W9-BGH-121

No. 1184
$19.95

The Master Guide To Electronic Circuits

by Thomas M. Adams

FIRST EDITION

FIRST PRINTING—JANUARY 1980

Printed in the United States of America

Library of Congress Cataloging in Publication Data

Adams, Thomas McConnell, 1913-
The master guide to electronic circuits.

Includes index.
1. Oscillators, Electric. 2. Electronic circuits.
I. Title.
TK7872.07A33 621.3815'33 79-25412
ISBN 0-8306-9971-6
ISBN 0-8306-1184-3 pbk.

Preface

The objective of this five-section book is to provide an understanding of *how* standard electronic circuits operate. Fundamentally, there are only a few basic actions involving electrons. And although these actions occur over and over in all circuits, each has acquired a multiplicity of names, which are not usually descriptive of action but of application. This nonconformity of names and terminology for electronic actions is a formidable barrier to the study of electronics.

Like many that have gone before it, this book is represented as a "nonmathematical" treatment. However, much of the terminology of electronics is of necessity mathematical and quite abstract. Thus, the omission of complex formulas in a text does not make it "nonmathematical" if it still contains words and phrases understandable only to those having the necessary mathematical background. It is practically impossible to write a book on electronics without using many such words and phrases, but in every instance the terms used in this book have been clarified through the use of illustrations which portray all identifiable currents and voltages in a somewhat unique manner.

The schematics show what takes place during successive half-cycles (or successive quarter-cycles) during the operation of each circuit under consideration. Every electron current is identified and discussed in detail, so that you can easily follow its path through all parts of the circuit. Frequently, more than one current will flow simultaneously through a component. An engineer or mathematician may rightly be interested in the *total* current through such a component, and would arrive at this figure mathematically. On the other hand, the student aspiring to become a technician or engineer must know more than the exact amount of total current or voltage; he must also be able to "visualize" each of the significant currents and voltages in order to learn how one affects the other. This ability to visualize what is happening inside a circuit can be accomplished by almost anyone, even without a mathematical background.

Electron currents in the various circuits have been treated much like "moving parts" of a machine. If we wanted to know how a piece of machinery worked, the simplest way would be to observe a cutaway model as it goes through several cycles of operation. Through the use of illustrations in this book, we can "look inside" the individual components, and "see" what is happening. Thus, the circuit diagrams used herein are more than just abstract drawings of circuit *connections*—they are concrete "working models" of circuit *actions* in diagram form.

It is our hope that this guide will find acceptance at each of the three important educational levels now concerned with the teaching of electronics—high schools, technical and vocational schools, and undergraduate levels in college engineering. The material is neither too advanced for high schools, nor too elementary for colleges. Students at any of these educational levels have one great desire in common—to achieve an early understanding of *how circuits operate.* The ability of an electronics student to visualize what is happening inside electrical components is an essential building block for later studies directed either toward design engineering, or toward maintenance, repair, and usage of electronic equipment.

The same approach and analysis is used in describing the circuits in all sections of this book. This approach consists first of clearly and unmistakably identifying every electron current at work in a circuit, and then of carrying out a detailed discussion of the movement of each current as well as the significance of these movements.

The circuit diagrams in this book treat electron currents as "moving parts." In any mechanical device, movements can be visualized and the motions of each part related to those of adjacent parts. The fact that these physical actions can be seen makes it easier to understand how the device works. Surprisingly, the same analytical technique can be adopted to explain electronic circuit actions. The ability to visualize the movements of electron currents is basic to a genuine understanding of all circuit operations. In addition, it will provide the background needed to understand the more advanced forms of circuit theory.

There is an interesting parallel between learning electronics and learning a foreign language. Even if a person never goes to school a day in his life, he usually learns to say enough words to get by. Contrast this with a diligent, intelligent high school student studying a foreign language. After two or three years of learning the rules of grammar, declension, spelling, etc., he may be capable of ordering an omelet in his newly acquired tongue—unless the waiter asks, "Chicken or duck?" At this point, the whole communications process may break down, simply because the fundamentals were neglected somewhere along the line.

In the study of electronics (or a foreign language), it is difficult to define "fundamental." This volume does not attempt to settle the question. Instead, the text has been written with a single objective in mind—teaching or explaining how circuits operate. At the same time, certain theories such as Kirchhoff's, Ohm's, and Coulomb's are also explained, wherever they apply.

None of the sections is considered more elementary or advanced than the others. Hence, read them in any order. In fact, it is hoped that almost any chapter can be read without reference to previous chapters.

The writer is indebted to many individuals in the educational field for their guidance and help in reviewing some of the original drawings for this series. Among those most helpful are Professors G. F. Corcoran, Joseph Weber, Henry Reed, and J. H. Rumbaugh of the University of Maryland; Professor J. C. Michalowicz of Catholic University; Professors E. R. Welch, H. R. Branson, W. K. Sherman, and K. T. Chu of Howard University; Professor N.T. Grisamore of George Washington University; Lou Godla of the Fairfax County (Va.) high-school system; Donald Ingraham and John Johnson of the Arlington County (Va.) high-school system; Keith Johnson of the District of Columbia high-school system; E. H. Reitzke, president of Capital Radio Engineering Institute, and Charles Devore and Latti Upchurch of the same Institute; Jamie Cruz of Emerson Electronics Institute, and the Institute staff; and David Lien of the Grantham School of Electronics. Also very helpful and encouraging were Dr. Hilary J. Deason of the American Association for the Advancement of Science, Commander David O. Mann, USN, and my long-time friend and associate, Captain Walter H. Keen, USN, Office of Naval Research.

I am also thankful for the encouragement and assistance rendered in countless ways by my wife, to whom this work is gratefully dedicated.

Thomas M. Adams

Contents

SECTION 1

OSCILLATOR CIRCUITS

Chapter 1

INTRODUCTION

An oscillator is any nonrotating device which generates a sig-
al at a frequency determined by the constants in the circuit.
scillators can be placed into almost as many categories as there
e oscillators. One method of classification is by the *type* of wave-
rm produced. A second is by the *frequency* of the waveform.
Oscillator waveforms can be broadly classified into one of two
pes—the *sinusoidal, or harmonic;* and the *nonsinusoidal, or*
nharmonic. A sinusoidal waveform, more familiarly known as
sine wave, is nothing more than a graphical representation of
mple harmonic motion. Most oscillators in everyday life are
nusoidal, or harmonic. Examples are the pendulum of a clock,
e spring of a watch, a child's swing, or the piston of a gasoline
steam engine. All the oscillations mentioned must be rein-
rced or replenished periodically, or they will die out because
natural losses. The pendulum of a clock, for instance, is re-
forced once each cycle by a small amount of energy released
the appropriate time by the clock mechanism.
The harmonic oscillators discussed in this volume are further
assified according to the frequency range within which they
erate—either the audio- or radio-frequency range. These cir-
its are as follows:

Radio-frequency
- Crystal oscillator.
- Hartley oscillator.
- Colpitts oscillator.
- Tuned-plate–tuned-grid oscillator.
- Electron-coupled oscillator.

Audio-frequency
- Phase-shift oscillator.

Nonharmonic oscillators, also called *relaxation* oscillators, are discussed in this volume as follows:

Thyratron sawtooth generator.
Blocking oscillator (pulse generator).
Multivibrators.

The term *relaxation oscillator* is derived from a particular property of resistance-capacitance combinations: A capacitor is "charged" with a positive or negative voltage and then permitted to "discharge" through a resistance. This process of discharging a capacitor through a resistive path gave rise to the descriptive terms *relaxing* and *relaxation.*

There is a fundamental difference between relaxation oscillators and the harmonic oscillators mentioned previously: All harmonic oscillators must have energy fed back from output to input in order for oscillation to continue, whereas relaxation oscillators need no such energy feedback. Therefore, oscillators can also be classified as either *feedback* or *relaxation* type.

BASIC ELECTRICAL FUNDAMENTALS

All electronic circuit actions consist of free electrons moving about under controlled conditions in the three main types of components—inductors, capacitors, and resistors. Added to these three components are switching or regulating devices known as vacuum tubes (and in more recent years, solid-state devices such as diodes and transistors, which are taking over many of the functions previously performed only by tubes).

The three circuit components are each closely related to certain fundamental electrical phenomena from which their names have been derived. Before these three phenomena—inductance, capacitance, and resistance—can be understood, it will be necessary to understand the principle of *electron drift.*

Electron drift is the foundation of all electric currents (or more accurately, of all electron currents)—except the electron flow within a battery and the electron flow across the evacuated space within a vacuum tube. The first is a chemical reaction, and the second is adequately described in practically every other chapter of this book and need not be belabored here.

Free Electrons and the Electron-Drift Process

In this day and age, almost every school child is aware of the atomic/molecular structure of matter. Each atom bears an astonishing resemblance to our solar system. At the center is the *nucleus,* where most of the atom's mass is concentrated. Orbiting around each nucleus are electrons, the number differing according to the type of material.

In our solar system the revolving planets are kept from flying off into space by the gravitational attraction between each planet and the sun, which is at the center, or nucleus, and contains most of the mass of the system. In each atom the planetary electrons are kept from flying off into space (or into the surrounding material) by an electrical rather than a gravitational attraction.

This electrical attraction exists because each electron carries within itself one negative unit of charge. The atom as a whole is electrically neutral, but its neucleus will always contain the same number of positive charge (called *protons*) as there are negative charges (meaning *electrons*) orbiting around it. For example, an atom of material which has eight orbiting electrons will have a nucleus with a positive charge of eight units; so the two charges will cancel each other.

There are several methods of setting electrons free from their positions in orbit or, in other words, "stealing" these tiny planets from their solar systems. Heat is one method; it is used in most electronic tubes. Chemical reaction, another common method, is the one used in batteries.

There would be no point in setting electrons free from their orbits unless we had some use for them. The immensity of our requirements for free electrons in everyday life is suggested in part by the definitions of electric current and voltage.

Definitions

Electric *current* consists of free electrons in motion.

Free electrons in concentration constitute a *negative voltage.*

A concentration of positive ions, caused by a *deficiency* of free electrons from their planetary positions, constitutes a *positive voltage.*

CONDUCTORS AND RESISTORS

Some materials release electrons from their orbits with the greatest of ease; these materials are classed as *conductors* of electricity. Other materials hold their electrons in orbit very tightly and release them only with the utmost difficulty; these materials are classified as *insulators.* Examples of both types abound in everyday life. Gold, silver, copper, and various other metals and their alloys are examples of good conductors. Glass, wood, and rubber are three familiar insulating materials.

Under the influence of an external force such as an applied voltage, electrons will move through a conductor. This movement is known as *electron drift*. As an example, if each end of a conductive wire is connected to the two terminals of a battery, electrons will leave the negative terminal and move through the wire toward the positive terminal, pushing other electrons into the positive terminal. A continuous flow from the negative to the positive terminal will exist as long as the circuit is connected (and the battery lasts).

No single electron completes the entire journey around even such a simple closed circuit as this one. Trillions upon trillions of electrons are in motion, even in a unit of current as small as one ampere. Also, the number of atoms which actually release free electrons is an infinitesimal fraction of the number of atoms within the material itself. Hence, in moving from the negative battery terminal and into the wire, each electron flows only the tiniest fraction of a millimeter before finding its path impeded by a planetary electron from another atom in the material.

Since each electron carries a like charge, the two repel each other. As a result, the planetary electron is dislodged from its orbit and starts down the wire in the same direction as—and probably ahead of—the approaching electron. The newly dislodge electron will in fact pick up some of the velocity of the other electron. The latter will be lowered down accordingly, and it then becomes an easy prey for being "recaptured" by the atom which has just lost an electron. The original electron "falls" into the orbit just vacated and again takes up planetary motion.

So we still have what we started with—an electron moving through the conductor and driven by an applied negative voltage. The interchange just described is electron drift. It will occur countless quadrillions of times for even the tiniest measurable electron current to flow—even a micromicroampere (one-millionth of one-millionth of an ampere!). The difference between a good conductor and a poor one is the relative ease or difficulty with which its electrons can be dislodged from their orbits and recaptured.

Resistors are aptly named, for they are inserted at certain points in the circuit to *resist* the flow of current. The value selected for each resistor is nothing more than a measure of the difficulty with which electron drift occurs within the resistor.

CAPACITANCE

Electron drift can also help us understand the important electrical characteristic known as capacitance, and from this to acquire a deeper appreciation of the role that capacitors play in electronic circuits. Consider again the situation as a moving electron approaches a planetary electron. There is a space between them, however small, when the planetary electron finally yields to the repulsive force of the approaching electron and jumps out of orbit. It does not really matter to the two electrons whether the space which separates them at that moment is the open space between copper atoms, air atoms . . . or the open space between atoms of any other material—either conductor or insulator.

What this suggests is that the *movement* of one electron can be transferred to another electron—irrespective of whether both are contained in the same conductive material, or *in two different but adjacent conductive materials.* As a clarifying example, suppose a conducting wire is severed between the points where the two electrons will be situated when the second one breaks away from its orbit. Suppose also that the two cut ends of the wire are separated by an infinitesimal distance, perhaps a millionth of an nch. Electrically this is an "open" circuit, through which current annot flow in the normal sense. Nonetheless, drift action can be made to occur between the two electrons. How do we explain his paradox?

We can do so by saying that the two ends of the wire *have capacitance toward each other.* It might be more descriptive, and also truer, to say that the two *electrons* have capacitance toward each other. Under these conditions, electron drift can be made o occur—*provided the first electron can be made to flow up to he break point in the wire.*

This is a fairly large proviso, of course, but it is subject to a simple laboratory demonstration which is also one of the standard capacitor checks performed by all technicians. The capacitor in question is connected across a DC voltage source such as a battery, and an ammeter is placed in the line to indicate the current flow.

The instant the switch is closed to activate the circuit, the ammeter needle will deflect, indicating an initial flow of electrons current) into the capacitor. (This current flow exists on both ides of the capacitor.) The ammeter needle quickly drops back o zero, indicating no further current flow. We say that the capacitor is "charged" to the value of the applied voltage. (If the mmeter needle remains in a deflected position—indicating some urrent flow—the capacitor is "leaking" and is defective. If no nitial deflection occurs at all the capacitor is "open," also a defect. n either case, the capacitor should be discarded.)

Fig. 1-1A shows the current condition for such a capacitor test he moment immediately after the switch is closed. An initial urge of charging current flows around the circuit and builds up he voltage across the capacitor, as shown in Fig. 1-1B, with negative electrons amassed on the upper plate. Since the electrons had to flow into the capacitor before they could become concentrated there, the current into a capacitor is said to *lead* (or precede) the voltage across it.

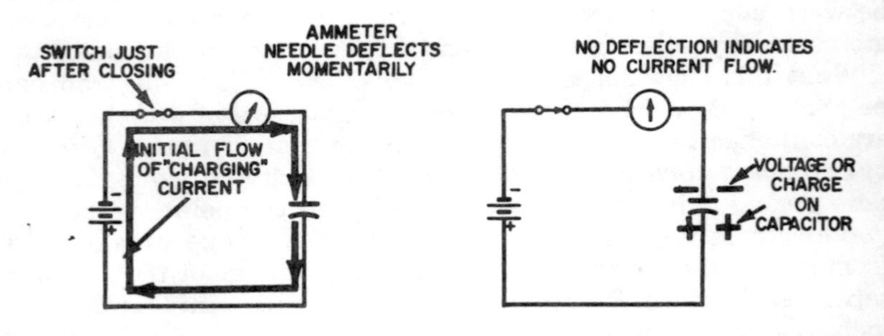

(A) Initial flow of current from DC source into capacitor. *(B) Current has stopped, but voltage exists across capacitor.*

Fig. 1-1. Current and voltage conditions for capacitor test.

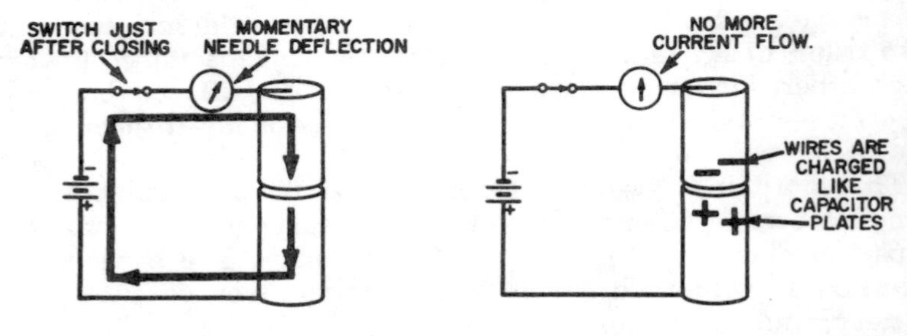

(A) Initial flow of current. *(B) Condition after initial current surge.*

Fig. 1-2. Enlarged view of conductor with a small break or opening.

Fig. 1-2 shows the hypothetical example discussed previously, in which the two ends of a severed wire have capacitance toward each other, so that an initial charging current flows around the circuit but a continuous current does not.

When a capacitor plate has charged to the value of the applied voltage as in Figs. 1-1B and 1-2B, no more electrons can be driven onto the negative plate and the flow of current around the circuit stops.

Although capacitors will not "pass" a pure direct current, the *initial surge* of electrons into the capacitor (or up to the break point in the wire, which resembles a capacitor) is not a smooth, steady flow and hence not a pure direct current. Capacitors will "pass" an alternating current because electrons are continuously being driven onto one plate of the capacitor (charging it), and hen pulled back off (discharging it) as the applied voltage reverses. Thus, electron current can be made to flow back and orth regularly on each plate of a capacitor, and in this sense he capacitor is said to be passing an alternating current.

The ability to visualize the action of a capacitor is absolutely essential to a quick and easy understanding of electronic circuit action. For this reason, wherever capacitors are mentioned in this book, their action has almost always been redescribed to fit the conditions for the circuit under consideration.

Capacitor Construction

A capacitor (sometimes referred to as a condenser) is a manufactured device for making use of two important electrical properties associated with capacitance. These are the ability of a capacitor to:

1. Store an electric charge, either negative electrons or positive ions.
2. Pass an alternating current from one plate to the other. This property is frequently stated as the ability of a capacitor to oppose any change in applied voltage.

The *amount* of capacitance exhibited by any capacitor depends on several interrelated factors which can be tied together with he following formula:

$$C = \frac{22.35\ KA\ (n - 1)}{10^8 d}$$

where,

C is the amount of capacitance in microfarads,
K is the dielectric constant (the dielectric is explained in a later paragraph),
A is the area of one plate in square inches (it is assumed all plates are the same size and shape),
n is the number of plates,
d is the distance between plates (thickness of the dielectric, as explained a little later).

It can be seen from the formula that the amount of capacitance varies *directly* with the area of the capacitor plates, and *inversely* with the distance between them. These considerations become highly significant at radio frequencies, where unwanted capacitive coupling may occur between components and cause interference.

For example, two wires passing close to each other will be capacitively coupled, the amount being determined by their diameters and by the distance between them. Also within vacuum tubes each electrode will exhibit some capacitance to the others, although the resulting *interelectrode capacitance* may have advantages as well as disadvantages.

The material between the plates of a capacitor is called the *dielectric*. It may be air or any other insulator, and its insulating ability is known as the *dielectric constant*, or K. The dielectric constant of air is unity, or 1. Mica has a dielectric constant of 5.5, and ordinary glass, slightly over 4.

Of equal if not greater importance is the dielectric strength of a material, or how high a voltage it can withstand before breaking down. The dielectric strength is expressed in volts per unit of distance. For air it is about 80 volts per .001 inch. Mica is 25 times stronger, or 2,000 volts per .001 inch.

Assuming two capacitors have equal plate area and separation, this means the breakdown voltage of one with a mica dielectric will be 25 times greater than one with an air dielectric.

The dielectric strength of the material (and its thickness) determine the working-voltage rating of a capacitor. This is the rating used by the manufacturer to indicate the maximum voltage a capacitor can safely withstand without breaking down.

Capacitive Reactance

Capacitors offer a certain opposition to the passage of alternating currents. The amount varies inversely with the frequency of the current and the value of the capacitor, in accordance with this formula:

$$X_C = \frac{1}{2\pi fC}$$

where,

X_C is the capacitive reactance (opposition to current flow) in ohms,

f is the frequency of the applied current in cycles per second,

C is the capacitance in farads.

INDUCTANCE

The third important circuit characteristic is *inductance*, and the manufactured components most directly associated with it are called *inductors*, which include all coils and transformers. We will find an endless variety of such devices, each manufactured for a specific application. As a group they cover the entire frequency spectrum, are called on to carry tiny currents or huge ones, and often must be able to withstand enormous voltages without rupturing or otherwise failing. Regardless of the conditions under which they are used or the circuit functions they are expected to fulfill, all inductors—coil, choke, transformer, etc.—take advantage of the electrical property known as inductance.

Inductance is the electrical equivalent of mechanical inertia. Stated more simply, inductance *is* electrical inertia. Let us compare it briefly with mechanical inertia.

A given mass possesses inertia, and for this reason requires a certain amount of force to set the mass in motion, or to speed it up or slow it down. Thus, a rolling ball—in the absence of friction—will continue to roll in the same direction and at the same speed. The same ball at rest will remain at rest.

The concept of electrical inertia is easy to visualize if the foregoing analogy is kept in mind. Remember that although the mass of one lone electron seems insignificant, its contribution becomes significant when we consider that there are trillions upon trillions of electrons in a single conductor!

A given quantity of electrons, flowing through any conductive material, will tend to keep flowing in the same quantity. The reason is that the electrical property known as inductance will always operate to support this natural law. Some common descriptive statements which apply to inductance are:

1. An inductance always tries to keep the total current constant.
2. An inductance always opposes any change in the current.
3. In an inductance, the voltage "leads" the current; or conversely, the current "lags" the voltage.
4. Inductors generate a back electromotive force which opposes the applied voltage.

These statements can be better understood from Fig. 1-3. In Fig. 1-3A the two adjacent closed circuits are "connected" only by the mutual coupling between the two wires placed side by side in the center. Even though shown as straight wires, they can be considered as two windings of a transformer, since all conductors exhibit some inductance toward all other conductors. The instant the switch in the left circuit is closed, a current will flow from the upper (negative) terminal of the battery, through the switch, *downward* through the wire we have designated as one winding of a transformer, and on around the circuit to the lower (positive) terminal.

Simultaneously, in the right circuit another current begins to flow *upward* through the other transformer winding, which we will call the secondary. With a sensitive ammeter or galvan-

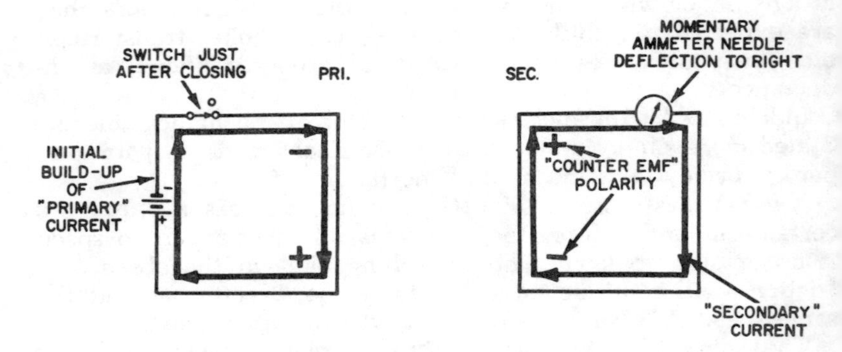

(A) Initial current flow when switch is closed.

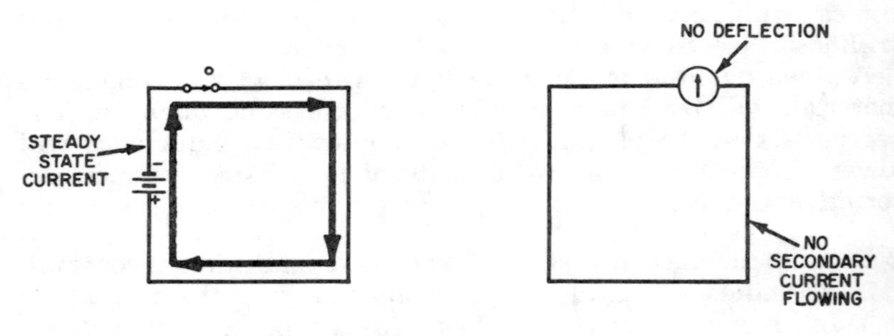

(B) Steady-state (DC) conditions.

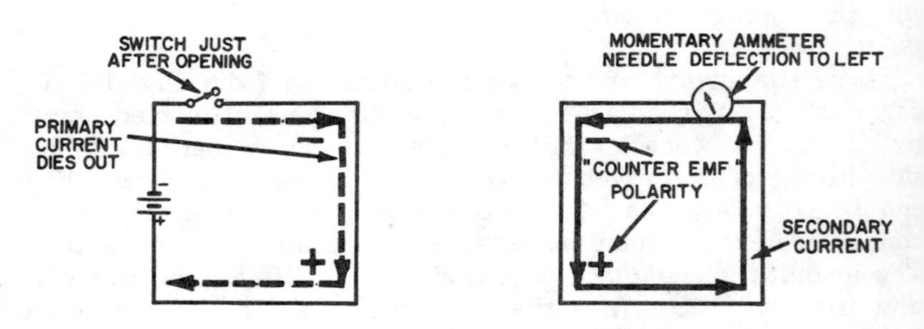

(C) Decrease in current flow when switch is opened.

Fig. 1-3. Current flow in an inductive circuit.

ometer in the circuit as shown, the needle will be deflected to the right slightly.

This secondary current has been shown at left. Where did it come from? Once we answer this question we can understand much about inductance.

This secondary current is nature's way of keeping the total current constant. Before the switch was closed, there was zero current flowing in both windings. The instant after the switch is closed, this zero total-current condition must be maintained, even though a substantial current begins flowing downward through the primary winding. As the electrons which make up the primary current begin to accelerate in the downward direction, the negative charge carried by them causes other electrons in the secondary winding to be accelerated upward. When the current flowing upward is subtracted from the one flowing downward, the remainder is zero *for the tiniest fraction of a second.*

This secondary current cannot flow forever—it is sustained only by *changes* in the amount of primary current, and the rate of change becomes less and less as the primary current approaches its steady-state value. The secondary current dies out gradually, and when the primary current finally reaches its full steady-state value, the secondary current will drop to zero as shown in Fig. 1-3B.

Fig. 1-3C depicts the conditions immediately after the switch in the primary circuit has been opened. When the primary-current flow is cut off, it must rapidly decrease (decelerate) until it reaches zero. Just as an increasing current flow in the primary winding will set up a current flow in the secondary winding, so too will a decreasing current. But this time, the current in the secondary winding will flow in the opposite direction (downward) to keep the total current through the two windings constant. A sensitive galvanometer in the secondary circuit will actually give a momentary deflection in the opposite direction from that shown in Fig. 1-3A, indicating a current reversal has occurred.

This secondary current lasts only while the primary current is changing. As soon as the primary current reaches zero, the secondary current will drop to zero also.

For every electron flow, or current, there must be a companion voltage to supply inertia. In the primary circuit of Fig. 1-3A, for example, the applied voltage is the companion and motivating force for the primary circuit. Because of the way the battery is connected, this applied voltage is negative at the top of the primary winding and positive at the bottom, causing the current to flow downward through the primary winding.

Suppose you were asked this question:

"What polarity of voltage must exist across the secondary winding in order for the secondary current to flow upward through the winding?"

There can be only one answer, since electrons *always* flow from a point of negative to a point of positive voltage—*never* in the opposite direction.

In Fig. 1-3A we can see that the polarity of the secondary voltage is positive at the top and negative at the bottom, and that the applied voltage in the primary is negative at the top and positive at the bottom. As a result, the voltage generated in the secondary "opposes" the applied voltage in the primary. This voltage associated with secondary currents in transformers and other inductors has been given the name of *counter electromotive force*, or more simply, *counter emf*.

We can also see that the applied voltage actually "leads" the resulting current. The current referred to here is the *total* transformer or inductor current, which is not achieved until sometime after the primary voltage is applied, as shown in Fig. 1-3B. Thus, attainment of the final current value has "lagged," or fallen behind, the voltage producing it.

In Fig. 1-3C we see that the polarity of the counter emf has reversed itself—now the negative voltage is at the top and the positive voltage is at the bottom. This is the only possible polarity consistent with the downward flow of secondary current at this moment. The fact that this counter emf is opposing the applied voltage may be a little difficult to visualize, since the two are of the same polarity. However, by opening the switch we have removed the negative voltage previously applied at the top of the primary winding. This is actually equivalent to applying a positive voltage at that point. In this sense, the applied voltage and counter emf are opposite in polarity.

The current in the primary winding flows in only one direction, and yet we have seen that it can cause a two-way current—better known as an alternating current—to flow in an adjacent winding. The reason is that it is a unique form of direct current known as *pulsating* DC. There will be several examples of its use in later chapters, such as in the Hartley and blocking oscillators and in the grid circuit of an electron-coupled oscillator

Self-Inductance

Fig. 1-4 shows how induced currents can be made to flow and a counter emf generated *within a single conductor* (shown enlarged in Fig. 1-4 for clarity). As before, the sudden application or removal of an applied voltage is required—or more specifically, a *change* in the applied voltage across the circuit is necessary for a self-induced current to flow.

As soon as the switch in Fig. 1-4 is closed and battery voltage is applied across the circuit, the primary current will attempt to rise to its full value. However, because all circuit components have self-inductance, the *total* current does not do so immediately, but remains at zero for an instant. The reason is simple—as the primary current tries to increase, the "electrical inertia" of the electrons in it will cause *other* electrons in the circuit to be accelerated in the opposite direction. Hence, the two currents cancel each other at first.

As the primary current approaches its full value, its *rate* of increase is steadily dropping, permitting the induced current to die out. The counter emf of Fig. 1-4A no longer exists when the steady-state condition shown in Fig. 1-4B has been replaced.

Fig. 1-4C shows the conditions an instant after the switch is opened. The primary current will now try to drop to zero. However, *all inductances act to keep the total current from changing*. Thus, as the electrons of the primary current start to decelerate, they induce other electrons in the same conductor to be accelerated in the same downward direction. Therefore, instead of the zero current we would expect to find upon first opening the switch, we find the full current flowing! Once again, a counter emf has come into existence, only this time its polarity is opposing the polarity we tried to apply by opening the switch to remove the negative applied voltage.

Inductor Construction

The entire art of inductors is built upon this property of electrical inertia. In theory, a coil of wire with many thousands of closely spaced turns is still the same straight wire shown earlier in Fig. 1-4. The object of the additional turns is to provide more inductance (by multiplying the effects of electron inertia).

The opposition offered by inductors to the flow of alternating current varies directly with the frequency of the applied current and with the size of the inductor, in accordance with this formula:

$$X_L = 2\pi f L$$

where,

X_L is the inductive reactance (opposition to current flow), in ohms,

f is the frequency of the applied current in cycles per second,

L is the inductance of the coil in henrys.

Fig. 1-5 shows cross-sectional views of the two common types of coils. Their *approximate* inductances can be calculated from the following formulas:

For a single-layer solenoid (Fig. 1-5A):

$$L = \frac{r^2 \times N^2}{9r + 10l}$$

where,

L is the inductance in microhenrys,
N is the number of turns,
l is the length of the coil in inches,
r is the radius of the coil in inches.

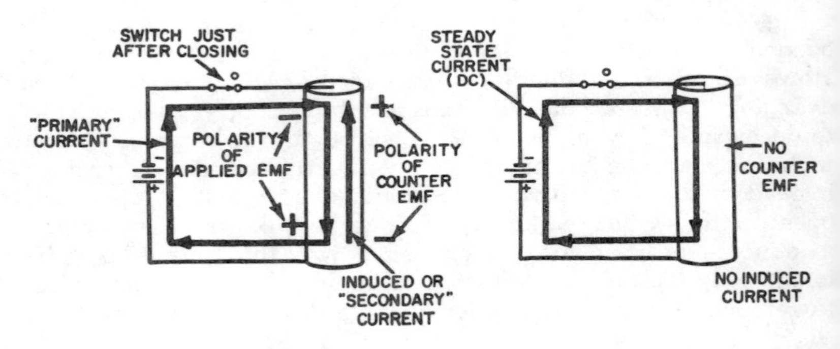

(A) Current flow when switch is closed.

(B) Steady-state (DC) current.

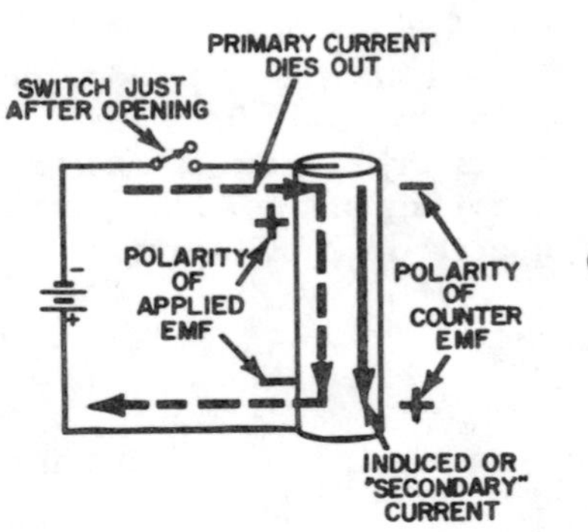

(C) Current flow when switch is opened.

Fig. 1-4. Self-inductance in a single wire (enlarged view).

For a multilayer coil (Fig. 1-5B):

$$L = \frac{r^2 \times N^2}{6\,r + 9l + 10t}$$

where t, the only new term, is the thickness of the coil winding in inches.

It will not be necessary for you to refer to these formulas in order to understand the circuit actions discussed in the text. But like the capacitance formula offered previously, this one shows the correlation between physical size and configuration of a coil and its resultant inductance.

THE INDUCTANCE-CAPACITANCE COMBINATION

Inductance and capacitance in combination possess the unique characteristic of *resonating* at a particular frequency. This im-

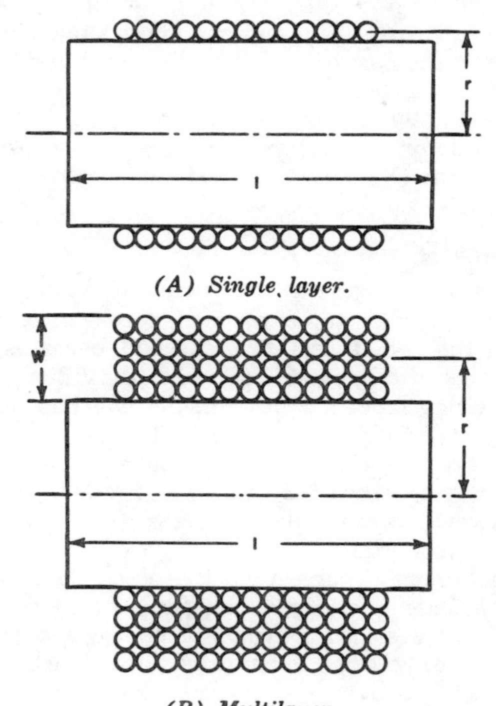

(A) Single layer.

(B) Multilayer.

Fig. 1-5. Cross-sectional views of single-layer and multilayer coils.

portant and basic circuit action is employed in all types of electronic equipment. Tuned inductance-capacitance (L-C) circuits are used in practically every piece of electronic equipment—and oscillators are no exception, as we shall learn in later chapters.

Fig. 1-6 shows the conditions at the ends of the four quarter-cycles of an oscillation between an inductor L and a capacitor C. At the end of the first quarter-cycle (Fig. 1-6A), electrons are amassed on the upper plate of the capacitor (thereby constituting a negative voltage), and there is a *deficiency* of electrons (positive voltage) on the lower plate. In other words, no current is flowing through the inductor yet.

The voltage across the capacitor represents stored electrical energy. Energy in storage is potential energy. So, during the second quarter-cycle the electrons from the top plate will start to flow downward through the inductor in an effort to redistribute themselves equally between the two plates and thus neutralize the electric field. The self-inductance of the coil prevents an instantanous build-up of current by generating a counter emf of the opposite polarity. This action is not shown in Fig. 1-6B, but because of it, the current is delayed exactly one quarter-cycle in reaching its peak. So, at the end of the second quarter-cycle the current is finally flowing at its maximum rate *downward* through the inductor. At this moment, the charge has been completely redistributed—no voltage remains across the capacitor or tuned tank.

Magnetic Lines of Force

As the current through the coil increases from zero to maximum, lines of magnetic force come into existence and expand outward from the coil. In essence, the coil becomes an electromagnet, and since the lines of force in Fig. 1-6B are entering at the top and exiting from the bottom, the coil has its north pole at the top and its south pole at the bottom.

These, then, are the conditions at the start of the third quarter-cycle—zero voltage across the tank, maximum electron current flowing downward through the inductor (coil), and maximum lines of force surrounding it.

The potential energy represented by the charged capacitor has now become kinetic energy, in the form of moving electrons. The lines of force are also a means of energy storage. So when the current tries to collapse during the third quarter-cycle, these lines of force will collapse and thereby drive additional current downward through the coil. The lines of force are trying to accomplish the effect we already know occurs with inductors—it is trying to keep the current from changing value.

As a result of this sequence, almost all the electrons which were on the upper plate of the capacitor will be delivered to the lower plate during the third quarter-cycle, and the charge distribution will look like Fig. 1-6C. As in Fig. 1-6A, this charge distribution again represents potential energy in storage, except now it has the opposite polarity.

At the start of the fourth quarter-cycle, the electrons massed on the lower plate must again redistribute themselves in order to neutralize the voltage across the capacitor. So they begin to flow upward through the inductor, and another familiar sequence begins: A counter emf with a polarity opposite that of the capacitor comes into existence. An induced current then flows downward in the coil and bucks the upward-flowing current.

A new set of magnetic lines of force (also known as *flux lines*) develops and expands outward as long as the total current in the coil is expanding. These lines of force have a different direction from the one they had in Fig. 1-6B, because in Fig. 1-6D the main current through the coil is flowing in the opposite direction. Now the lines of force are entering at the bottom and exiting at the top, again making the coil a small electromagnet but with its north pole at the bottom and its south pole at the top.

These, then, are the conditions at the start of the first quarter-cycle—zero voltage across the capacitor and tank, maximum upward flow of electrons through the inductor, and maximum number of magnetic lines of force around it.

The collapse of these lines of force during the first quarter-cycle tries to keep the current from dying out by drawing additional electrons upward through the inductor. The end result is that almost as many electrons are delivered to the upper plate as were amassed on the lower plate a half-cycle earlier.

Thus, we see that one cycle of an oscillation is characterized by (1) a periodic changing of the voltage across the tank from plus to minus and back to plus; (2) periodic up-and-down alternations of the current through the inductor; and (3) periodic expansion and contraction of the magnetic lines of force around the inductor, the lines changing direction every half-cycle.

In brief, an electric oscillation consists of a cyclic interchange of energy between an electric and a magnetic field. The electric field is represented by the voltage across the capacitor in Figs. 1-6A and 1-6C; and the magnetic field, which is associated with a high current flow, is represented by the expanded lines of force in Figs. 1-6B and 1-6D.

The number of cycles occurring each second is called the *frequency*. Tuned circuits of the type shown in Fig. 1-6 are used in the generation of radio frequencies ranging from a few thou-

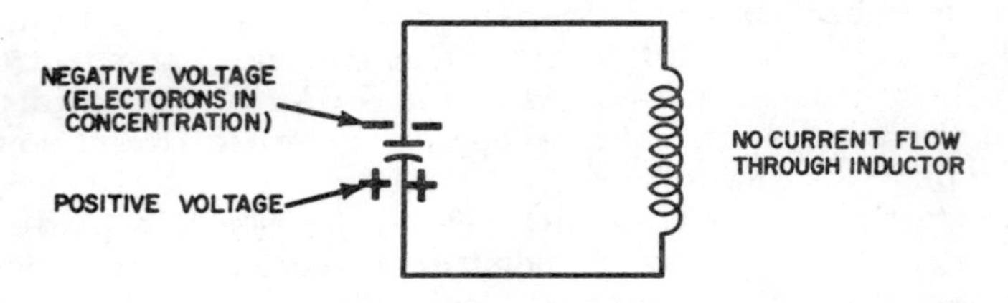

(A) At end of first quarter-cycle.

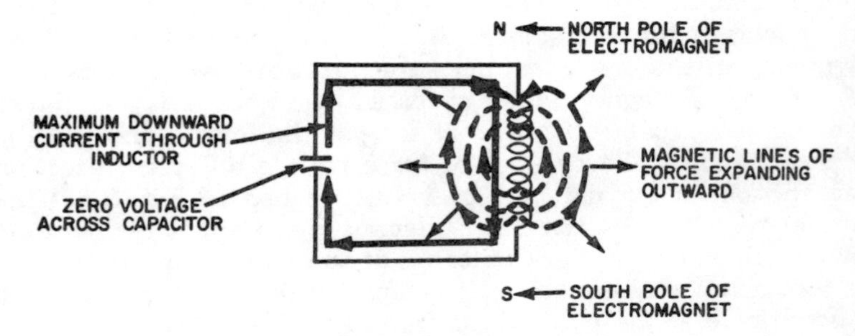

(B) At end of second quarter-cycle.

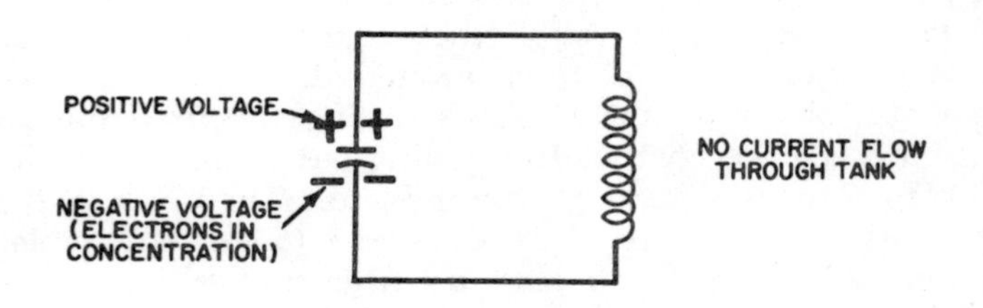

(C) At end of third quarter-cycle.

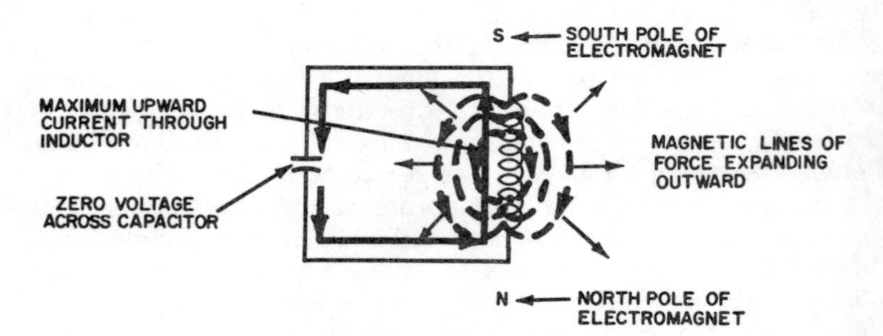

(D) At end of fourth quarter-cycle.

Fig. 1-6. Oscillation in an L-C circuit.

and cycles per second, up to hundreds of kilocycles and even negacycles (a megacycle is one million cycles per second).

The values of the capacitor and inductor determine the basic operating frequency of any tuned tank, in accordance with the tandard formula which states that:

$$f = \frac{1}{2\pi\sqrt{LC}}$$

vhere,

f is the frequency in cycles per second,
L is the inductance in henrys,
C is the capacitance in farads.

A later chapter shows how this formula is derived from the wo simple reactance formulas for inductance and capacitance.

The strength of an oscillation is measured by the amount of oltage (*amplitude* of the voltage peaks) across the tuned tank. Once started, an oscillation could continue indefinitely at the ame amplitude, if it were not for the inevitable *losses* due to vire resistance and other effects. Consequently, unless an oscillation is replenished from an outside source of energy, each suceeding cycle will be a little weaker than the preceding one until ventually the oscillation will die out entirely. (This "dying out" rocess is called *damping* of the oscillation).

The quality, or *Q*, of a tuned circuit is a comparative measure f its freedom from the losses which damp out an oscillation. Once started, a high-*Q* circuit will oscillate for many thousands f cycles, whereas an oscillation in a low-*Q* circuit will die out n a relatively few cycles. A convenient means of visualizing the neaning of *Q* is to consider it a ratio between the number of lectrons in oscillation and the number which drop out each ycle due to losses. Mathematically, circuit *Q* is stated by several ifferent formulas, two of which are:

$$Q = \frac{2\pi \times \text{Energy in Storage}}{\text{Energy Lost Each Cycle}}$$

r,

$$Q = 2\pi\frac{L}{R}$$

here,

L is the coil inductance in henrys,
R is the circuit resistance in ohms.

Fig. 1-7A shows a lightly damped waveform for an oscillation n a high-*Q* tuned circuit, where losses are low. The heavily

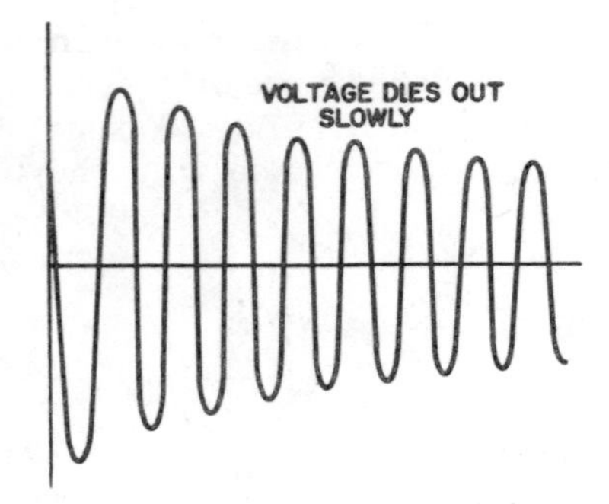

(A) *Light damping from high-"Q" circuit.*

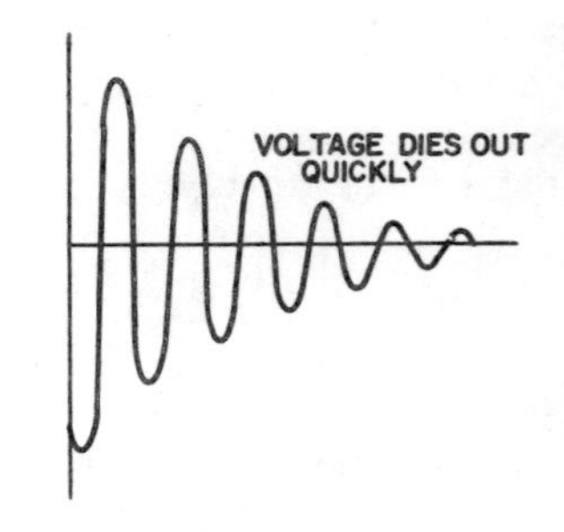

(B) *Heavy damping from low-"Q" circuit.*

Fig. 1-7. Effect of circuit "Q" on oscillatory waveform.

damped voltage waveform in Fig. 1-7B is for an oscillation in a low-Q tuned circuit, where losses are heavy.

THE RESISTOR-CAPACITOR COMBINATION

The action between resistors and capacitors in combination is the last in our discussion of the five basic actions taking place in electronic circuitry. These combinations are most easily classified into one of two broad groups, depending on (1) the values of the resistor and capacitor, and (2) the frequency of the current/voltage combination to which they must respond in the particular circuit. These two groups are the long time-constant and the short time-constant combinations. The same R-C combination will provide a long time-constant at one applied frequency and a short time-constant at a lower frequency.

Almost all circuits discussed in later chapters have at least one long time-constant R-C combination. For this reason, it is certainly worth your while to achieve an early mastery of the basic action in a resistor-capacitor combination. Of the five, it is probably the easiest to understand.

The drawings in Figs. 1-8 through 1-12 are based on an interesting set of analogies. Notice that the capacitance of a capacitor is likened to the capacity of a water tank, the resistance of a resistor is likened to the resistance of a water pipe, water level or pressure is compared to "electron level" or pressure (meaning voltage) and water flow or current is compared to electron flow or current.

Each of the five combinations in Figs. 1-8 through 1-12 shows two periods (or half-cycles) of operation—the charging half-cycle, and the "other" half-cycle. Just to keep things simple, we

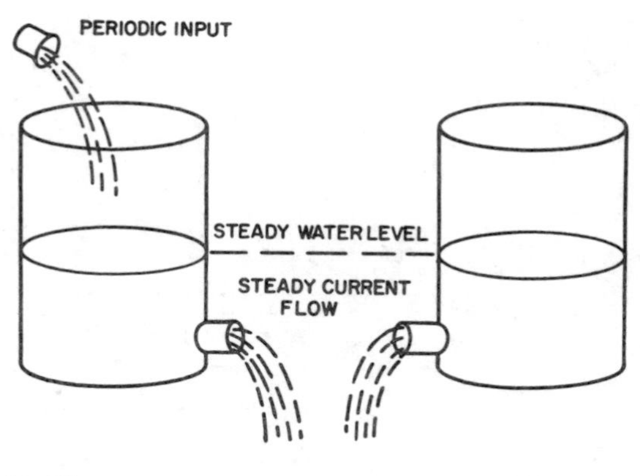

(A) Charge. *(B) Discharge.*

'ig. 1-8. A water tank with a fluctuating input and steady output is comparable to a "long time-constant" resistor-capacitor combination.

vill assume the amount of water added during each one of the .ve charging half-cycles is always the same—say, one bucketful. Moreover, let us keep the frequency the same throughout by add- ıg water to each tank at the rate of one bucketful per second. Hence, we can say that each one of the five combinations operates t a frequency of one cycle per second.

In Fig. 1-8 the tank is large enough, and the output nozzle small nough, that the addition of one bucketful of water each second oes not significantly raise the water level. Consequently, the vater pressure does not change during the charging half-cycle

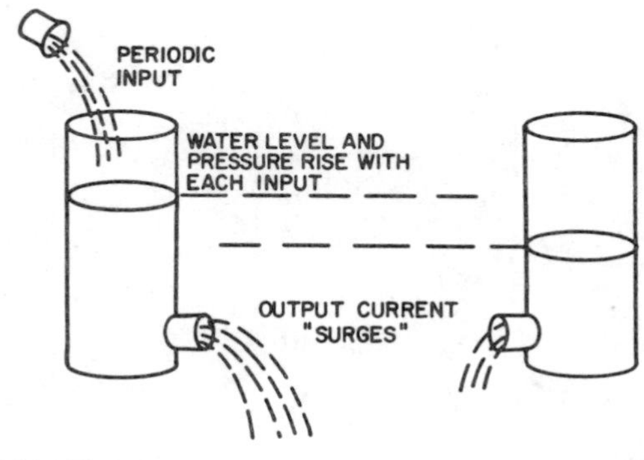

(A) Charge. *(B) Discharge.*

ig. 1-9. If tank is too narrow and thus has too little "capacity," the output urrent will surge. This is comparable to a "short time-constant" resistor-capacitor combination.

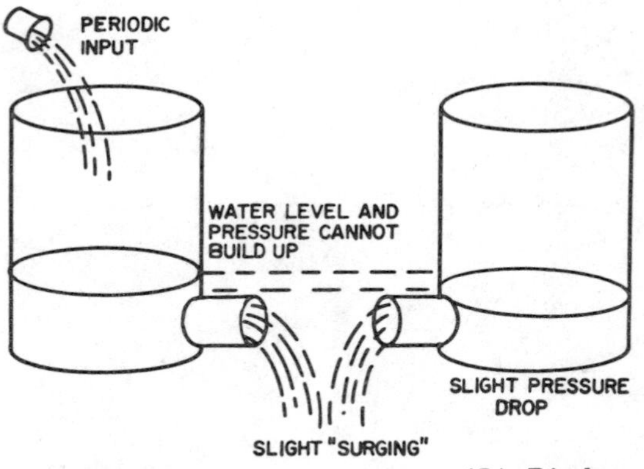

(A) Charge. *(B) Discharge.*

Fig. 1-10. With a low-resistance outlet (wide, short nozzle), pressure cannot build up and the water runs out as fast as it is added. This compares to a "short time-constant" R-C combination.

depicted in Fig. 1-8A, because the same amount of water flows out through the nozzle during the discharging half-cycle of Fig. 1-8B. In electronic circuitry, this is analogous to a long time-constant R-C combination.

Fig. 1-9 depicts the same conditions as before, only with a smaller water tank. Now a single bucketful of water *does* make a difference in the water level and consequently the pressure. Because the pressure rises as each bucketful is poured in, more water flows out through the nozzle during the charging than the

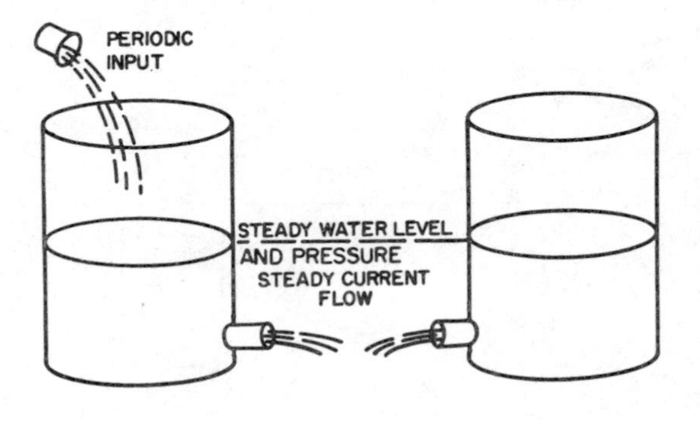

(A) Charge. *(B) Discharge.*

Fig. 1-11. With a high-resistance outlet (narrow, long nozzle), a high water level and pressure will build up and current will be steady. This is also a long time-constant combination, but the same amount of input will produce a higher pressure than in Fig. 1-8.

discharging half-cycle, since rate of flow (current) depends on water pressure. The same is true of electric current, which depends on electrical pressure, or voltage. Whenever the pressure level and consequently the amount of current in a circuit changes significantly during one complete cycle, the combination has a short time-constant at the particular frequency.

In Fig. 1-10 we have restored the tank to its original size but enlarged the outlet nozzle (thus lowering its resistance). Now it is difficult to build up water pressure, since the water flows out almost as fast as it flows in. Because the water level fluctuates, the current through the nozzle is pulsating rather than steady. Thus, as in Fig. 1-8, we have a short time-constant combination here.

In Fig. 1-11 we have narrowed the outlet nozzle down and thereby substantially *increased* its resistance. Thus, it will impede the flow, and more water will be added each cycle than can flow out, until the water level builds up to the point where there is sufficient pressure to force out the same amount that is added each cycle. However, since the water level does not decrease during the discharging half-cycle, there is no surging—the flow will gradually build up and then remain constant. Hence, this combination can be classified as a long time-constant.

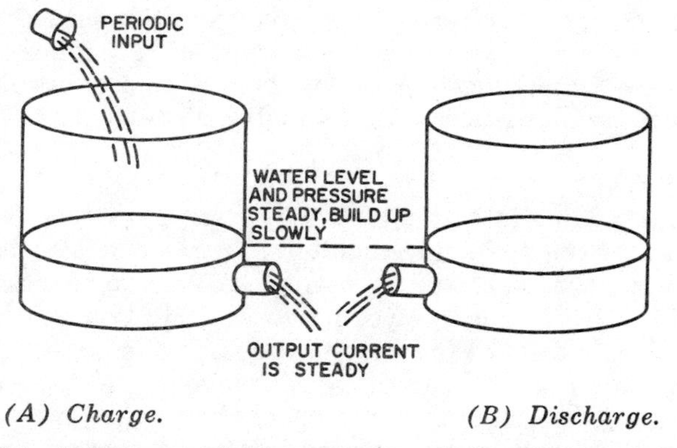

(A) Charge. *(B) Discharge.*

Fig. 1-12. If tank has excessive capacity, it will require more water and more time to build up to the same pressure as in Fig. 1-8. This level will be achieved eventually, however; therefore this is also a long time-constant combination. The current output is steady from half-cycle to half-cycle, but will increase as the pressure does.

Fig. 1-12 shows the effect of greatly increasing the tank capacity. One bucketful of water each cycle will not change the water level substantially; many cycles will be required to build up a pressure equal to that of Fig. 1-8. The moral is that the water tank is probably bigger than necesary, if its function is merely to maintain a constant water pressure and thus a constant water flow through the outlet nozzle. Anyway, this combination is a long time-constant one.

Time-Constant Formula

From the previous analogies it is possible to demonstrate the underlying meanings of several important formulas you will encounter in later chapters—in fact, in all electronic-circuit applications. The first of these is the time-constant formula.

The *time constant* of a resistor-capacitor combination is the time it requires to complete 63.2% of its charging or discharging action. The formula states that this time is equal to the *product* of the component values as follows:

$$T = R \times C$$

where,

T is the time in seconds,
R is the resistance of the charging path in ohms,
C is the capacitance in farads.

When we consider the water tanks and their outlets, it is evident that the emptying, or discharge, time from any given water pressure or level (voltage) will vary directly with the size, or capacity, of the tank and with the amount of resistance the nozzle offers to the water flow. (A wide, short nozzle offers low resistance; a narrow, long nozzle offers high resistance.)

Coulomb's Law

Coulomb's law states another important principle referred to repeatedly in the following chapters, and it can also be demonstrated with the water-tank analogies. This law says that the attraction or repulsion between two electric charges is proportional to the product of their magnitudes, and is inversely proportional to the square of the distance between them. Hence, from this we can deduce that the amount of charge (negative electrons or positive ions) stored in any capacitor is proportional to the voltage across the capacitor. These three quantities are related arithmetically by the following formula:

$$Q = C \times E$$

where,

Q is the quantity of charge in coulombs (1 coulomb = 6.25×10^{18} electrons),

C is the capacitance in farads,
E is the voltage, or electrical pressure, in volts across the capacitor as a result of the stored charge.

From the water-tank analogies it is evident that an increase in the amount of water stored in the tank will raise the water level and consequently the pressure. It is also apparent that a relatively small amount of water will fill up a narrow tank with a small capacity, thereby creating a high water level and pressure. This same quantity of water, when transferred to a wide tank, however, will barely cover the bottom of the tank, and the water level and pressure will be very low.

Ohm's Law

Another important formula whose meaning can be demonstrated with the tank analogies is Ohm's law, which states that the voltage developed across a resistor is proportional to the current flowing through it. These three quantities are related arithmetically as follows:

$$E = I \times R$$

where,

E is the voltage across the resistive path in volts,
I is the current through the resistor in amperes,
R is the resistance of the path in ohms.

If the water level in one of the tanks is raised, it should be clear that the higher pressure will force more water (current) through the nozzle. Also, if the water level is not changed but a larger nozzle is substituted, its resistance will be lower and more water will flow out than before.

Conclusion

Each of these three formulas is elaborated on whenever an R-C combination appears in one of the later chapters and it seems necessary to do so in order to clarify circuit operation. There is a truly enormous body of literature making up all the mathematics of electronics. However, it is not necessary to master it all to understand how electronic circuits operate. A working understanding of the preceding three formulas, plus the two for capacitive and inductive reactance and the frequency formula for tuned circuits, is adequate in helping you *visualize* the action of any circuit. The reason is that all circuit actions can be placed into one of seven major categories. These might be called basic actions, of the types described briefly in this chapter. They are:

Resistor.
Capacitor.
Inductor.
Inductor-capacitor combination (both at resonance and off-resonance.)
Resistor-capacitor combination.
Resistor-inductor combination (which enjoys only limited usage).
Vacuum tubes (which provide the necessary regulation).

Chapter 2

CRYSTAL OSCILLATORS

The crystal-controlled oscillator depends on the *piezoelectric* effect of certain crystals for their ability to generate electric oscillations at radio frequencies. The piezoelectric effect can be visualized from Fig. 2-1. One peculiar quality of such a crystal is its ability to oscillate structurally as well as electrically. Like all oscillations, this one must be sustained by adding energy from an outside source or it will die out. Using Fig. 2-1 we will discuss the condition of the crystal during four successive quarter-cycles of an oscillation.

CRYSTALS

Fig. 2-1A shows the normal physical configuration of a typical crystal. Crystals have various shapes, a flat rectangular plate being the most common. The upper and lower plates are the electrical ones, and the right and left plates the mechanical ones. When a crystal is oscillating, positive and negative electric charges will alternate between the two electric faces, and mechanical distortion will be evident on the two mechanical faces.

Fig. 2-1B shows the concentration of a negative charge on the upper plate of the crystal. Although not shown, positive ions are concentrated on the bottom plate. Fig. 2-1D shows the electrical conditions of the crystal a half-cycle later, when the positive charge is concentrated on the upper plate and the negative charge on the lower one.

Fig. 2-1B also shows the mechanical distortion of the crystal. Here the right and left plates are drawn *inward,* while in Fig. 2-1D they are expanded.

The arrows in Figs. 2-1A and 2-1C point in the direction these two plates are moving at the end of the first and third quarter-cycles.

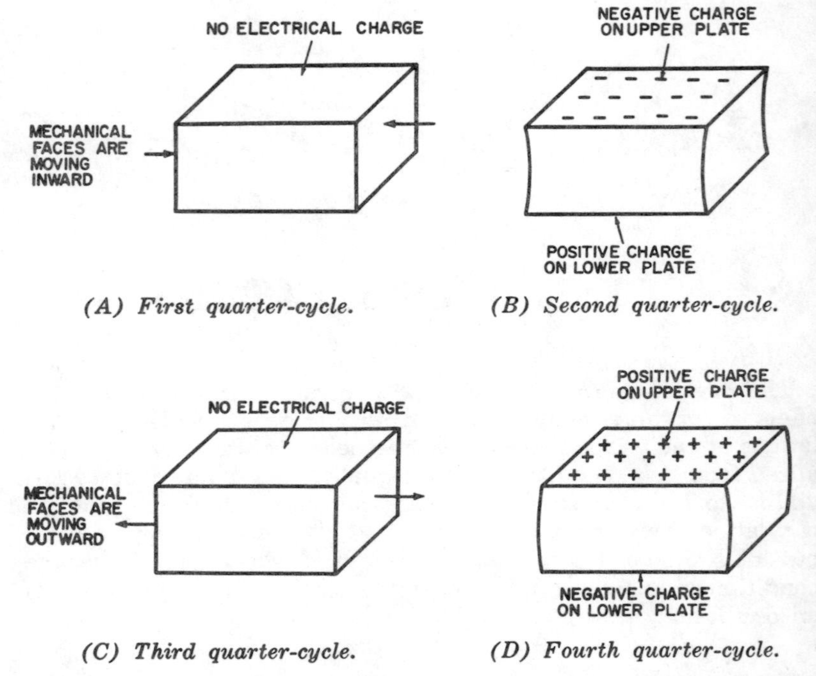

(A) First quarter-cycle. *(B) Second quarter-cycle.*

(C) Third quarter-cycle. *(D) Fourth quarter-cycle.*

Fig. 2-1. Piezoelectric effect on a typical crystal for one cycle of operation.

The thickness of the crystal, the type of material it is cut from, and the orientation of the cut to the crystal material, all have a direct bearing on the frequency at which the crystal will oscillate. A typical quartz crystal, oscillating in the region of 500 kilocycles per second (near the low end of the broadcast band), might have the following *approximate* dimensions:

Thickness—0.25 inch
Width—1.3 inches
Length—1.0 inch

This describes a crystal about one inch square and one-fourth inch thick.

The type of material used and the orientation of the cut to the original axis of the crystal material are of major significance to ne crystal grinder and circuit designer, but not to the average echnician. Hence, this type of information is omitted here, since ne emphasis is more on conveying a general understanding of ne *electrical actions* occurring in a series of widely-used standard ircuits.

CRYSTAL-OSCILLATOR CIRCUIT

Figs. 2-2 and 2-3 show a typical crystal-oscillator circuit using standard triode tube. The necessary circuit components, in ddition to the crystal, are:

R1—Grid-drive and -return resistor.
V1—Triode tube.
C1—Plate tank capacitor.
L1—Plate tank inductor.
M1—DC power supply.

This oscillator requires energy to be fed back from the output plate circuit) to the input (grid circuit) in order to sustain scillation. Feedback is accomplished via interelectrode capaciance from plate to grid of the tube. There is a distinct similarity etween the electrical actions in this circuit and those in the uned-plate–tuned-grid oscillator circuit discussed in a later chaper. The main difference between the two is that the quartz crysal here takes the place of the tuned inductance-capacitance cominaton in the TPTG oscillator.

The electron currents at work in this circuit are:

Grid-driving current, driven by the crystal voltage.
A small amount of grid-leak current.
Plate current of the tube.
Electrons in oscillation in the plate tank.
Feedback electron current.

When the cathode is heated by the filament current (not hown) and plate voltage is applied, the tube will begin to conuct electrons from cathode to plate. This current (P. 38) ows along a closed path from cathode to plate, through the uned L-C circuit and the power supply, and into a common round connection where it can readily return to the cathode. As is true for all tube currents, *a closed path must be available ack to the cathode.*)

The sudden arrival of plate current at the tuned circuit will nevitably cause the circuit to begin oscillating at its natural frequency. Electrons (shown on p. 37) will then move back and forth through the inductor, between the two plates of tank capacitor C1. Fig. 2-2 shows the electrical conditions at the end of the first half-cycle, when the oscillating current has delivered the maximum number of electrons to the top plate of C1. The voltage at the top of C1 is now reduced to a low positive value, as shown by the single plus sign at the top plate and the two at the bottom.

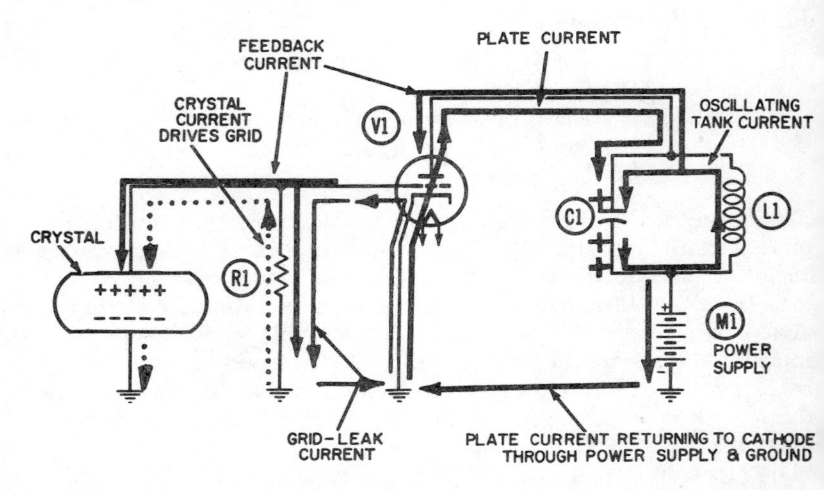

Fig. 2-2. Operation of a simple crystal-oscillator circuit—first half-cycle.

Since it is connected to the top of the tuned circuit, the plate of the tube will always have the same voltage as the top of the tuned circuit. The flow of a small amount of current between the two points serves to transfer the voltage from one to the other. This flow, as shown above, has been labeled the *feedback current*. During the time the upper plate of C1 is less positive than the lower one, the electrons which make up the feedback current will flow *toward* the plate of the tube and drive an equal number of electrons away from the grid and into the grid circuit. This occurs because of interelectrode capacitance within the tube, wherein the plate and grid act as the two plates of a small capacitor. The natural function of any capacitor is to "pass an alternating current." That is, as electrons are driven onto one of its plates, an equal number are driven from the opposite plate. The flow of electrons away from the grid is labeled "feedback current" in Fig. 2-2.

Likewise, whenever the upper plate of C1 is more positive than the lower one, the feedback current will flow in the opposite direction—*away* from the plate of the tube and toward the tuned circuit. The withdrawal of these electrons from the tube plate now draws an equal number of electrons through the grid circuit, *toward* the grid. This condition is depicted in Fig. 2-3.

Now that we have succeeded in getting feedback current to flow in the grid circuit, we have satisfied the condition that en-

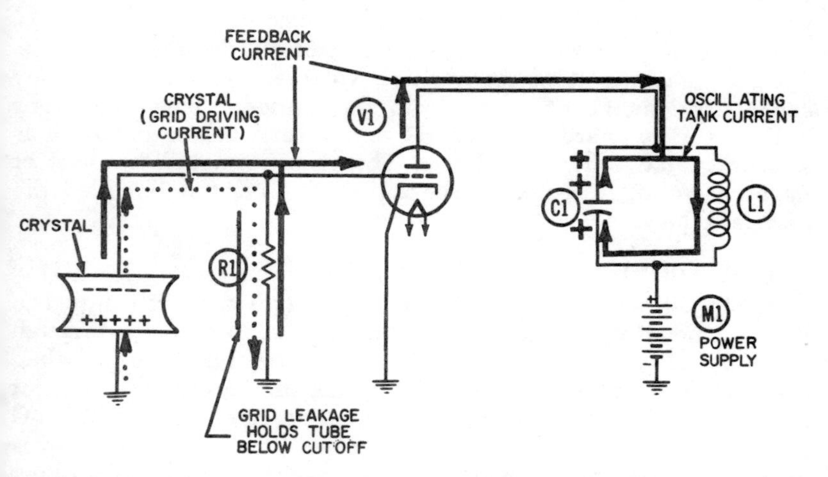

Fig. 2-3. Operation of a simple crystal-oscillator circuit—second half-cycle.

ergy in some form must be coupled back from output to input of the circuit. This coupled energy must also be of the appropriate phase so it can turn the control grid on and off at the proper times in order to sustain the oscillation. Let us now consider the action occurring in the grid circuit.

The first surge of feedback current should be sufficient to start the crystal oscillating. Fig. 2-2 depicts the half-cycle of oscillation when the upper face of the crystal is positively charged. This positive voltage draws a small amount of current (labeled "crystal current" and shown as a dotted line) *upward* through grid resistor R1, performing the function recognized as "driving the grid."

During the same half-cycle shown in Fig. 2-2, the feedback current (shown above) flows *downward* through R1. Thus it appears that the feedback and driving currents, and the respective voltages developed across R1 by them, are clearly out of phase.

In Fig. 2-3 both currents have reversed direction and so they are still out of phase. These conditions are still not conducive to oscillation. (We will return to this point shortly.)

The polarities of the crystal voltage, as shown in Figs. 2-2 and 2-3, are such that they properly support the plate tank oscillation. When the top of the crystal is positive, as in Fig. 2-2, the control grid will release maximum plate current through the tube. This current will reach the top of the tuned circuit (the upper plate of capacitor C1) at the moment the voltage at this point is least positive and, in effect, will further reduce this voltage. Thus, the oscillation is sustained by strengthening it during each cycle.

Upon reaching its most positive voltage, the grid will draw some grid-leakage electrons from the cathode. These electrons will flow downward through R1 and return to the cathode, as shown by the solid thin lines. The resulting small amount of negative voltage created at the top of this resistor is called the *grid-leak bias.*

Class-C Operation

In Fig. 2-3, when the crystal current drives the control grid to its most negative voltage, minimum plate current flows through the tube. This current reaches the top of the tuned circuit when the voltage produced at this point by the oscillating tank current is most positive. If the tube is being operated under Class-C conditions, then no plate current at all will flow when the grid is negative. (The definition of Class-C operation is that plate current shall flow for less than half of each cycle..)

The circuit here would most likely be operating under Class-C conditions. The amount of bias voltage developed by the grid-leakage electrons flowing through R1 is sufficient to cut the tube off during most of the individual cycle. The voltage swing developed by the crystal can be quite substantial, 3 to 5 volts being fairly normal with most amplifier tubes. When higher-power tubes are used, more powerful oscillations are developed in the plate tank circuit. Also the feedback impulses become stronger and may drive a crystal to oscillations of 20 to 25 volts or more. In fact, it is possible to shatter a crystal by driving it with too strong a feedback impulse.

TIME CONSTANTS AND VOLTAGE STORAGE

In order for the grid-leakage current to create a bias voltage which will persist during that part of the cycle when the tube is not conducting, there must be some method of storing these electrons at the top of the resistor. This is normally performed by a grid-coupling capacitor. It is omitted in this circuit because the crystal and its electrodes to which the electrical faces are connected have sufficient capacity to store the electrons, or charge.

These inherent capacitances need be only a few micromicrofarads in order to provide the necessary storage. The size of the grid resistor is determined to some extent by the amount of capacitance, in accordance with the time-constant formula for resistors and capacitors. This formula tells us that the time required to discharge 63.2% of any stored voltage is equal to the *product* of the components values involved—namely, the capacitance of he unit where the voltage (electrons or positive ions) is stored, and the resistive path over which the discharge must occur. If we assume a grid-resistor value of 300,000 ohms and an inherent crystal plus holder capacitance of 10 micromicrofarads, we can solve for time as follows:

$$T = 300{,}000 \times 10^{-12}$$
$$= 3 \times 10^{-6} \text{ second, or 3 microseconds.}$$

If a crystal has a natural frequency of oscillation equal to 500 kilocycles per second, then one cycle of oscillation will occur in microseconds. Thus, we see that the time constant of 3 microseconds is equal to one and a half cycles of oscillation.

Under these assumptions the circuit might oscillate satisfactorily, although when successful circuit operation depends on maintenance of a grid-leak bias voltage, it is usually desirable or this voltage to discharge *much more slowly* than is indicated here. In fact, it is more normal for the R-C combination to have time constant equal to five or ten cycles. When we consider the quation T= RC, it is evident that the time constant can be made onger by increasing either the resistance or the capacitance in he circuit.

If we increase the value of the grid resistor to, say, 1 or 2 megohms we may overcome one objection, only to create another ne. This problem, known as *squegging,* occurs as the oscillation uilds up, causing successively larger amounts of grid-leakage urrent to flow each cycle. Eventually this current becomes great nough to cut off the oscillation, and it will remain cut off for many cycles (perhaps even hundreds of them) until the leakage lectrons can escape from storage by flowing back to ground hrough the grid resistor.

A much simpler means of lengthening the time constant of the rid-discharge circuit is to add a small capacitance in parallel with the crystal and grid resistor. This permits using grid resisors of nominal size, on the order of 25,000 or 50,000 ohms.

PHASE RELATIONSHIPS

Now to return to the phasing which should exist between the feedback and crystal voltages. The crystal oscillation must be sustained electrically by receiving one voltage impulse, or "kick," each time the feedback voltage reverses polarity. However, when the plate circuit is tuned to the *exact* crystal frequency, the condition in the previous sentence will be impossible to attain because the plate tank voltage and resulting feedback voltage will always be out of phase with the crystal voltage. Under such conditions, oscillation cannot occur.

By detuning the plate tank circuit slightly, it is possible to shift the phase of the feedback voltage enough that a tiny fraction of each cycle of feedback voltage will be in phase with the crystal voltage. This momentary in-phase condition is sufficient to reinforce the crystal oscillation. Such a condition is difficult to show, either with diagrams or from the waveform. However, it is subject to intricate mathematical demonstration, which we will forego in this text.

If the plate circuit is detuned toward a higher frequency, the circuit is then said to be slightly inductive—an often-used descriptive term which merits considerable elaboration.

Suppose the plate circuit is tuned to a somewhat higher frequency than that of the plate-current pulses resulting from grid-circuit action. Because these pulses are arriving at the lower frequency, they will not "see" a resonant tank circuit waiting for them. Instead, two reactive paths—one capacitive and the other inductive—present themselves. The pulses will be divided between the two in proportion to the amount of reactance presented by each path. The one offering the *lesser* reactance will get the *larger* share of current, and the circuit will also bear its name. The smaller share of current goes over the higher-reactance path. If the two reactances are the same, the pulses will be equally divided, of course, and the circuit is said to be *resonant*.

In fact, the formula for the resonant frequency of a circuit is derived from the formulas for inductive and capacitive reactance. For example:

$$X_L = 2\pi fL$$

and,

$$X_C = \frac{1}{2\pi fC}$$

Since, at resonance, the inductive reactance must equal the capacitive reactance, the resonant-frequency formula is derived as follows:

$$2\pi fL = \frac{1}{2\pi fC} \qquad f^2 = \frac{1}{2\pi L \times 2\pi C} \qquad f = \frac{1}{2\pi\sqrt{LC}}$$

vhere,

X_L is the inductive reactance in ohms,
X_C is the capacitive reactance in ohms,
f is the frequency in cycles per second,
L is the inductance in henrys,
C is the capacitance in farads.

It is obvious, from the equations for X_L and X_C, that if a cur-ent is being supplied at the resonant frequency of an L-C com-ination, and then if the frequency of this current is lowered, he inductive reactance will be less for this lower frequency. Sim-larly, the capacitive reactance will be higher. Such a condition xists when the plate tank circuit is tuned above the crystal fre-uency—the inductor will have a lower reactance (opposition o electron flow) and therefore claim a greater share of plate urrent (also called line current). Consequently, the tank is said o be tuned inductively.

It is well to keep clearly in mind exactly which frequency is eing discussed. For example, in addition to the frequency of the pplied current, there may be a resonant circuit frequency, and he two could be entirely different in value.

DETERMINING OPERATING FREQUENCY

If the crystal oscillates at a different frequency from that of he plate tank circuit, which circuit determines the over-all oper-ting frequency? The answer: the circuit with the higher *Q*.

The *Q*, or quality, of a tuned circuit was described in Chapter 1 s the ratio between the electrons in oscillation and those which lrop out each cycle because of circuit losses. A "clean" tuned ircuit, consisting of inductance and capacitance plus a small mount of internal resistance, may have a *Q* of over a hundred. Such a circuit can be considered fairly "high quality." A crystal, n the other hand, may have a *Q* of over a thousand, and in spe-ial cases as high as a half million. The crystal used in the cir-uit of Figs. 2-2 and 2-3 could have a *Q* of two or even three housand.

Therefore, in this circuit the natural frequency of the crystal letermines the operating frequency of the circuit, and a tuned late-tank circuit must be operated off-resonance to sustain os-illations.

Chapter 3

THE HARTLEY OSCILLATOR

The Hartley oscillator shown in the accompanying diagrams ranks high in popularity when radio-frequency oscillator circuits are mentioned. Recall that the function of any oscillator circuit is to generate an alternating current at the desired frequency. This circuit will be explained by assuming the desired frequency to be within the broadcast band—for example, 1,000 kilocycles (1,000,000 cycles) per second.

Figs. 3-1 through 3-4 show the operation of the Hartley oscillator during each quarter-cycle. The necessary circuit components are as follows:

C1—Radio-frequency tank capacitor.
L1—Radio-frequency tank inductor (also acts as an autotransformer).
C2—Grid-coupling capacitor.
R1—Grid-leak biasing resistor.
V1—Oscillator tube.
M1—Power supply.

Here are the currents at work in this circuit.

1. Oscillating tank current .
2. Plate current .
3. Grid-leak biasing current.
4. Grid-driving current.
5. Feedback current.

Feedback occurs in the Hartley oscillator as current is drawn to the cathode through the tapped portion of coil L1. As more and more tube current (or plate current) flows, autotransformer action occurs between the tapped portion of L1 and the whole coil, causing another component of current to flow *down-*

ward through the coil. This is the feedback current, shown by the dotted lines.

How the voltage polarities and current movements in the tuned tank are correlated with those in the cathode circuit of the tube, so that the feedback will have the appropriate phase to support the oscillation, will be discussed later. First, let us look at the individual currents in detail.

OSCILLATING TANK CURRENT

Twice each cycle, the oscillating current (or tank current) moves from plate to plate of tank capacitor C1, via coil L1. The values chosen from C1 and L1 are such that these two com-

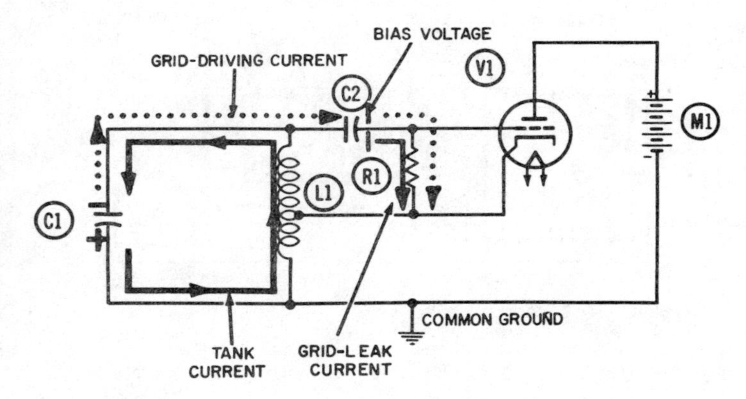

Fig. 3-1. Operation of the Hartley oscillator—first quarter-cycle.

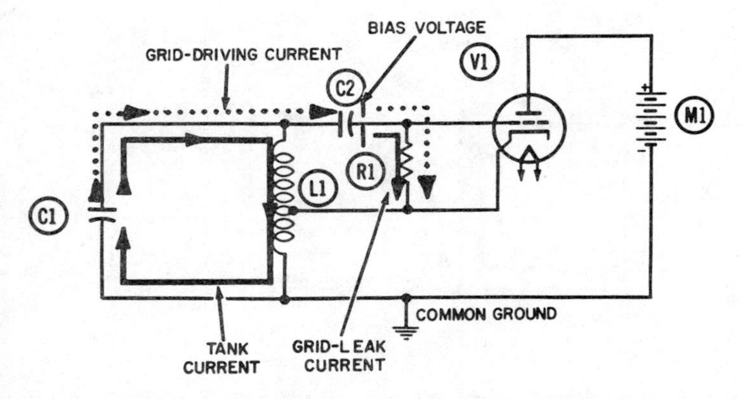

Fig. 3-2. Operation of the Hartley oscillator—second quarter-cycle.

ponents will be resonant at the desired frequency—1,000,000 cycles per second in our example. These values can be determined from the standard formula for finding the resonant frequency:

$$f = \frac{1}{2\pi\sqrt{LC}}$$

Fig. 3-1 shows the first quarter-cycle, during which the current flows from the lower plate of capacitor C1 to the bottom of coil

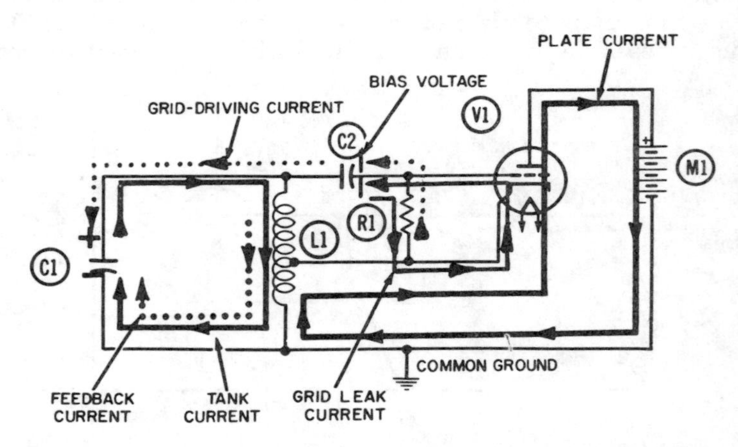

Fig. 3-3. Operation of the Hartley oscillator—third quarter-cycle.

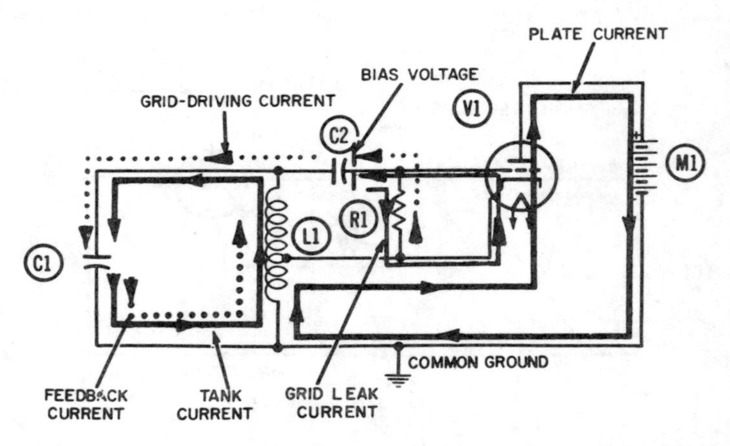

Fig. 3-4. Operation of the Hartley oscillator—fourth quarter-cycle.

L1 and through it, onto the upper plate of C1. The surplus of electrons makes the upper plate of C1 more negative than the lower one, as indicated by the minus sign.

This charging of the upper plate of tank capacitor C1 also delivers electrons to the left plate of coupling capacitor C2, which becomes charged to the same negative voltage. The number of electrons flowing onto the left plate of coupling capacitor C2 is accompanied by an equal number flowing *away* from the opposite plate. The resulting current, grid-driving current, develops the grid-driving voltage across resistor R1.

The circuit diagram for the first quarter-cycle (Fig. 3-1) shows conditions as the grid-driving voltage becomes more and more negative. Electrons are flowing downward through R1, so we know it is more negative at the top, since electrons always flow from a more negative to a less negative area.

The second diagram (Fig. 3-2) denotes the second quarter-cycle, when the tank current has reversed its direction. There is no electron flow in the tank at the start, but once the flow starts, it builds up rapidly and reaches maximum at the end of the quarter cycle. The electrons formerly concentrated on the upper plate of tank capacitor C1 and the left plate of coupling capacitor C2 are distributed equally between both plates of C1 *the moment the quarter-cycle ends.* Now the tank voltage (measured at the top of the tank) is zero.

The waveform diagrams in Fig. 3-5 show this relationship. In interpreting sine-wave diagrams of current and voltage, it would be wise to clarify some basic assumptions. In the waveform representing tank current, the half-cycles *above* the reference line are arbitrarily chosen to represent electron current flowing *from* the upper plate of tank capacitor C1 *to* the lower plate (through tank coil L1, of course).

The half-cycles below the reference line represent electron current flowing in the opposite direction, or *upward* through L1. Whenever the sine waves cross the reference line, current is flowing in *neither* direction (meaning there is zero current). At this time the current is changing direction. As an example, at the *end* of the first quarter-cycle, electrons have been flowing upward and charging the upper plate of C1 with a negative voltage. During the second quarter-cycle, this current reverses and begins flowing downward.

By the end of the second quarter-cycle, the current is flowing downward through L1 at the maximum rate, and the negative voltage on the upper plate of C1 has discharged to zero. During the third quarter-cycle (Figs. 3-3 and 3-5), this downward electron flow continues, building up a peak negative voltage on the lower plate, and a peak positive voltage on the upper plate, of capacitor C1.

At the start of the fourth quarter-cycle, the electron current in the tank again reverses direction, as shown in Figs. 3-4 and

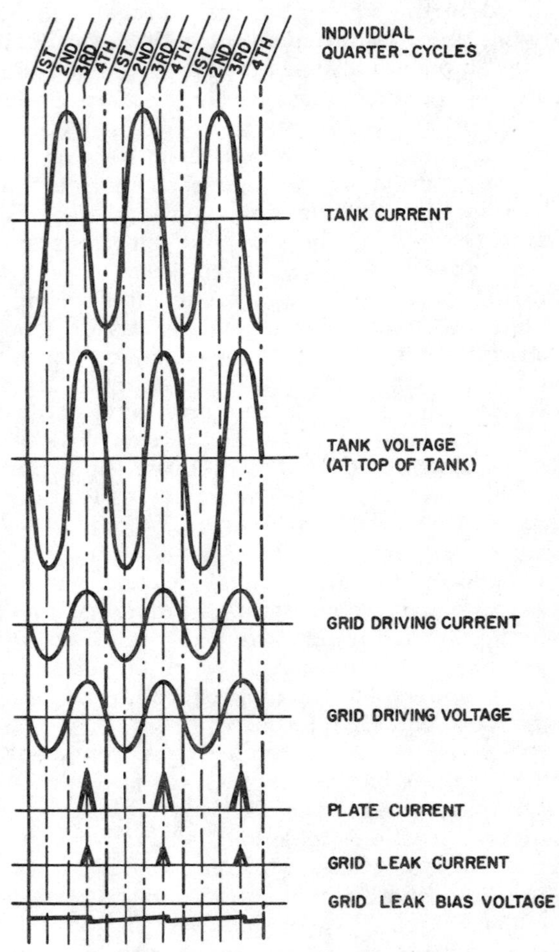

Fig. 3-5. Hartley-oscillator waveforms.

3-5, and begins flowing *upward* through coil L1. This action begins discharging the negative voltage on the lower plate of C1. So, at the end of the fourth quarter-cycle, the voltage on both capacitor plates is again equalized and the voltage at the top of the tank is zero.

During the entire first two quarter-cycles, no current can flow through the tube itself because of the combination of permanent negative bias and instantaneous grid-driving voltages. The origin of the negative bias voltage becomes apparent when we study the action during the third quarter-cycle, depicted in Figs. 3-3 and 3-5. The tank current has now completed its journey through the inductor to the lower plate of the tank capacitor, charging the plate negatively and at the same time leaving a deficiency of electrons on the upper plate (which we know to be a positive voltage). Electrons are also being drawn away from the left plate of coupling capacitor C2 and, in turn, away from the grid area of the tube and upward through grid resistor R1. (This electron flow is shown as a dotted line to differentiate it from the grid-leak current shown as a solid line.) An instantaneous positive voltage is created at the grid and momentarily exceeds the permanent negative bias voltage. The result is that current is permitted to flow from cathode to plate at the end of the third quarter-cycle.

The plate current (right-hand solid line) flows during the latter part of the third and first part of the fourth quarter-cycles, as shown in Fig. 3-5. This current will flow whenever the grid voltage is less negative than the cutoff voltage of the tube.

Grid-leak current (left-hand line) flows for an even shorter time. This occurs at the end of the third and the start of the fourth quarter-cycles, when the instantaneous grid voltage is positive enough that the grid wires will attract and capture the negative electrons.

It is during the third quarter-cycle that the control-grid voltage is raised above the cutoff value and tube current flows. For the remainder of the cycle, the control-grid voltage becomes less and less negative as the tank current completes its journey through the tank inductor to the lower plate of the tank capacitor. As the grid voltage becomes less negative, the current through the tube continues to increase. The complete path for this tube current (or plate current) includes the lower portion of tank inductor L1.

When tapped in this fashion, an inductor is called an *autotransformer*. The action of an autotransformer is identical to that of a conventional transformer, where the windings are separated. The tube current, or plate current, can be considered the *primary* current in this transformer action, and the current it sets up in the tank circuit (lower left corner) is the *secondary* current. The latter is appropriately labeled the feedback current, since it transfers energy from the output (plate circuit) to the input (grid circuit).

Feedback in an oscillator circuit hinges on two essentials: One is a means for transferring energy from output to input. The other is that this energy must be in the proper phase to reinforce the oscillation in the grid tank circuit. The following discussion will show how the second essential is achieved.

The tube current in the cathode circuit flows in one direction only—from the common ground to the cathode, through the tube to the plate, then through the power supply and back to ground. However, transformer action will occur just as readily whenever a direct current increases or decreases as it will for outright reversals in the direction of current flow.

In Fig. 3-5 we see that plate current begins to flow during the third quarter-cycle, and that it builds up to maximum at the end of this quarter-cycle. Since all inductors act to oppose *any change* in current, this build-up will cause another current to flow in the opposite direction and buck the tube current as it passes through the lower portion of coil L1. This bucking current is called the counter emf (also, the back emf or back voltage) of the inductor.

As the third quarter-cycle comes to a close, this counter-, or secondary, current (lower left corner) also stops flowing in the tank circuit. During its brief life it has been flowing in the same direction as the tank current, thereby being at least partially in phase with the tank current and thus reinforcing it. The fact that such reinforcement has occurred can also be deduced from the following observation: At the end of the third quarter-cycle, both currents have independently delivered electrons to the lower plate of the tank capacitor and increased its negative voltage.

This secondary current both *starts and stops* during the latter part of the third quarter-cycle. Therefore, as depicted in Fig. 3-6 by the distorted sine wave representing feedback current, an entire half-cycle has occurred in less time than a quarter-cycle of tank current.

At the beginning of the fourth quarter-cycle, the tank current begins to flow upward through tank inductor L1, tending to neutralize the positive voltage on the upper plate of C1. An examination of the sine-wave relationships for this circuit (Fig. 3-5) will reveal that the tank voltage, plus the resultant grid-driving current and its attendant voltage developed across resistor R1, are all essentially in phase. Consequently, during the fourth quarter-cycle the positive tank-voltage decreases to zero and the positive grid voltage also disappears.

A reduced grid voltage in turn lowers the plate current flowing through the tube and, of course, through the lower portion of

:he inductor. This decrease in what we earlier identified as ‘primary” current results in transformer action between the whole winding and its lower portion. As before, the transformer action will oppose any change in current, so a secondary current starts flowing *upward* through the winding in a steadily increasing amount to compensate for the corresponding drop in the primary current.

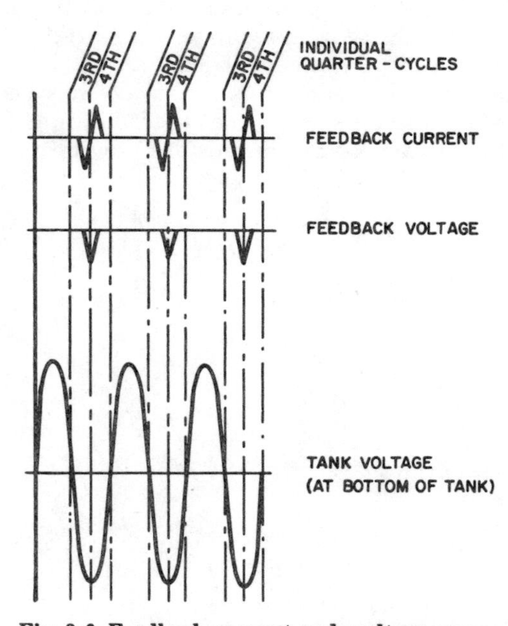

Fig. 3-6. Feedback-current and -voltage waveforms in the Hartley oscillator.

The plate current is cut off midway in the fourth quarter-cycle, and this secondary current also quickly drops to zero. During its brief span the secondary current flowed in the same direction as the tank current and the two can be considered roughly in phase. Again, the secondary current thus reinforces the oscillation in the tank. Fig. 3-6 shows that the feedback current will again go through half a cycle in less time than the tank current and voltage require for a quarter-cycle. It is also fairly obvious that the whole cycle of feedback current is a very distorted sine wave. Although not shown on its sine wave, the tank voltage (and current) will in turn be distorted, resulting in higher positive than negative peaks. An even more important phenomenon resulting from this and most other feedback is a shift in the tank frequency. In a Hartley oscillator, this shift is toward the lower end of the spectrum; therefore the circuit oscillates at a somewhat lower frequency than that computed using the standard frequency formula.

The feedback current drives additional electrons to the lower plate of tank capacitor C1. As a result, a negative pip (pulse) of voltage occurs at the end of each third quarter-cycle, as shown in Fig. 3-6. This extra component is then added to the tank voltage. Fig. 3-5 shows the changes in voltage, measured at the top of the tank, during each quarter-cycle.

For the sake of convenience, this sine wave has been inverted in Fig. 3-6 to show the tank voltage at the bottom of the tank. As you can see, the feedback-voltage pips are now in phase with the tank voltage and will reinforce it each cycle.

Chapter 4

THE COLPITTS OSCILLATOR

The Colpitts oscillator is another widely used circuit whose function is to generate a continuous alternating current at a fixed radio frequency. The Colpitts oscillator is usually operated under Class-C conditions, which means plate current flows during less than half of each cycle. Figs 4-1 through 4-4, depicting successive quarter-cycles, serve to explain how this circuit operates. The essential circuit components and their functional titles are as follows:

L1—Radio-frequency tank inductor.
C1 and C2—Radio-frequency tank capacitors.
C3—Grid coupling capacitor.
R1—Grid-leak biasing resistor.
V1—Oscillator tube.
C4—Feedback and blocking capacitor.
L2—Radio-frequency choke.

Each of the currents at work in the circuit is shown for easier identification and analysis. The currents are:

1. Radio-frequency tank current.
2. Unidirectional plate current.
3. Grid-leak biasing current.
4. Grid-driving current.
5. Feedback current.

Fig. 4-5 shows representative waveforms of the currents and voltages in this circuit. Two complete cycles are included, with dividing lines between each quarter-cycle so that momentary points on the waveforms can be related to their respective current directions and voltage polarities.

As in the Hartley oscillator, the feedback impulse is applied at the end of the tank opposite the grid connection. When the top of the tank is negative, the bottom is positive, and vice versa. Figs. 4-1 and 4-2 show the tank current flowing in a direction which delivers electrons to the lower plate of tank capacitor C2. At the end of the first quarter-cycle there is no voltage on the plates of C1 and C2, although maximum tank current is flowing between them. Proper interpretation of the tank-current sine wave of Fig. 4-5 tells us that peak amplitude is also the moment of maximum current. This particular sine wave has been drawn so that any displacement *above* the reference line means electrons are flowing downward through the tank coil, or *away* from the upper plate of C1, resulting in a positive voltage at this point.

By the same token, any displacement of the sine wave *below* the reference line means just the opposite—electrons are flowing *upward* through the tank coil, toward the upper plate of C1, resulting in a negative voltage there. The tank voltage waveform of Fig. 4-5 is the waveform appearing at this point.

At the end of the second quarter-cycle, the tank current has stopped flowing; this is also the moment the current sine wave crosses the reference line. Electrons have been delivered to the lower plate of C2, giving it maximum negative voltage. Likewise, the upper plate of C1 has an equivalent deficiency of electrons, or a peak positive voltage. These two voltage peaks are indicated in Fig. 4-2 by appropriate minus and plus signs on C2 and C1 respectively.

The grid-driving currents, shown on pps. 56 and 57 in Figs. 4-1 through 4-4, are driven by the oscillating voltage in the tank circuit and constitute the main load on it. The grid driving current is external from the tank current/voltage combination and, as such, is in phase with the voltage at the bottom of the tank. The voltage which this current develops across grid resistor R1 will be in phase with the driving current at all times (because current and voltage in a resistive path are always in phase with each other). Consequently, the grid-driving voltage will be in phase with the tank voltage when measured at C2, and exactly out of phase with the tank voltage when measured at the top of the tank, as portrayed in Fig. 4-5.

The directions indicated in Figs. 4-1 through 4-4 for the grid-driving current are the result of electron concentration (negative voltage) or deficiency (positive voltage) on the lower plate of tank capacitor C2. For instance, during the second and third quarter-cycles, there is always some negative voltage on C2, so the grid-driving current is repelled from it. But during the first and fourth quarter-cycles, the positive voltage on C2 draws the grid-driving current toward this point.

As explained previously, current does not actually flow through grid coupling capacitor C3—the electrons flowing onto one side of the capacitor drive an equal number off the other side. Conversely, whenever electrons are drawn off one plate of a capacitor, an equal number will be attracted to the other plate.

The grid-driving current will develop a positive-going voltage across resistor R1 during the fourth quarter-cycle. Midway through this quarter-cycle, the positive voltage cancels out the more permanent negative voltage resulting from grid-leak biasing action. The grid voltage is now less negative than the cutoff voltage, and plate current will start to flow through the tube. (The cutoff voltage is a negative value below which tube current cannot flow, and above which it can.)

Plate current begins to flow during the fourth quarter-cycle, as indicated in Fig. 4-5, and steadily increases until the grid voltage reaches zero just before the fourth quarter-cycle ends.

The conventional path for unidirectional plate currents is through the tube from cathode to plate; then through any load, where the essential work of the circuit is done; and finally, through the power supply to a common ground which provides the necessary return to the cathode. This plate circuit is no exception, as you will note by the solid path in Fig. 4-4. The load is made up of feedback capacitor C4 and radio-frequency choke L2. It is necessary that the plate current flow onto the right plate of C4, and the radio-frequency choke is placed in the circuit to make sure it does.

A radio-frequency choke, as its name implies, will stifle the passage of radio-frequency currents. A choke is nothing more than an inductor, and the universal property of any inductor is that it will oppose any change in current. The *amount* of opposition depends on the inductance and on the *frequency* of the current trying to flow, as related by the mathematical formula for inductive reactance:

$$X_L = 2\pi fL.$$

While the plate current is increasing, it is unable to enter the choke, even though the normal path is through the choke to the power supply. Momentarily succumbing to this opposition, the plate current is diverted down the only other path available to it and heads for the right plate of feedback capacitor C4. This influx of electrons onto one plate of C4 drives an equal number away from the other plate and toward the tank circuit.

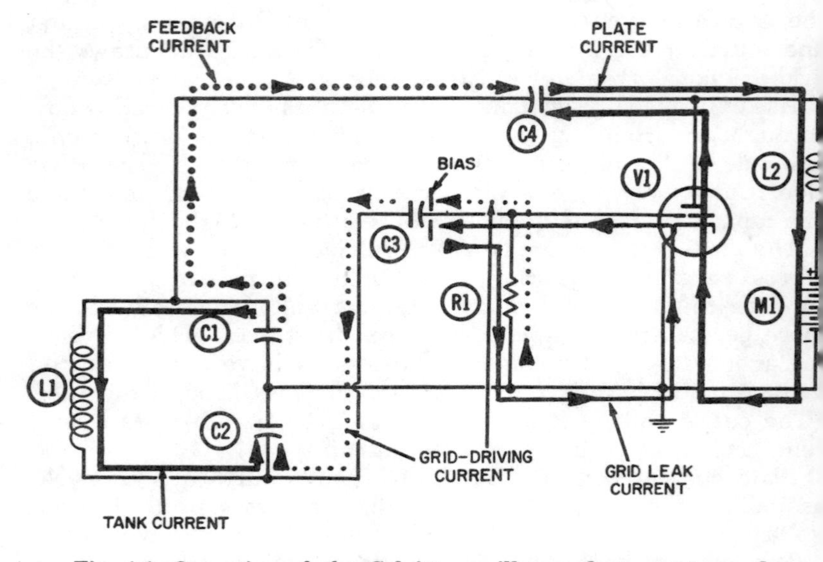

Fig. 4-1. Operation of the Colpitts oscillator—first quarter-cycle.

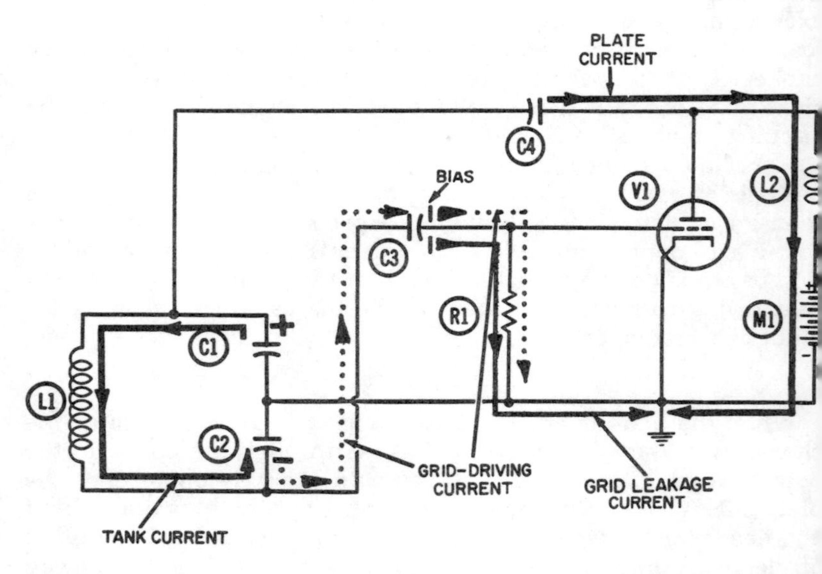

Fig. 4-2. Operation of the Colpitts oscillator—second quarter-cycle.

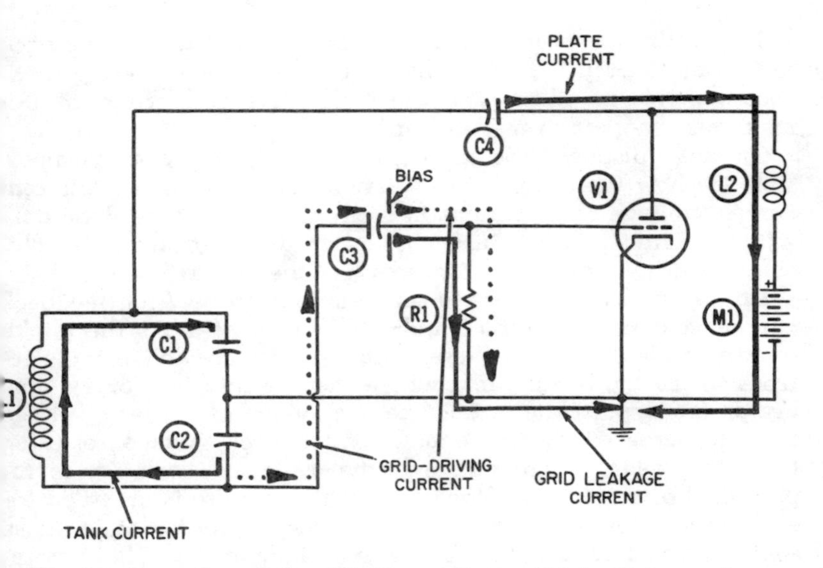

Fig. 4-3. Operation of the Colpitts oscillator—third quarter-cycle.

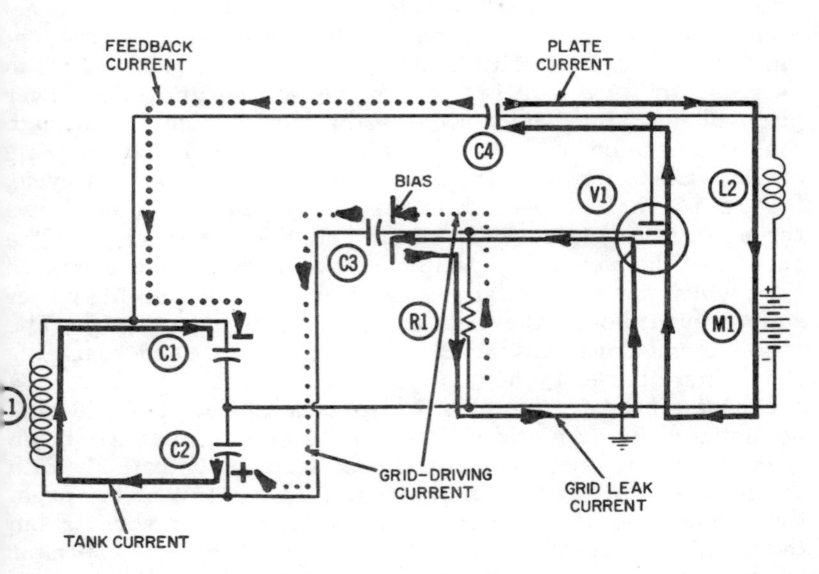

Fig. 4-4. Operation of the Colpitts oscillator—fourth quarter-cycle.

This is the way a feedback pulse is delivered from the output to the input circuit. The feedback current, shown in large dots, should arrive in phase with the tank current in order to offer maximum support to the oscillation.

On static displays like circuit diagrams, it is obviously impossible to show the *exact* phase relationships between all different current-voltage combinations in a circuit. This can be done only with animated drawings, although it can be demonstrated graphically by using waveform diagrams like the one in Fig. 4-5.

An approximation of the phase relationship between feedback current and tank current is indicated in Fig. 4-5. As the fourth quarter-cycle draws to a close, both currents are delivering electrons to the top of the tank, where they go into temporary storage on the upper plate of C1. The peak value of the tank voltage is thus increased by the amount of feedback voltage resulting from the feedback current, and the cycle of oscillation is replenished or reinforced. Since one feedback pulse is provided for each cycle of oscillation and it is in the same approximate phase each time, the oscillation will continue indefinitely. The inevitable losses due to wire resistance, interaction with nearby objects, etc., are compensated for by the regular feedback pulse.

As soon as the next quarter-cycle begins (which corresponds again to Fig. 4-1), the tank current begins its journey from the upper to the lower side of the tank, through the inductor. This action begins delivering electrons to the lower plate of tank capacitor C2. The high positive tank voltage at this point begins to decrease, in turn reducing the voltage at the grid. The lower grid voltage immediately reduces the electron stream through the tube, as indicated by the waveform in Fig. 4-5, and the plate current falls to zero about midway through the first quarter-cycle.

Figs. 4-1 through 4-4 shows that a continuous current flows through radio-frequency choke L2 and the power supply. The choke and the feedback capacitor form an inductance-capacitance filter which maintains a fairly steady current flow into the power supply, even though the plate current is arriving in spurts. The tube current flows only at the end of the fourth and beginning of the first quarter-cycle.

During a brief portion of this plate-current flow, the grid-tank oscillation drives the grid positive, and grid-leak current (shown a solid line) begins to flow. It originates at the cathode with the plate current, but exits at the grid as a result of the negative electrons striking the momentarily-positive grid wires. From the tube, the current flows onto the right plate of grid capacitor C3, forming a pool of electrons (or negative voltage) there until they can drain off ("leak") back to ground and the cathode through grid resistor R1. This leakage goes on continuously, as indicated by the solid thin lines on all four circuit-operation diagrams.

The negative voltage created at the grid by these grid-leak electrons is known as the grid-bias voltage, or more simply as the bias voltage. The instantaneous grid voltage is always the product of the permanent bias and grid-driving voltages, the current of which is shown in small dots. The bias voltage very

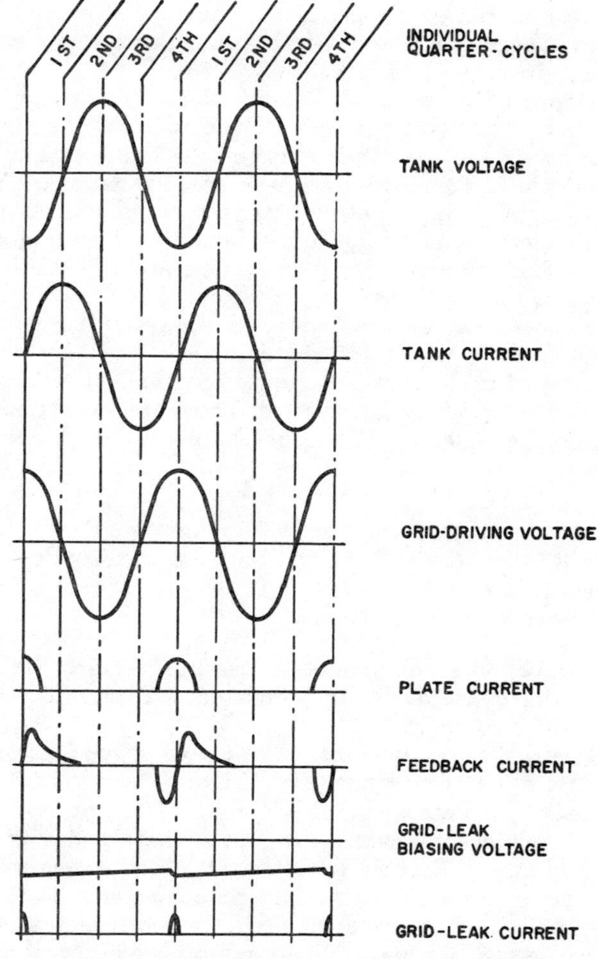

Fig. 4-5. Voltage and current waveforms in the Colpitts oscillator.

quickly stabilizes at an average value. When this value has been reached, we know that during each cycle, the number of electrons flowing into the "electron pool" on the right plate of C3 must equal the exact number flowing out. Otherwise, the bias voltage would not be stable. If more electrons come in during the fourth quarter-cycle than are drained down through the resistor during the whole cycle, the negative bias voltage will increase and act like the closing of a throttle to the electron stream through the tube. Now, fewer electrons will also be attracted from the electron stream to the control-grid wires, and the number coming into the electron pool on the capacitor will likewise be reduced. Therefore, the grid-leak voltage is automatically prevented from increasing indefinitely.

It can happen that more electrons will be drained away from the capacitor each cycle than will come in from the tube. This will occur when for any reason the oscillation is weakened, reducing the value of the grid-driving–current-voltage combination. When this happens the negative bias voltage must decrease until the outgoing electrons no longer exceed the incoming ones, at which time the bias voltage will again be stabilized. Thus, we see that natural limitations existing in either direction prevent the unlimited growth or decay of the grid-leak bias voltage, so that the oscillator tends to be self-stabilizing.

The values of grid-resistor R1 and capacitor C3 are important in the design of a grid-leak system. They are regulated by the time-constant formula, explained in Chapter 1, which states that:

$$T = R \times C$$

where,

T is the time constant of the combination in seconds,
R is the resistance in ohms,
C is the capacitance in farads.

It is usually more convenient to use microfarads rather than farads for the capacitance, in which case T will be in microseconds instead of seconds.

If, in this example, the tank current were suddenly stopped, the grid-driving–current-voltage combination would also die, and no more leakage electrons would flow out of the tube to the grid capacitor. The electrons which are stored there and make up the grid-bias voltage will immediately begin discharging through the grid resistor to ground. During one time-constant, 63.2% of this discharge will have been completed . . . after another equal period has passed, 63.2% of the remaining electrons will have discharged to ground . . . and so on.

Theoretically, it would take an infinite number of periods for any voltage to completely discharge itself. Practically speaking, as few as ten time-constant periods are more than sufficient, for in this length of time a voltage will discharge down to about a millionth of its original value!

In choosing suitable values of R and C for providing the grid-leak bias, we try to select ones whose product (in other words, time constant) is at least five times the period required for one cycle of the basic oscillation frequency. In other words, five or more cycles of oscillation (and grid leakage) will occur during each time constant. This assures us that the bias voltage will have relatively little chance to discharge to ground before another "shot" of electrons arrives from the tube. When this condition is satisfied, the combination of resistor and capacitor is identified as a long time-constant combination. That is, it is a long time constant *with respect to the particular frequency under consideration*—the basic oscillator frequency.

There is still another important consideration in choosing the size of the resistor, and that is the amount of load the grid circuit places on the oscillating tank current. It is desirable for this loading to be as small as possible (consistent with getting the job done). The task being performed here is driving the grid at the tank frequency, and the grid-driving current (shown in small dots) performs this task, as discussed previously. If allowed to become too large, this current will overload the tank oscillation. The grid-driving current can be kept small by increasing the value of the grid resistor. If the latter has an extremely low resistance, it will constitute a heavy load on the tank voltage because it will permit an extremely large grid-driving current to flow—so large, in fact, that the tank voltage could never reach the desired peak value. On the other hand, if the grid resistor has an extremely high value, it will constitute a very light load on the tank voltage because it will allow only a very small grid-driving current to flow.

The *amount* of grid-driving voltage will always equal the amount of tank voltage. The former can be developed by either a small current flowing through a large resistor, or a large current through a small resistor. This is in accordance with Ohm's law, which tells us the amount of voltage developed across a resistor is proportional to the amount of current flowing through it, and also to the size of the resistor.

Chapter 5

THE TUNED-PLATE – TUNED-GRID OSCILLATOR

The tuned-plate–tuned-grid (TPTG) oscillator, as its name implies, utilizes tank circuits in both the plate and grid circuits. Feedback from output to input, a necessary function in any self-sustained oscillator, is accomplished by using the interelectrode capacitance between the plate and grid.

Figs. 5-1 and 5-2 show the circuit and the currents flowing for each half-cycle of operation. The circuit components are as follows:

L1—Grid tank inductor.
C1—Grid tank capacitor.
C2—Grid coupling capacitor.
R1—Grid-leak bias resistor.
V1—Oscillator tube.
L2—Plate tank inductor.
C3—Plate tank capacitor.
C4—Output coupling capacitor.
C5—Decoupling capacitor.

The output voltage of the oscillator is capacitively coupled, via C4, to the next stage. Capacitor C5 acts as a decoupling filter to keep pulses of plate current from entering the power supply and affecting its output voltage.

There are three main groups of electron currents whose movements, if analyzed, will lead to understanding the operation of this oscillator circuit. These groups might be labeled as follows:

1. Alternating radio-frequency currents.
2. Unidirectional radio-frequency currents.
3. Pure direct currents.

ALTERNATING RADIO-FREQUENCY CURRENTS

There are two tank currents in the alternating RF category. One is in the grid circuit and is shown a dotted line; the other is in the plate circuit and is shown a solid line. The feedback currents, which are driven by the plate tank current, appear in the figures; and the output current, which provides the driving

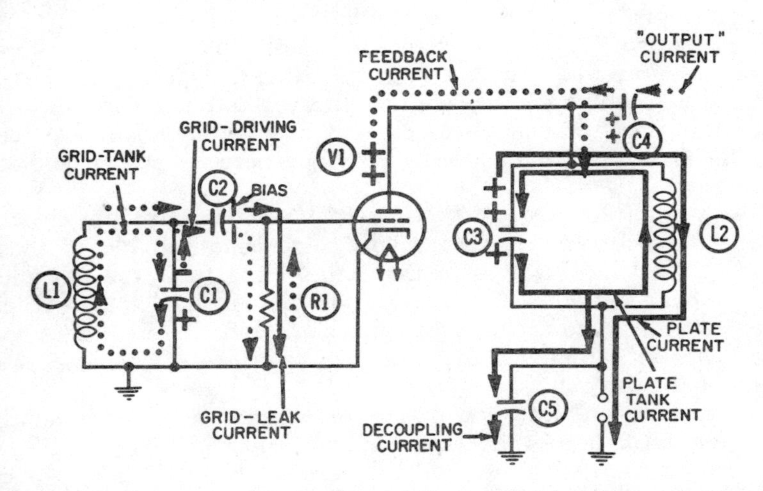

Fig. 5-1. Operation of a tuned-plate–tuned-grid oscillator—first half-cycle.

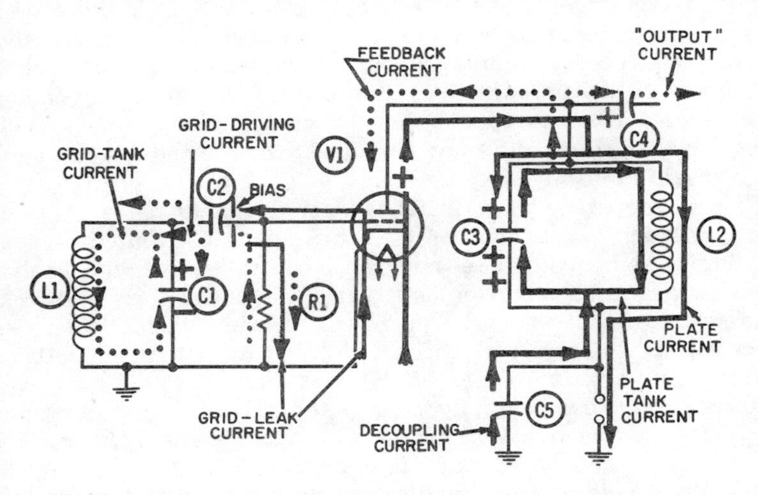

Fig. 5-2. Operation of the tuned-plate–tuned-grid oscillator—second half-cycle.

current and consequently the driving voltage for the next stage, is shown.

For successful self-oscillation, the two tank circuits must be tuned to approximately the same frequency. When the tube is first turned on, the initial surge of plate current sets up oscillation in the plate circuit at its natural, or tuned, frequency. Recall that even a single surge of current or a sudden voltage change will cause any tuned circuit to oscillate at its natural frequency. Even though the oscillation is not sustained by further voltage or current changes, it will continue for several cycles before the initial energy is expended. The purpose of oscillator circuitry is to continue the oscillation indefinitely by providing such a voltage or current change, usually once each cycle. This repetitive action is provided here by pulses of plate current.

As the initial surge of tube current reaches the plate circuit, its voltage-current conditions will correspond roughly to those in Fig. 5-2. The voltages at the plate of the tube, the entrance to coupling capacitor C4, and the top of the tuned tank will all be positive, as indicated by the plus signs, but these voltages will be lower in value than the supply voltage present on the other side of the tank.

Once this uneven distribution of current and consequently voltage exists across the tuned tank circuit, the charge will redistribute itself in an attempt to overcome the unbalance. Since the lower plate of tank capacitor C3 is more positive than the upper plate, current will flow from top to bottom. If there were no inductance or resistance in the current path, this redistribution would occur instantaneously. However, the primary characteristic of any inductance is that it tends to oppose any change in current: If no current is flowing, an inductance tends to oppose any build-up; and once current flows, the inductance tends to prevent it from decaying.

These properties of inductance should enable us to see why an oscillation is set up when electrons are redistributed. Instead of taking place instantaneously, the current requires the equivalent of a quarter-cycle of oscillation to build up from zero to maximum. After maximum current is flowing, the inductive effect will try to keep it from dying out. At this instant the voltages are the same on each side of the tank capacitor, meaning both charges have been equally distributed. The current, however, requires another quarter of a cycle to decay to zero. At the end of this cycle the charge again is unevenly distributed, but in the opposite direction. The flow of electrons from top to bottom during the next quarter cycle—*after* the voltages on the two capacitor plates have been equalized—has charged the bottom plate to a lower positive voltage than the top plate.

The first half-cycle is shown in Fig. 5-1. The greater number of plus signs on the upper plate indicates that midway in the first half-cycle the upper plate is more positive than the lower one. This is confirmed by the sine-wave representation of tank voltage in Fig. 5-3, which shows the voltage at the *top* of the tank—in this case, also the point where output voltage is taken off.

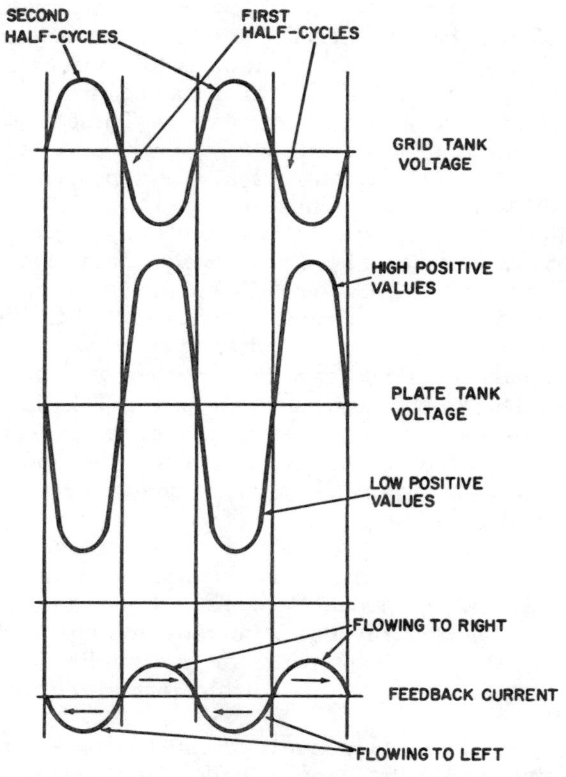

Fig. 5-3. Current and voltage waveforms in the tuned-plate–tuned-grid oscillator.

Fig. 5-2 depicts current-voltage conditions during the second half-cycle of oscillation. The voltage unbalance across the tank capacitor will again attempt to neutralize itself in the following manner: Since there is an excess of positive ions on the upper plate, current will be drawn from the lower plate, through tank inductor L2. Again the inductor will oppose both the build-up and the decay of electron current. As a result, redistribution of the electric charge will again go too far—the voltage unbalance across the tank will be reversed a second time, with the top plate now less positive.

In order to see how this oscillation sustains itself indefinitely, we must now consider what has been happening in the rest of the circuit. During the single cycle just described, certain inevitable losses will have occurred. Hence, at the end of the first complete cycle, the voltage difference between the two plates of tank capacitor C3 will be smaller than it was at the beginning. Another way of visualizing this condition is to consider that fewer electrons will complete the cycle than started it, some dropping out because of internal circuit resistances. These losses, which must be replenished before the next cycle begins, are supplied by "turning on" the plate current at the appropriate moment. Let us see how this is accomplished.

The oscillating current (shown in a solid line) in the plate tank circuit feeds three external paths, or loads, These paths are decoupling filter capacitor C5 (also shown in solid line), output coupling capacitor C4 (top right-hand), and the feedback path to the plate and grid, shown in dotted lines. (The interelectrode capacitance between these two elements couples the feedback to the grid.) During the first half-cycle in Fig. 5-1, this feedback current draws electrons away from the plate. In Fig. 5-2 the polarity of the oscillating voltage is reversed and feedback current is driven back toward the plate, as shown by the arrow on the dotted feedback line.

There are three alternate paths in the grid circuit, and current flows in each one, in response to the feedback current in the plate circuit. These grid components are also shown a dotted line to help tie them in with the feedback current, and also to differentiate them from the grid-driving current (in small dots) and the grid-leak current (a thin line) which are also flowing in the grid circuit.

Note how all three components of the feedback current are flowing in unison. During the first half-cycle they are all drawn to the *right,* and during the second half-cycle they are all flowing to the *left*—both times being driven by the plate tank voltage. These current components are said to be *in phase* with the plate tank voltage.

The components of current to the left of the grid capacitor actually "deliver" the feedback pulse to the grid tank circuit and thereby set up an oscillation of appropriate phase to support the plate-circuit oscillation. The oscillation current in the grid tank has been shown a dotted line .

The oscillation in the tuned-grid circuit will build up to a maximum strength determined by:

1. The strength of the feedback pulses from the plate circuit.
2. The amount of losses during each cycle.

The grid-circuit oscillation will have internal losses due to electrical resistance, dielectric leakage, etc. Normally they will be very small, only a fraction of one per cent each cycle. Thus, this is a "high-*Q*" circuit as explained previously.

In addition, the oscillation actually drives the control-grid voltage to its two extremes by sending the grid-driving current (shown in small dots) up and down through grid resistor R1. This current flows in phase with the oscillating voltage in the grid tank and thus acts as a load on the oscillation by adding to its total losses during each cycle.

The oscillation will build up until the total losses during each cycle are equal to the energy supplied by the feedback pulse. When energy lost equals energy supplied, the grid oscillation will become stabilized.

UNIDIRECTIONAL CURRENTS

Currents which flow essentially in only one direction are classified as direct, or unidirectional, currents. In our circuit they are the pulsating DC of the plate current (shown a solid line), and the grid-leakage current (shown a thin line). The plate current replenishes each cycle of oscillation, and the grid-leak current provides the grid voltage (also called operating bias) for the tube.

In any vacuum tube, electrons in the tube stream will tend to strike the control grid whenever it is more positive than the cathode. We have already seen how the grid voltage is made positive, midway in the second half-cycle, by the grid tank oscillation. Electrons (shown a thin line) will now be attracted to the grid wires and leave the tube via the control grid. Thus, three separate electron currents are flowing in the grid circuit. Shown in a thin line, large dots, and small dots, they represent the grid-leak, feedback, and grid-drive currents respectively.

Fig. 5-3 shows the waveform for the feedback current as being in phase with the plate tank voltage. This means that during the first half-cycle (Fig. 5-1) the plate tank voltage, being at its most positive value, draws the electrons to the *right,* or toward the high positive voltage. Conversely, during the second half-cycle the plate tank voltage has its lowest positive voltage and repels the electrons of the feedback current, moving them to the left as in Fig. 5-2.

The grid-leakage electrons accumulate on the right plate of the grid capacitor and build up a permanent negative voltage there, as indicated by the solid thin minus signs. These electrons drain continuously downward through grid resistor R1, the amount of current depending on the quantity of electrons in storage, the size of the grid capacitor, and the resistance of grid resistor R1.

This current flow through the resistor is pure DC; consequently, it is represented by a solid thin line in both Figs. 5-1 and 5-2. Grid resistor R1 and grid capacitor C2 form a conventional long time-constant R-C combination so the grid-leak voltage will remain steady in the face of the pulsating electron current coming to it from the tube. This current enters the grid capacitor during the second half-cycle only, when the control grid has been driven positive. (During the first half-cycle the control grid has been driven negative and no grid-leak electrons can leave the tube.)

The amount of grid-leak voltage can be computed from two separate formulas. The first one, known as Coulomb's law, states that:

$$Q = C \times E$$

where,

Q is the quantity of the charge in coulombs,
C is the value of the capacitor in farads,
E is the voltage in volts.

The second formula—much more widely used—is Ohm's law, which states that:

$$E = I \times R$$

where,

E is the voltage across a resistor in volts,
I is the current flowing through the resistor, as a result of that voltage, in amperes,
R is the resistance of the resistor in ohms.

The presence of this fixed biasing voltage at the grid accounts for the fact that the grid voltage is always lower than the grid tank voltage driving it. The voltage at the top of the grid tank fluctuates around zero as a reference point, whereas the voltage at the grid fluctuates around the negative biasing voltage, represented by the solid minus signs. The control-grid voltage momentarily becomes positive in the middle of each half-cycle and allows leakage electrons to leave the tube via the control grid.

The plate current will flow for a longer part of each cycle than the grid current. For every value of plate voltage there is a negative grid voltage below which plate current cannot flow and above which it can.

Note that the tube conducts electrons when the plate voltage is at or near its lowest value. In the middle of the second half-cycle, for instance, we see the lowest concentration of plus signs—representing positive ions—at the plate and at the top of the tuned tank. This condition is brought about by the oscillating electrons in the plate tank circuit of course, and can be confirmed from the sine waves of voltage in Fig. 5-3. Each pulse of plate current arrives at the top of the tank and adds to the oscillating electrons concentrated there, thus replenishing the oscillation. The amount of this reinforcement must compensate exactly for the internal resistance losses and the output and feedback loads faced each cycle by the oscillation.

POWER-SUPPLY DECOUPLING

In Figs. 5-1 and 5-2, capacitor C5 is placed in parallel with the power supply to sidetrack, or decouple, large fluctuations in current before they reach the power supply. Otherwise, in flowing through the power-supply filters, these currents could cause corresponding voltage fluctuations which would be reflected into other vacuum-tube stages.

It was shown previously that the oscillating electrons in the plate tank will flow out along any available path, such as the line to the power supply. When a capacitor is placed in parallel with this line (as C5 has been), the oscillating current will choose this alternate path because of its lower impedance. Thus, most of the current fluctuations are diverted harmlessly into C5.

This decoupling current is shown a solid line so you can see its relationship to the oscillating tank current driving it. The decoupling network constitutes one more load, or loss, for the oscillating voltage, along with the feedback and output currents described previously.

A small decoupling resistor is often added in the power-supply line to provide additional filtering. Even without it, the power-supply impedance and the filter capacitor constitute an effective filtering combination.

TANK-CIRCUIT TUNING

In this oscillator circuit the plate tank must be tuned "slightly inductive" with respect to the grid tank. In other words, the plate tank should have a somewhat lower resonant frequency. One way of accomplishing this is to add more inductive reactance to the plate tank by increasing the inductance of L2.

However, it is also possible to lower the resonant frequency of a tuned circuit by adding capacitance and thereby lowering the capacitive reactance. Now the inductive reactance is greater, and the circuit will again be tuned slightly inductive as before.

Conversely, tuning a circuit "slightly capacitive" means to *increase* its natural frequency. Here the capacitance must be lowered in order to increase the capacitive reactance of the circuit. As before, the same result would be obtained by lowering the inductance to decrease the inductive reactance in the circuit.

Figs. 5-1 and 5-2 give no hint of a frequency difference between the two tanks. If the phase relationships were exactly as shown in these diagrams, the circuit would be unable to support its own oscillation for these reasons: The plate current reaches the plate tank at the precise moment it can give the most support to the oscillation in the tank. However, the feedback current from the plate tank will deliver a pulse to the grid circuit at the wrong instant to support the grid-tank oscillation. In fact, the oscillation will be dampened because the current-voltage combination in the grid circuit is always exactly out of phase with the grid tank voltage. The dotted arrows in Figs. 5-1 and 5-2 represent the feedback current in the grid circuit. As you can see, it is flowing in the opposite direction from the external grid-driving current produced by the grid tank oscillation. It is likely, under these phase conditions, that the oscillation in the grid tank would not be allowed to build up at all.

When the oscillation in the plate tank is lower than the resonant frequency of the grid tank circuit, the feedback current will "see" two different impedances in the grid tank—one in the direction of the tank capacitor, and the other in the direction of the tank inductor. At the lower feedback-current frequency, the inductor will have much lower reactance than the capacitor, so most of the feedback current will be shunted into the inductor path.

Because the two tanks have different natural frequencies, neither operates at its own resonant frequency. The pulses of plate current, released once each cycle at the grid-tank current frequency, will arrive slightly early for maximum reinforcement of oscillation in the plate tank. This will shorten each cycle of oscillation by hastening the end of one and the beginning of the next one. Instead of being a true harmonic waveform, each cycle will be somewhat distorted, and the plate oscillation will occur at slightly *higher* than the resonant frequency of the plate tank.

By similar reasoning, the grid oscillation is slightly *lower* than the resonant frequency of the grid tank. This oscillation is sustained by the feedback pulses coupled from the plate to the grid via interelectrode capacitance. The feedback current, a dotted line, is driven by the tank voltage in the plate circuit and must stay in phase with it at all times.

It is impossible to show, in Fig. 5-1 and 5-2, how the phase relationship of the feedback current is able to support the grid oscillation. This can be demonstrated graphically and mathematically, but it would require extraordinarily complex waveforms and computations.

It is sufficient to say that the feedback current has a slightly lower frequency than the grid oscillation and can thus provide sufficient "kick" to sustain oscillation during each cycle. In the process the feedback current, itself driven by a distorted current waveform, manages to distort the oscillating current waveform in the grid circuit. Also, the feedback current lengthens each cycle so that oscillation will occur slightly below the resonant frequency of the grid tank.

Chapter 6

THE ELECTRON-COUPLED OSCILLATOR

The name *electron-coupled oscillator* (ECO) is derived from the way the oscillation in the plate tank circuit is supported or replenished by fluctuations in the electrons streaming through the tube. Upon reflection we will realize that the name *electron coupling* does not describe something unique to this circuit, since the oscillation in *any* plate tank circuit is likewise replenished by fluctuations in the electron stream.

Figs. 6-1 and 6-2 show the operation of the electron-coupled oscillator for each half-cycle. Inspection of the circuit will reveal a Hartley-type oscillator between the grid and cathode. Since the Hartley oscillator was covered in detail in Chapter 3, its mode of operation will be reviewed only briefly here. The necessary components of an electron-coupled oscillator include:

C1—Grid tank capacitor.
L1—Grid tank inductor (used as an autotransformer).
R1—Grid-leak bias resistor.
C2—Grid coupling capacitor.
V1—Tetrode vacuum tube.
C3—Screen-grid bypass or filter capacitor.
R2—Variable resistor used for adjusting the screen-grid bias voltage.
C4—Plate tank capacitor.
L2—Plate tank inductor.
M1—Power-supply or other voltage source.

These components form convenient *combinations:*

C1 and L1 form a tuned oscillatory circuit.
C2 and R1 form a conventional RC filter with a long time-constant.
C3 and the upper part of R2 form another long time-constant RC filter.
C4 and L2 form a second tuned oscillatory circuit.

The currents at work in this circuit include:

1. Grid tank current.
2. Feedback current.
3. Plate and screen-grid currents.
4. Grid-leak bias current.
5. Screen-grid "biasing" current, which might also be considered a voltage-divider current.
6. Plate-tank oscillating current.

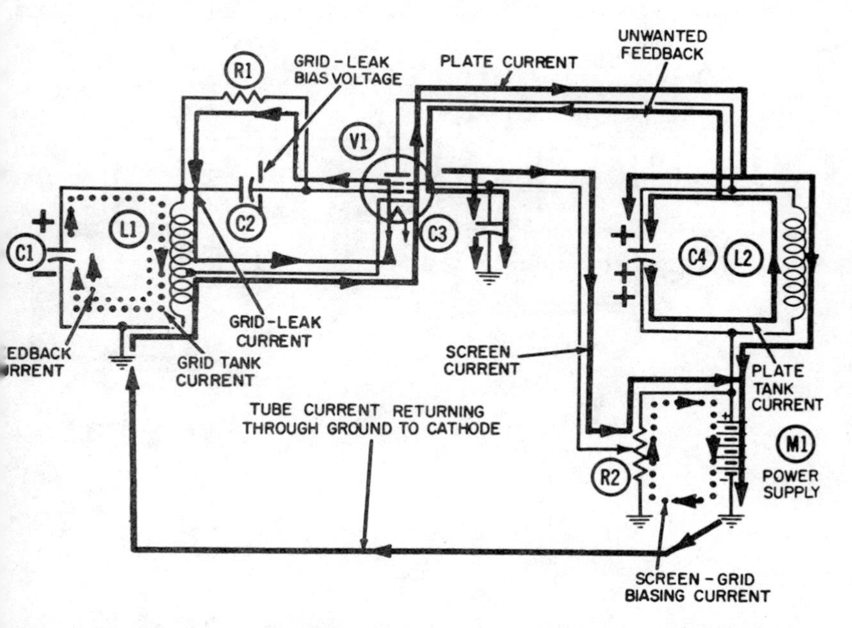

Fig. 6-1. Operation of the electron-coupled oscillator—positive half-cycle.

The grid tank current (left-hand dots in Figs. 6-1 and 6-2), oscillates between the upper and lower plates of capacitor C1 through inductor L1, alternately driving the grid negative and positive. To keep the tank current oscillating, it is periodically replenished or strengthened by the feedback current (dotted thin line). The latter in turn is driven by the plate current through the lower portion of inductor L1.

Fig. 6-1 depicts the positive half-cycle of operation. Grid tank electrons are amassed on the lower plate of capacitor C1, making the upper plate positive—and also the control grid, since it is coupled to the top of the tank through capacitor C2. A pulse of plate current is released through the tube and arrives at the top of the tuned tank (L2-C4) at the moment its voltage is least positive (depicted by the single plus sign on the upper plate of C4 in Fig. 6-1). Being composed of negative electrons, the plate-current pulse lowers the already low positive voltage and thereby replenishes the oscillation in the plate tank.

Some of the current exits from the tube at the screen grid and flows through the upper portion of potentiometer R2, through the power supply to ground, and back to the cathode.

Another current leaves the tube at the control grid in the form of grid-leak current, shown by the solid thin lines. This occurs

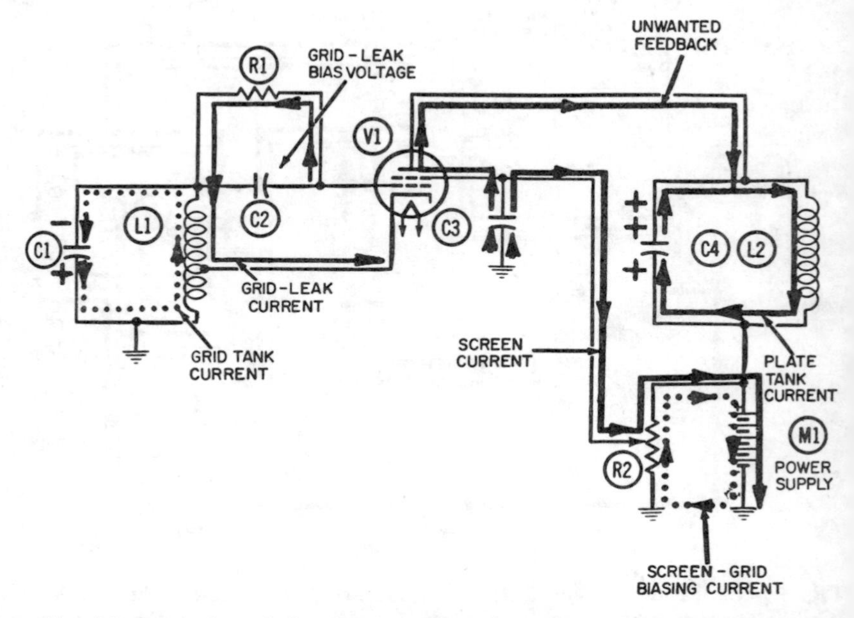

Fig. 6-2. Operation of the electron-coupled oscillator—negative half-cycle.

only once each cycle, when the control-grid voltage becomes momentarily positive. Because of the high resistance of grid resistor R1, these electrons cannot immediately return to the cathode, but will accumulate on the right plate of grid capacitor C2, building up the negative grid-leak bias voltage. Throughout the entire cycle, there has been a slow and continuing drain of electrons back to the cathode, through the grid resistor and the upper portion of L1.

As with plate and screen currents, a closed path back to the cathode must be available. Otherwise, enough electrons will accumulate on the grid capacitor and on the grid itself that the flow of tube current will be cut off entirely (known as "grid blocking").

Fig. 6-1 also shows a feedback current (dotted thin line) flowing in the grid tank. This current is produced by the autotransformer action of the inductor. (For a fuller treatment of the phase relationships between the grid-tank, plate, and feedback currents, refer to the chapter on the Hartley oscillator.)

This electron-coupled oscillator is a special tuned-plate–tuned-grid configuration. Its distinguishing feature is that the grid and plate oscillations are isolated from each other by the screen grid. Hence, there can be no feedback from plate to grid. An unwanted feedback current (solid heavy line in Figs. 6-1 and 6-2) flows from the top of the plate tank and back to the plate, where it is coupled into the screen-grid circuit by interelectrode capacitance. In the screen circuit we see it being bypassed harmlessly to ground through capacitor C3.

There is nothing mysterious about this coupling from plate to screen by means of interelectrode capacitance, nor about the bypassing the feedback to ground. The same electrical principle is employed for both—namely, the natural ability of a capacitor (including two objects having a capacitance toward each other) to pass an alternating current. Fig. 6-1 shows a half-cycle of this unwanted feedback current flowing from the plate tank to the plate, from the screen grid to the upper plate of capacitor C3, and finally from the lower plate of C3 to ground. In the negative half-cycle of operation of Fig. 6-2, these directions are reversed.

You may wonder why feedback from the plate tank is unwanted in this circuit; in others such as the crystal or TPTG, the continuance of the oscillation *depends directly* on feedback between the output (plate) and input (grid) circuits. Obviously, the reason is the Hartley oscillator in the grid circuit. As explained in Chapter 3, it needs no feedback from the plate, since it generates its own between the cathode and grid circuits.

One of the virtues of an electron-coupled oscillator circuit is its frequency stability, made possible by the isolation between the load circuit (plate tank) and the basic oscillation in the grid tank.

The oscillating current in the plate tank is generated for the sole purpose of getting it to perform some useful function. But in order to do so, it must first be coupled out of this circuit and into another one. One way is by transformer coupling between L2 and another inductor (not shown). Another is to connect the coupling circuit directly to the normal output point at the top junction of C4 and L2.

Whichever means is used, it is inevitable that a new current will flow in the coupling circuit. Driven by the current/voltage combination in the plate tank, it will have the *same frequency* as, and its strength will be *proportional* to, the plate tank oscillation driving it. Thus, it constitutes a load on the plate tank oscillation and both weakens and detunes it. The detuning can perhaps be better visualized by considering the frequency formula:

$$f = \frac{1}{2\pi\sqrt{LC}}$$

and then recognizing that the proximity of additional coupling components—whether they be capacitors, inductors, resistors, or any combination—will change the effective values of L and C, and consequently the frequency.

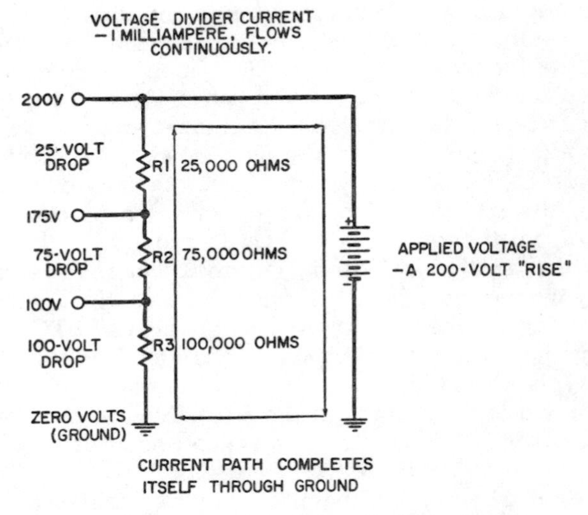

Fig. 6-3. A simple voltage-divider network.

The purpose of the screen grid and its bypass capacitor is to shunt the feedback current harmlessly to ground before it reaches the grid. The strength and frequency of the feedback will vary with changes occurring in the plate tank oscillation as a result of loading. Hence, if allowed to reach the grid circuit, such feedback will change the basic oscillator frequency being generated in the grid tank.

A pentode tube is often used in place of the tetrode shown in Figs. 6-1 and 6-2. Because its plate current is relatively unaffected by variations in plate voltage, a pentode will make a significant contribution to the *amplitude* stability of the plate tank oscillation. In simpler terms, any variation in coupling or loading between the plate tank circuit and the next stage or circuit may modify the strength of the oscillation, and accordingly, the strength of the voltage peaks at the top of the tank. However, these variations in plate voltage will cause no significant change in the plate current coming through a pentode. It is true that an unvarying plate current cannot correct existing deviations in amplitude or strength—but at least it will not be the source of new ones. In this respect, the pentode has the advantage over the less stable tetrode. If the tetrode of Figs. 6-1 and 6-2 is replaced by a pentode, all other currents will remain relatively unchanged.

The screen-grid biasing current (right-hand dots) deserves special mention. Driven by the power-supply voltage, this current flows from the negative terminal of the power supply (the ground or neutral point, in other words), upward through potentiometer R2, and back into the positive terminal of the power supply. As a result of this current, a voltage is developed across R2 and a partial voltage exists at any point along this resistor, the amount depending on the distance between the two segments on each side of the contact point. Calculation of the voltage at any point is a straight Ohm's-law problem.

When a potentiometer is connected across a voltage source as is done here, this combination becomes a special form of voltage divider. Resistor R2 might be considered as being made up of five or even ten smaller resistors, all connected in series across a voltage source. Each resistor serves to divide the available voltage into smaller ones which will always add up to the applied voltage. This is the meaning of Kirchhoff's law that all voltage drops and all voltage increases around a closed circuit must add up to zero. (Watch those polarity signs!) The following paragraph makes the meaning a little clearer.

In interpreting Kirchhoff's law, an applied voltage is considered a voltage increase and given a positive sign, whereas voltages developed across resistive loads are considered drops and given an opposite, or negative, sign. Thus, when a certain number of positive units are added to an equal number of negative units, the sum is zero of course.

Fig. 6-3 shows a sample voltage-divider circuit closely resembling the one used in the screen circuit of Figs. 6-1 and 6-2. Three resistances in series—R1, R2, and R3—have replaced the potentiometer. Their values are 25,000, 75,000, and 100,000 ohms, so the series resistance of the combination is 200,000 ohms. With an applied voltage of 200 volts, we can calculate the current flowing in this closed circuit by using Ohm's law, as follows:

$$I = \frac{E}{R}$$

$$= \frac{200 \text{ volts}}{200{,}000 \text{ ohms}}$$

$$= .001 \text{ ampere}$$

$$= 1 \text{ milliampere.}$$

Since this same current flows through each resistance, the voltage developed—or dropped—across each resistor can also be calculated from Ohm's law.

Call the voltage across R1 E_1:

$$E_1 = IR$$

$$= .001 \text{ ampere} \times 25{,}000 \text{ ohms}$$

$$= 25 \text{ volts.}$$

Call the voltage across R2 E_2:

$$E_2 = .001 \text{ ampere} \times 75{,}000 \text{ ohms}$$

$$= 75 \text{ volts.}$$

Call the voltage across R3 E_3:

$$E_3 = .001 \text{ ampere} \times 100{,}000 \text{ ohms}$$

$$= 100 \text{ volts.}$$

Note that the sum of these voltage drops is 200 volts, the amount of voltage rise represented by the power supply.

Voltage dividers are widely used in electronic circuitry for obtaining an infinite variety of partial voltages from a single source such as a power supply. They are usually the essence of simplicity. Paradoxically, they can be quite difficult to recognize on a schematic, particularly for someone with an untrained eye. First of all, they are rarely labeled voltage dividers. Also, the resistors frequently are widely separated from each other and from the applied voltage, often by a full page of the schematic.

It is evident from Fig. 6-1 that the same positive voltage that exists at the tap on the potentiometer will be applied to the screen grid. It is also evident that the screen-grid current must flow through the upper portion of R2 in order to reach the power supply. The voltage at R2 will be altered somewhat when this happens. It is possible, by adjusting the tap, to compensate for variations in amplitude of the plate tank oscillation (caused by corresponding variations in the loading or coupling current from circuits which come after the plate circuit). This self-compensation enables the tetrode to exhibit some of the amplitude

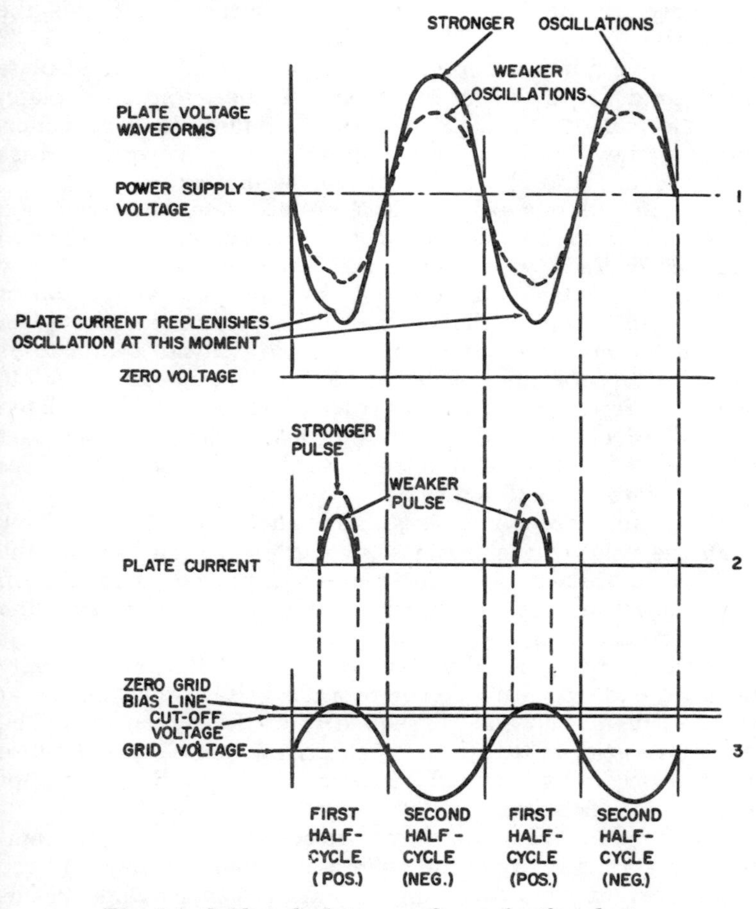

Fig. 6-4. Grid and plate waveforms in the electron-coupled oscillator.

stability previously attributed to the pentode. Now let us consider an example.

Fig. 6-4 shows one cycle of plate voltage for an electron-coupled oscillator. (The plate voltage is the same as the oscillating voltage in the plate tank.) The positive half-cycle of Fig. 6-4 corresponds to the current/voltage conditions depicted in Fig. 6-1. That is, the grid voltage (line 3 of Fig. 6-4) reaches its most positive voltage in the middle of the half-cycle, releasing a pulse of plate current (solid curve in line 2 of Fig. 6-4). The current arrives at the top of the plate tank when the latter is least positive (solid curve of line 1 of Fig. 6-4) and thereby reinforces the oscillation.

The negative hump in the middle of the first half-cycle of plate voltage (solid curve of line 1) represents the addition of plate-current electrons to the voltage across the plate tank capacitor. Arriving as they do during this half-cycle, the electrons *increase* this voltage and thereby reinforce the oscillation.

Continuing with our example, assume this plate tank oscillation is weakened by, say, a variation in the load current being driven by the tank current. Also assume the dotted curve in line 1 now represents the voltage waveform at the plate and at the top of the tank. Since the oscillation has been weakened, the plate voltage at the center of the first half-cycle will not swing to *as low a positive value* as before. So now a larger pulse of plate current can be drawn across the tube. This condition, depicted by the dotted curve of line 2 in Fig. 6-4, stems from the fact that the plate voltage of a tetrode (unlike the pentode) *does* affect the plate current somewhat.

As the plate current increases, more and more electrons flow through the positively charged screen grid on their way to the plate. Thus, more and more electrons are captured by the screen grid and flow through the upper portion of R2, increasing the voltage drop across the top of the potentiometer.

Subtracting this greater voltage drop from the power-supply voltage, we end up with a lower positive voltage than before at the potentiometer tap and consequently at the screen grid. The lower screen-grid voltage now weakens the plate-current pulse which had originally lowered the screen voltage. So, a form of self-compensation exists.

These cumulative actions do not necessarily *correct* the amplitude variation already in the plate oscillation, but they do keep the plate current from following the changes in loading. In this sense they contribute to the over-all stability of the circuit—including the frequency stability of the oscillation in the grid circuit, because any change in the plate current drawn by the tube will also change the strength of the feedback in the cathode-to-grid tuned circuit. As is true for any tuned-circuit oscillator, any such variations in the feedback, or in loading, coupling, etc., will raise or lower the basic frequency.

Chapter 7

PHASE-SHIFT OSCILLATOR

The phase-shift oscillator circuit is widely used in laboratories for generating an audio frequency, and for other testing where precision is a must. A special combination of resistors and capacitors are employed to achieve regenerative feedback and thus permit the circuit to generate a continuous alternating current. Unlike other oscillators, the phase-shift oscillator has no tuned circuit and hence is not as susceptible to detuning by occasional stray capacitances and inductances. Consequently, the output remains fairly stable at the desired frequency.

Figs. 7-1 and 7-2 show one of the several possible configurations for this special application of the R-C oscillator. The circuit components and their functions are as follows:

V1—Conventional pentode amplifier tube.
C1—Cathode bypass capacitor for preventing degeneration.
R1—Cathode biasing resistor.
C2—Screen-grid bypass or filter capacitor.
R2—Screen-grid voltage-dropping resistor.
R6—Plate-load resistor.
C3, C4, and C5—Feedback, coupling, and phase-shifting capacitors.
R3, R4, and R5—Phase-shifting resistors.
M1—DC power supply.

There are five electron currents at work in this circuit, and each one needs to be analyzed before you can understand how the circuit operates. These currents are:

1. Cathode-to-plate current (conventional tube current).

2. Screen-grid current.
3. Current in the first R-C combination (R3 and C3).
4. Current in the second R-C combination (R4 and C4).
5. Current in the third R-C combination (R5 and C5). Its associated voltage is applied to the control grid as the feedback voltage.

CURRENT PATHS

Fig. 7-1 shows the current flow during the first half-cycle, which begins when the grid is most negative and ends when it is most positive. Conditions are just the opposite for the second half-cycle in Fig. 7-2—it begins when the grid is most positive and ends when it is most negative. In both illustrations, significant voltage polarities are shown as they exist *at the end* of the half-cycle.

At the start of the first half-cycle, you will recall that the grid voltage is most negative. However, in this circuit (unlike the ones discussed previously), plate current must flow at all times. Therefore, the most negative grid voltage merely restricts the flow of plate current, instead of cutting it off completely. A low plate current always leads to a high positive plate voltage, particularly when this current must flow through a resistive load such as R6 on its journey to the power supply.

Line 1 of Fig. 7-3 shows a sine wave of plate voltage. At the start of the first half-cycle this plate voltage has its maximum positive value.

Line 4 of Fig. 7-3 shows two sine waves of grid voltage. The one in dashed lines is developed across R5 by the current shown as a thin line in Figs. 7-1 and 7-2, and represents the actual voltage applied at the grid. The solid line is its hypothetical value if the signal could pass through the R-C network unattenuated.

It is of course significant that the grid-voltage waveforms on line 4 and the plate voltage on line 1 of Fig. 7-3 are 180° out of phase with each other—or more simply, "out of phase," or "of opposite phase." This out-of-phase condition is necessary for the oscillation to continue. Now let us see how this phase shift between output and input can be achieved from the combination of resistors and capacitors making up the phase-shifting network in Figs. 7-1 and 7-2.

Fig. 7-1 shows the current conditions at the end of the first half-cycle. Prior to this, the grid voltage has been increasing steadily from negative to positive. Consequently, the plate curent, shown solid , has also been steadily increasing, until it eaches maximum. Now it is drawn across the tube by the posi- ive voltage applied to the plate from power supply M1. (Although hown as a large battery, the power supply can be a vacuum- ube rectifier or any other device capable of furnishing DC).

Ohm's law tells us the voltage developed, or "dropped," across resistor by the current flowing through it is proportional to he amount of that current. Consequently, the voltage drop across esistor R6 will be much larger at the end than at the beginning f the first half-cycle. Since the voltage at the plate is always he supply voltage *minus* the voltage developed across the plate oad resistor, we can see why the plate voltage is lowest when he grid voltage is at its highest.

At the end of the first half-cycle, when maximum plate current s flowing, it is convenient to visualize the consequent reduction n plate voltage resulting from the excess of plate-current elec- rons flowing out of the tube, into load resistor R6, and on to he power supply. Until these electrons can enter the load re- istor, they are in a sense "dammed up" at the junction of R6 nd coupling capacitor C3. For this reason, during the first half- ycle they are shown flowing *onto* the left plate of C3.

Conversely, during the second half-cycle, when the flow of late current is restricted by the negative grid voltage, it is onvenient to equate the drop in number of plate-current elec- rons with the corresponding rise in plate voltage. This is done y picturing it as current being drawn out of the left plate of apacitor C3 and into load resistor R6, to the power supply.

Fig. 7-1 shows current (solid thin line) flowing upward hrough grid-driving resistor R5 during the first half-cycle. Note that if there were no 180° shift in phase as the voltage ulse from the plate passes through the three R-C combinations, he current in R5 would flow *downward* instead.

Likewise, during the second half-cycle, the normal downward low of current through R5 would be reversed if it were not for his 180° phase shift.

Suppose this phase shift did not occur? Then, when electrons lowed onto the left plate of capacitor C3, other electrons would low in unison and in the same direction through all these ca- acitors, much as they would through a straight wire. Also, when lectrons flowed out of the left plate of C3, other electrons would low in the same direction through all three capacitors.

How, then, does a shift in phase occur in an R-C network? Jnfortunately, it is impossible to show the voltage pulse actually hanging phase. The dashed diagonal lines in Fig. 7-3 are meant o strengthen your conviction that such a phase shift does occur, even though not clarifying exactly how. Let us now look more closely at these waveforms, to see what makes them do so.

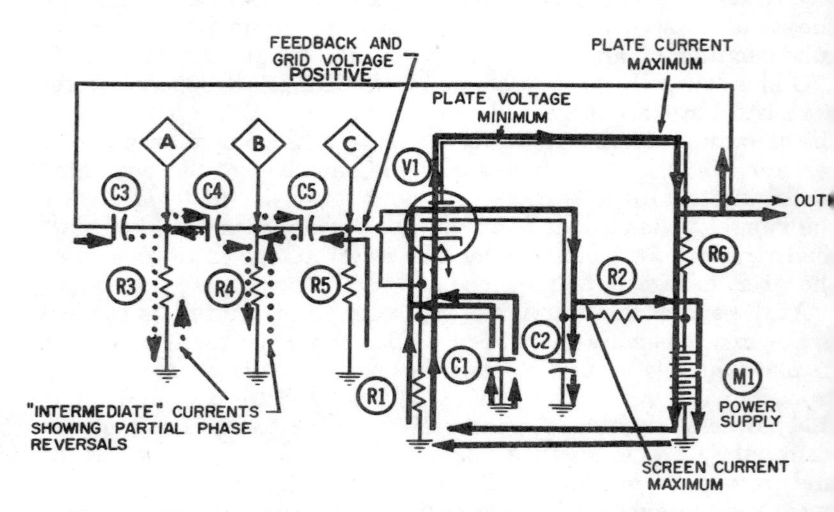

Fig. 7-1. Operation of the phase-shift oscillator—positive half-cycle.

PHASE-SHIFT NETWORK

Suppose resistor R3, R4, and R5 each have the same resistance and C3, C4, and C5 the same capacitance. Then it is safe to assume that each of the R-C combinations (R3-C3, R4-C4, and R5-C5) will produce equal portions of the total phase shift of 180°; in other words, the applied voltage will be shifted 60° by each R-C combination.

What would happen if no phase shift occurred in any R-C combination during the entire first half-cycle of Fig. 7-1? Then the electron current shown by the dotted thin line would flow down through R3 and reach its maximum value at the same instant the plate voltage drops to its minimum at the end of the half-cycle. However, the voltage peak across R3 must occur 60°, or one-sixth of a cycle, later. For this reason, the green arrows in Fig. 7-1 point in both directions through R3. Admittedly this is a crude presentation of what is happening within the resistor and will require help from your imagination.

Fig. 7-3 shows the voltage waveforms at the plate and at feedback points *A*, *B*, and *C*. Each successive waveform is displaced 60° in phase, and consequently in time, from the preceding one. As a result, a positive-voltage peak at point *C* occurs 180°, or half a cycle, after the identical peak at the plate. Since the grid is connected directly to point *C*, this positive-voltage peak will release maximum plate current through the tube. We already know that in a vacuum tube a high plate current usually coincides with a low plate voltage because of the voltage drop across plate load resistor R6.

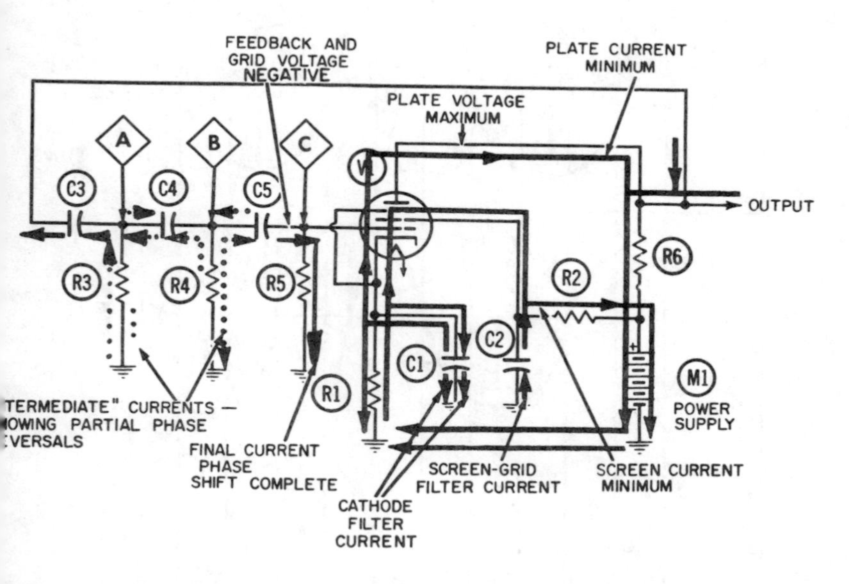

Fig. 7-2. Operation of the phase-shift oscillator—negative half-cycle.

Also in Fig. 7-3, we see that a negative-voltage peak occurs at point *C* a half-cycle, or 180°, after the plate voltage reaches its minimum. This peak, applied to the control grid via the feedback line, drastically reduces the flow of plate current. Minimum plate current usually coincides with maximum plate voltage, because the lower current through load resistor R6 reduces the voltage drop across it.

From the two preceding paragraphs and Fig. 7-3 we may conclude that the feedback has the appropriate phase to control the current flow in the tube and deliver an alternating current or oscillation at the output point. Several important approximations have been made in arriving at Fig. 7-3. The signal voltage will of course be attenuated as it passes through the R-C network. The solid curves of lines 2, 3, and 4 represent the theoretical voltage waveforms at point *A*, *B*, and *C*, respectively, if no attenuation occurred. The waveforms shown in dashed curves indicate the actual reduction in strength at each point. The feedback voltage at point *C* is only a small fraction of the output voltage which causes it. This is normal for all oscillators—the

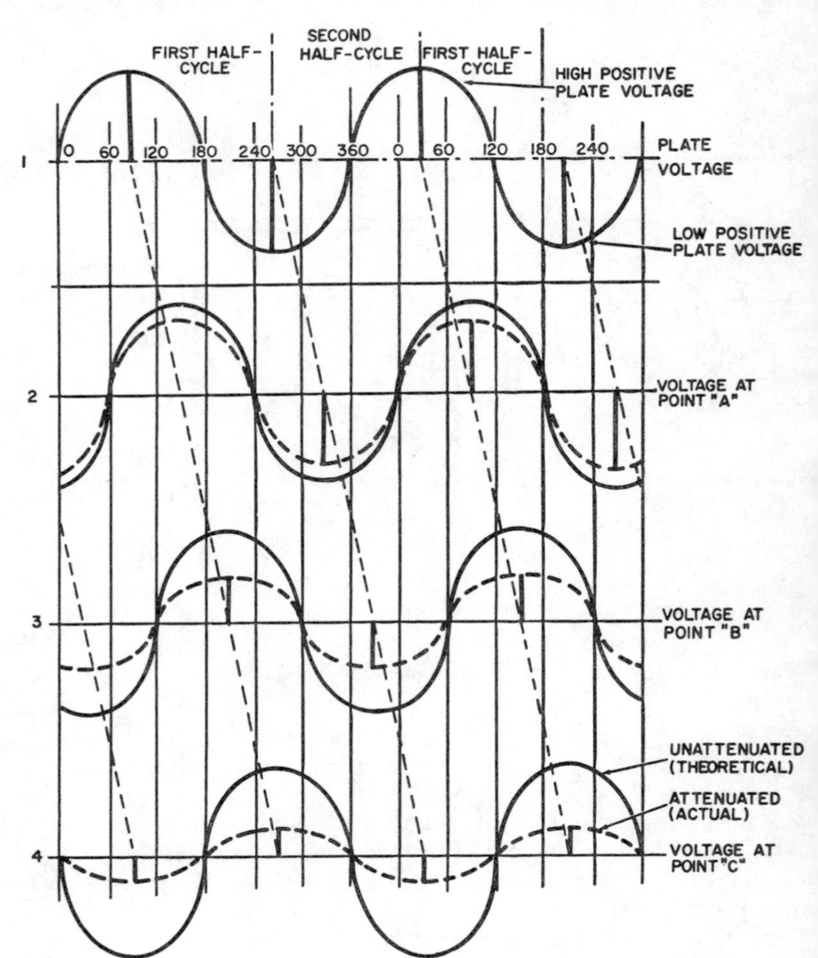

Fig. 7-3. Voltage waveforms at four key points in the phase-shift oscillator.

tube acts as an amplifier; hence, only a small grid-driving voltage is required to control a much larger plate current through the tube and thereby contribute to large swings of plate voltage.

The values of resistors R3, R4, and R5 and capacitors C3, C4, and C5 determine the basic frequency of the voltage delivered to the output point. As an example, if each capacitor has a value of .01 microfarad, and each resistor a value of 10,000 ohms, the circuit will oscillate in the region of 700 or 800 cycles per second, well within the range of the human ear.

Let us consider each R-C combination separately (an approximation which the actual currents and voltages refuse to recognize). It is often said of capacitors that the current "leads" the voltage, frequently cited as unassailable evidence that some other related condition exists or will occur. Recall that in inductors the opposite is true—the current lags the voltage. (This latter property has been adequately described in earlier chapters.) Let us give some thought now to the significance of the axiom that the current leads the voltage in capacitors.

This statement is directly related to another widely quoted truth that capacitors will oppose any change in *voltage*. The existing voltage across the plates of a capacitor can be changed only by adding or withdrawing charged particles (electrons) and these electrons must enter or leave the capacitor *first*, in order for the voltage to change.

By using trigonometry it is possible to calculate the exact number of degrees the voltage developed across the capacitor will lag the current which produces it. The resultant is called the *vector relationship* between reactances and resistances, or between reactive and resistive voltages.

The maximum voltage developed across R3 should occur one-sixth of a cycle after the maximum current flow into C3. Hence, the reactive voltage across the capacitor will be 1.73 times the resistive voltage. The only frequency which satisfies these conditions turns out to be 920 cycles per second.

The reactance and resistance at the given frequency are arrived at from the trigonometric relationship existing between two legs of a right triangle with an included angle of 60°, as follows:

$$C = .01 \text{ microfarad } (10^{-8} \text{ farads}),$$

$$R = 10{,}000 \text{ ohms } (10^{4} \text{ ohms}),$$

$$X_C = 1.73R$$

We know the formula for capacitive reactance is:

$$X_C = \frac{1}{2\pi fC}$$

Therefore, substituting this formula for X_C we have:

$$\frac{1}{2\pi fC} = 1.73R$$

Rearranging for f:

$$f=\frac{1}{2\pi \times C \times 1.73 \times R}$$

Substituting the known values and solving:

$$f=\frac{1}{6.28 \times 10^{-8} \times 1.73 \times 10^{4}}$$

$$=\frac{1}{6.28 \times 1.73 \times 10^{-4}}$$

$$=\frac{1}{10.86 \times 10^{-4}}$$

$$=920 \text{ cycles.}$$

If each of the three R-C combinations could be considered separately, a frequency of 920 cycles per second would satisfy the requirement that the reactive voltage lead the resistive voltage by 60°. This is another way of saying that the capacitor current should lead the resistor current by 60°, since the current through a resistor is always in phase with the voltage developed across the resistor. The three R-C combinations would then cause a 180° phase shift between the input current and feedback voltage.

This solution is not 100% correct because all the current driven through capacitor C3 does not flow downward through R3, but divides between C4 and R3 in proportion to the impedance offered by each path. Likewise, when there is a negative-voltage peak at point *A*, the current it drives into C4 does not all flow downward through R4 (the current shown by the dotted heavy line), but again divides between R4 and C5 in proportion to the impedance offered by each path.

The significance of these multiple-current paths is that the amplitudes and phase shifts of the voltages achieved at points *A*, *B*, and *C* cannot be calculated separately as was done earlier. Also, these amplitudes and phase shifts will be modified by the type of circuit which follows.

Summary

1. Current must first flow into or out of a capacitor before the voltage across it can be changed. Therefore, the current leads the voltage in a capacitor.
2. In an R-C combination, maximum current will flow through the capacitor before it will through the resistor.

3. The voltage developed across a resistor by a current flowing through it is in phase with that current. Consequently, the voltage across a resistor lags the capacitor current.
4. A capacitor opposes the flow of electron current, the amount varying inversely with the frequency of the applied current, in accordance with the standard reactance formula.
5. The current in a capacitor leads the resulting voltage and the resistor current by the same number of degrees, which can be calculated by triangulation. The capacitance and resistance are considered two legs of a right triangle. When these two values are known, the third, or phase, angle can then be determined.

SCREEN-GRID AND FILTER CURRENTS

Little has been said so far about the remaining currents in he circuit. The screen-grid current, shown a solid line, follows closed path within the tube, from cathode to screen grid. Here t exits and flows through screen-dropping resistor R2 and the ower supply, then through common ground and cathode biasing esistor R1, and back to the cathode.

During the first half-cycle, the control-grid voltage becomes nore and more positive, and the plate- and screen-grid currents lso steadily increase. This demand is satisfied by the electrons n the upper plate of cathode filter capacitor C1, and their leparture from the top plate draws an equal number (labeled 'cathode filter current") from ground to the bottom plate.

The increase in screen-grid current during the first half-cycle lso drives an excess of electrons onto the upper plate of screen-rid filter capacitor C2. Coincidentally, an equal number of lectrons are driven from the lower plate to ground. This is the creen-grid filter current.

In Fig. 7-2 we see that during the second half-cycle, when the ontrol-grid voltage goes negative and reduces the plate and creen currents, both filter currents flow in the opposite direction. The upper plate of C1 becomes positively charged and draws lectrons to it through R1, permitting an equal number to flow ack to ground from the lower plate. Therefore, the cathode filter urrent will always have the same frequency as the basic oscil-ator frequency. The cathode filter capacitor acts as a "shock bsorber"—it keeps the voltage at the cathode from changing as he tube current changes.

The reduction in screen current during the second half-cycle ermits the excess of electrons on the upper plate of C2 to be rawn off through R2 and the power supply. An equal number will then be drawn upward from ground to the lower plate of C2. Thus this capacitor also acts as a shock absorber by keeping the screen-grid voltage from changing. It does this by maintaining the flow of electrons constant through R2 so that the voltage drop (or rise) across it will likewise be steady. Thus, the screen-grid filter current will also remain at the basic oscillator frequency.

Chapter 8

BLOCKING OSCILLATORS

The blocking oscillator is one of the most widely used relaxation oscillators. It is often employed in the vertical section of a television receiver, as well as in the timing circuit of many radar sets, oscilloscopes, and similar electronic equipment. A typical blocking-oscillator circuit, along with the current flows during each part of the cycle, is given in Figs. 8-1 through 8-3. The circuit components and their functions are:

T1—Pulse transformer, which operates within a specific range of pulse-repetition frequencies (PRF's).
V1—Triode amplifier tube.
C1—Input and blocking capacitor.
R1 and R2—Grid-leak bias and grid-drive resistors (R2 is a potentiometer used for varying the basic repetition frequency).
C2—Output capacitor.
R3—Plate-load and isolating resistor.
M1—Power supply providing a fairly high positive voltage to the plate of V1.

CIRCUIT OPERATION

You will recall from Chapter 1 that feedback in the normal sense is not required to sustain oscillations in a relaxation circuit. Let us consider that a cycle of operation begins when the grid voltage first permits plate-current electrons to pass through the tube.

This plate current, a solid line on the circuit diagrams, flows through the primary winding of transformer T1 and induces a current of equal value but opposite polarity in the secondary winding.

The polarity of the two windings is such that the induced current in the secondary makes the grid voltage more and more positive as the plate current increases in the primary. These are the conditions depicted in Fig. 8-1. In fact, the effect is cumulative—as soon as the slightest amount of plate current flows into the primary winding, transformer action responds to the changing current and begins to drive the grid positive. As the grid becomes more positive, more plate current flows, causing another positive rise in the grid voltage and a still larger plate current. Soon, maximum plate current is flowing, and any rise in plate voltage will no longer increase the current. When this happens, the tube is said to be saturated.

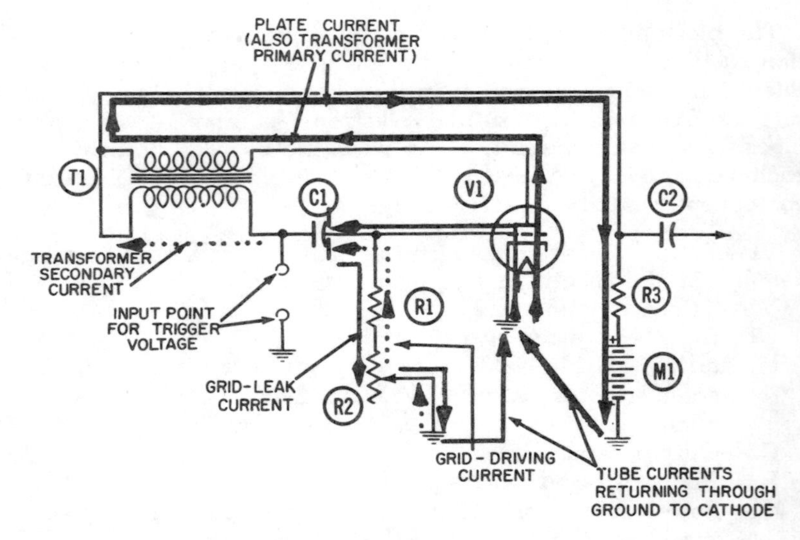

Fig. 8-1. Operation of the blocking oscillator—first part of conduction period.

When saturation occurs, the primary current remains steady; therefore, transformer action between the primary and the secondary (which is dependent on a changing current) ceases. There is a temporary stoppage of the secondary current, and also of the grid-driving current flowing upward through R1 and R2. Both currents are shown in dotted green lines, since they are so directly related to each other.

Meanwhile, early in the conduction cycle, the grid is actually driven slightly positive and attracts a substantial number of electrons from the plate-current stream. These grid-leak elec-

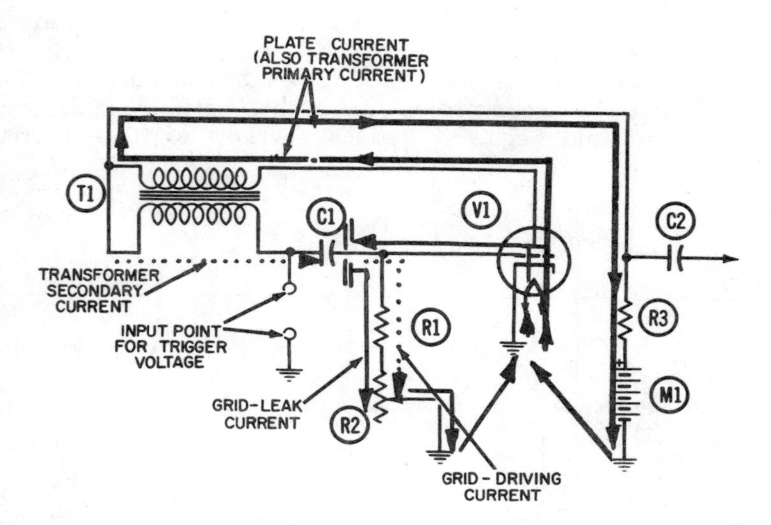

Fig. 8-2. Operation of the blocking oscillator—second part of conduction period.

trons, as they are called, leave the tube via the grid and accumulate on the right plate of capacitor C1. (This flow is shown by the solid thin line in Fig. 8-1.) Because of them, the grid is prevented from rising to its full positive voltage during the transformer action. The dotted lines in Fig. 8-4 show how positive

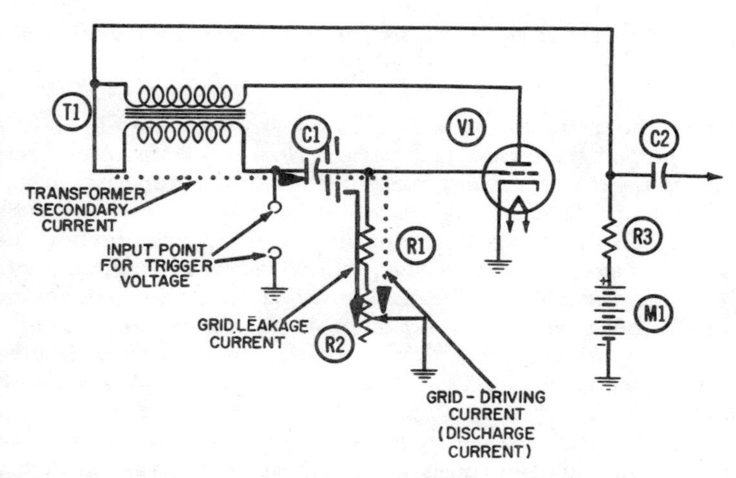

Fig. 8-3. Operation of the blocking oscillator—nonconducting period.

the grid would be driven, were it not for this accumulation of grid-leak electrons.

As soon as plate-current saturation stops the transformer action, the negative voltage represented by these grid-leak electrons "takes over" and begins to reduce the flow of plate current

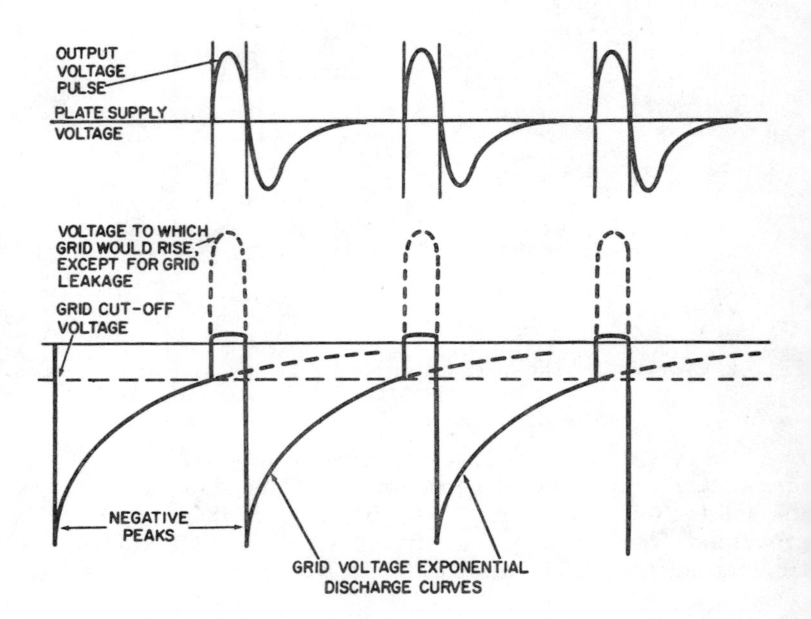

Fig. 8-4. Plate and grid waveforms during free-running operation of the blocking oscillator.

through the primary winding. This leads us into Fig. 8-2, the second part of the conduction period. Now the transformer action is reversed—the secondary current flows in the opposite direction and makes the grid voltage *more negative.*

These two events are also cumulative—less plate current leads to a more negative grid voltage, which reduces the plate current still more, and so on. The net result is that the plate current through the tube is cut off almost instantaneously. There is nothing to prevent the grid from being driven very negative by the transformer action, as it is by the negative peaks in Fig. 8-4.

During the much longer nonconduction period (Fig. 8-3), there are two currents flowing in the grid circuit. Both of them—the grid-leakage and the grid-driving current—discharge through the grid resistors.

The discharge of grid-driving electrons, which accounts for the sudden negative peak in the grid voltage at the start of each nonconducting period, is completed almost instantaneously.

The flow, or "discharge," of grid-leakage electrons through the grid resistors occurs at an exponential rate which is governed

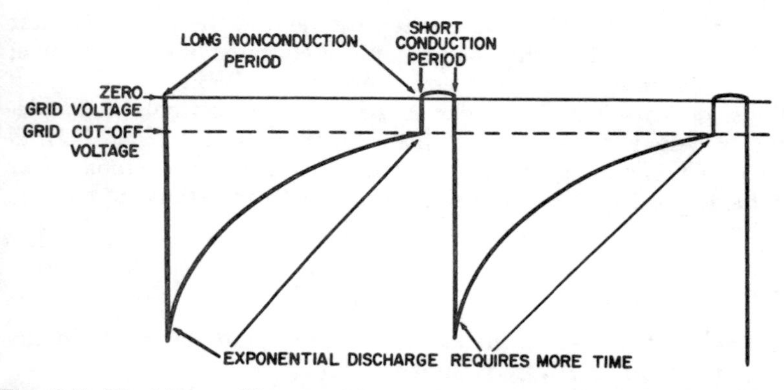

Fig. 8-5. Blocking-oscillator grid waveform, showing decrease in pulse frequency when grid resistance is increased.

by the time constant of the R-C circuit formed by capacitor C1 and resistors R1 and R2.

The exponential curve of the grid voltage during the nonconduction periods in Fig. 8-4 represents the voltage at the grid as the leakage-current electrons discharge to ground. This grid voltage rises towards zero until it reaches the cutoff value (the point where the grid voltage is no longer negative enough to keep the plate current from flowing). With a 6SN7 tube and a plate-supply voltage of about 250 volts, the grid-cutoff voltage will be in the region of −10 volts. The cumulative effects associated with Fig. 8-1 then begin, and a new cycle of oscillation is underway.

As long as the values of C1, R1, and R2 do not change, the oscillation frequency of a blocking oscillator should be fairly stable. However, the number of electrons coming out of the tube and onto the grid capacitor may vary from cycle to cycle because of slight changes in plate voltage, heater voltage or temperature, etc. Consequently, each discharge period may have a slightly different negative voltage than its predecessor, and will thus require more or less time to reach the cutoff voltage and start a new cycle. When a blocking oscillator is running unsynchronized in this manner, it is said to be "free-running." Fig. 8-4 shows the grid-voltage waveform in such an oscillator.

The waveform in Fig. 8-5 has a longer discharge time, resulting in a lower pulse-repetition frequency (PRF). This condition is achieved by moving the tap downward on potentiometer R2, so that the discharging grid-leakage electrons will encounter *more* resistance as they flow to ground. In other words, the time constant of the grid circuit is increased.

The opposite effect can be created by moving the potentiometer arm *upward* so that the leakage path contains *less* resistance; this will shorten each cycle and *increase* its PRF.

Thus, potentiometer R2 acts as a frequency control (in a television receiver it is labeled the Vertical or Horizontal Hold control). Its application will be more clearly understood after you have studied the discussion on synchronization of the oscillator.

Summary

1. The grid voltage reaches cutoff and plate current begins to flow through the primary winding.
2. This increasing current in the primary winding is coupled to the grid circuit by transformer action. The transformer is connected so that this rising current causes the grid voltage to become less and less negative, thereby increasing the plate current even further.
3. The plate current reaches saturation, and can no longer increase.
4. The grid voltage becomes positive, and the grid draws grid-leak current from the tube.
5. The plate current in the transformer primary collapses and, the current induced in the secondary drives the grid highly negative, cutting off the tube.
6. Grid-leakage electrons leak off until the grid voltage is again raised to the cutoff point. Plate current again flows, and the cycle is repeated.

SYNCHRONIZING A BLOCKING OSCILLATOR

To synchronize a blocking oscillator, positive-going voltage "spikes" from an external source are applied to the grid. Here they add to the existing voltage and raise it enough to trigger the cycle prematurely. Fig. 8-6 shows the sequence. Reading from left to right, you can see that the first cycle of oscillation begins before the trigger or synchronizing spike arrives. In the second and third cycles, the spike is applied to the grid when its voltage is very negative. Notice, however, that the combined voltages still are not sufficient to raise the grid voltage to cutoff. It is evident from Fig. 8-6 that the oscillator is running *faster*, or at a *higher* frequency, than the spike frequency, and cannot be synchronized.

Fig. 8-7 shows the opposite condition. Here the oscillator is running *slower*, or at a *lower* frequency, than that of the synchronizing spikes. This is the desirable condition, because now the oscillator can be synchronized. Positive spikes of trigger voltage are applied to the grid *near the end of each nonconduction period*. Now the spike is sufficient to raise the grid voltage to the cutoff point. Thus, each new cycle starts the instant a trigger spike is applied.

In a television receiver these trigger spikes are known as sync pulses and are part of the received signal. At the transmitter, their frequency is rigidly held to the correct frequency (60-cps for the vertical-sync pulses, and 15,750 cps for the horizontal-sync pulses). At the receiver the pulses are separated from the picture information and amplified to the point where they are able to trigger the blocking oscillator. It in turn generates a much more powerful pulse, at the same frequency, which is further amplified and used to sweep the beam back and forth, and up and down, on the screen.

When horizontal or vertical sync is lost in a television receiver, it is most likely caused by an increase in the natural frequency of the corresponding oscillator. As a result, the picture information does not arrive in step with (is not synchronized with) the beam as it moves across the screen. The remedy is to decrease the natural frequency of the oscillator by lengthening the nonconduction period. This is accomplished by adding resistance to the grid-leak discharge path. In a television receiver, potentiometer R2 would thus be connected directly to an adjusting knob and labeled the Vertical Hold control.

A second (and less likely) reason for loss of horizontal or vertical sync is a *decrease* in oscillator frequency, of such magnitude that the trigger pulses arrive too early during the nonconduction period (too far down on the exponential curve), instead of near the end as they should. This is not too likely to happen, because a much greater deviation in natural frequency is required here before loss of sync occurs. When it does, the remedy of course is to *increase* the natural frequency of the oscillator by *shortening* the nonconduction period. This is accomplished by decreasing the resistance of potentiometer R2 in the grid circuit, in order to speed up the discharge time of the grid-leakage electrons so that the pulses will again trigger the oscillator.

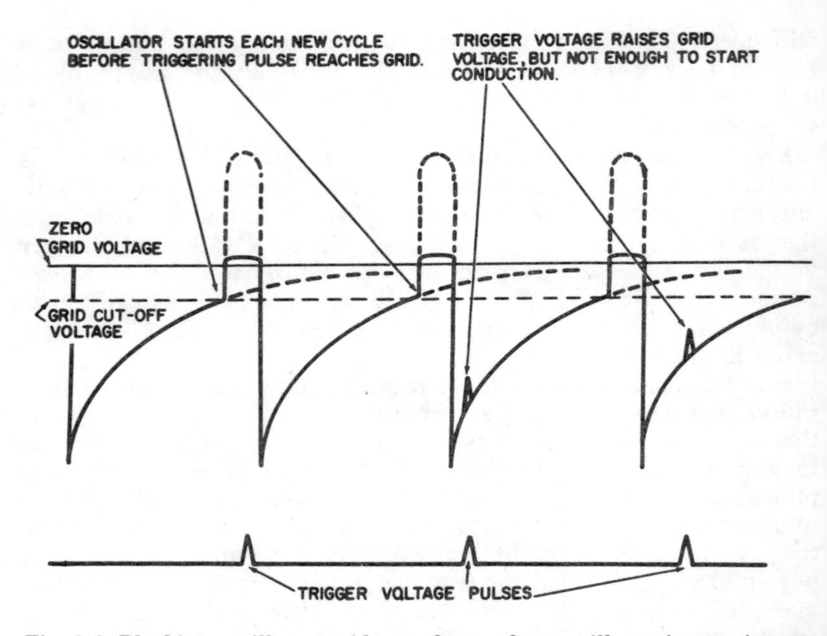

Fig. 8-6. Blocking-oscillator grid waveform when oscillator is running too fast to be synchronized.

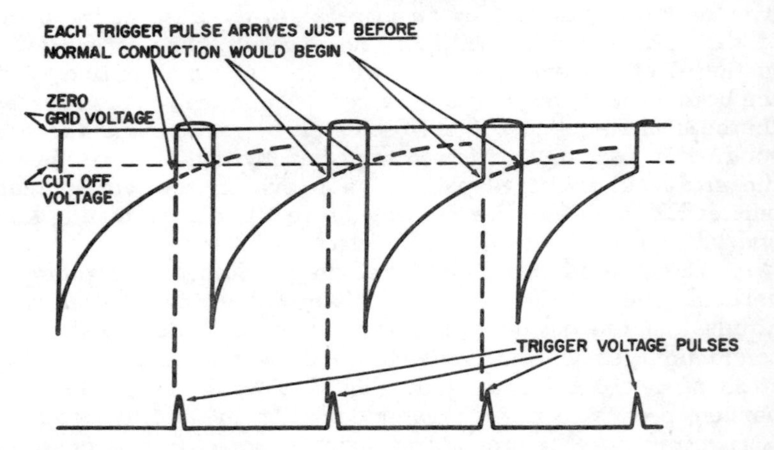

Fig. 8-7. Blocking-oscillator grid waveform when oscillator is being synchronized by applied trigger voltage.

OUTPUT COUPLING

Output voltage (shown in Fig. 8-4) can be coupled out of the blocking-oscillator circuit via coupling capacitor C2. The output point is connected to the junction of the transformer primary and secondary windings and thus exhibits the voltage at this point. In Fig. 8-1, which shows conditions during the first part of the conduction period, the current flows from right to left in the secondary winding. This can only mean that the secondary voltage is more positive at the left side winding, accounting for the initial rise in output voltage shown by the waveform in Fig. 8-4.

Toward the end of the conduction period, as depicted in Fig. 8-2, the secondary current reverses direction and flows from left to right, in accordance with the concept of counter emf discussed in Chapter 1. Now the voltage at the left side of the winding is negative. Actually, it does not reach its most negative value until shortly after the nonconduction period begins. This accounts for the negative peak in the output waveform of Fig. 8-4.

The positive output pulse can easily have a magnitude of a hundred volts or more. Even without additional amplification, it is thus able to provide a variety of functions in electronic circuits.

Chapter 9

MULTIVIBRATORS

The multivibrator circuit is widely used for providing the timing function in television, computers, radar, and many other applications. The basic free-running multivibrator circuit is given in Figs. 9-1 and 9-2. Its components are:

R1—Plate-load resistor (about 20,000 ohms) for V1.
R2—Grid-biasing and -driving resistor (about 2 megohms) for V1.
R3—Plate-load resistor (about 20,000 ohms) for V2.
R4—Grid-biasing and -driving resistor (about 1 megohm) for V2.
C1—Grid coupling and blocking capacitor (800 micromicrofarads) for V1.
C2—Grid coupling and blocking capacitor (also 800 micromicrofarads) for V2.
V1 and V2—Triode amplifier tubes.
M1—Power supply.

There are six currents at work in this circuit, and their movements must be understood in order to comprehend how the circuit operates as a whole. Each of the current paths is shown in Figs. 9-1 and 9-2. These currents are:

1. Plate current for V1.
2. Grid-leak biasing current for V1.
3. Grid-driving current for V1.
4. Plate current for V2.
5. Grid-leak biasing current for V2.
6. Grid-driving current for V2.

The voltage waveforms resulting from the current flow at the plate and grid of each tube are given in Fig. 9-3.

CIRCUIT DESCRIPTION

As mentioned previously, the circuit in Figs. 9-1 and 9-2 is connected to operate as a free-running multivibrator. Its pulse-repetition frequency is determined by the values of the resistors and capacitors, and by the conduction characteristics of the two vacuum tubes.

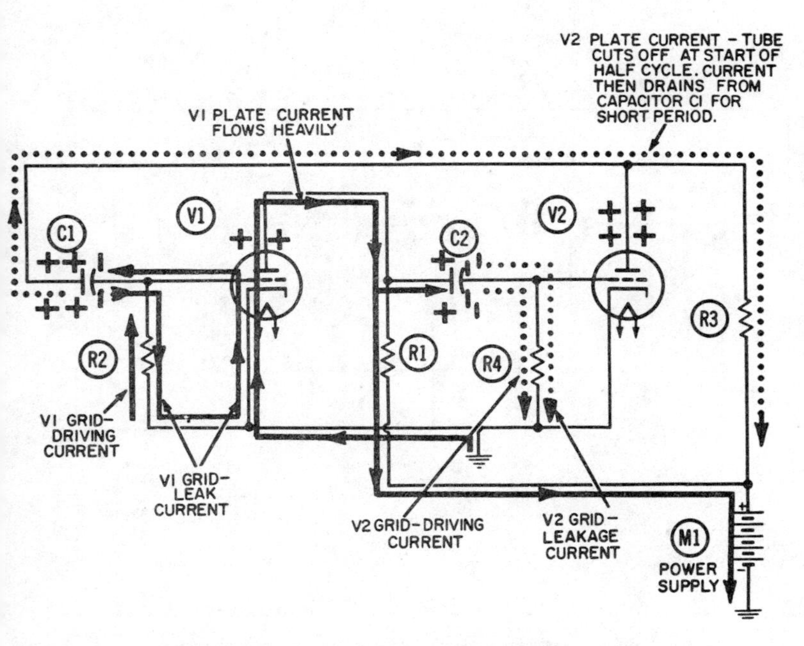

Fig. 9-1. Operation of free-running multivibrator—first half-cycle.

Operation begins when plate voltage is applied to V1 and V2 simultaneously. As each tube conducts electrons from cathode to plate, the voltage at each plate will drop—but not by the same amount. If both tubes conducted exactly the same number of plate-current electrons (which they never do) and if the two plate-load resistors were *exactly* the same value (which they never are), then both would be identical. But such an assumption is purely hypothetical—the two drops will never be the same.

Let us assume that V1 conducts the greater amount of plate current. As a result, its plate voltage will drop to a lower positive value than V2's. Let us also assume that it drops to +100 volts, and that the supply voltage is +250 volts.

It is true to say that capacitors oppose any change in voltage, or another way of saying the same thing: "it is impossible to instantaneously change the voltage across a capacitor." The fundamental meaning of these statements will be made clearer by now observing the action within capacitor C2. Before tube V1 conducts, the voltage on the left plate of C2 (and also on the

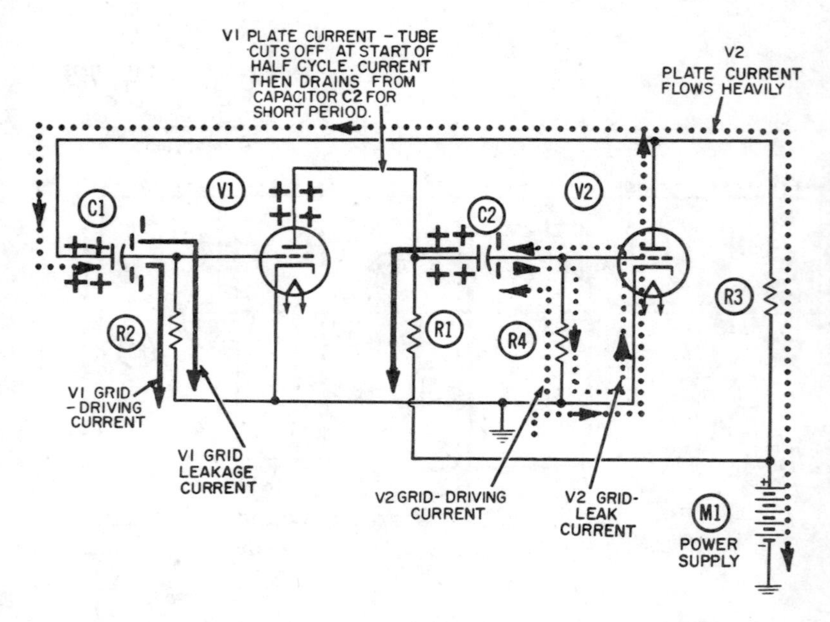

Fig. 9-2. Operation of free-running multivibrator—second half-cycle.

plate of V1) is the +250 volts from the power supply, while the voltage on the right plate is essentially zero since it is connected to ground through grid resistor R4. Thus, the moment before conduction begins, the voltage *across* the two plates of the capacitor is +250 volts.

We know that the voltage across C2 does not change at the same instant the tube conducts and the plate voltage drops. It will remain at +250 volts momentarily, and then decrease to +100 volts as the electrons flow onto the left plate of C2. As they do, an equal number will flow from the right plate and downward through R4. This grid-driving current, shown a dotted line in Fig. 9-1, produces a voltage drop across R4. Now the voltage at the junction of R4 and C2 (hence, the grid voltage) is −150 volts. Line 4 of Fig. 9-3 shows the voltage waveform at the grid of tube V2. At the start of the first half-cycle this voltage suddenly goes from zero to −150 volts, and then returns toward zero at an exponential rate determined by a corresponding exponential decay in the grid-leak current (shown in small dots) flowing downward through R4 to ground.

The sudden drop immediately cuts off the plate current through V2 and thus allows its plate voltage to rise toward the value of the supply voltage. It will stay at 250 volts until the grid voltage has gone from −150 volts to the cutoff value of −10 volts. The first half-cycle has now ended.

The second half-cycle begins when the rise in plate current through tube V2 causes a drop in its plate voltage, similar to the action in tube V1. Since plate-load resistors R1 and R3 are both equal in value, the voltage at the plate of V2 drops from +250 to +100 volts, just as the plate of V1 did.

Capacitor C1 starts this second half-cycle with 250 volts across it, since its one plate is connected to the plate of tube V2 and its other plate is connected to ground through grid resistor R2. C1 will also oppose any change in voltage across its plates. The voltage on the right plate, which is also the voltage applied to the grid of tube V1, immediately goes from zero to −150 volts as current flows downward through grid resistor R2. This current decays exponentially to zero, and the resultant voltage across it (which is also the voltage applied to the grid of tube V1) will likewise decay toward zero at the same exponential rate, as shown in Fig. 9-3.

It is important to note why the second half-cycle is so much longer than the first. You will recall that C1 and C2 have equal values of 800 micromicrofarads; but R2 is 2 megohms, or twice the value of R4. Using the formula ($T = R \times C$) explained earlier, we can calculate the time constant for the C1-R2 combination as follows:

We know:

$$R = 2 \text{ megohms} = 2 \times 10^6 \text{ ohms},$$

$$C = 800 \text{ micromicrofarads} = 800 \times 10^{-12} \text{ farads},$$

$$T = R \times C.$$

Therefore,

$$\begin{aligned} T &= 2 \times 10^6 \times 800 \times 10^{-12} \\ &= 2 \times 800 \times 10^{-6} \\ &= 1{,}600 \times 10^{-6} \text{ seconds} \\ &= 1{,}600 \text{ microseconds} \end{aligned}$$

Similarly for C2-R4, where the only new value is R4 which is 1 megohm (1×10^6 ohms), the time constant is:

$$\begin{aligned} T &= 1 \times 10^6 \times 800 \times 10^{-12} \\ &= 800 \times 10^{-6} \text{ seconds} \\ &= 800 \text{ microseconds} \end{aligned}$$

Thus, the time constant of C1-R2 in the grid circuit of V1 is twice as long as the one for C2-R4 in the grid circuit of V2.

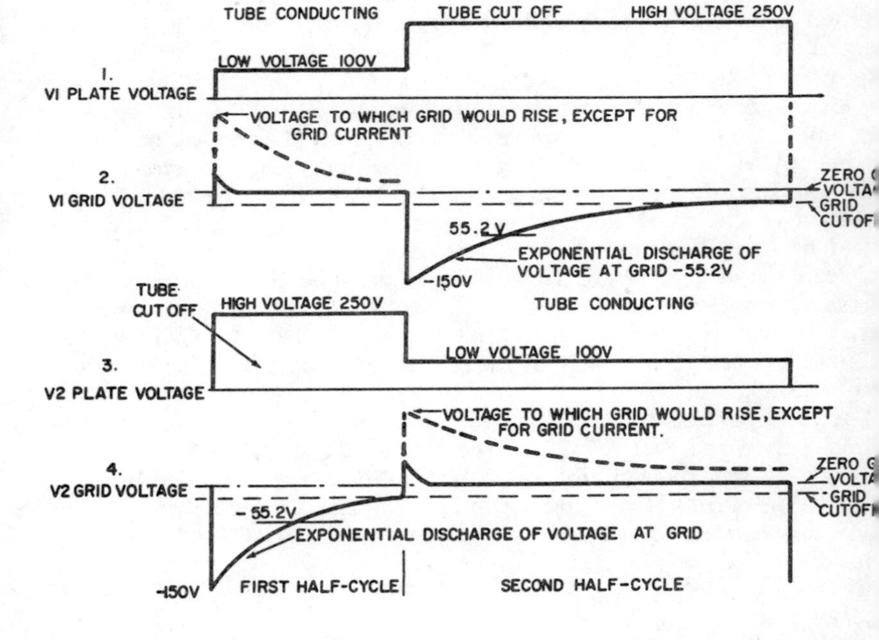

Fig. 9-3. Grid- and plate-voltage waveforms for free-running multivibrator.

Remember that the time constant of an R-C circuit is the length of time it will take the capacitor to discharge to 63.2% of its original value. Therefore, during the first 800 microseconds after the first half-cycle starts, the grid voltage on V2 will discharge to about −55.2 volts, or 36.8% of its original −150 volts. This is shown in line 4 of Fig. 9-3.

After three time periods of 800 microseconds each, the grid of V2 will have discharged to about −7.5 volts. Since its cutoff value is −10 volts, V2 will start conducting before the third time eriod is completed, ending the first half-cycle and starting the second one.

Now the grid of V1 will also discharge to −7.5 volts after three time periods of 1,600 microseconds each. Slightly before the end of the third time period, V1 begins conducting and the first half-cycle starts all over.

Thus, we see that the first half-cycle is slightly less than 3 × 800 microseconds, or 2,400 microseconds; and that the second half-cycle is twice as long, or not quite 4,800 microseconds.

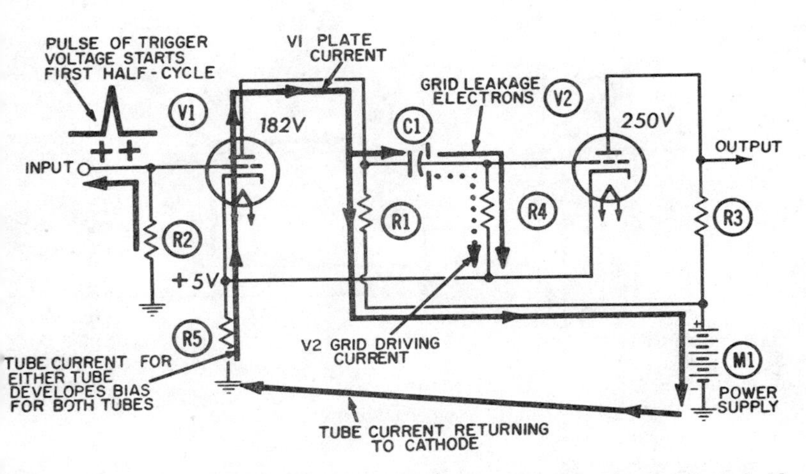

Fig. 9-4. Operation of a modified Eccles-Jordan multivibrator—first half-cycle.

The waveforms at the plates of V1 and V2 will be rectangular, as shown by lines 1 and 3 of Fig. 9-3, because of the unequal half-cycles. If the two half-cycles were equal—that is, if R2 and R4 had identical resistances—the waveforms would be square, and the multivibrator could then be considered a crude form of square-wave generator. (Actually, they would only be approximations of square waves. However, the term *square wave* is adequate for conveying a general understanding of how the circuit operates.)

Note in lines 2 and 4 of Fig. 9-3 that the grid voltage rises exponentially until it reaches −10 volts (its cutoff value). At this point, it almost immediately jumps to a slightly positive value, then drops back to the zero-voltage line and hovers there during the remainder of the half-cycle that the tube is conducting.

What prevented the grid from being driven positive instead of remaining at (or near) zero? Whenever the control grid is driven positive (as it was here), it will attract electrons from the plate current. These grid-leak electrons, sometimes more loosely described as "grid current," collect on the control grid and drive the positive-going grid voltage back to zero.

The dotted portions of lines 2 and 4 in Fig. 9-3 show how high the grid voltage would rise, were it not for the grid-leak current. At the start of the first half-cycle (Fig. 9-1), for example, the current through V2 is suddenly cut off, and the voltage at the

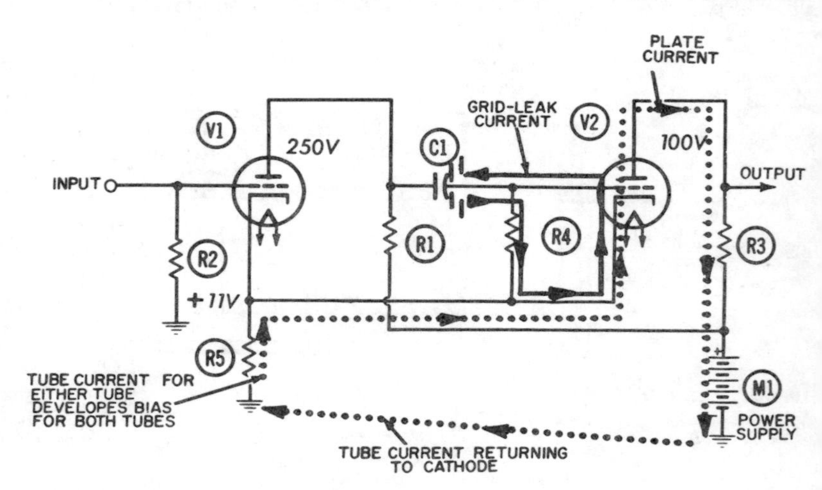

Fig. 9-5. Operation of a modified Eccles-Jordan multivibrator—second half-cycle.

plate rockets from +100 volts toward the supply voltage of +250 volts. Capacitor C1, connected to the plate of V2, opposes this sudden change in voltage across *its* plates—it tries to drag the V1 grid 150 volts in the positive direction, from −10 volts to +140 volts, but is prevented from doing so by the grid-leak current flowing out of V1 and onto the left plate.

A series of similar events also occurs at the start of the second half-cycle (Fig. 9-2) and prevents the V2 grid voltage from rising any further after it passes the zero-voltage point.

During the first half-cycle, the grid-leak and -driving currents flowing through R2 travel in opposite directions. The driving current predominates and "triggers" V1 to start the first half cycle. Through R4, however, the two currents flow in the same direction. (Remember that grid-leakage electrons always flow downward, in an attempt to discharge from their storage points on the grid sides of the grid capacitors.)

In Fig. 9-2 the situation has been reversed—the currents flow in the same direction through R2, but in opposite directions through R4. The driving current is more temporary than the leakage current, since it flows only while the plate voltage in the other tube is rising or falling.

Summary

Both multivibrator tubes will begin to conduct plate current when the supply voltage is applied. One of the tubes inevitably will conduct slightly more current, even though the two circuits are identical. When this happens, its plate voltage will fall and, by capacitive coupling, reduce the voltage at the grid of *the second tube.*

The lower voltage will reduce the current through the second tube and raise its plate voltage. This rise is capacitively coupled to the grid of the first tube, increasing the plate current and reducing the plate voltage of the first tube still further. The effects are cumulative, so that the first tube conducts more and more heavily until it cuts off the second tube.

The grid voltage of the second tube will have been driven considerably below cutoff during this sequence; and because of the grid-leakage electrons it has accumulated, the second tube requires considerable time to discharge this negative voltage to ground before it can conduct again.

When this does happen, the earlier sequence is repeated, only this time in reverse. The plate voltage of the second tube begins to drop because of the rising plate current. Since the plate is capacitively coupled to the grid of the first tube, its voltage drops accordingly, reducing the plate current and raising the plate voltage of the first tube. By capacitive coupling the grid voltage of the second tube is raised still higher, and so on, in another cumulative series of events. The first tube will soon be cut off, and will remain off until its negative grid voltage can again discharge to the cutoff value.

Shifting conduction from one tube to the other in this manner, the multivibrator circuit will oscillate indefinitely. For this reason it has been aptly named a "flip-flop" circuit.

THE ECCLES-JORDAN MULTIVIBRATOR

The modified Eccles-Jordan multivibrator (Figs. 9-4 and 9-5) is also known as a "one-shot" or "trigger" multivibrator. Unlike in the free-running multivibrator described previously, only one of the tubes has its plate coupled back to the grid of the other. This feature prevents the circuit from cycling repeatedly. Also, both tubes use a common cathode-biasing resistor to drive the voltage at both cathodes in the positive direction, in a manner which will be explained later. The grid of V1 is returned to ground, whereas in V2 it is returned to the cathode. Since ground

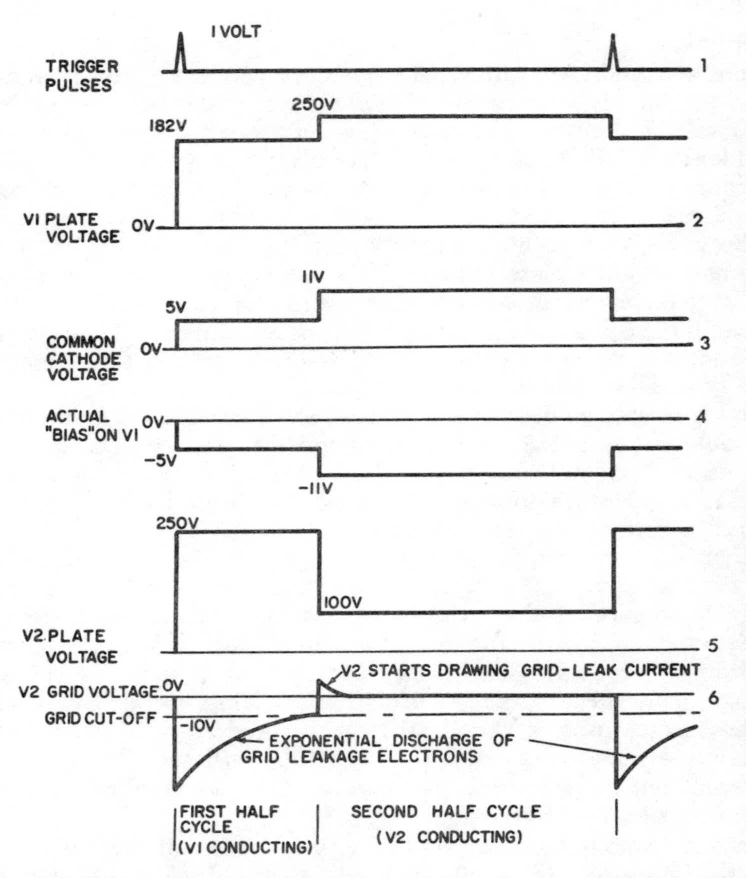

Fig. 9-6. Voltage waveforms during one cycle of operation of the Eccles-Jordan multivibrator.

is at a lower voltage than the cathodes, normally V1 remains cut off and V2 conducts.

The components of this circuit include:

R1—Plate-load resistor (about 20,000 ohms) for V1.
R2—Grid driving resistor for V1.
R3—Plate-load resistor (about 20,000 ohms) for V2.

R4—Grid blocking and driving resistor (about 1 megohm) for V2.
R5—Common cathode-biasing resistor (about 1,000 ohms) for V1 and V2.
C1—Grid coupling and blocking capacitor (about 800 micromicrofarads) for V2.
V1 and V2—Triode amplifier tubes.

The currents at work in this circuit appear in Figs. 9-4 and 9-5:

1. Grid driving current for V1.
2. Plate current for V1
3. Grid driving current for V2.
4. Plate current for V2.
5. Grid-leak current.

The first half-cycle begins with the arrival of a positive triggering pulse at the control grid of V1. This pulse, a solid line in Fig. 9-4, draws a small electron current upward through grid resistor R2 and into whatever coupling device is at the left of V1.

V1 is normally cut off (nonconducting) because its grid is connected to ground and thus is at zero voltage, whereas its cathode is at the positive voltage produced by the current flowing upward through cathode biasing resistor R5. On the other hand, V2 normally conducts a substantial amount of plate current.

Plate current for either tube will flow upward through resistor R5. Since both tube cathodes are connected to the top of R5, the two will always have the same positive voltage on the cathode, irrespective of which tube is conducting. Let us assume it is +11 volts.

Tube V1 is normally cut off because its control grid is connected to ground, or zero voltage, through resistor R2. Hence, the grid is at zero volts also, and we said earlier that the cathodes are at +11 volts. This leads us into *bias*, an often misused term.

Bias, in the correct sense, refers to the *difference* in voltage between the grid and cathode. In this circuit we would normally say the cathode is 11 volts more positive than the grid. But it would be just as correct to say the grid is 11 volts more negative than the cathode. This 11-volt difference in voltage between the grid and cathode is the bias. (If the cathode were at zero volts instead and the grid were at −11 volts, the bias would still be the same −11 volts.)

If we assume that the grid cutoff voltage is −10 volts, V1 will remain cut off as long as its bias is −11 volts. Nothing happens until a trigger pulse exceeding +1 volt arrives at the grid and raises its voltage above the cutoff value. When the pulse arrives, plate current begins to flow through V1, marking the beginning of the first half-cycle (Fig. 9-4).

Now the V1 plate voltage begins to fall. This drop in voltage is coupled from the plate of V1, via C1, to the grid of V2. The flow of plate current through V2 is cut off, causing the plate voltage to rise rapidly toward the value of the supply voltage.

Unlike in the free-running multivibrator discussed previously, this rise in plate voltage is not coupled back to the grid of V2. Consequently, the grid voltage of V1 will not rise as high, and V1 will conduct less plate current, than it would if connected as a free-running multivibrator.

This reduction in plate current through V1 (compared with that which would flow in the free-running multivibrator) has two important effects. First, the plate voltage of V1—and hence the voltage at the grid of V2—will not be reduced to as low a value. Secondly, the cathode-bias voltage developed across R5 will be lower during the first half-cycle. Line 3 of Fig. 9-6 shows this reduction in cathode voltage.

Let us assume the cathode voltage across R5 drops from +11 volts during the second half-cycle to +5 volts during the first one. This has the effect of reducing the total grid bias from −11 to −5 volts.

Line 4 of Fig. 9-6 is essentially an inversion of Line 3; it shows the actual bias on V1. Once the trigger voltage has started this tube conducting, the grid returns fairly quickly to zero volts and the bias stabilizes at −5 volts—enough to permit a reasonable amount of plate current to flow.

Line 5 of Fig. 9-6 shows the plate-voltage waveform for tube V2. At the start of the first half-cycle when the plate current is cut off, the plate voltage rises to the value of the supply voltage, and remains there until the plate current is turned on again at the start of the second half-cycle.

Line 6 of Fig. 9-6 is quite similar to the grid-voltage discharge curves shown in Fig. 9-3 for the free-running multivibrator, and will therefore not be reanalyzed in detail here. Tube V2 will remain cut off as long as the difference in voltage between its grid and cathode is greater than the cutoff value of −10 volts. The discharging of capacitor C1 consists of grid-leakage electrons flowing downward through grid resistor R4. The rate of flow, which decreases exponentially, determines the voltage developed across R4.

The first half-cycle ends when V2 begins to conduct. This cuts off the flow of plate current through V1, but in a different manner than in the free-running multivibrator. The increase in total urrent through cathode resistor R5 raises the voltage at both athodes, immediately restricting the flow of plate current hrough V1. As a result, the plate voltage of V1 begins to rise. 'his increase is coupled via C1 to the grid of V2, raising the grid oltage and hence the plate current of this tube still further. 'his increase in V2 plate current, flowing through common-athode resistor R5, further reduces the plate current through V1, utting the tube off very quickly.

The second half-cycle, once begun, will continue indefinitely nless retriggered by an outside voltage source. Thus, it will sually last longer than the first half-cycle. The grid voltage f V2 will rise momentarily to a slightly more positive value han the cathode, but will then begin to acquire some grid-leak lectrons from the tube, preventing it from rising any further. 'hroughout the second half-cycle, the two voltages will remain lose to the same value.

The different amount of plate current through each tube is eflected by the waveform voltages at the two plates. When not onducting, each tube plate will rise to the +250-volt supply oltage. However, since V2 conducts more than twice as much urrent as V1, and since R1 and R3 are equal in value, it follows rom Ohm's law that the resulting voltage drop across R3 will e more than twice the one across R1. Consequently, approxi-nately 182 volts has been indicated at the plate of V1 during ne first half-cycle, as opposed to 100 volts for V2 during the econd half-cycle.

Chapter 10

THYRATRON SAWTOOTH GENERATORS

A thyratron differs from a vacuum triode in that it is filled with an easily-ionized gas, rather than being evacuated. Under certain conditions the gas will become ionized, when grid and plate voltages are applied, and permit a very heavy electron current to flow between cathode and plate.

An important characteristic of the thyratron is the inability of its control grid to cut off the plate current—once a thyratron begins to conduct, the only way it can be cut off is for the plate voltage to be lowered almost to zero—15 or 20 volts is not abnormally low.

The schematic symbol is the same for a vacuum tube or thyratron except for a dot within the tube envelope, as shown for the thyratron in Figs. 10-1 and 10-2.

Thyratrons are capable of generating sawtooth-voltage waveforms (so named because of their shape) which are used for sweeping the electron beam across the face of a cathode-ray tube in an oscilloscope, radar set, or other application where it is desired to gradually move the beam across the face of the tube and then quickly return it to the starting point. Fig. 10-3 shows the waveform generated at the plate.

The only significant current in this circuit crosses from cathode to plate and could perhaps be called "plate current." However, its movement during the conducting and nonconducting period of each cycle is quite different from the plate-current flow in a vacuum tube. Let us start at the beginning, when the power supply is turned on and applies a positive voltage to the plate.

In order for the plate voltage to rise from zero to the power-supply voltage, free electrons must be drawn away from the late of the tube, the upper plate of capacitor C1, and from any omponent connected to the output. This is the charging process nd the amount of time it takes depends on the values of plate-oad resistor R1 and capacitor C1 in Fig. 10-1. R1 is very large, everal hundred thousand ohms; and C1 is 1,000 micromicro-arads or so. The plate-voltage curve begins at zero and rises xponentially toward the power-supply voltage, as shown in 'ig. 10-3. (An exponential curve is one which does not rise or all at the same rate from left to right.) The dashed-line portions f the curves in Fig. 10-3 represent the voltage which would xist at the plate if the charging process were completed.

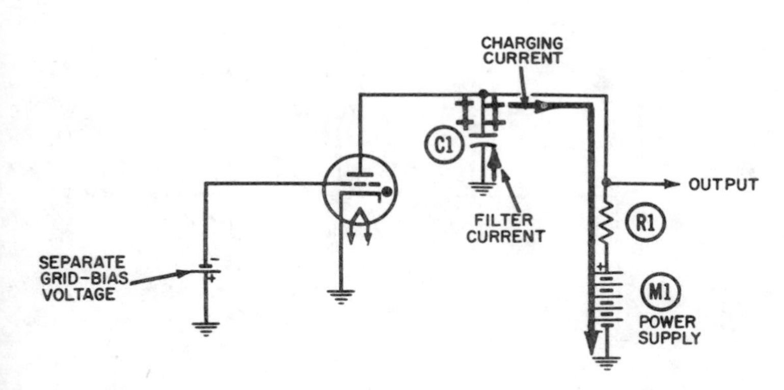

'ig. 10-1. Operation of the thyratron sawtooth-voltage generator—charging half-cycle.

In a thyratron tube, the value of plate voltage at which the as "breaks down" or becomes ionized is called the *firing voltage*. Vhen this happens, the molecules of gas becomes ionized and elease their free electrons, and the interior of the tube becomes illed with both positive ions and negative electrons. Attracted y the positive voltage at the plate, the negative electrons flow here in a steady stream and are replaced by other free electrons mitted by the heated cathode.

The value of the firing voltage depends partly on the applied rid-bias voltage, shown as a separate battery in Figs. 10-1 nd 10-2. In a typical thyratron tube, this grid bias is on the rder of −5 volts, and the firing voltage at the plate is 75 to 00 volts.

Once the tube begins to conduct, it delivers electrons to the late much faster than high-value plate-load resistor R1 can ass them. The electrons accumulate on the upper plate of ca-acitor C1 and reduce its charge to a very low positive voltage, as depicted in Fig. 10-2. The charged capacitor is now said to be "discharging through the tube." Strictly speaking, the capacitor is not discharging through the tube, but its positive *voltage* is being discharged because the tube supplies a large number of free electrons (in the form of plate current) to the capacitor. This discharging process occurs in only a tiny fraction of the time required for the charging process, as shown by Fig. 10-3.

Once the gas has been ionized, the grid can no longer control the flow of current. Now it can only be stopped by reducing the *plate* voltage to a very low value—on the order of 15 or 20 volts. This might be called the "cutoff" voltage for a thyratron.

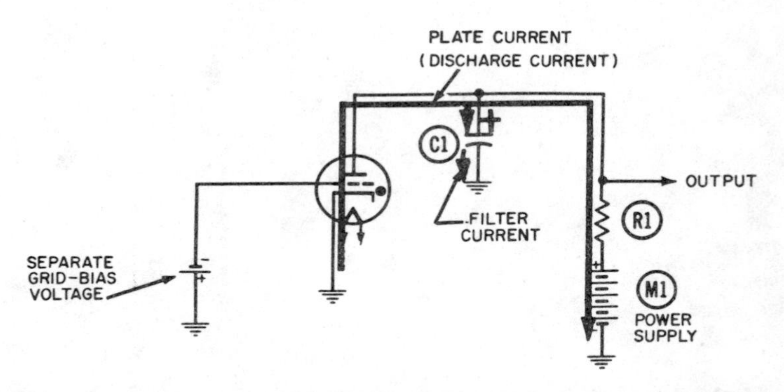

Fig. 10-2. Operation of the thyratron sawtooth-voltage generator—discharging half-cycle.

The exponential charging curve is almost a straight line in the low plate-voltage regions, between the cutoff and firing voltages. This area, referred to as the linear portion of the curve, is where the sawtooth voltage is generated.

As in any tube, the current must have a closed path leading back to the cathode. In Figs. 10-1 and 10-2, the complete path is from cathode to plate, through the load resistor and power supply to ground, and back to the cathode.

The much shorter discharge half-cycle makes the thyratron an ideal sweep generator. After the beam has been swept across the face of the screen, it must be retraced as quickly as possible or the viewer will see a flickering picture. The short discharge period is normally used for retrace, so that the beam will be out of action as briefly as possible.

A filter current, a thin line, flows between the lower plate of capacitor C1 and ground. In Fig. 10-1 the current flows upward because of normal capacitor response to the voltage changes at

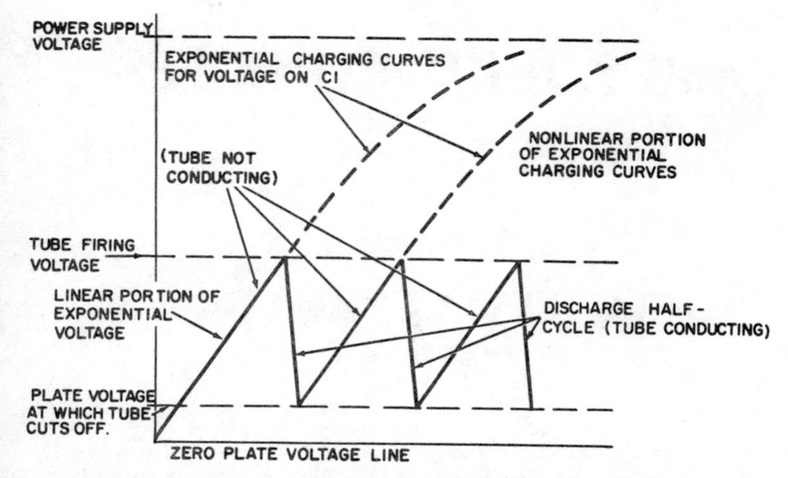

Fig. 10-3. Sawtooth output voltage generated at plate of thyratron generator.

the plate of the tube, but is driven downward in Fig. 10-2 by the plate-current electrons flowing onto the upper plate.

Obviously, a high positive value of plate-supply voltage contributes to greater linearity of the charging half-cycle, whereas a lowered value would lead to greater curvature in the discharge curve.

SECTION 2

AMPLIFIER CIRCUITS

Chapter 1

BASIC VACUUM-TUBE ACTIONS

An elementary understanding of the actions within a vacuum tube is essential to an understanding of those which occur outside the tube. Most of this book is devoted to a study of these common circuit actions, all of which are regulated in some manner within the tube. These basic phenomena which occur over and over include:

1. Free electron emission from a heated cathode.
2. Easy passage of free electrons through a vacuum.
3. Attraction of electrons to positive voltage points.
4. Repulsion of electrons from negative voltage points.

DIODE TUBES

The simplest, and also the earliest, vacuum tube is the diode. The discovery of the Edison effect by Thomas Alva Edison led to the invention of the diode. Edison had observed that when two conductors were terminated inside an evacuated glass envelope and one was heated, a small electric current would flow between conductors under certain other conditions of applied voltage. Although he recorded his observations, he made no immediate attempt to explain or apply them.

The phenomenon he observed can be fully explained by examining the operation of the diode vacuum tube, which is in wide use today. Fig. 1-1 reveals its essential construction details in symbolic form. The two elements, called electrodes, are the *cathode* and the *plate* (sometimes referred to as the anode).

Three conditions must be met in order for electric current (meaning electrons) to flow across the open space within the tube:

1. The cathode must be made hot enough that free electrons will actually "jump" off the cathode.
2. A positive voltage must be applied to the plate in order to attract the negative electrons which have left the cathode.
3. The space between the electrodes must be evacuated of all air molecules, so that there will be nothing within the tube to hinder the movement of electrons.

The relative ease or difficulty with which electrons will leave the heated cathode is called the *work function* of the cathode material. When any material is heated, a tremendous increase in molecular motion occurs. This molecular agitation results in the releasing, or "shaking off," of many of the electrons from their orbit around the nuclei of the various molecules. Once it is released from its planetary orbit within a molecule, such an electron becomes what is known as a "free" electron. Some electrons from the molecules immediately adjacent to the surface will not only jump free of their parent molecules, but may jump free of the surface of the metal also. Any electron which has done the latter becomes available for making up *space current* (the name applied to the electron current drawn across vacuum tubes, from cathode to plate).

This process of releasing electrons into the tube by heating the cathode can be likened to the familiar pot of boiling water. If you observe the boiling water closely, you will be able to see tiny droplets jumping off its surface. This analogy gives rise to the fairly common expression that electrons "boil off the cathode."

Gravity normally causes the droplets of water to fall back onto the surface. Electrons, however, do not respond to gravitational force but to electric forces. Since all electrons carry one negative electric charge, they are repelled by each other, and hence by any negative voltage, however small. By the same token, they are all attracted by positive voltages. It thus becomes a simple matter to attract electrons across the vacuum tube by applying a suitable positive voltage to the plate. The amount of positive voltage determines to some extent the number of electrons that will cross the tube, and the size and shape of the cathode itself is another factor.

Fig. 1-2 shows a common graphical representation of the voltage-current characteristic of a 2X2A diode tube. A curve of this nature reveals the *amount of current* that will cross the tube for any particular plate voltage. As an example, a plate voltage of 100 volts (which is read on the horizontal scale) indicates that approximately 15 milliamperes of current will flow across the tube (this being read on the vertical scale as shown by the dashed line). Another dashed line shows that if 200 volts is applied to the plate, about 35 milliamperes of current will flow.

Literally hundreds of different diode tubes are available, all operating in the same fundamental fashion described. These tubes will differ in construction, depending on the amount of plate voltage it is necessary or desirable to apply and on the amount

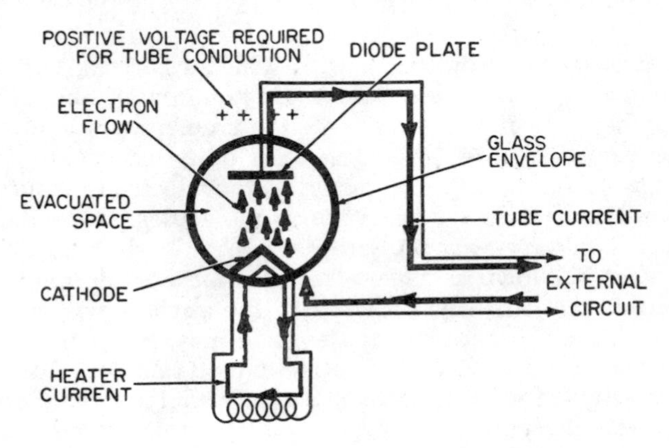

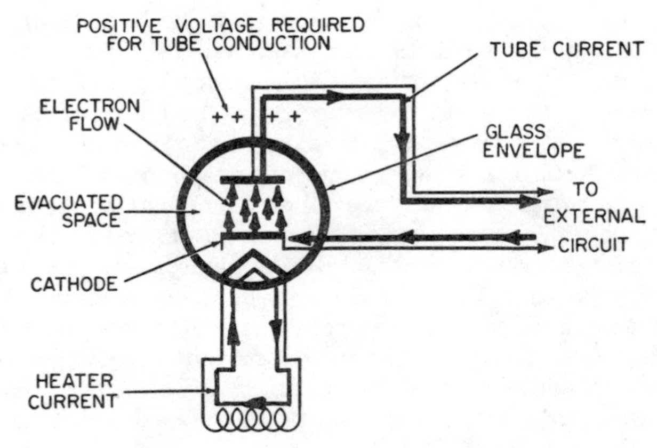

Fig. 1-1. The diode vacuum tube.

of plate current (tube current) desired. The amount of current required is dictated by external circuit considerations, some of which are discussed in later chapters.

Diode vacuum tubes are used extensively as rectifiers and detectors. The term *rectifier* is usually applied to a circuit for converting low-frequency alternating current into direct current.

The term *detector,* on the other hand, is usually applied to a circuit employed to *demodulate* a high-frequency (usually a radio-frequency) alternating current. That is, a detector separates the low-frequency component of a signal from the high-frequency component it is superimposed on. A detector cannot demodulate without first rectifying the alternating current. Therefore, only unidirectional currents will flow as a result of this rectifying action. (A unidirectional current is one that, although varying in amplitude, always flows in one direction.)

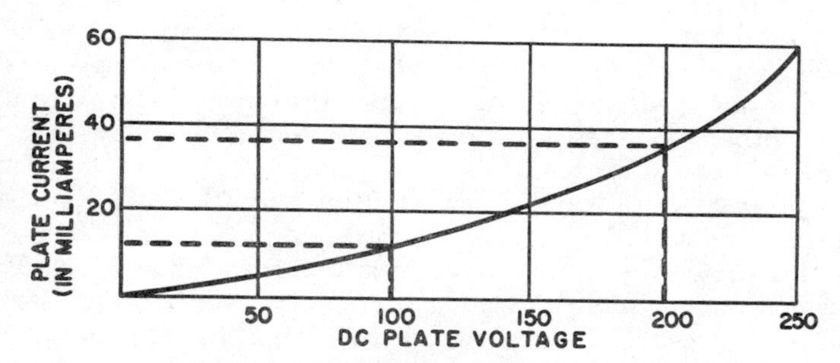

FILAMENT HEATING VOLTAGE — 2.5 VOLTS.
STEADY STATE PEAK CURRENT — 60 MA.
TRANSIENT PEAK CURRENT — 100 MA.

Fig. 1-2. Plate voltage-current characteristics of a typical diode

TRIODE TUBES

Fig. 1-3 shows the commonly accepted symbol for a *triode* vacuum tube. The triode differs from the diode in that a third element, called the *control grid,* has been added between the cathode and plate. The control grid is constructed of very fine mesh wire, and because it is between the cathode and plate, any electrons which flow from one to the other must pass through the mesh itself. The control grid is usually much closer to the cathode than to the plate. The wire-mesh construction and this closeness to the cathode make it possible to regulate, or control, the electron stream through the tube by varying the voltage of the control grid.

The control grid, invented early in the twentieth century by Dr. Lee de Forest, made it possible to *amplify* small signals—perhaps the most important single function performed by electronic circuitry. Tubes used as amplifiers are so constructed that a small charge in voltage at the grid will cause a relatively large change in the current flowing through the tube. The changes in tube current flowing through the external load will then develop a correspondingly large variation in the voltage present at the plate of the tube.

To understand how a tube amplifies, it is necessary for you to understand its characteristic curves. There are three quantities in question, and a change in any one will affect the amount of amplification attainable. These three quantities are the grid voltage, plate voltage, and plate current.

(The term *grid voltage* is sometimes taken to mean the *difference* in voltage between the grid and cathode. However, a more correct term for this voltage is *grid bias*. If the cathode is maintained at zero voltage (ground), then the actual voltage at the grid is also the grid bias. However, when the cathode is maintained at some other voltage, as is more frequent, the voltage at the cathode must be subtracted from the voltage at the grid to arrive at the proper "grid voltage" to be used with the characteristic curves of the tube.)

Fig. 1-4 shows the plate characteristic curves for a 6BN4. This is a triode tube which is used primarily as an RF amplifier in television receivers. The abscissa, or horizontal scale, is measured in volts and represents the possible range over which the plate voltage of the tube may be varied. The ordinate, or vertical scale, is measured in milliamperes and represents the range over which the plate current will vary for the various grid voltages during operation.

The curves across the face of the graph represent particular values of grid voltage (grid bias) which is designated as E_c. The curve farthest to the left represents a grid voltage of zero. Each point on this line tells us how much plate current will flow through this tube, for the particular value of plate voltage, when the grid voltage is zero.

As an example, the vertical line which represents 100 volts of plate voltage intersects the zero grid-voltage curve at point *A*. The dashed horizontal line, which also intersects point *A*, indicates that approximately 19 milliamperes of plate current will flow through the tube for this combination of grid and plate voltages. As long as the combination of 0 volts on the grid and 100 volts on the plate is maintained, 19 milliamperes of current will flow through the tube.

Voltage amplification cannot be demonstrated adequately from a single example. By taking a second example, however, we can come closer to describing its meaning. Consider now what would happen if we should change the grid voltage from zero to −1 volt, without changing the plate voltage. The grid-voltage line for −1 volt intersects the plate voltage line for 100 volts at point *B*.

Reading horizontally from point *B* to the left edge of the graph, we see that approximately 9 milliamperes of plate current will flow through the tube for this new combination of grid and plate voltages.

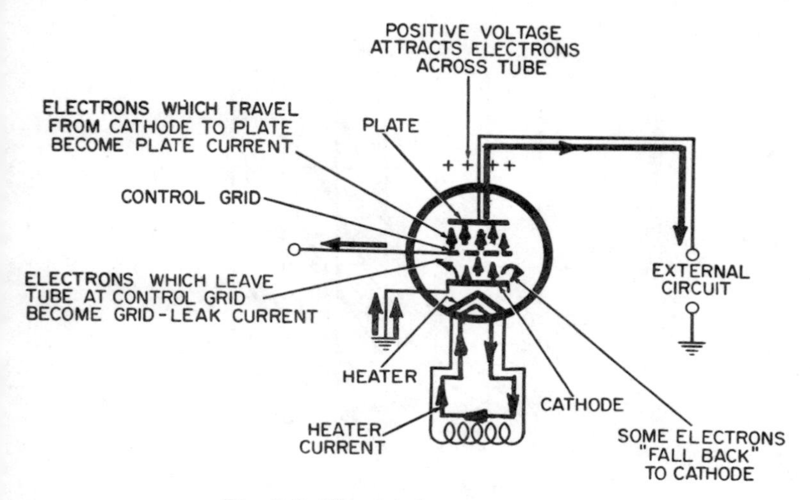

Fig. 1-3. The triode vacuum tube.

The most important point to grasp from these two examples is this—by reducing the grid voltage (meaning to make it more negative), we can reduce the amount of plate current which flows through the tube. By the same token, when we increase the grid voltage (meaning to make it less negative—or more positive), we can increase the amount of current through the tube.

These two examples enable us to arrive at a definition of the *mutual transconductance* (g_m) of a tube. Mutual transconductance is defined as the ratio between a small change in the plate current and the corresponding change in grid voltage which caused it, *when the plate voltage is maintained constant*. Transconductance (the name it usually goes by) is measured in mhos or micromhos. (From Ohm's law we know that the ratio of voltage to current is a measure of resistance, which is measured in ohms. Therefore, since transconductance is just the opposite—the ratio of current to voltage—it has been given the unit of measurement of a *mho* which is ohm spelled backwards.)

Since plate current is normally measured in milliamperes, it becomes more convenient to measure transconductance in millionths of a mho, or micromhos. Values of several thousand micromhos are normal; the 6BN4 tube used in this example has a transconductance of 7700 micromhos.

Many tube testers have their meters calibrated directly in micromhos. One of the commonest symptoms of vacuum-tube failure is a decrease in the transconductance; this automatically implies a decrease in the total emission of electron current within the tube.

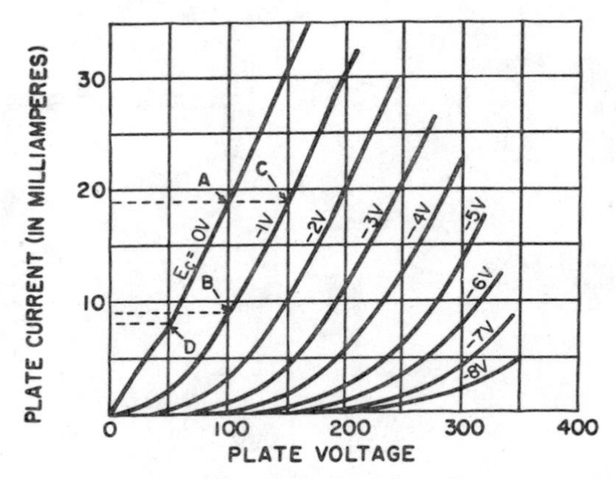

Fig. 1-4. Plate characteristic curves of a typical triode.

The second important rating factor of vacuum tubes is the *amplification factor*. It is the ratio between changes in grid and plate voltages, *when the plate current is maintained constant.* Fig. 1-4 can also be used to describe this rating factor. Suppose a line is drawn horizontally through point *A* until it intersects the next grid line, which is for a grid voltage of −1 volt. The point of intersection with this grid-voltage line has been labeled point *C*. It is evident, from the bottom scale for plate voltage, that point *C* represents a plate voltage of 150 volts. This is 50 volts greater than is represented by point *A*. Thus, a 1 volt change in grid voltage corresponds to, or has the same effect on plate current as, a 50-volt change in plate voltage. The amplification factor is the ratio of these two amounts, or:

$$\frac{50 \text{ volts}}{1 \text{ volt}} = 50$$

In other words, a tube with these values is said to have an amplification factor (μ or *mu*) of 50. This value of *mu* tells us the grid voltage is 50 times more effective than the plate voltage in controlling the flow of plate current.

The third important rating factor for vacuum tubes is a mythical quantity known as the *plate resistance,* the symbol for which is r_p. It is the resistance (in ohms) of the path between the cathode and the plate of the tube, which, of course, is not measurable with a meter. It is equal to the ratio between a small change in plate voltage and the corresponding change in plate current, *when the grid voltage is maintained constant*. This can also be understood from Fig. 1-4. The two points used to illustrate this quantity are *A* and *D*, both of which are on the same zero grid-voltage line. The change in plate current between these two points is evidently 19 − 8, or 11, milliamperes. This is read from the vertical scale at the left.

The change in plate voltage, read from the horizontal scale at the bottom, is evidently 100 − 50, or 50 volts.

These two quantities enable us to compute an approximate value of plate resistance for the particular operating conditions, as follows:

$$r_p \text{ (in ohms)} = \frac{\text{change in plate voltage (in volts)}}{\text{change in plate current (in amperes)}}$$

$$= \frac{50 \text{ volts}}{11 \text{ milliamperes}}$$

converting milliamperes to amperes, we have

$$r_p = \frac{50}{.011}$$

$$r_p = 4545 \text{ ohms}$$

Because of the curvature of the grid-voltage lines in Fig. 1-4, the locations of points *A* and subsequently B, C, and D will have some bearing on the exact amounts calculated for the three quantities. If point *A* had initially been positioned on a different grid-voltage line (or for a different combination of plate current and plate voltage), the *changes* in these quantities—represented by moving to points *B, C,* and *D*—will of course be different. Likewise, the calculated values of transconductance, amplification factor, and plate resistance will be changed somewhat.

Each of these quantities—*mu*, g_m, and r_p—is of great significance in the mathematics of electronic tube actions. (These mathematical considerations are beyond the scope of this book.) Transconductance (g_m) probably has the greatest everyday significance to the technician, since some tube testers are calibrated directly in transconductance. Also, any decrease below a specified level will instantly identify tubes which have "failed" (cannot deliver the proper amount of electron current from cathode to plate).

TETRODE TUBES

A refinement in vacuum-tube functions is made possible by the addition of a second grid between the control grid and plate as shown in Fig. 1-5. This screen grid, as it is called, is also a fine wire mesh. All electron current which eventually reaches the plate must pass through the grid wires.

The screen grid has two important functions:

1. Its screening action between the plate and control grid serves to isolate the voltage changes at the plate from those at the control grid, thereby preventing undesirable feedback between plate and grid.
2. It speeds up, or "accelerates," the plate-current stream flowing through the tube. In this way, electrons are prevented from accumulating around the cathode and impeding the plate-current flow. This accumulation is called *space charge*. In addition, the amount of plate current flowing is almost independent of the plate voltage, thus making a higher amplification factor possible.

To understand the screening action, it is first necessary for you to understand the nature and effects of interelectrode capacitance between the plate and control grid. In a tube, these two elements act like plates of a capacitor (capacitor action is covered in subsequent chapters). Suffice it to say here that whenever a radio-frequency voltage exists in the plate circuit, a radio-frequency current at the same frequency will also be present. During each cycle this current will drive a small quantity of electrons on and off the plate. As electrons are driven onto the plate, from its external circuit, an equal number will be driven off the control grid and into the grid circuit. Likewise, as electrons are drawn off the plate, an equal number of electrons will be drawn onto the grid from its external circuit. Thus, this feedback action induces an RF current/voltage combination in the grid circuit, with results that are generally undesirable.

Primarily because it is closer to the plate than to the control grid, the screen grid will exhibit much more capacitance to the plate. Consequently, it acts as an electrostatic shield between the plate and control grid. That is, radio-frequency voltages at the plate will in effect be isolated from the control grid. As a result, a similar RF current/voltage combination will be set up in the screen-grid circuit, where it can easily be diverted, or bypassed, to ground.

The screen grid normally is connected to a B+ supply voltage, which is about the same value as the plate voltage; or through a

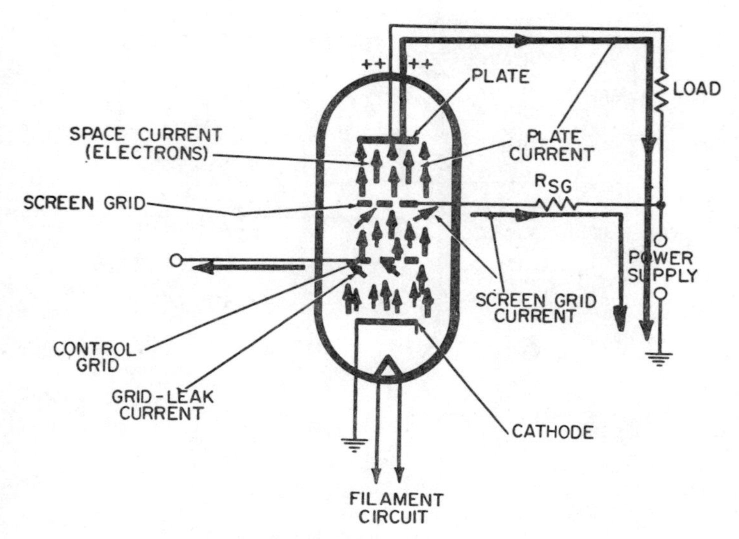

Fig. 1-5. The tetrode vacuum tube.

resistor, which results in a somewhat lower screen voltage. Because it is between the plate and the cathode, and has a high positive voltage on it, the screen grid provides an extra attraction to the electrons leaving the cathode. Thus, they are drawn away from the cathode area more easily, and in greater quantity, than if the screen grid were not there.

PENTODE TUBES

Fig. 1-6 shows the conventional symbol for the five-element tube called the *pentode.* (The filament and the connecting pins to it, required for heating the cathode, are not considered elements.) The pentode tube has three different grids between its cathode and plate. In addition to control and screen grids, (the functions of which were described for triode and tetrode tubes), the pentode has a third grid, called the *suppressor grid,* between the screen and plate.

Recall that when a high positive voltage is applied to it, the screen grid draws the electron stream toward it from the cathode, at a much higher velocity than if there were no screen grid. Most of this electron stream passes through the mesh of the screen grid and is then in position to be attracted by the high positive voltage of the plate.

Traveling at such a high velocity, these electrons will bombard the plate and discharge, or "knock off," other "secondary" elec-

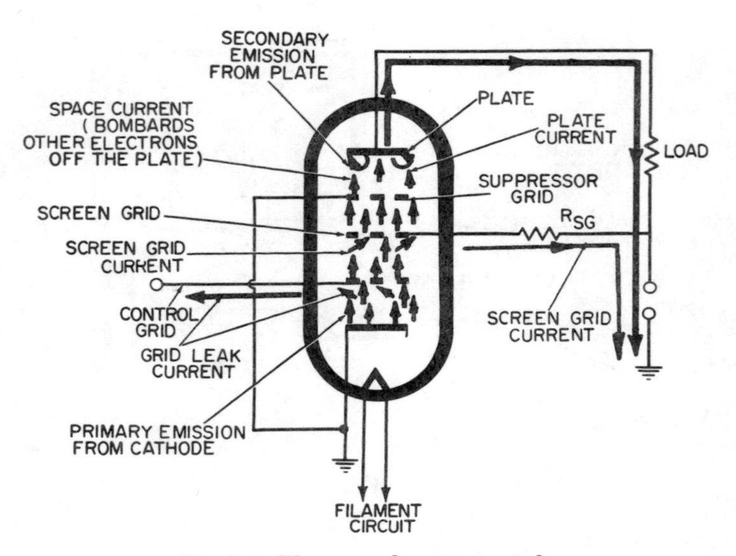

Fig. 1-6. The pentode vacuum tube.

trons. Once they have entered the open space between plate and screen grid, these secondary electrons are subjected to the positive voltages of the screen grid and plate. Under these conditions, a large percentage of the secondary electrons would probably be attracted to the screen grid and exit from the tube via it.

This phenomenon of electrons being knocked off the plate and re-entering the electron stream is called *secondary emission*. It is disadvantageous in two ways:

1. It will increase the total amount of screen-grid current, but in such an unpredictable manner that precise operation of the tubes and their circuits is very difficult.
2. It will decrease the total plate current by an equal amount and thereby largely nullify one of the principal reasons for adding a screen grid in the first place—namely, to increase the amount of plate current through the tube.

These disadvantages are circumvented by the suppressor grid between the screen and plate. The prime function of this grid is to suppress secondary emission of electrons from the plate of the tube. It is usually connected to the cathode and exhibits its same low voltage. Secondary electrons from the plate, upon entering the space between the plate and suppressor grid, will now "see" this low voltage instead of the high positive voltage of the screen grid. Therefore, the high positive voltage on the plate will exhibit a much stronger attraction for these electrons than the low suppressor grid voltage. As a result, most of them will return to the plate.

The plate-current stream must of course pass through the wire mesh of the suppressor grid. However, the electrons in this stream will be relatively unaffected by the low voltage on the suppressor grid, because their own inertia carries them beyond it. Once these electrons have moved beyond the suppressor grid, the high positive plate voltage quickly draws them to the plate.

Chapter 2

R-C COUPLED AF VOLTAGE AMPLIFIERS

The R-C coupled AF voltage amplifier circuit is one of the simplest amplifier circuits available. It is widely used for amplification of signals in the audio range, up to 10,000 or 20,000 cycles per second. The circuit goes by the name of RC amplifier, which refers to the combination of resistive load and coupling capacitor. It also is frequently referred to as a resistance-coupled amplifier.

CIRCUIT DESCRIPTION

When a signal voltage of a given amplitude (strength) is impressed on the control grid (input circuit), a signal voltage of larger amplitude will be delivered in the plate circuit, or output circuit. The accompanying diagrams portray the essential circuit actions occurring in two successive half-cycles. Fig. 2-1 depicts current (electron) flow conditions *during* an entire half-cycle of the signal which will be referred to here as a negative half-cycle; and also the instantaneous-voltage conditions at the midpoint of that same half-cycle.

Fig. 2-2 depicts the same current flow for the *entire* second half-cycle, and the voltage conditions at the midpoint of the second half-cycle. This, we will call the positive half-cycle.

The circuit components for the resistance-coupled AF voltage amplifier are:

R1—Grid load resistor for V1.
R2—Cathode biasing resistor.
R3—Plate load resistor.
R4—Grid load resistor for following stage.
C1—Cathode bypass capacitor.
C2—Coupling capacitor to following stage.
V1—Triode amplifier tube.
M1—Power supply.

There are three main currents at work in the circuit. Each has been shown in Figs. 2-1 and 2-2. They are:

1. Grid-driving current for stage.
2. Plate current of the tube.
3. Grid-driving current for following stage.

The first current, flows in the input circuit and is called the grid-driving current (solid). It is directly associated with the *grid-driving voltage* (also a solid line), a more common term than *grid-driving current* in discussing tube operation.

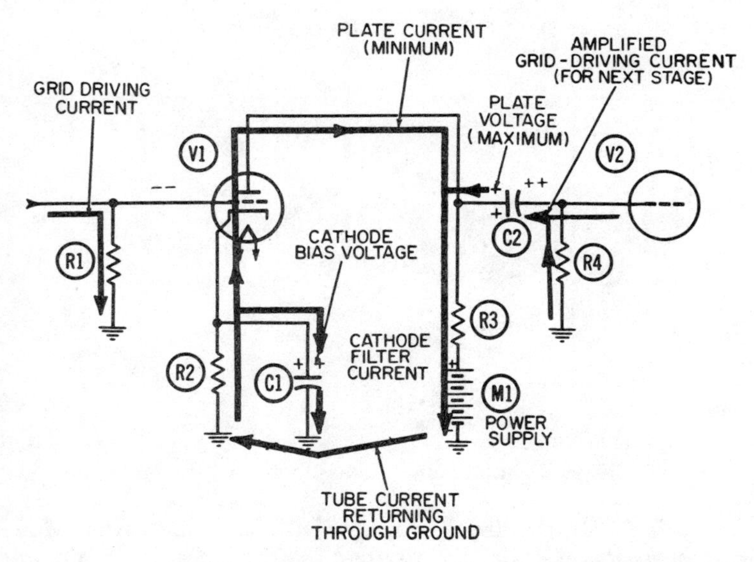

Fig. 2-1. The R-C coupled audio amplifier—negative half-cycle.

The second current is the unidirectional tube current, shown in and called the *plate current*. The current which flows into and out of output coupling capacitor C2 and the cathode filtering current are also shown because of their dependence on the unidirectional tube current.

The final current, which is actually flowing in the next tube circuit and is not intrinsically a part of the RC amplifier, is the grid-driving current for the next tube stage. It is shown here (thin line) to make easier the comparison between input and output voltages and thereby further clarify the term *amplification*.

Undoubtedly the most important single characteristic of vacuum tubes having control grids is their ability to amplify voltages. The valving, or throttling, action of the control grid within the tube makes it possible for small voltage changes impressed on the control grid to cause fairly large changes in the amount of electron current flowing through the tube. These large changes in tube current can then be made to generate larger voltage changes in the plate circuit. The output voltage should faithfully reproduce

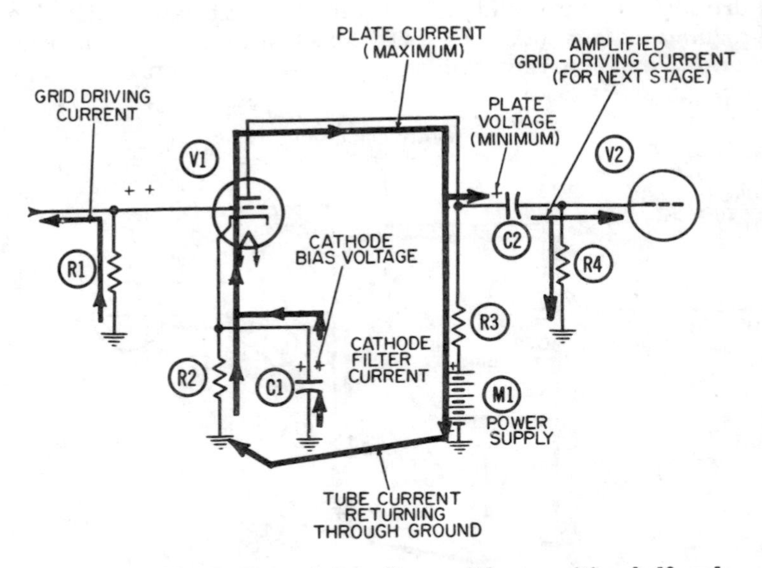

Fig. 2-2. The R-C coupled audio amplifier—positive half-cycle.

the waveshape of the input voltage, with no distortion. However, this goal is not always attained. The degree to which this is achieved determines the fidelity of the circuit.

CURRENT FLOWS

In Fig. 2-1, a small electron current is shown, as solid , flowing downward through the grid resistor R1. The control grid will be at its maximum negative voltage when maximum current is flowing downward through the grid resistor. Since Fig. 2-1 represents an entire half-cycle, the moment of maximum negative grid voltage occurs roughly near the midpoint of this half-cycle. The unidirectional tube current, a solid line, will be throttled down to its minimum value by this negative grid voltage.

The second diagram shows this grid-driving current flowing upward through resistor R1, in response to the positive voltage mpressed from the external circuit. The maximum flow rate, and he resultant maximum positive voltage at the grid, will occur oughly at the midpoint of the second half-cycle. The unidirecional tube current will then flow at its maximum value.

If no signal voltage were applied to the grid, the tube current plate current) would flow in a steady stream through the path ndicated—that is, up from ground through cathode resistor R2, hrough the tube, downward through plate-load resistor R3 and he power supply to ground, then back to the cathode resistor. This is the *quiescent,* or static, operation of the tube. When a small ignal voltage imposes fluctuations on the electron stream through he tube, these fluctuations develop larger voltage variations in he plate circuit, and the input signal is considered to be amplified.

This amplified signal voltage can perhaps best be explained by eferring to Fig. 2-2. The increase in plate current (due to the positive grid voltage) is causing electrons to be dammed up at the entrance to plate-load resistor R3. These electrons are shown pouring onto the left-hand plate of the coupling capacitor C2, hereby reducing the positive voltage stored on that plate. At the same time, on the other side of this coupling capacitor, electrons are being driven away from the capacitor and downward through R4, the grid resistor for the next tube stage. (This is the current, a thin line). Thus, we can see that a *phase shift* has occurred n this amplifier stage, for when the first grid is at maximum positive voltage, the second grid is at its maximum negative voltage.

The voltage on the left side of coupling capacitor C2 does not actually become negative, but varies between low and high values of positive voltage. To indicate this, the two plus signs at this point in Fig. 2-1 have not been replaced by minus signs, but are merely changed to a single plus sign to signify the reduction in voltage caused by the additional electrons from the plate-current stream.

Simultaneously with this action, more electrons will flow through plate-load resistor R3 toward the power supply, causing an increased voltage drop across this resistor. Since the voltage at the lower end is fixed (being connected directly to the power-supply voltage), this increased voltage drop across the load will be evidenced by a decrease in the positive voltage at the top of resistor R3.

Thus we have two means of associating a drop in plate voltage with an increase in plate current. One is by the inflow of electrons into coupling capacitor C2, neutralizing an equal number of positive ions there. The second is by the increased electron flow through load resistor R3.

Fig. 2-1 depicts the half-cycle when the tube current is restricted by the negative grid voltage. Minimum plate current is indicated during this period. This decrease in current affords an opportunity for the low positive voltage on the left plate of C2 to build up again. Electrons will now be drawn out of this capacitor and toward the power-supply voltage. As they depart, there is an attendant increase in the number of positive ions remaining at this point and the voltage increases to a higher positive value.

This departure of electrons from the left plate of the coupling capacitor draws an equal number of electrons upwards through the grid resistor and onto the right plate.

Thus, we have two means of visualizing that the voltage changes imparted to the grid of the next tube will duplicate, or *follow,* the voltage changes in the plate circuit of the amplifier stage. When electrons flow downward through R4 (shown in Fig. 2-2), we know that the top of R4 is more negative than the bottom, or ground. This action coincides with the period of increased tube current which, as mentioned, has reduced the positive plate voltage.

CATHODE CIRCUIT

In the cathode circuit, capacitor C1 functions as a filter to prevent degeneration. The resulting filtering current is being driven by the fluctuations in demand for electrons at the cathode. In the Fig. 2-1, the control grid is negative, so the demand for electrons to enter the tube is not great. However, since the upper plate of the capacitor is already charged to a positive voltage, this voltage will draw electrons upward through cathode resistor R2. During the first half-cycle they are shown flowing onto the upper plate of the capacitor and of course neutralizing some of the positive ions there which make up the positive voltage on the cathode.

During the second half-cycle (Fig. 2-2) when the control grid is positive, the demand for electrons in the plate current stream is greatly increased. This bigger demand is supplied by free electrons from the upper plate of the capacitor. The departure of these electrons naturally creates additional positive ions on the capacitor plate.

These changes in the quantity of positive ions naturally would indicate a change in the positive voltage at the cathode. However, normal amplifier operation requires an unchanging cathode voltage. The answer to this apparent paradox lies in the *amount* of voltage change which occurs because of this regular inflow and outflow of electrons. To determine the amount of voltage change, a factor known as the time constant must be considered.

Time Constant

The time constant of a circuit is the time required by a resistance-capacitance combination to complete 63.2% of its charging or discharging action. Numerically, the time constant is equal to:

$$T = R \times C$$

where,

T is the time in seconds,
R is the resistance in ohms,
C is the capacitance in farads.

Any RC combination is defined as a "long time-constant" circuit when the value of T is "long" (five or ten times as long), compared with the time which elapses during one cycle of the frequency under consideration, and as a "short-time constant" circuit if significantly (five or ten times) shorter than the frequency under consideration.

Thus, if we make the time constant of the resistor-capacitor combination in the cathode circuit (C1 and R2 in Figs. 2-1 and 2-2) long, it will provide a relatively constant voltage at the cathode. If the signal frequency is 1,000 cycles per second, then the time required for one cycle is .001 second. Thus the time constant of this combination should be at least .005 seconds. Any combination of resistor and capacitor values whose product exceeds .005 second would theoretically meet this requirement. However, one additional consideration, which virtually dictates the size of the resistor, is the amount of positive voltage wanted at the cathode (cathode biasing voltage). Since all the tube current is going to pass through the cathode resistor, we can calculate by Ohm's law what size of resistor we would need, with a given optimum tube current, to achieve the desired cathode bias voltage.

Ohm's Law

Ohm's law states that the voltage developed across a resistor is proportional to the current through it, or:

$$E = I \times R$$

where,

E is the voltage developed, in volts,
I is the current in amperes,
R is the resistance in ohms.

Any capacitor plate which is charged to a negative voltage can be looked upon as an electron pool, since electrons in concentration constitute a negative voltage. Likewise, when charged to a positive voltage (as is the upper plate of C1 in this example), a capacitor plate can be considered an ion pool.

The amount of voltage in any such charge is directly proportional to the quantity of electrons or ions which are so concentrated. This is expressed by Coulomb's law, which states that:

$$Q = C \times E$$

where,

Q is the quantity of charge in coulombs,
C is the capacitor size in farads,
E is the voltage.

One coulomb equals slightly more than half a trillion *trillion* electrons (6.28×10^{23}, to be exact!). A coulomb of positive charge equals the same number of positive ions.

Filter Action

As indicated before, the actual number of electrons entering the tube is determined primarily by the construction of the tube, as is the size of the variations in this electron stream which are imposed by the alternating signal voltage. The size of the tube-current variations determines the quantity of electrons which actually flow onto the upper plate of capacitor C1 during the first half-cycle and off it during the second. If the quantity of positive ions already stored there (which represent the positive cathode voltage) is substantially greater than the number of electrons which flow in during the first half-cycle, the positive cathode voltage will not be appreciably affected. But if the electrons flowing in and out of the capacitor during each cycle is a significant fraction of the total number of ions stored there, the ion level will rise and fall, and the positive cathode voltage, being dependent on the ion level, will rise and fall too.

As an example, suppose the desired cathode voltage is +10 volts and the number of electrons flowing in during the first half-cycle is exactly 1% of the number of positive ions stored there. Then the cathode voltage would be reduced by 1%, or 0.1 volt, thereby making its value 9.9 volts on this half-cycle. On the second half-cycle, with the same number of electrons leaving the capacitor, the stored voltage would again increase to the full 10 volts.

Filtering is completed by allowing current to flow between the lower plate of the filtering capacitor and ground. This current, a solid line, is driven downward during the first half-cycle by the inflow of electrons to the top plate, and upward during the second half-cycle by the outflow of electrons going toward

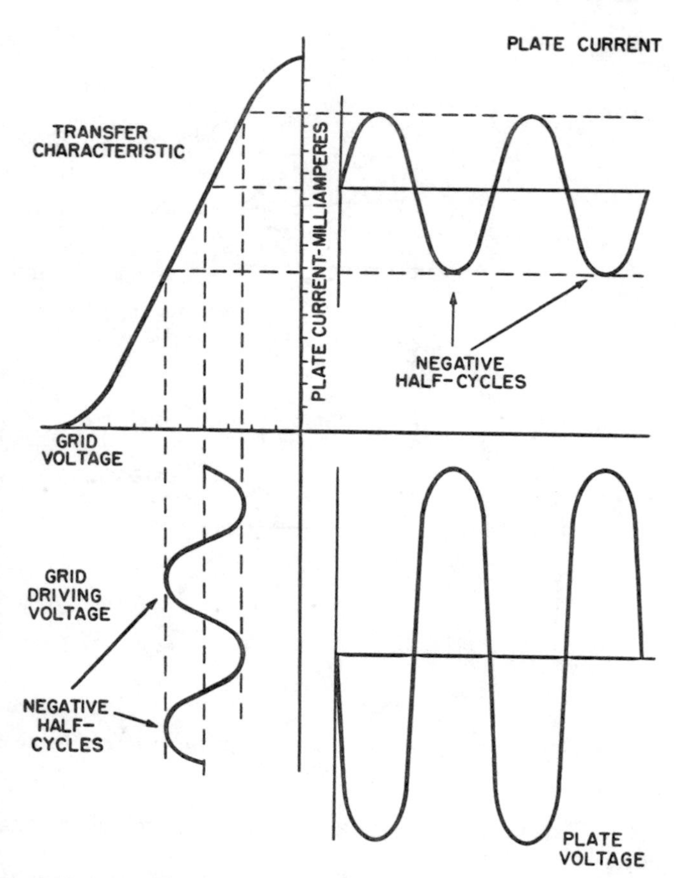

Fig. 2-3. Grid and plate waveforms for an R-C coupled audio amplifier.

the cathode. The electrons flowing along this path will always equal in quantity the electrons flowing to or from the other side of the capacitor. If this current were restricted somehow from flowing, filtering could not be done.

The variation in cathode voltage, as a result of the changes in tube current, is a measure of the degeneration which occurs. In other words, degeneration reduces the total amplification the tube can deliver.

WAVEFORM ANALYSIS

Fig. 2-3 is the conventional waveform diagram for relating the grid-driving voltage, the biasing or reference voltage, and the

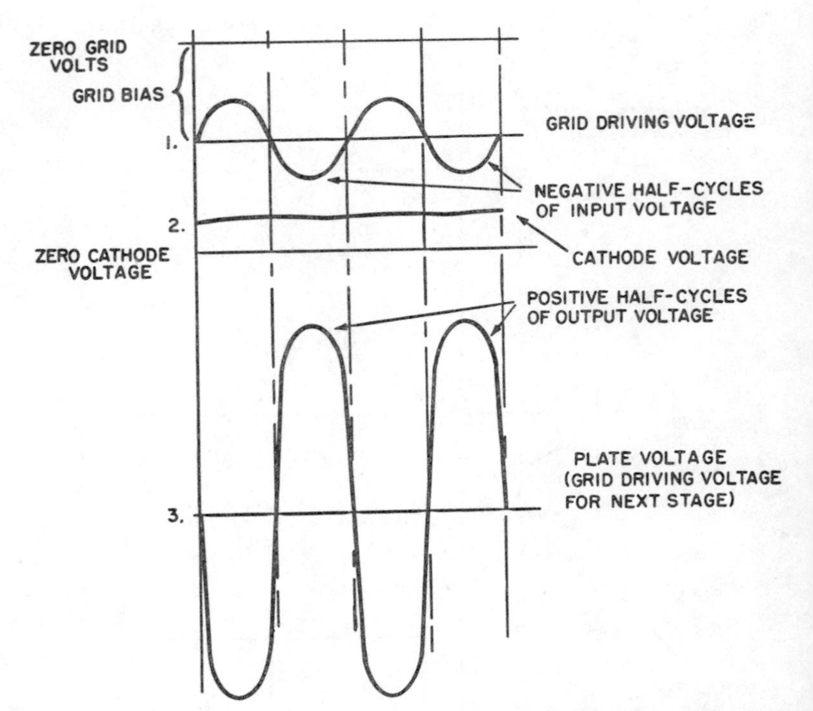

Fig. 2-4. Relationship between grid, cathode, and plate voltages in an R-C coupled audio amplifier.

plate current. This relationship is achieved by means of the transfer characteristic curve of the tube.

For any instantaneous grid voltage, a line projected vertically to the transfer characteristic curve and thence horizontally to the plate-current scale will indicate in milliamperes the resulting plate-current flow.

The plate-voltage sine wave is given below the one for the plate-current. Of special interest is the fact that when plate current is maximum, plate voltage is minimum, and vice versa. This illustrates the phase shift which occurs between grid and plate voltages in most vacuum tubes. When the grid voltage is most negative, the plate voltage is most positive; when the grid voltage is least negative, the plate voltage is least positive.

Fig. 2-4 shows this phase relationship between input and output voltages more clearly. The cathode voltage has also been shown in Fig. 2-4. Notice that it is essentially a flat line, increasing very slightly during those half-cycles when the tube conducts most heavily.

Response Curve

Fig. 2-5 shows a conventional frequency-response curve for a resistance-coupled amplifier. Below 100 cycles per second (cps). the response of the amplifier falls off because the signal is attenuated, or "consumed," by coupling capacitor C2. Capacitive reactance varies inversely with the frequency of the applied signal, in accordance with the standard formula:

$$X_C = \frac{1}{2\pi fC}$$

where,

X_C is the capacitive reactance in ohms,
f is the frequency in cycles per second,
C is the capacitance in farads.

Thus, at very low frequencies the coupling capacitor will couple, or "pass," only a small portion of the available signal from the plate circuit of the tube to the grid circuit of the next tube, and will attenuate most of the signal as it passes through.

The response of an amplifier circuit is a measure of how well it amplifies a voltage at any particular frequency. The response

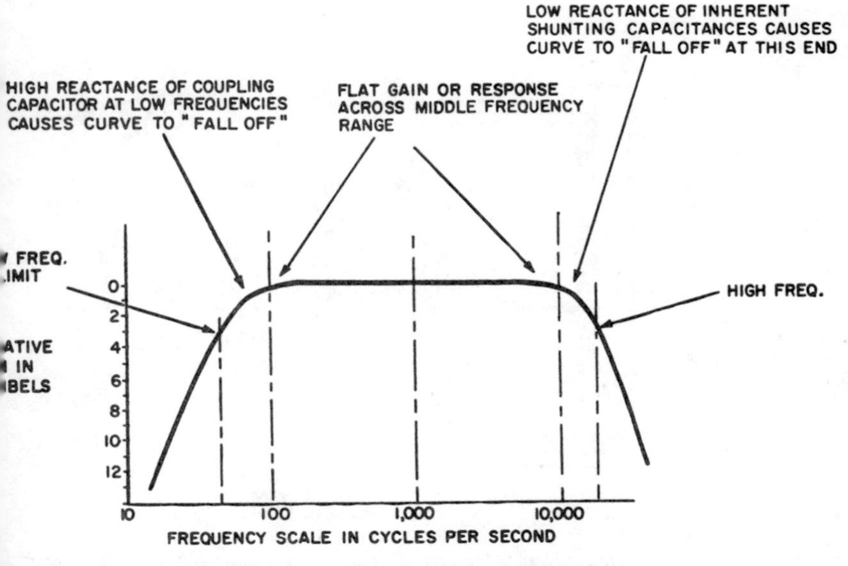

Fig. 2-5. An R-C coupled audio amplifier response curve.

curve is a means of comparing how well it responds to, or amplifies, voltages at any frequency within the range of the amplifier. The curve will also indicate the frequency limits within which the amplifier is designed to operate. It is very important in design consideration that an amplifier response curve be flat and that the sides be as steep as possible. A flat response curve indicates that the circuit will provide equal amplification for applied voltages of any frequency within its range. An amplifier which did otherwise would provide very poor sound reproduction indeed.

The high-frequency limit of operation for a resistance-coupled amplifier is determined by the interelectrode capacitances of both tubes and by the distributed or wiring capacitances of the entire circuit. Fig. 2-6 shows the equivalent circuit of the RC amplifier in Figs. 2-1 and 2-2. The cathode, grid, and plate of the next succeeding triode stage have been added in Fig. 2-6 because they have a definite bearing on the high-frequency limit of operation.

Three different currents are shown passing simultaneously through the amplifier tube:

1. Low-frequency current.
2. Medium-frequency current.
3. High-frequency current.

Let us define these terms. The low-frequency current is one having a frequency *lower* than the low-frequency limit of the

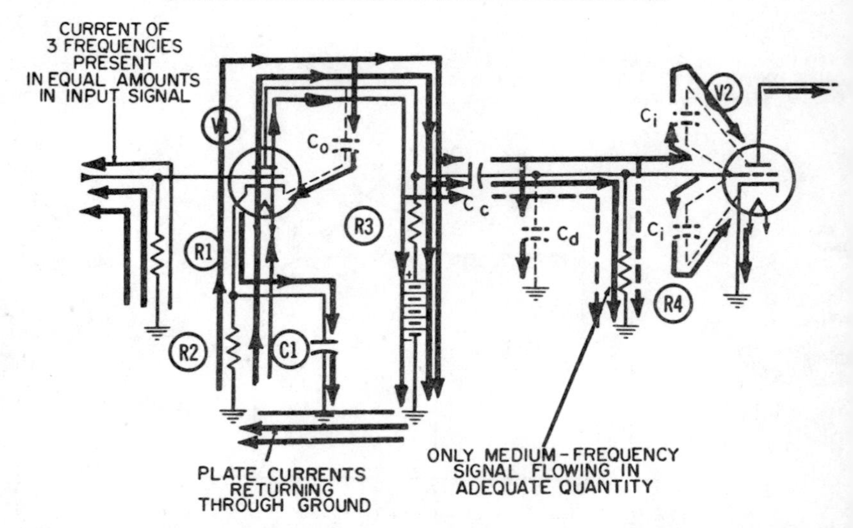

Fig. 2-6. Current flow at low, medium, and high frequencies in the R-C coupled audio amplifier—positive half-cycle.

response curve in Fig. 2-5, and the high-frequency current is one having a frequency *higher* than the high-frequency limit. Each of these limiting points is defined as being the point at which the amplifier response has fallen 3 decibels from the response achieved across the flat area of the curve.

It should be assumed that equal amplitudes or amounts of each of these three frequency components are applied at the input point, meaning the control grid of the first tube, V1. These three components of current are all shown as flowing upward through grid driving resistor R1 in Fig. 2-6. Since the frequencies of these three currents are drastically different, the time durations, or "periods," for single cycles at the three frequencies will differ greatly from each other. Consequently, the actual condition depicted in Fig. 2-6, where the three currents appear to be flowing in phase with each other through the grid resistor, would be achieved only rarely. There is an infinite variety of combinations when three such currents of three widely separated and variable frequencies can be somewhat out of phase with each other. But there is only one combination when they can all reach their maximum amplitude, in the same flow direction, at the same instant, thereby being truly in phase with each other as depicted in Fig. 2-6.

As an example, consider the following possible frequencies and resulting periods for one cycle of each of the three currents. The period of a sine wave is related to the frequency by the formula:

$$T = \frac{1}{F}$$

where,

T is the time for one cycle in seconds,
F is the frequency in cycles per second.

Current	*Frequency, F*	*Time for One Cycle, T*
Low Frequency	50 cps	.02 sec.
Middle Frequency	1,000 cps	.001 sec.
High Frequency	20,000 cps	.00005 sec.

In addition to the normal "manufactured" circuit components in Fig. 2-6, there are inherent characteristics which place the upper limit on the frequency response of a resistance-coupled amplifier like this one. These inherent characteristics include:

1. The output capacitance of amplifier tube V1. This is shown in dashed lines between the plate of V1 and its cathode, and is labeled C_o.

2. The distributed capacitance between all the wiring of the circuit and the nearest ground points. This is normally shown in equivalent circuits as a single, or "lumped," capacitance and is labeled C_d.
3. The input capacitance of the next amplifier stage. This consists of the inherent interelectrode capacitance between the control grid, and the cathode and plate of the next amplifier stage (designated C_i in Fig. 2-6).

Let us consider now what effects these inherent capacitances may have on the passage of currents at the three chosen frequencies—low, intermediate, and high.

Low-Frequency Limitation

The primary limitation on the passage of low-frequency currents is the reactance of the main coupling capacitor, labeled C in Fig. 2-6. Because of the excessive reactance which all capacitors exhibit at low frequencies, only a very small portion of the available low-frequency current can enter this capacitor. An equally small portion of current, at the same low frequency, will be driven out of the opposite plate and be available to flow downward through the grid-driving resistor for the next stage labeled as R in Fig. 2-6).

The amount of grid-driving voltage developed at any frequency depends on the amount of current, at this same frequency, which can be made to flow through the grid-driving resistor. In Fig. 2-6 the low-frequency current flowing in the circuit to the right of capacitor C_c is in dashed thin lines to indicate the severe attenuation it suffered in getting through the coupling capacitor.

The classical method of verifying the extent of this attenuation is to calculate the reactance of the capacitor at the particular frequency, and to compare it with the resistance of the grid driving resistor, since these two components comprise the complete path of this current. At the low frequency of fifty cycles per second the reactance of capacitor C_c will be *twenty times as great* as it is at 1,000 cps, and *two hundred times as great* as it is at 10,000 cps. (Reactance, of course, is the measure of a capacitor's *opposition to electron flow.*) At 50 cps the coupling reactance so greatly exceeds the grid resistance that most of the low-frequency signal is lost in the capacitor.

Middle-Frequency Current Passage

The middle-frequency current, a solid line in Fig. 2-6, can have any frequency in the entire middle range between the low and high-frequency limitations shown in Fig. 2-5. In passing through the coupling capacitor, the current has only insignificant losses. Most of it is available to flow up and down through R4 (on alternate half-cycles, of course) and develop grid-driving voltage for tube V2. The inherent capacitances (C_d and C_i) have small values, on the order of 10 micromicrofarads each; consequently, their reactances are prohibitively large throughout the entire range of low and middle frequencies, and little or none of the signal currents bleed off at these frequencies.

High-Frequency Limitations

The range of higher audio frequencies is where the inherent capacitances of vacuum tubes and their circuitry limit the operation of audio-frequency amplifiers. The reason is that excessive quantities of the available current are bled off, leaving little or none to flow through grid resistor R4 and develop driving voltage for the next stage. The high-frequency currents are indicated in in Fig. 2-6. We see that some portion of the high-frequency currents are bypassed back to the cathode of tube V1 by the output capacitance, C_o, of the tube. At the high frequencies this action reduces the amount of signal current available for coupling across C_c to the next stage.

The signals suffer no intrinsic loss at the higher frequencies as they pass through coupling capacitor C_c, since the reactance of all capacitors decreases as the frequency increases. It is thus possible to say that there is negligible voltage drop, or only slight attenuation of its voltage, as the signal is passed through the coupling capacitor.

That portion of higher-frequency current which succeeds in getting through coupling capacitor C_c, finds four alternate paths available to it, all in parallel. Consequently the current will divide into four parts, the amount going to each part being inversely proportional to the impedance offered by each particular path. Only the current which manages to flow up and down through grid resistor R4 will develop driving voltage for the next stage; the other three are losses.

The distributed, or wiring, capacitance of the circuit has been represented by the simulated capacitor C_d. In Fig. 2-6 we see the high-frequency current flowing freely into this simulated capacitor and thus being bypassed to ground. This is the first of the three loss currents referred to in the previous paragraph.

The second loss current flows directly into the interelectrode capacitance between the control grid and cathode of tube V2. In Fig. 2-6 a component of high-frequency current is shown flowing between the cathode and the ground connection. This component is driven by the current flowing toward the control grid from the coupling capacitor, but contributes nothing to the operation of amplifier V2.

The third loss current associated with this amplifier stage flows in the interelectrode capacitance between control grid and plate. In Fig. 2-6 a component of this loss current is shown flowing *away* from the plate, since it is being driven by the electrons flowing toward the grid from the coupling capacitor. This current also contributes nothing to the operation of the amplifier stage.

Fig. 2-7 shows the equivalent circuit of Fig. 2-6, but redrawn with only the high-frequency currents flowing, since the inherent

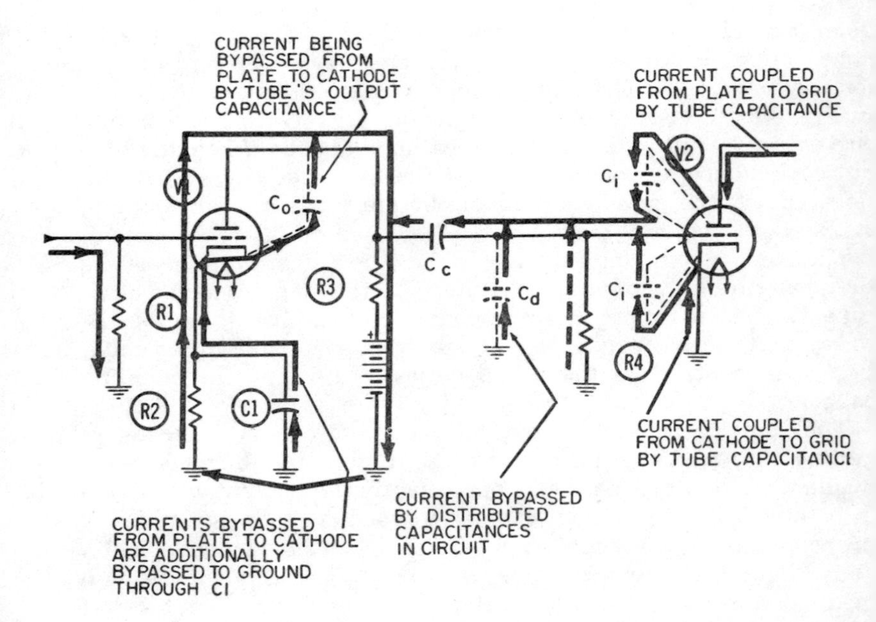

Fig. 2-7. Equivalent circuit at high frequencies in the R-C coupled audio amplifier—negative half-cycle.

capacitances discussed earlier are significant at the higher frequencies only. The flow directions of currents in Fig. 2-7 are the reverse of those in Fig. 2-6. Fig. 2-6 represents conditions during a positive half-cycle of operation, and Fig. 2-7 represents a negative half-cycle.

Therefore, in Fig. 2-7 the high-frequency signal current is flowing *downward* through resistor R1, making the control grid of tube V1 negative. As a result, the plate current through tube

V1 is reduced and the plate voltage rises. Some portion of the high plate-voltage peak, which would otherwise have been obtainable, will be lost. The reason is that the shunting effect of output capacitance of the tube permits some instantaneous electron current to be drawn *upward* from the upper plate of C_0. These additional electrons reduce the positive plate-voltage peak somewhat, and consequently are another loss current when the circuit is operated at high frequencies.

As the plate voltage rises (Fig. 2-7) and falls (Fig. 2-6), this loss current flows alternately up and down fairly freely, bypassing some of the high-frequency voltage back to the cathode. From here it has an easy bypass path through cathode filter capacitor C1 to ground.

The rise in plate voltage in Fig. 2-7 also draws current onto the right plate of coupling capacitor C_c. If there were no shunting capacitances to worry about, all of this current would be drawn upward through grid resistor R4 and a maximum positive voltage peak for application to the control grid of tube V2 would be developed across R4. In reality, some current will be drawn through each of the inherent capacitances and only a small remainder will flow upward through the grid resistor. This explains why all three loss currents are now flowing *toward* coupling capacitor C_c, instead of away from it as in Fig. 2-6.

FEEDBACK

Often, the output waveform will be distorted, that is, it will not be an accurate reproduction of the input waveform. Feedback is often employed to compensate for any distortion introduced by the circuit and to extend the frequency response of the amplifier.

Figs. 2-8 and 2-9 depict two successive half-cycles in the operation of a two-stage resistance-coupled amplifier which utilizes both negative and positive feedback. The positive feedback is achieved by coupling the plate voltage from the second stage back to the control grid of the first stage. The negative feedback is achieved through the degeneration process, by leaving the cathode resistor for the tube unfiltered, or unbypassed. Let us examine these two important actions in more detail. In order to do so, it will prove desirable to review the actions occurring in the entire circuit.

The essential components of this two-stage amplifier are:

R1—Grid input resistor for V1.
R2—Grid input and feedback resistor.
R3—Voltage-divider resistor used in feedback path.

R4—Cathode resistor used for developing negative feedback.
R5—Plate-load resistor for tube V1.
R6—Grid driving, or input, resistor for V2.
R7—Cathode biasing resistor for V2.
R8—Plate-load resistor for V2.
R9—Grid-driving, or input, resistor for next stage.
C1—Coupling capacitor between stages.
C2—Cathode filter capacitor for V2.
C3—Feedback capacitor.
C4—Output coupling capacitor to next stage.
V1 and V2—Triode amplifier tubes.
M1—Common power supply for both tubes.

There are seven electron currents operating in this circuit:

1. Three grid driving currents.
2. Two plate currents.
3. One positive feedback current.
4. One cathode filter current.

Analysis of Operation

By inspection of Fig. 2-8 we see that at the end of the first half-cycle, the control grid of V1 is positive, the control grid of V2 is negative, and the control grid of the next succeeding amplifier stage is positive. These voltage polarities reflect the normal 180° phase shift between the control-grid and plate voltages in an R-C vacuum-tube stage. Since the control grid of the second tube, V2, reaches its most negative value at the end of the first half-cycle, the plate current through this tube will be reduced to its minimum value and its plate voltage will rise to its maximum positive value. This rise in plate voltage is coupled into the feedback network by drawing electrons off the right plate of feedback capacitor C3. As these electrons flow out of the capacitor, toward the power supply, they draw an equal number upward from ground and through voltage divider R2-R3.

This upward flow of electrons makes the voltage at the bottom of resistor R1 positive with respect to ground; this instantaneous feedback polarity is indicated by a heavy plus sign. The instantaneous grid voltage for the first tube, V1, is also positive as a result of the grid driving current. Consequently, the sum of these two positive voltages will be a higher voltage at the grid than the input-signal voltage can provide by itself. Because these two voltages are in phase, the feedback voltage (solid) *reinforces* the signal voltage (thin line) and the feedback is said to be positive.

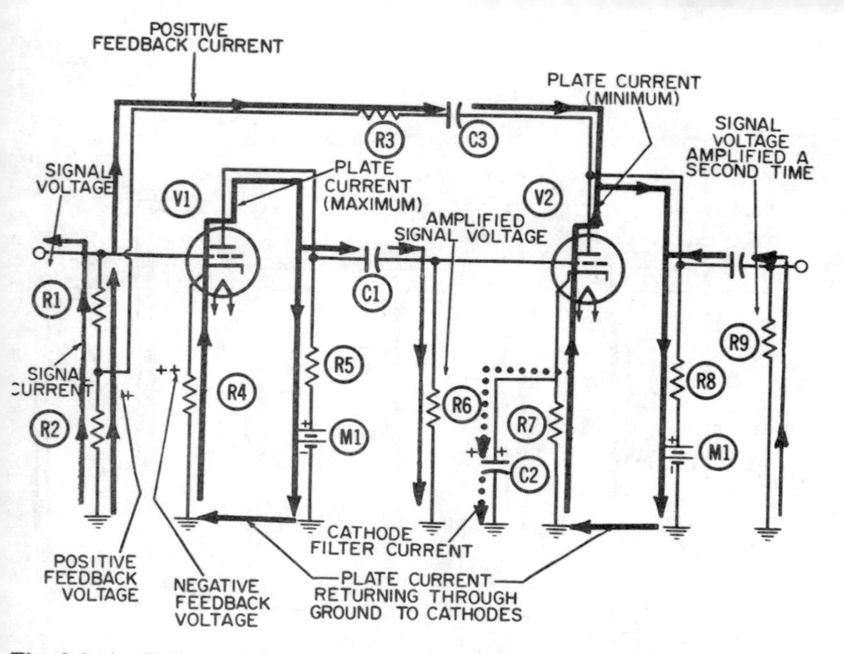

Fig. 2-8. An R-C coupled audio amplifier utilizing both positive and negative feedback—positive half-cycle.

This type of feedback is called voltage feedback, since it is in parallel with the plate voltage and in effect is driven by changes in the latter.

During the second half-cycle (Fig. 2-9), the control grid of V2 becomes positive, releasing a large amount of plate current through that tube and thus lowering its plate voltage. This reduction in plate voltage is coupled into the feedback network. It is convenient to visualize this action as consisting of the excess plate-current electrons pouring *onto* the right plate of capacitor C3, until they can be drawn through load resistor R8 to the power supply. As the electrons enter C3 they drive an equal number off the left plate of the capacitor, and downward through voltage-divider network (R3-R2) to ground. This downward flow of electrons tells us the resulting voltage at the top of resistor R2 must be negative with respect to ground, since electrons will always flow *away* from a negative voltage, to a less negative or a more positive voltage. This instantaneous feedback-voltage polarity has been indicated by a heavy minus sign.

The alternating grid voltage for V1 is also negative at this instant, as a result of the downward flow of grid driving current

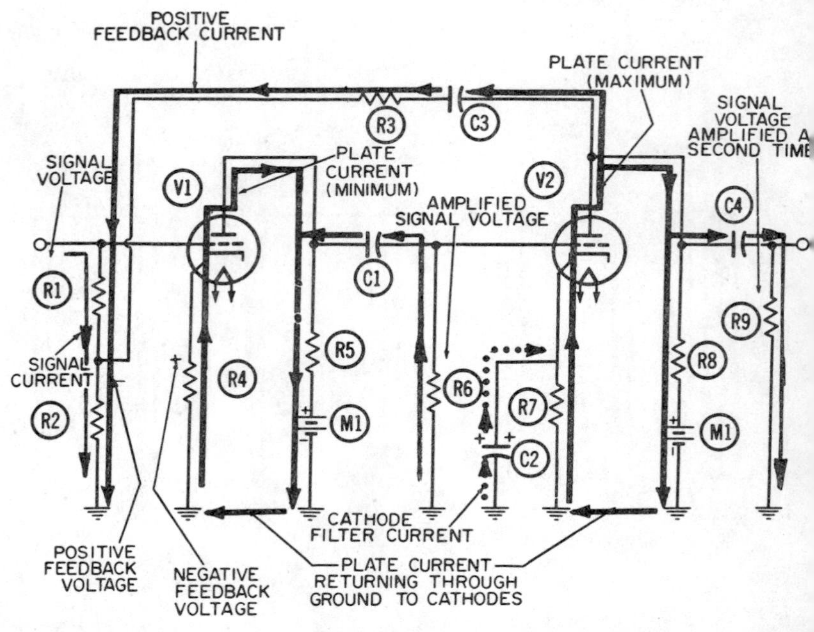

Fig. 2-9. An R-C coupled audio amplifier utilizing both positive and negative feedback—negative half-cycle.

through resistors R1 and R2. Since these two voltages are now both negative, their sum is a higher negative voltage at the grid than the driving signal can achieve alone. Looking at the results from both half-cycles, we see that the feedback voltage reinforces the signal voltage during the whole cycle. This logically gives rise to its designation as *positive feedback.*

Current Feedback

During the first half-cycle in Fig. 2-8, the positive grid voltage on V1 releases a large amount of plate current through this tube. This plate current flowing upward through resistor R4 to the cathode produces positive voltage (indicated by the plus signs) at the cathode.

When the control grid V1 is negative (Fig. 2-9) the plate current through the tube decreases. This causes a smaller voltage drop to be produced across the cathode resistor, and thereby lowers the positive voltage at the cathode. The result of these two actions is a continual *fluctuation* in the cathode voltage—rising when the grid voltage goes positive, and falling when it goes negative.

An *increase* in the positive voltage at a cathode has the same effect as a *decrease* in grid voltage—namely, the amount of plate current flowing across the tube is restricted. The reason for this behavior is that the cathode "recaptures" some of its own electrons after initial emission occurs. This recapturing process comes about in the following manner: When an electron is first emitted into the tube, it "sees" or "feels" (is subject to) whatever voltage exists on any electrode. The high positive plate voltage will try to attract the electron across the tube. A positive voltage at the grid, indicated in Fig. 2-8, will encourage this process, by allowing a large number of electrons to flow from cathode to plate, resulting in a heavy plate current. However, as the cathode voltage becomes more positive, it will tend to reattract the emitted electrons after emission. The cathode has two advantages over the other electrodes—(1) the instant after emission, the electron is *closer* to the cathode than to any of the other electrodes, and (2) the electron has not yet had the opportunity to build up any velocity as it travels away from the cathode. Consequently, even a slight increase in positive voltage at the cathode causes many of the emitted electrons to fall back onto the cathode, making fewer electrons available for plate current.

In the example of Fig. 2-8, the increase in the cathode voltage nullifies part of the effect of the positive voltage applied to the control grid by the input signal. Recall that the definition of instantaneous grid bias is the instantaneous *difference* in voltage between the grid and cathode of a tube. It is this difference between the two voltages that determines what portion of the emitted electrons will be permitted to cross the tube and become plate current.

In Fig. 2-9 we find an opposite set of conditions. The negative voltage at the grid of a tube will tend to repel most of the emitted electrons back to the cathode. However, since the positive cathode voltage has now been reduced, the cathode exerts less attraction for these electrons and so partially counteracts the negative grid voltage.

This is classified as a form of negative feedback since these continuing changes in the voltages at grid and cathode are "out of phase" with each other. Here the term "out of phase" refers to the *effects* of changing each of the two voltages, rather than to changing their polarities. Viewed in this light, a rise in the positive grid voltage is *out of phase* with a rise in the positive cathode voltage.

This type of negative feedback has the more familiar title of *degeneration*. The obvious result of degeneration is some loss in the amplifying capability of the circuit. Degeneration is usually avoided by the addition of a filter capacitor of suitable size across the cathode resistor, to make a long time-constant combination. This is why capacitor C2 and resistor R7 are in this circuit.

This type of feedback is also classified as *current feedback,* because the feedback voltage is developed directly by the plate-current stream flowing through the cathode resistor. It is an interesting anomaly, and one which may provide some confusion, that a separate "feedback current" does not exist as such. In the voltage-feedback example discussed previously, a feedback current has to flow through the voltage-divider network (R2-R3) in order for the necessary feedback voltage to be developed. Often, voltage feedback is used for the negative feedback circuit, too. Such a feedback circuit is discussed in the next chapter in conjunction with the transformer-coupled AF amplifier and could be used equally well in an R-C coupled circuit.

There is nothing unique about the remaining operations in the circuit. Each of the grid driving currents (a thin line) develops a larger, or more amplified, version of the signal voltage than the one at the preceding stage. That is, the alternating voltage across resistor R9 is greater than the voltage across R6, and the latter in turn is greater than the voltage developed across R1 and R2.

AMPLITUDE DISTORTION

Fig. 2-10 is the transfer characteristic curve of a typical triode tube. Note that it is similar to the curve of Fig. 2-3. Two cycles of input (grid) voltage are shown, along with the resulting two cycles of output (plate) current. Cycle A, which is similar to the grid driving cycle of Fig. 2-3, causes the tube to operate along the linear portion of the curve. Being a fairly faithful reproduction of the input waveshape, the resulting plate current is considered to be distortionless.

Cycle B is large enough to make the tube operate at *both* ends of the transfer characteristic curve. Because the curve is nonlinear at both ends, each half-cycle of Cycle B is badly distorted. Likewise, if *either* half-cycle of grid voltage drives the tube into one of the nonlinear portions of the curve, the resulting distortion would occur on that half-cycle only, but would still be unacceptable. This could happen if the grid-bias voltage were too small for Cycle A. Then, the grid-bias line would shift to the right, and the positive half cycles of plate current would be distorted. Alternatively, the negative half-cycles would be distorted if the grid-bias voltage were too large (shifting the grid-bias voltage line to the left).

PHASE DISTORTION

Fig. 2-11 shows two voltage sine waves of different frequencies, both before and after passing through an amplifier (such as the resistance-capacitance amplifier of Fig. 2-9). The purpose in showing these sine waves is to demonstrate the meaning of phase distortion.

Recall that for each stage of amplification, the signal should be inverted; that is, the output waveform should be 180° out-of-phase with the input. Therefore, in the two-stage amplifier of Fig. 2-9, the signal should be inverted twice or shifted 360°—a complete cycle. Hence, the output voltage waveform should be exactly in phase with the input waveform. Often, in passing through an amplifier, some frequencies will not be shifted exactly 180° in phase.

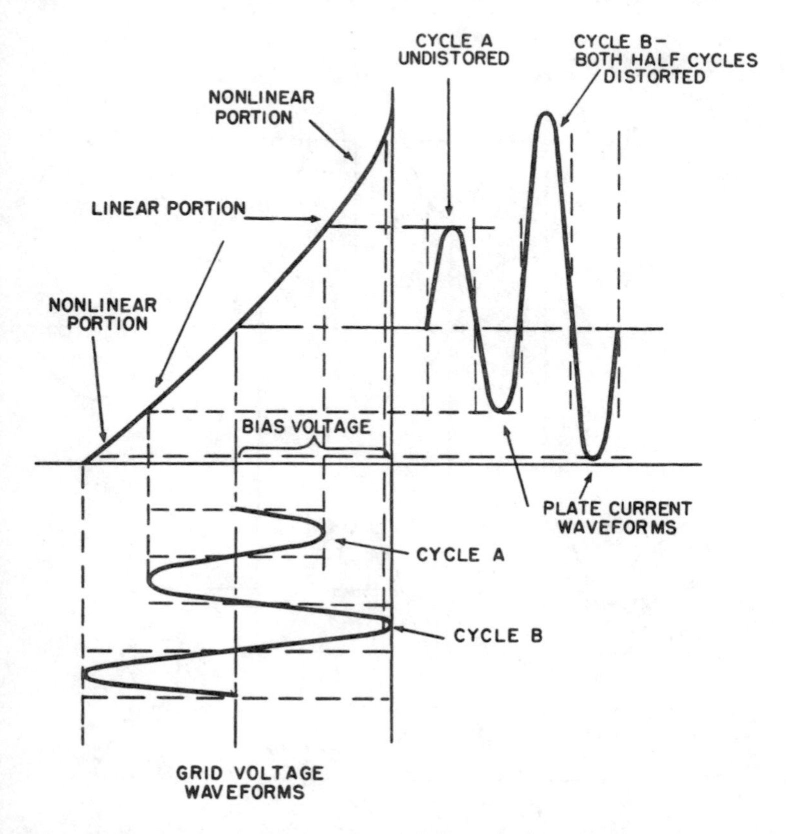

Fig. 2-10. How amplitude distortion is introduced when tube is operated on nonlinear portion of transfer characteristic curve.

Phase distortion occurs in an amplifier when currents or voltages of some frequencies suffer only a negligible variation from the normal phase inversion, whereas currents or voltages of other frequencies suffer a somewhat greater shift in phase.

In lines 1 and 3 of Fig. 2-11, the high-frequency voltage "in" has the same phase as the high-frequency voltage "out." This means the two voltages achieve their peak values, pass through zero, etc., at the same moment. These conditions tell us that no phase shift has occurred in this particular voltage during amplification. However, lines 2 and 4 show that the low-frequency voltage "in" is not in phase with the low-frequency voltage "out"; the latter has been shifted 90°, or a quarter of a cycle, in phase. This means the output voltage achieves its peak value a quarter of a cycle later than the input voltage.

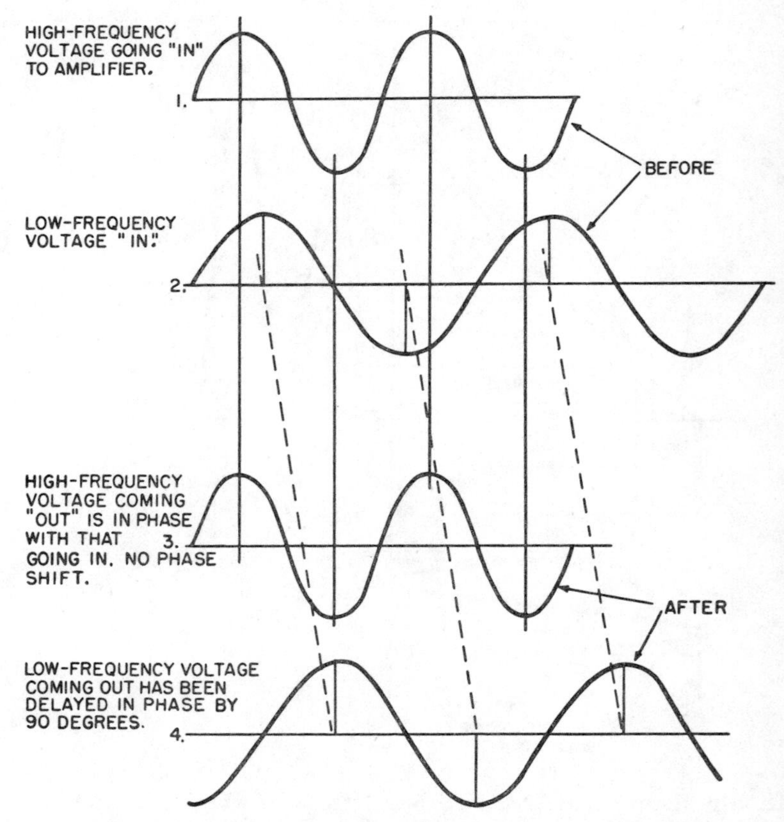

Fig. 2-11. The effect of phase distortion on a waveform.

Phase distortion normally is not a serious problem in the reproduction of sound because the ear cannot detect it. A moment's reflection will clarify why this is so. When we listen to a sustained chord of music we are listening to several tones of different pitches, each of a different frequency. Assume the low-frequency sound represented by the voltage waveforms of lines 2 and 4 is middle *C*, which has a frequency of 256 cycles per second. A 90° shift in phase means we will now hear each peak *about a thousandth of a second later*. Obviously, much greater phase differences, amounting to many whole cycles, can be created by the artist without noticeably degrading the over-all result.

Phase distortion is a more serious problem in the amplification of complex waveshapes. These usually represent the algebraic sum of many simple shapes, such as sine waves of many frequencies. If some (but not all) of these subordinate waveshapes have undergone shifts in phase, the resultant waveshape in the output is likely to look entirely different from the input waveshape.

Chapter 3

TRANSFORMER-COUPLED AF VOLTAGE AMPLIFIERS

Transformer coupling, while not as popular as R-C coupling discussed in the previous chapter, is sometimes used between AF voltage amplifiers. The chief disadvantage of transformer coupling is the high cost of a transformer which will provide the required frequency response.

CIRCUIT DESCRIPTION

Figs. 3-1 and 3-2 show two successive half-cycles in the operation of an audio voltage amplifier, in which two adjacent amplifier stages are linked together by means of an audio frequency transformer. The components which make up this circuit are as follows:

R1—Grid-driving and grid-return resistor.
R2—Cathode biasing resistor for V1.
R3—Cathode biasing resistor for V2.
C1—Cathode filter capacitor for V1.
C2—Cathode filter capacitor for V2.
T1—Audio-frequency coupling transformer.
V1 and V2—Triode amplifier tubes.
M1—Power Supply.

This is essentially a simple circuit. The relatively few currents are :

1. Grid-driving current for tube V1 .
2. Plate current for each tube .
3. Transformer secondary current .
4. Cathode filter current for each tube .

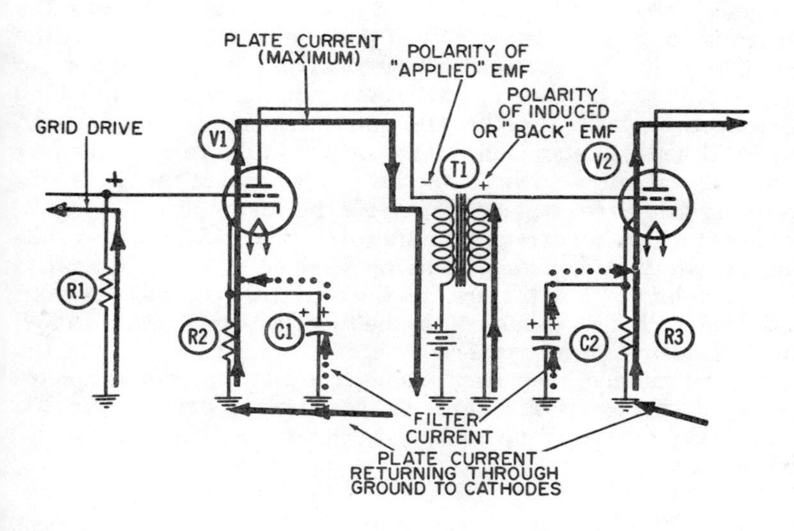

Fig. 3-1. The transformer-coupled AF amplifier—first half-cycle.

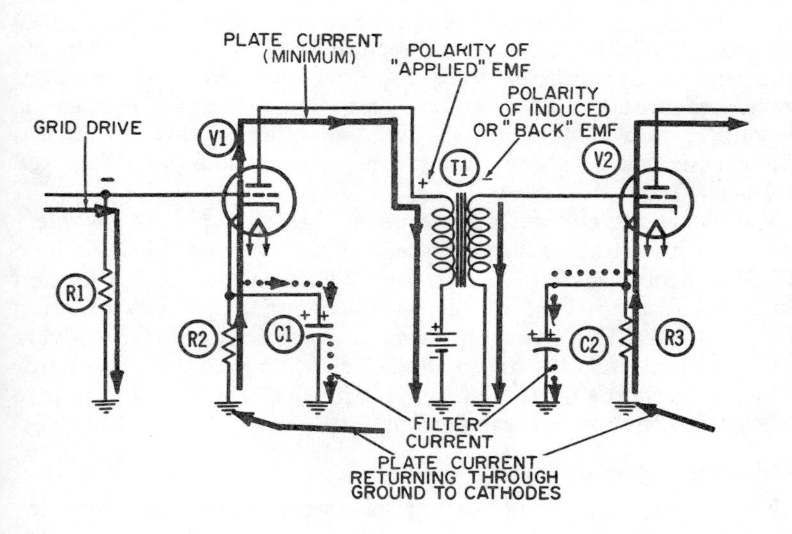

Fig. 3-2. The transformer-coupled AF amplifier—second half-cycle.

Fig. 3-1 is termed a positive half-cycle of operation because the control grid of V1 is being driven positive. The plus sign at the top of resistor R1 represents the instantaneous value of applied grid voltage and is consistent with the upward flow of grid-driving current through the resistor. The movements of this current, and the resulting instantaneous voltages at the grid, are controlled by preceding circuitry such as another voltage amplifier. It is this resulting grid-drive voltage we are seeking to amplify.

Maximum plate current through tube V1 coincides with maximum positive grid voltage. This heavy flow of plate current is indicated by the line traversing the conventional plate-current path from cathode to plate within the tube and downward through the transformer primary. From here it can be drawn into the positive terminal of the power supply. The plate current completes its round trip by passing through the power supply to common ground. From here, it travels upward through cathode resistor R2 to the tube.

TRANSFORMER ACTION

Fig. 3-3 shows the graphical relationships between grid current and voltage, plate current and voltage, and transformer secondary voltage. Line 1 of this figure can be used to represent both the grid-driving current and the grid-driving voltage, since the current through a pure resistance is always in phase with the voltage across it. The reference line in the center of line 1 thus represents zero conditions for both current and voltage. When the current waveform crosses the reference line, this means no current is flowing *at that instant* and no voltage is developed across R1. These conditions occur midway during each half-cycle depicted in Figs. 3-1 and 3-2.

Whenever the current sine wave of Fig. 3-3 is below the reference line, the current it represents is flowing *downward* through R2. This occurs during the last half of the second half-cycle and the first half of the first one. Or we could say that it occurs during the fourth and first quarter-cycles of operation, and that during these same periods the grid voltage has to be negative. The latter is indicated by the portion of the voltage sine wave below the zero reference line during these periods.

Sine-wave Symbology

The use of voltage and current sine waves raises an interesting point. No other symbology available can present as much detailed information on a single cycle or on many cycles of operation as the sinusoidal, or sine-wave, symbology can. However, there are two important discrepancies in the way they are used which are traps for the unwary, as we shall soon see.

The definition of a sine wave is "a graphical representation, along a time reference line, *of simple harmonic motion.*" The world is full of voltage sine waves, but the interesting fact is that voltage does not have motion . . . voltage does not move. Only the current has motion, and the simplest and broadest

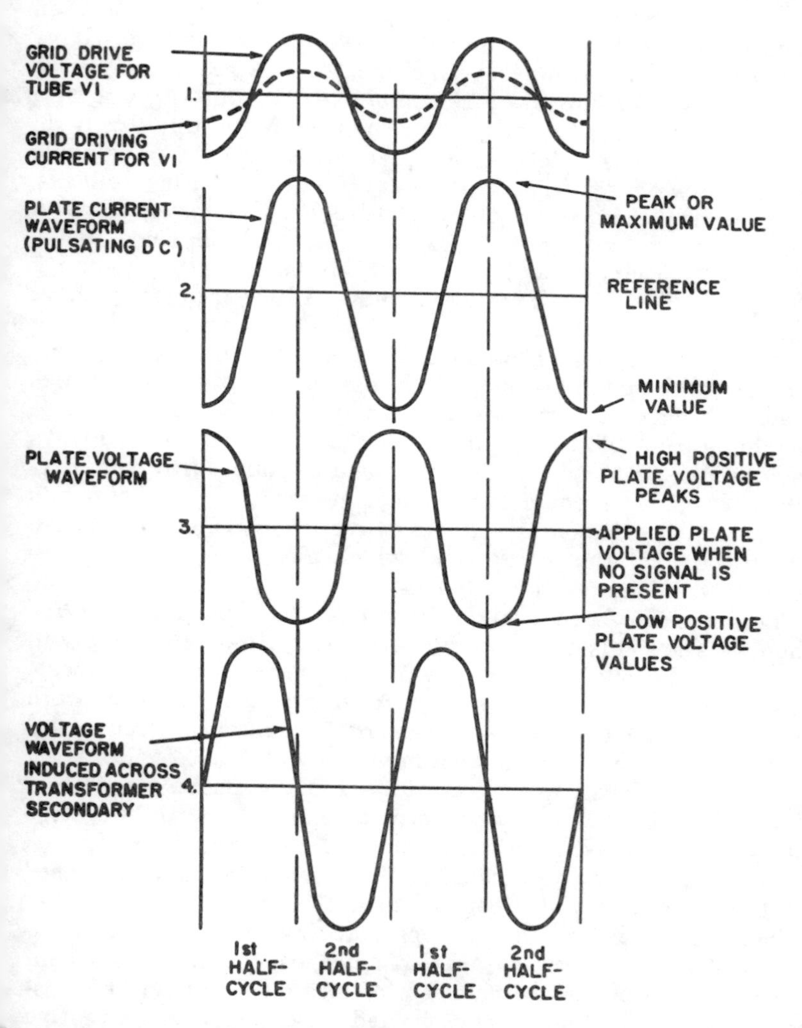

Fig. 3-3. Voltage and current waveforms in the transformer-coupled AF amplifier.

definition of electric current is that it consists of *electrons in motion*. A negative voltage is defined as *electrons in concentration.* Concentrations of electrons do not move around in a circuit; rather, the electrons flow from point to point. While electrons are moving they constitute a current. when they *accumulate* at any point, they constitute a negative voltage. When there is *deficiency* of electrons at any point, it constitutes a positive voltage.

So the discrepancy in a voltage sine wave is the implication, based on the definition of the sine wave, that all voltages have some kind of mysterious wave motion. For this reason, any time you see a voltage sine wave, remember that you are looking at a sine wave of quantity rather than motion. With this firmly in mind, you will be much better qualified to use the invaluable tool of sine-wave graphical representations of alternating voltages.

Since currents have motion, it is entirely proper to represent an alternating current by a simple harmonic-motion since wave. However, currents do not have polarity in the sense that a voltage has. Rather, they have directions of flow. All too frequently there is the tacit implication that any portion of a current sine wave above the reference or center line is "positive," whereas any portion below the line is "negative." The assignment of such polarities to current may have mathematical significance but is otherwise meaningless, particularly to the neophyte. Since current is always made up of electrons in motion, its flow direction during any portion of a cycle is important because it reveals (1) the points toward which electrons are being delivered to build up negative voltages, or (2) the points from which electrons are being depleted to build up positive voltages.

There has also been a tendency to refer to current as *negative* when we know it is composed of electrons flowing indisputably in one certain direction (such as the plate current of vacuum tubes). This is a holdover from olden days when negative electrons were not yet recognized as the prime current carriers in electric and electronic circuits. Current was universally assumed to flow from a positive to a negative voltage source. Later with the recognition of negative electrons as the current carriers, the term "negative current" was coined to describe them, the "positive-to-negative" or "conventional" current came to be known as "positive current" and this terminology still persists today.

There can be little genuine hope that this convention of "positive current" will ever disappear, since early researchers constructed much of the mathematics of electronics under the mistaken assumption that current flowed from positive to negative rather than vice versa. In fact, millions of words have been written, based on this early assumption. When you get right down to it, though, it is easier for *you* to remember to "change the sign" that to rewrite all that literature!

The alternating current in the primary winding is really a pulsating direct current. Changes in this primary current will induce a voltage in the secondary winding of the transformer, and the induced voltage will be maximum when the primary current is changing at its maximum rate. Referring to the plate-current sine wave in line 2 of Fig. 3-3, you can see that the plate current is not changing whenever it is passing through its maximum and minimum values (at the end of the first and second half-cycles respectively).

A true alternating current is changing at its maximum rate whenever it is changing direction—in other words, whenever its graphical representation is crossing the reference line. Since plate current is a pulsating direct current, it always flows in the same direction (clockwise in the usual circuit diagram), unlike a true alternating current which flows in both directions. For this reason it is necessary to treat pulsating direct current *as an alternating* current. The additional convention devised to accomplish this is to look upon the plate current as consisting of two separate but related currents—a pure direct current (the value of which is indicated by the reference line in line 2), and an *alternating component superimposed on the direct current.*

The *reference value* of plate current is the amount which will flow through the tube when no signal voltage is applied to the grid. This occurs only twice each cycle. Midway in the first half-cycle, the application of a positive voltage to the grid will increase the plate current. A negative voltage will decrease the plate current, midway in the second half-cycle.

The voltage polarities shown at the tops of the primary and secondary windings in Fig. 3-1 are chosen to coincide with that instant when the plate current is *increasing at its maximum rate*—midway in the first half-cycle. We see from line 4 of Fig. 3-3 that the voltage induced across the secondary winding has its maximum positive value at this instant. This voltage is applied to the control grid of vacuum tube V2. As a result, maximum plate current flows from cathode to plate in V2.

These voltage polarities are normally referred to as *applied emf* or *back emf.* In Fig. 3-2 these polarities are reversed—because the plate current is *decreasing* at its maximum rate—in the middle of the second half-cycle. The plate voltage is considered to be *increasing* in the positive direction, also at its maximum rate. This is the reason for the plus sign at the top of the primary winding in Fig. 3-2; it indicates the positive polarity of the applied emf across the primary winding *at this instant.*

Both tubes in this circuit employ the type of self-bias known as cathode biasing. Note the small resistor in series between the cathode and ground; plate current must flow through it before reaching the tube. When cathode-biasing is used, the plate current must flow at all times. Otherwise the positive cathode bais voltage would fall toward zero. This would change the grid-to-cathode voltage difference (called *grid bias*) in such a direction that more plate current would flow. Consequently, the use of cathode biasing automatically signifies that a tube is being operated under Class-A conditions—by definition, a tube in which plate current is flowing throughout the entire cycle.

The process of bypassing the cathodes with filter capacitors C1 and C2 has been indicated on the circuit diagrams of Figs. 3-1 and 3-2. The filter currents which flow in and out of ground below these capacitors are indicated in dotted red lines. Since there is absolutely no difference between these filtering processes and the ones in the resistance-capacitance coupled amplifiers of the previous chapters, the cathode filtering process will not be redescribed here.

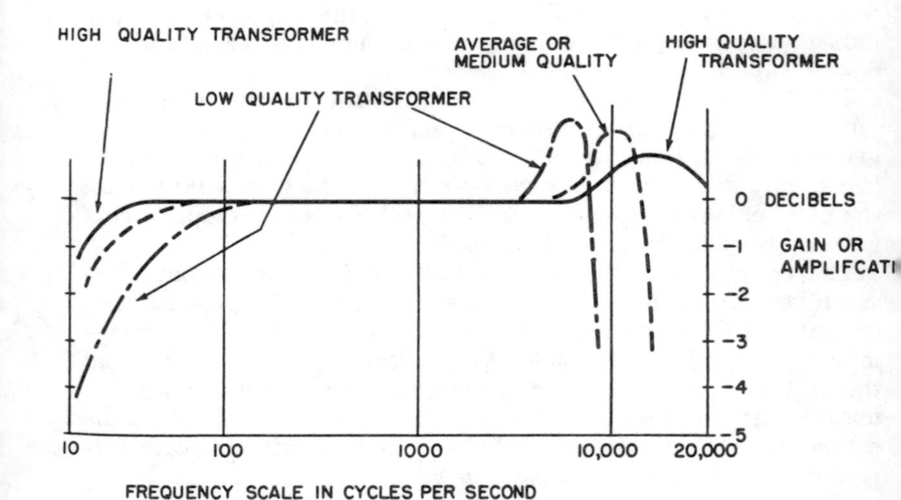

Fig. 3-4. Response curves of transformer-coupled AF amplifiers.

FREQUENCY RESPONSE

Fig. 3-4 shows some conventional frequency-response curves for an audio-frequency voltage amplifier using transformer coupling. The horizontal scale indicates that its range of usefulness is approximately the same as for the resistance-capacitance coupled amplifier previously discussed. Low-frequency operation is limited here by the extremely low inductive reactance which the transformer primary winding presents to the changing plate current. The applied emf developed across the primary winding is proportional to the inductive reactance of the primary winding through which the plate current must flow; inductive reactance is of course proportional to the frequency. Voltage, current, inductive reactance, and frequency are related by these two common formulas:

$$E = I \times X_L$$

where,

E is the applied emf in volts,
I is the AC component of plate current in amperes,
X_L is the inductive reactance in ohms;

and:

$$X_L = 2\pi fL$$

where,

X_L is the inductive reactance in ohms,
f is the frequency of the applied AC component of plate current in cycles per second,
L is the inductance of the primary winding, *plus* any inductance coupled from the secondary, in henrys.

The average voltage amplifier using transformer coupling will have a good frequency response down to about 50 cycles per second. Below 50 cps the response deteriorates rapidly.

The high-frequency response of a transformer-coupled amplifier is limited by many inherent circuit capacitances. Figs. 3-5, 3-6, and 3-7 are adaptations of the equivalent circuit normally used in arriving at mathematical solutions for a transformer-coupled amplifier. In addition to the inherent capacitances, an equivalent circuit also indicates the presence of such other inherent circuit constants as resistances and inductances.

A combined listing of the actual, or manufactured, circuit components, along with the inherent characteristics *which act like components,* would read as follows (refer to Figs. 3-5, 3-6 and 3-7):

C1—Output capacitance of tube (inherent).
C2—Shunting effect of capacitance between adjacent turns of primary winding (inherent).

C3—Capacitance between primary and secondary windings (inherent).

C4—Shunting effect of capacitance between adjacent turns of secondary winding (inherent).

C5—Input capacitance of next tube (inherent).

R1—Theoretical plate resistance of tube (inherent).

R2—DC resistance of primary winding (inherent).

R3—Resistance representing eddy-current losses in primary winding (inherent).

R4—Resistance representing hysteresis losses in primary (inherent).

R5—Resistance representing direct-current losses in secondary winding (inherent).

R6—Input resistance of next stage (inherent).

L1—Leakage inductance of primary winding (inherent).

L2—Actual primary-winding inductance (manufactured).

L3—Actual secondary-winding inductance (manufactured).

L4—Leakage inductance of secondary winding (inherent).

V1—Triode amplifier tube (manufactured).

Fig. 3-5 shows the currents in the three audio-frequency ranges —low (thin), middle (solid), and high (solid)—in their passage through the amplifier circuitry. All currents are drawn as if they are momentarily in phase. Of course three currents of such widely different frequencies will rarely be in phase even momentarily. This has been done for illustrative purposes only; in general, the amplifier reacts independently to each current.

The plate current of medium frequencies passes through the amplifier and receives maximum amplification. Like all others, this current must flow through all inherent circuit constants in its series path before passing through the power supply and back to ground. This path includes the theoretical plate resistance of the tube, the inherent DC resistance of the primary winding, the inherent leakage current of the primary, the theoretical resistance associated with hysteresis losses, and the primary winding of the transformer. In passing through the transformer primary, the plate current induces another current (solid line) in the secondary winding. This secondary current also must flow through the three inherent circuit characteristics which lie in its series path—the secondary leakage inductance, the DC resistance of the secondary winding, and the theoretical input resistance of the next amplifier stage.

Note that no significant portion of this middle-frequency current flows into any of the four inherent shunting capacitances which lead to ground; nor into the capacitance C3, which would

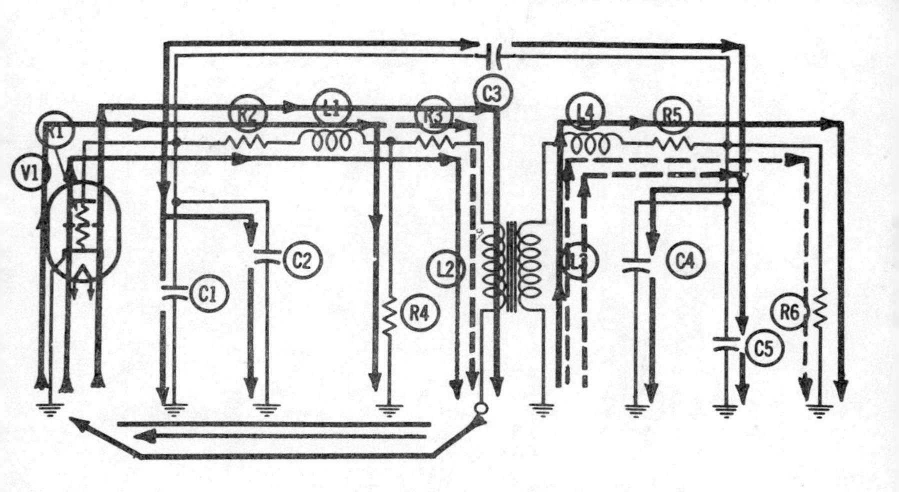

Fig. 3-5. Equivalent circuit of transformer-coupled AF amplifier—first half-cycle.

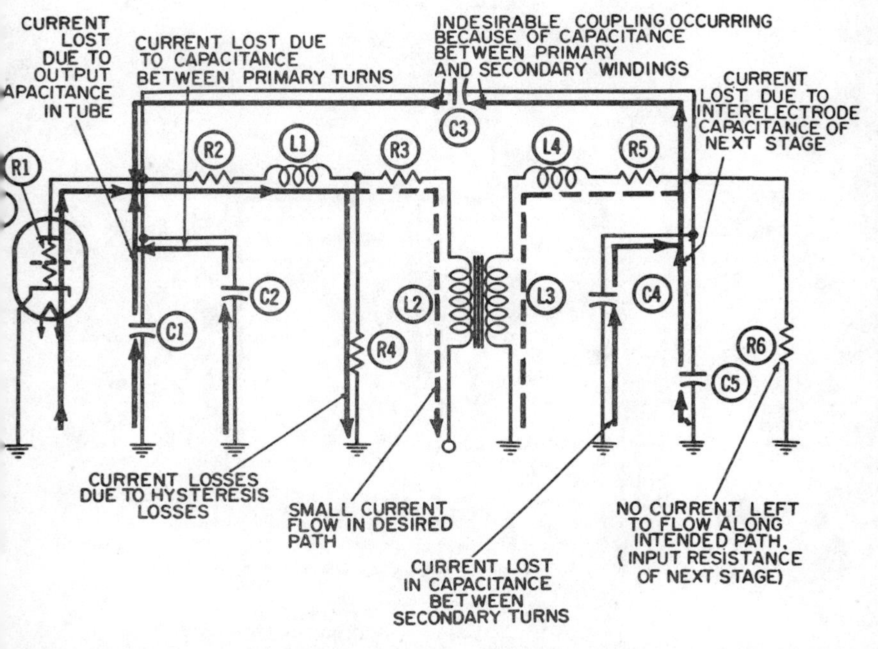

Fig. 3-6. Equivalent circuit of transformer-coupled AF amplifier at high frequencies—second half-cycle.

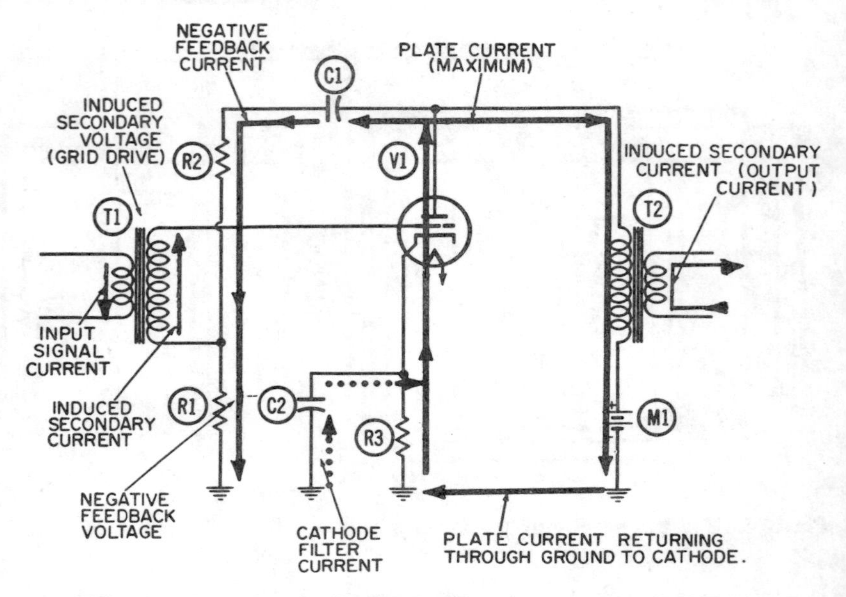

Fig. 3-7. Negative feedback amplifier circuit—first half-cycle.

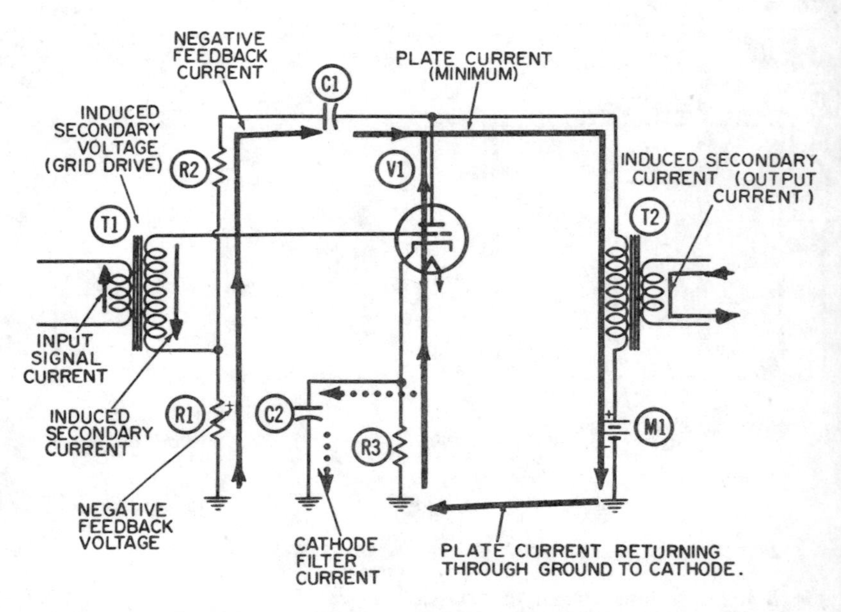

Fig. 3-8. Negative feedback amplifier circuit—second half-cycle.

couple it directly to the next stage rather than via normal transformer action. All these inherent capacitances are so small in value (10 or 20 micromicrofarads) that they have an extremely high reactance at low and middle frequencies. For this reason they do not bleed off any currents at these frequencies and consequently have no adverse effect on circuit operation.

The low-frequency current (thin line) follows the same path through the amplifier as the middle-frequency current. Because transformer action is inadequate at very low frequencies, the low-frequency current induced in the secondary winding is seriously weakened. Therefore, it is drawn as a dashed rather than a solid line, to indicate this attenuated condition.

Like the middle-frequency current, none of this low-frequency current flows into any of the inherent shunting capacitances, again because of their small size, and hence, high reactance at the low frequencies. As stated previously, the sole limitation on low-frequency operation of a transformer-coupled amplifier is the inability of the transformer, *unless especially designed for the task,* to induce sufficient secondary current at extremely low frequencies.

The high-frequency current, a solid line in Fig. 3-5, has a rough time getting through the transformer-coupled amplifier. Portions of it flow into every one of the inherent shunting paths. These shunted portions must be counted as losses, since they no longer are available to flow into the primary winding of the transformer and induce a current in the secondary winding.

Fig. 3-5 is considered a positive half-cycle; at this time, the control grid (not shown) becomes positive and maximum plate current is released into the tube and external plate circuit. This accounts for the flow directions shown. The excess electrons in the plate current are driving electron current away from the plate and along the three unwanted shunt paths available to them. These paths are downward into C1, (the output capacitance of the tube), downward into C2 (the capacitance between adjacent turns of the primary winding), and to the right into C3 (the inherent capacitance between adjacent turns in the primary and secondary windings of the transformer). None of these current components are available to do their primary task of flowing through the primary winding and inducing a current in the secondary-winding.

The high-frequency current in Fig. 3-5 is also shown flowing downward through R3 which represents the eddy-current losses in the primary winding. There will be some eddy currents induced in the transformer core at all frequencies; consequently, currents of all frequencies will incur some eddy-current losses. However, the amount of eddy-current losses increases as the *square* of the frequency. Hence, losses become most significant at the higher frequencies.

The small portion of current which succeeds in flowing through the primary winding is shown as a dashed line, to indicate the seriousness of the reduction. This small current induces a small current and voltage combination in the secondary winding—also shown as a dashed line. This secondary current is further reduced by the two inherent shunting capacitances, C4 and C5, and practically nothing is left to flow into input resistance R6.

Fig. 3-6 represents a negative half-cycle of the same high-frequency current. During this half-cycle the currents previously driven downward in the four shunting capacitances—C1, C2, C4, and C5—are now being drawn upward from ground. These currents all flow very freely over these unwanted capacitive paths to ground as the frequency increases, leaving less and less current to do the necessary work.

The current coupled directly to the secondary via the inherent capacitance (C3) between adjacent turns of the two windings also reverses its direction each half-cycle. In Fig. 3-6 we see it flowing to the left, attracted of course by the inevitable rise in plate voltage whenever the plate current falls.

At first thought, this current does not really appear to be "lost" —after all, it *is* coupled to the secondary circuit and is therefore available to drive the control grid of the next amplifier stage, despite its unorthodox journey up to that point. This assumption is fallacious for several reasons, including the important one that current which reaches the control grid via this path will be *out of phase* with a current of the same frequency which traveled over the conventional path. Consequently, there would be two components of what was originally a single current, each attempting to drive the control grid independently of the other. As a result, the two would reach their peaks at different moments in the same cycle and partially cancel each other.

FREQUENCY DISTORTION

Of the three broad types of distortion—amplitude (nonlinear), frequency, and phase distortion—encountered in tube operation, it is possible to visualize at least one of them by reference to Fig. 3-5. Frequency distortion occurs when an amplifier provides unequal amplification of currents at different frequencies.

All currents whose frequencies fall within the band covered by the flat portion of the response curve (Fig. 3-4) will be amplified equally and passed equally by the coupling network. Hence, no frequency distortion occurs within this frequency band which is shown in solid Fig. 3-5. However, the low-frequency currents (thin line) are amplified very poorly because of transformer limitations, and the high-frequency currents suffer because of all the losses caused by undesired capacitances. Thus, the low, medium, and high-frequency components are not equally amplified. This is defined as frequency distortion.

Examples of amplitude and phase distortion were discussed in Chapter 2 and will not be repeated here since they are the same for all amplifiers.

NEGATIVE VOLTAGE FEEDBACK

Figs. 3-7 and 3-8 illustrate two half-cycles in the operation of a commonly used feedback circuit. In this circuit a portion of the plate voltage is coupled, or fed back, to the input (grid circuit). Since the plate voltage is normally out of phase with the grid voltage, this type of feedback is considered to be negative.

The circuit used here is a simple transformer-coupled voltage amplifier for audio frequencies. The only difference between this circuit and the preceding one (other than the feedback connection from plate to grid, of course) is that the control grid is driven by a transformer rather than a resistor.

The necessary components of this circuit include:

R1—Portion of voltage-divider network used for developing feedback voltage.
R2—Remaining portion of voltage-divider network.
R3—Cathode biasing resistor.
C1—Blocking and coupling capacitor.
C2—Cathode filter capacitor.
T1—Input transformer of the voltage step-up type.
T2—Output transformer of current step-up type.
V1—Triode voltage amplifier tube.
M1—Power Supply.

Identification of Currents

The operating currents in this circuit which need to be identified include:

1. Input transformer primary and secondary currents.
2. Plate current of tube.
3. Feedback current.

4. Output current in output-transformer secondary.
5. Cathode filter current.

Details of Operation

Fig. 3-7 shows what is called a positive half-cycle, because the driving signal delivered through the input transformer to the control grid becomes progressively more positive throughout the half-cycle. The polarity of this voltage is indicated by the plus sign at the top of the secondary winding of input transformer T1. This is the voltage induced in the secondary by the input current flowing in the primary winding. The *directions* of current flow in the primary and secondary are at all times related to each other and to the instantaneous polarity of the induced voltage. This is in accordance with the normal principles of transformer action.

The positive-going signal on the control grid in Fig. 3-7 releases a large surge of plate current into the tube. The normal path of plate current is from cathode to plate, then downward through the primary winding of output transformer T2 and into the positive terminal of power supply M1, through the power supply to ground and through it to the bottom of cathode resistor R3. From here the plate current flows upward and returns to the cathode.

With the feedback network (consisting of C1, R2, and R1) in the circuit, a portion of this surge of plate current electrons flows onto the right plate of capacitor C1. An equal amount of electrons are driven off the left plate of C1 and downward through R2 and R1; this current is shown as a solid line. When electrons are flowing downward through this path, we know that the instantaneous voltage at any point along the path will be more negative than the voltage at any other point farther down the path, since electrons always flow from negative to positive. Hence we can infer that the voltage at the junction of R1 and R2 will be negative with respect to the ground at the bottom of R1.

Since the bottom of the secondary winding of input transformer T1 is also connected to the junction of R1 and R2, the positive control-grid voltage resulting from the transformer action will be reduced by the amount of this negative voltage developed across R1. Since the driving and feedback voltages are out of phase, the feedback is considered negative.

The total voltage developed across R1 and R2 by the flow of feedback current is divided between the two in proportion to their resistances. This is a straight Ohm's-law relationship:

$$E_T = I \times R_T$$
$$= I\,(R1 + R2)$$

or,

$$E_T = E_1 + E_2$$

also,

$$E_1 = I \times R1$$
$$E_2 = I \times R2$$

where,

E_T is the total voltage drop across R1 and R2,
E_1 is the voltage drop across R1,
E_2 is the voltage drop across R2,
R_T is the sum of resistances R1 and R2, in ohms,
I is the effective value of the alternating current, in amperes.

Let,

$$R1 = 10{,}000 \text{ ohms.}$$
$$R2 = 90{,}000 \text{ ohms.}$$
$$I = 1 \text{ milliampere.}$$

Then,

$$E_1 = .001 \times 10{,}000$$
$$= 10 \text{ volts,}$$
$$E_2 = .001 \times 90{,}000$$
$$= 90 \text{ volts,}$$

and,

$$E_T = 10V + 90V$$
$$= 100 \text{ volts}$$

Thus, the total voltage across network R1 – R2 is divided in the ratio of 9 to 1, which is the ratio between the two resistances. The effective value of the feedback voltage is 10 volts.

Fig. 3-9 shows a simplification of the voltage waveforms in this circuit. Line 1 is the alternating voltage, which is normally applied to the control grid via the input transformer. During the first half-cycle, this voltage increases continuously in the positive direction. Because the cathode is biased positive by the R3 – C2 filter combination, the grid is automatically biased an equal amount in the negative direction (since the true definition of grid bias is "the *difference* in voltage between grid and cathode").

Under the conditions indicated, the peak voltage swing at the control grid is slightly less than this bias value. Hence, the control grid never becomes positive with respect to the cathode and no significant amount of grid-leakage electrons will flow.

Line 2 shows the sinusoidal waveforms for plate current and plate voltage. Note that they are 180° out of phase with each other; when the plate current reaches its maximum value, at the end of the first half-cycle, the plate voltage will be at its minimum value.

This alternating component of plate voltage can all be considered as being coupled into the feedback network. This coupling is accomplished by means of the feedback current (solid line in Figs. 3-7 and 3-8), which works in and out of feedback capacitor C1. This current suffers negligible losses and phase shift;

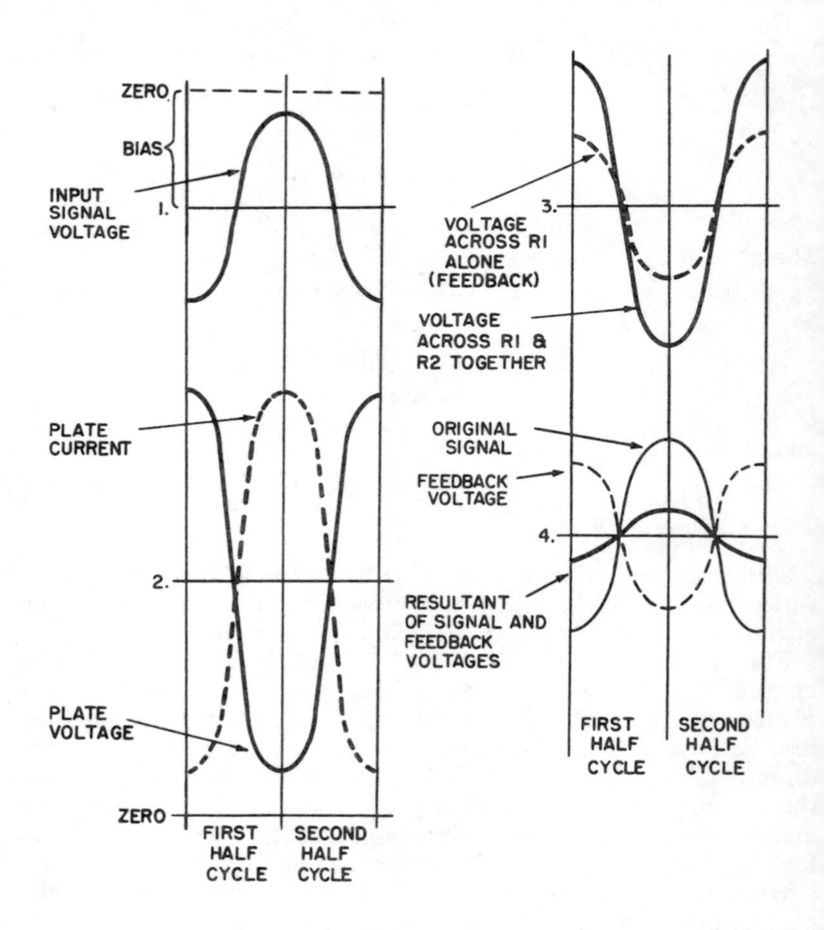

Fig. 3-9. Voltage and current waveforms in negative-feedback amplifier circuit.

so the alternating voltage developed across the two series resistors (line 3 of Fig. 3-9) is an exact replica of the alternating component of plate voltage. The reduced portion in dashed lines represents the amplitude and phase of the voltage developed across feedback resistor R1 and fed back to the grid from the plate.

Line 4 is the control-grid voltage redrawn with the feedback voltage superimposed on the same axis. Because the two voltages are exactly out of phase, they tend to cancel each other out. The values of the resistors must be so adjusted that the alternating voltage applied through the transformer to the control grid is at least slightly larger than the feedback voltage. Otherwise, the resultant voltage will not have sufficient amplitude to provide the necessary excitation.

It will be apparent that the grid is actually being driven by the smaller voltage indicated in line 4 as the "resultant grid voltage," rather than by the original alternating voltage indicated in line 1. This implies that *smaller* pulsations of plate current will occur. This will result in smaller swings of plate voltage, and also less feedback current through the two resistors. Hence, the feedback voltage will likewise be smaller. Thus, the indicated amplitudes of the four voltage waveforms must be considered gross approximations only.

During the second half-cycle (Fig. 3-8), the grid driving voltage goes negative. The greatly reduced plate current which results is accompanied by a rise in the plate voltage, as indicated by the waveforms in line 2 of Fig. 3-9. This is the opportunity for electrons to flow *off* the right plate of capacitor C1. An equal number of electrons are drawn *upward* through the voltage-divider network, making the feedback voltage *positive* at the junction of R1 and R2. Since this is also the same half-cycle the input transformer is trying to make the grid *negative,* we see that these two voltages still oppose each other. As indicated in line 4 of Fig. 3-9, the algebraic sum of these two voltages, at the end of the second half-cycle, is a small negative voltage.

This is called *voltage feedback* because it in effect is "tapped off" from the voltage changes occurring at the plate, even though the feedback voltage across the voltage divider owes its existence to the feedback current .

Although we have shown the feedback loop from the plate to the grid of the same tube, the feedback voltage can be (and often is) taken from other points. The only requirement is that the feedback voltage be 180° out-of-phase with the signal voltage. For example, the feedback voltage could be taken from the plate of the last tube in a three-stage amplifier and applied to the grid of the first stage. Thus, any variations introduced by any of the three tubes would be cancelled out.

Also, the two forms of feedback discussed in the previous chapter (positive voltage and negative current) can also be used with the transformer-coupled amplifier. Since operation is the same, they will not be discussed again here.

Cathode-Filter Circuit

The cathode filtering current, shown in large dots, reflects *withdrawals of electrons from* the ion pool, on the upper plate of C2 during each first half-cycle or *additions of electrons to* the pool during the second half. As electrons are added to or withdrawn from the plate of C2 an equal number flows to or from ground to the bottom plate.

Thus, capacitor C2 aids in keeping the cathode voltage constant, by supplying electrons when the tube current is maximum, and storing them when it is minimum.

Chapter 4

AUDIO-FREQUENCY POWER AMPLIFIERS

Audio-frequency power amplifiers are used in all public-address systems, radio and television receivers, record players, recorders, etc., to provide sufficient audio power to drive the speakers or other devices connected to its output. Speakers are usually *current-operated* devices requiring a fairly heavy electron current to flow at the audio frequencies. The power developed across any resistive load varies as to the square of the current flowing through that load, in accordance with the power formula which states that:

$$P \text{ (or W)} = I^2 \times R$$

where,

P is the power in watts,
I is the current in amperes,
R is the resistance of the load in ohms.

A tube used as a power amplifier is designed to deliver a large quantity of cathode-plate current. The current is normally delivered to a transformer in the load circuit. This enables the particular speaker load to be matched in impedance with that of the plate circuit. An exact match between the two impedances is necessary for optimum, or maximum, transfer of power from the plate circuit to the load circuit.

Since the impedance of the average speaker is fairly small (on the order of a few ohms), and since the plate-circuit impedance will be much larger, the impedance-matching process takes advantage of the turns ratio (which is a manufactured characteristic of transformers). The impedance-matching characteristic of a transformer varies in accordance with the *square* of the turns ratio between the primary and secondary windings. If a primary winding has twice as many turns as the secondary, the turns ratio—normally expressed as N1 ÷ N2—will be 2, or 2-to-1, and the transformer can match two impedances in the ratio of 4-to-1.

If a turns ratio is 20-to-1, the transformer can match two impedances which differ from each other by a factor of 400-to-1. Thus, if a speaker has an impedance of 8 ohms and the plate circuit an impedance of 3,200 ohms, the two impedances can be matched to each other by using a transformer having a 20-to-1 turns ratio between the primary and secondary.

Sample circuits have been chosen to depict circuit conditions and problems during audio-frequency power amplification. The first circuit uses a single power-amplifier triode. All use output transformers of the current step-up type. In the first example. the movement of the output current through the speaker coil, and the resultant movements of the speaker diaphragm, will be discussed in detail. The later examples represent typical push-pull circuits. This portion of the discussion has not been repeated here, since the principles are identical for all circuits.

TRANSFORMER-COUPLED POWER AMPLIFIER

Figs. 4-1 and 4-2 show two successive half-cycles in the operation of a conventional audio-frequency power amplifier. A transformer couples the plate circuit to the load, which is shown as a speaker. This type of circuit must be operated under Class-A conditions, meaning the plate current must not be cut off, or interrupted, during any portion of an individual cycle. Self-biasing is accomplished by using a cathode resistor and bypass capacitor. The power-amplifier tube is driven by the output from the preceding voltage amplifier.

The components which make up this complete circuit are as follows:

R1—Grid driving and grid-return resistor.
R2—Cathode biasing resistor.
C1—Cathode filter capacitor.
T1—Audio-frequency output transformer.
V1—Power-amplifier triode.
M1—Power supply.

Identification of Currents

The currents at work in this circuit include:

1. Grid driving current .
2. Plate current (pulsating DC).

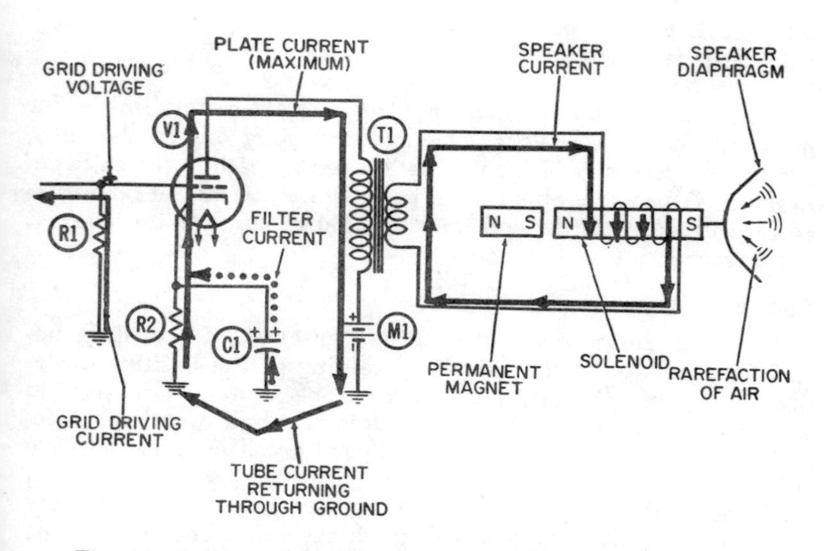

Fig. 4-1. Operation of an AF power amplifier—first half-cycle.

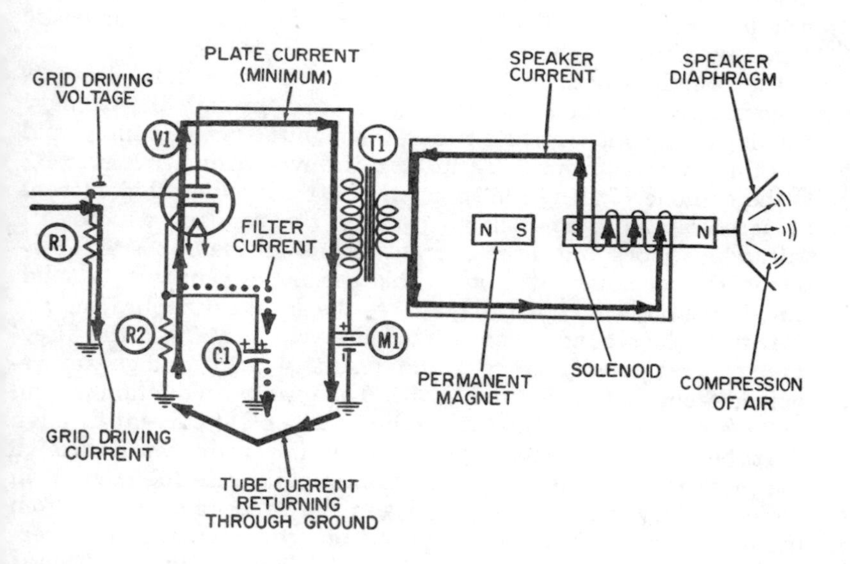

Fig. 4-2. Operation of an AF power amplifier—second half-cycle.

3. Speaker or output current.
4. Cathode filter current.

In addition to these currents, provision is always made for the possibility of grid-leakage current flowing out of the tube, and for the necessity of furnishing it a return path to the cathode. Resistor R1 provides this return path. Since the circuit operation as a whole does not depend on grid-leakage current, however, it is not shown in Figs. 4-1 and 4-2.

Details of Operation

Fig. 3-1 is considered the positive half-cycle of operation because the voltage at the grid is increasing in the positive direction throughout the entire half-cycle. The upward flow of grid driving current through R1 during this first half-cycle exists because electrons will always flow toward areas of more positive voltage.

As the grid voltage becomes more positive, more and more plate current will be released through the tube. Throughout the ranges of grid voltage and plate current in which this operation is carried out, it is desirable that the *amount* of plate current bear the same proportion to the amount of grid-cathode voltage at all times. When this condition is achieved, the tube is considered to be a linear amplifier. This term is derived from the characteristic curve for a triode (shown in Fig. 2-3).

The complete path for plate current (shown a solid line) is of course from cathode to plate, downward through the primary winding of the output transformer, through the power supply, and into the common ground. From here it flows through resistor R2 to the cathode. During that portion of a cycle when this current is *increasing* as it flows down through the primary winding, it will induce more and more current to flow in the opposite direction in the secondary winding. This secondary current, a solid line, is the one which actually drives the speaker diaphragm.

During the second, or negative, half-cycle in Fig. 4-2, the direction of grid driving current and resultant grid voltage are reversed from that shown in Fig. 4-1. The gradual reduction in grid voltage causes a continual reduction in the plate current through the tube. As this decreasing plate current flows downward through the primary winding, it will simultaneously cause the current in the secondary winding to *decrease* in the opposite direction from the plate current, which is actually an *increase* in the same direction (downward) as shown by the solid line in Fig. 4-1. This is how a pulsating direct current in one winding can actually cause an alternating current to flow in the other winding. Thus, from one half-cycle to the next, the output currents in the secondary winding and speaker coil alternately flow in the opposite direction.

Speaker Action

A transducer is broadly defined as a device for converting energy from one form into another. A speaker qualifies as a transducer because it converts the electrical energy represented by the output current into sound energy which can be heard. As shown in Fig. 4-1 and 4-2, this is accomplished with the aid of a permanent magnet, a movable solenoid, and a diaphragm which is connected to the solenoid and moves with it to set up the air vibrations we know as sound waves.

The solenoid operates on the same principle as the electromagnet. When electron current flows through a coil of wire, magnetic lines of force are established which pass through the coil in an axial direction. Furthermore, the direction (polarity) of these lines of force depends on the direction of the electron flow through the coil. The polarity of these or any magnetic lines of force is either north or south.

The magnetic lines of force (also called *lines of flux*) are greatly strengthened by placing an easily magnetized material, such as soft iron, within the coil. This has been done in the speaker of Figs. 4-1 and 4-2. The *polarity* of the magnetic field created, as a result of the electron current through the moving coil, has a north pole at the left end and a south pole at the right end of the coil, as shown in Fig. 4-1.

This combination of a coil of wire and a movable iron core becomes a solenoid with the addition of a permanent magnetic field to the temporary one. This permanent magnetic field is provided by the magnet in Fig. 4-1 and 4-2. Its permanent south pole is on the right, adjacent to the left end of the movable iron core. When the core is magnetized, as shown in Fig. 4-1, it will be drawn to the left, closer to the permanent magnet, since a south and a north magnetic pole will always be attracted to each other. The core pulls the flexible diaphragm with it, creating a "rarefaction" of the air in front of the speaker, which becomes one half of a single cycle of a sound wave.

When the direction of the output current in Fig. 3-2 has been reversed (Fig. 4-2) the resulting magnetic lines of force change direction. The movable iron core now has a south magnetic pole at its left end and a north magnetic pole at its right end. Since the two south poles repel each other, the entire core moves to the right, pushing the flexible diaphragm ahead of it. This diaphragm movement compresses the air in front of the speaker which becomes the other half of a single cycle of a sound wave.

A sound wave consists of these alternate compressions and rarefactions of air traveling through the atmosphere. When these air vibrations strike another transducer such as the human ear, they are "transduced" to appropriate nerve vibrations, bringing pleasure (presumably!) to the listener.

The frequency of the output current determines the frequency, or pitch, of the sound produced. As an example, if the current frequency is 400 cycles per second, the iron core will move back and forth 400 cycles each second and thus accurately reproduce the desired pitch.

The *volume* of sound produced depends on the distance the diaphragm travels back and forth each cycle. This will be the same distance as the one traveled by the iron core, since the two are rigidly joined together. The strength of the alternating current through the secondary winding determines the strength of the temporary magnetic field, which in turn regulates the movements of the iron core.

The Permanent Magnet

The permanent magnet shown in Figs. 4-1 and 4-2 to attract or repel the solenoid is used in the majority of speakers. However, in some speakers, a coil of wire is wound around an iron core and connected to a source of direct current. Thus we have a second electromagnet. Unlike the solenoid electromagnet, however, the magnetic field for this second electromagnet will remain constant and the poles unchanged because it is connected to a *DC source* instead of the AC through the solenoid coil.

How do we tell which pole is which in an electromagnet? It's simple! Just grasp the coil with the left-hand so that your fingers point in the direction the electrons are flowing through the coil. Then extend your thumb, and it will point toward the north magnetic pole.

Speakers employing permanent magnets are called permanent-magnet, or PM, speakers, while those with electromagnets are called electromagnetic or EM speakers. Operation of both types is the same—the only difference is the added connections for the extra coil, usually called the *field*. In practice, the field is usually connected to a DC source in the equipment's power supply, but sometimes a separate power supply is used.

Cathode Filtering

The cathode filter or bypass current, shown in large dots, flows back and forth into and out of capacitor C1. Its only purpose is to keep the voltage at the cathode constant. Remember that the grid bias of any tube is the instantaneous *difference* in voltage between grid and cathode. The actual voltage at the grid is changed by the flow of grid driving current up and down through the grid driving resistor, and it is desirable that the amount of plate current, at any instant throughout the cycle, always be proportional to the grid driving voltage. If we were measuring grid driving voltage with respect to a fixed value such as ground, this condition would be relatively simple to attain. However, we are really measuring the grid driving voltage with respect to the *cathode,* since the total instantaneous grid bias (which really determines how much plate current will flow) is the difference between the cathode and grid voltages. Therefore, the cathode should, insofar as possible, be held at some fixed value of positive voltage.

It is obvious that the cathode voltage is subject to fluctuation, since it is produced by the flow of plate current through cathode biasing resistor R2 and the plate current is varying over a fairly wide range. The addition of a filtering capacitor stabilizes this cathode voltage and thereby prevents such voltage fluctuations. The combination of resistor R2 and capacitor C1 is a familiar example of a long time-constant RC filter. A positive voltage builds up on the upper plate of capacitor C1. When the demand for plate current is very low (the condition depicted in Fig. 3-2), this positive cathode voltage continues to draw electron current upward through the cathode resistor. During these negative half-cycles the excess electron current "spills over" onto the capacitor rather than entering the tube. This action drives an equal number of electrons off the lower plate and into common ground.

During the positive half-cycles depicted by Fig. 3-1, the control grid is positive; hence the demand for the cathode to emit more electrons into the tube increases. If there were no cathode bypass capacitor, the additional plate current would have to be supplied directly from ground, below the cathode resistor. In flowing upwards through R2 this current would cause an increased voltage drop across the resistor, in accordance with Ohm's low. The larger drop would result in a higher positive voltage at the cathode.

When a suitable bypass capacitor is connected across the resistor, the additional demand for plate-current electrons will be supplied directly from the positive ion pool on the upper plate of the capacitor. *If the capacitor is large enough in value,* the ion pool will be large enough that an insignificant number of extra electrons will be given up into the tube, in comparison with the number of positive ions stored there. Hence the positive voltage of the ion pool will not be changed noticeably by the departure of the electrons.

The filtering current which flows between the lower plate of the capacitor and ground will flow *upward* during the positive half-cycle of Fig. 3-1, drawn by the departure of the extra electrons into the tube.

Thus we see that the filtering current flows up and down between the capacitor and ground at the frequency of the applied voltage.

ADVANTAGES OF PUSH-PULL OPERATION

Several important advantages accrue from using two power-amplifier tubes in push-pull connection. As discussed in earlier chapters, it is normally desirable to operate a tube on the linear portion of its transfer characteristic curve in order to avoid distortion. The distortion arising from operating a tube along nonlinear portions of this curve is classified as second-harmonic distortion. In the push-pull connection, the second-harmonic distortion caused by one tube is canceled out, in the output transformer, by second-harmonic distortion caused by the second tube. Consequently, the output is essentially distortionless. The fact that these two distortions can be canceled out permits each tube to be driven harder—in fact, into the nonlinear portions of the characteristic curve. The power output achieved from two tubes in push-pull is more than double the power that can be achieved with a single tube of the same type. This feature permits the use of two lower power tubes in push-pull, instead of one high power tube, to achieve the desired power outputs.

Another advantage of the push-pull circuit is that DC magnetization of the output-transformer primary winding is avoided. In the circuit just discussed, some plate current is flowing downward through the primary winding at all times. (This can be verified by examination of Figs. 4-1 and 4-2.) As a result of the continuous current through one winding, the iron core becomes magnetized and this detracts from its ability to operate efficiently as a transformer. This undesirable effect is eliminated in all the other circuits of this chapter, since any permanent magnetism which might otherwise be acquired as one plate current flows downward through part of the primary winding will be canceled out (neutralized) by the other plate current flowing upward through the other portion of the primary.

Still another advantage of a push-pull connection is that the sum of the two currents entering the power supply is very nearly constant, particularly when the tubes are being operated under Class-A conditions. This eliminates the necessity of decoupling the power supply with an additional filter capacitor to bypass small surges in current around the power supply to ground. Because a single power supply is normally used to supply many circuits, it is essential that no voltage or current surges be permitted to momentarily raise or lower the power-supply voltage, since the operating conditions of all other tubes connected to the power supply would be immediately affected.

Because of the symmetrical relationship between the two grid voltages and the resulting plate currents, a push-pull circuit is sometimes referred to as a balanced amplifier.

PUSH-PULL AMPLIFIER USING GRID-LEAK BIAS

Figs. 4-3 and 4-4 show two successive half-cycles of operation of a push-pull power amplifier for audio frequencies. A transformer is used in the input circuit to supply driving voltages to the two tubes. These driving voltages must be equal in amplitude but 180° out of phase with each other. There are several methods of meeting the two conditions. The transformer input is one.

This circuit has the following components:

R1—Grid-leak biasing resistor for both tubes.
C1—Grid-leak biasing capacitor for both tubes.
T1—Input transformer.
T2—Output transformer.
V1—First power-amplifier triode.
V2—Second power-amplifier triode.
M1—Common power supply for both tubes.

There are seven different electron currents at work in this circuit:

1. Input-signal current flowing in primary of transformer T1.
2. Grid driving current flowing in secondary of transformer T1.
3. Grid-leakage current for V1.
4. Grid-leakage current for V2.
5. Plate current for V1.
6. Plate current for V2.
7. Output current flowing in secondary winding of transformer T2.

Circuit Description

As indicated in Figs. 4-3 and 4-4, the secondary winding of transformer T1 has more turns than the primary. This indicates the transformer is a voltage step-up and current step-down type. This transformer is being driven by the input signal current, a solid line in Fig. 4-3. The primary current is shown flowing downward through the primary winding.

The action of any transformer or inductor is to oppose any change in the total current flowing in the two windings. Thus, if we consider that fraction of the first half-cycle when the input-signal current is *building up* as it flows downward through the primary winding of T1, we can immediately visualize a different current being caused to flow upward in the secondary winding. The latter also is increasing, so that the *total* current flowing in the two windings (the algebraic sum of the two currents) will be considerably less than if the secondary current did not flow.

This induced current, shown as a thin line in Fig. 4-3, flows upward in the secondary winding. Associated with this secondary current is the back emf or counter emf. Its polarity during this portion of the first half-cycle is indicated by the plus sign at the top and the minus sign at the bottom of the secondary winding. (It is easy to correlate these polarities with the upward flow of secondary current, since electrons will always flow *toward* a positive voltage and *away* from a negative one.)

The positive polarity also exists at the control grid of V1 during the first half-cycle. This positive grid voltage releases a large amount of plate current through V1. The complete path of the plate current (solid) is from cathode to plate, downward through the upper half of output transformer T2 to its center tap, then through common power supply M1 to ground, and back to the cathode.

Since the control grid of tube V2 is connected to the lower end of the secondary winding of input transformer T1, its grid voltage will always be 180° out of phase with the voltage applied to the grid of V1. This condition is indicated in Fig. 4-3 by the minus sign at the bottom of the secondary winding. The negative voltage at the grid of V2 drastically restricts the flow of plate current through V2 during this half-cycle. The complete path of plate current for V2 is from cathode to plate within the tube, upward through the lower half of the primary winding of output transformer T2, then through the common power supply to ground, and back to the cathode of V2.

Capacitor C1 and resistor R1 serve as a common grid-leak bias combination for both tubes. Leakage electrons which strike the grid of either tube must flow back to ground through the appropriate half of the secondary winding of transformer T1, and then through resistor R1. If R1 has a sufficiently high resistance, all grid-leakage electrons from both tubes will accumulate on the left plate of capacitor C1, and a negative voltage will build up and be applied equally to the control grid of each tube. This negative voltage, known as a *grid-leak bias voltage,* is characterized by an *intermittent* input of new electrons from each tube once each cycle, and by a continuous drain of electrons from the capacitor, through resistor R1 and back to ground.

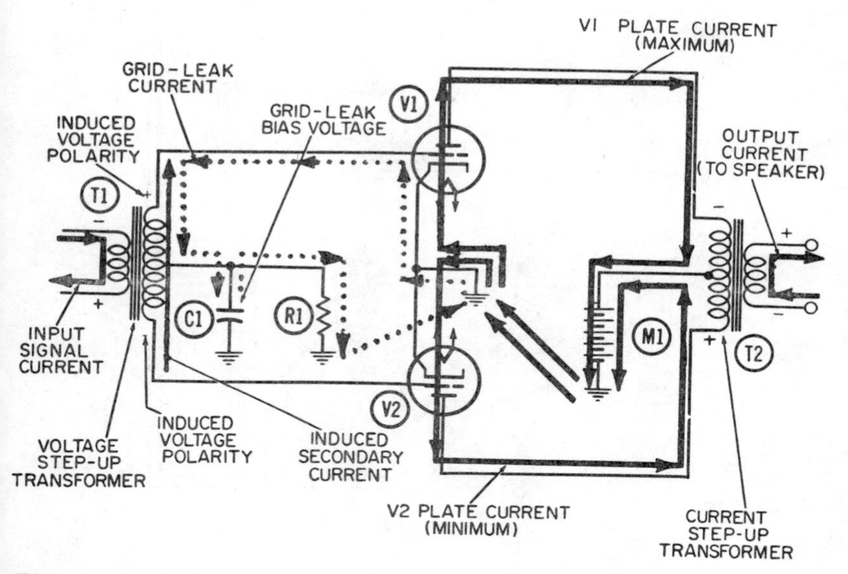

Fig. 4-3. Operation of a push-pull AF power amplifier using grid-leak bias—first half-cycle.

A circuit constructed in this manner—with a transformer in the input portion—will normally be operated Class-A, which means each tube is conducting some plate current during the entire audio-frequency cycle. Thus, in Fig. 4-3, when V1 conducts heavily, V2 will conduct lightly. During the second half-cycle, V2 will have a more positive grid voltage and will conduct heavily while tube V1 will have a negative grid voltage and consequently conduct only a small amount of plate current. The plate current in each tube is a pulsating direct current, and the pulsation from each tube will cause transformer action between the primary and secondary windings of T2. These two separate transformer actions will fortunately be in the appropriate phase to aid, or reinforce, each other. As a result, a very heavy secondary current (solid) will flow in the output circuit. This output current is a greatly amplified version of the small input-signal current, a solid line, which flows up and down in the primary winding of T1.

Notice, transformer T1 has more turns in its secondary than in the primary. Therefore, it steps-up the voltage (and steps-down the current) before applying it to the two control grids, since vacuum tubes are voltage controlled devices. Transformer T2, on the other hand, has more turns of wire in its primary than in its secondary winding. So it will act as a current step-up (and volt-

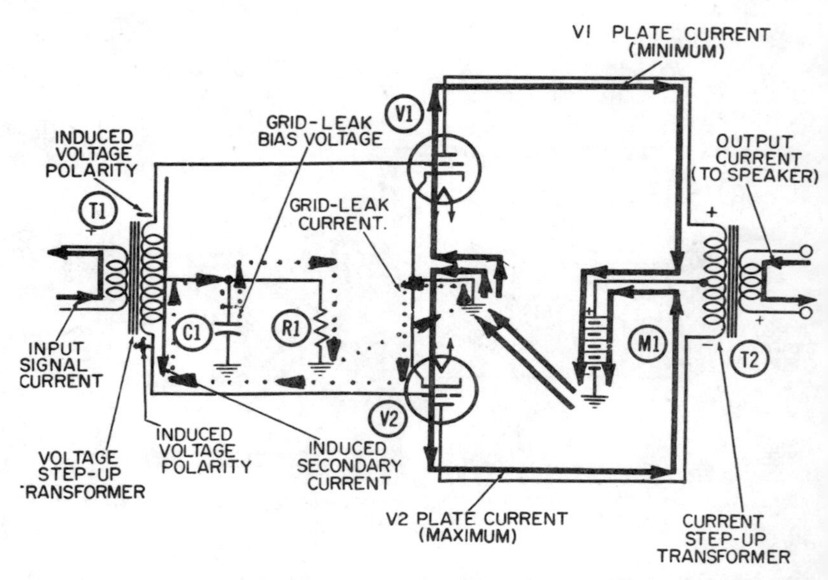

Fig. 4-4. Operation of a push-pull AF power amplifier using grid-leak bias—second half-cycle.

age step-down) transformer. This is advantageous for this reason—the transducer (speaker) to which the secondary winding of the output transformer is connected will usually be a current-operated device. Hence, the volume of sound available will depend directly on the strength, or amplitude, of the *current* pulsations flowing through the speaker coil. The speaker current, and the current through the secondary winding of transformer T2, are of course one and the same.

Transformer Action

Let us consider the transformer actions as a result of the two plate currents in Fig. 4-3. During the first half-cycle the plate current starts at its minimum value and *increases* continuously until it reaches its maximum value at the end. As this current flows *downward* through the upper half of the primary winding of T2 at an increasing rate, another current will be induced to flow *upward* in the secondary winding (also at an increasing rate) because the natural tendency of any transformer is to *oppose* any change in the total current flowing through the windings. Associated with these two currents are appropriate voltage polarities, the signs of which are indicated at the tops of the primary and secondary windings. An increase in plate current is normally associated with a reduction in plate voltage, because of the greater number of electrons in the plate area. This is the meaning of the minus sign at the top of the primary winding in Fig. 4-3.

By definition, a back emf always has a polarity opposite that of the applied emf, or voltage. Consequently, in Fig. 4-3, the resulting back emf in the secondary winding of T2 will have a positive polarity at the top. It is possible to correlate both voltage polarities with the directions of current flow indicated—more electrons flow *away* from the minus sign in the primary winding, and more electrons flows *toward* the plus sign in the secondary winding.

During the same first half-cycle (Fig. 4-3), current through V2 starts out at its maximum value and decreases to minimum at the end of the half-cycle. Associated with this decrease in plate current is the conventional rise in plate voltage. The latter can be symbolized by the plus sign at the bottom of the primary winding, which also designates the polarity of the applied emf across the lower half of the primary.

The resulting back emf across the secondary winding will have a negative polarity at the bottom of the winding, as indicated by the minus sign. Associated with this polarity will be a flow of electron current away from the negative region and *upward* through the winding.

Hence, during the first half-cycle two different components of secondary current are made to flow *in phase* with each other by the two streams of plate current, thus making a very heavy current available for operating the speaker.

During the second half-cycle (Fig. 3-4) most of the previous conditions are reversed. The direction of signal current in the primary of the input transformer T1 is upward, and the resulting back emf in the secondary winding is positive at the bottom. This causes a steady *increase* in the plate current through tube V2, with the resultant lowering in plate voltage and a negative polarity sign for the applied emf across T2. The resulting back emf across the secondary winding will have a positive polarity at the bottom, indicated by the plus sign, and during this half-cycle electron current will of course be flowing *downward* in the secondary winding at an increasing rate.

As the control grid of tube V1 becomes progressively more negative during this second half-cycle, the plate current through V1 decreases. This causes a rise in plate voltage, as indicated by the plus sign at the top of the primary winding of T2. The transformer action across T2 gives a back emf of negative polarity at the top of the secondary winding, as indicated by the minus sign. Therefore, electron current flows *downward* through the secondary winding.

Thus, the two transformer actions resulting from the two plate currents will drive currents in the secondary winding, which are in phase during each half-cycle of operation. This feature is of course the special attractiveness of the push-pull circuit.

Plate and Grid-Leak Currents

For the newcomer to electronics, the complete path of plate and grid-leakage currents will now be briefly reviewed.

Electrons which leave the cathode and reach the plate of tube V1 will flow through the upper half *only* of the primary winding of transformer T2 before being drawn into the positive terminal of power supply M1 and delivered through the power supply to ground and back to the cathode. It is of course the high positive voltage of the power supply which causes this entire sequence to occur, since this voltage provides the positive plate voltage which draws the electrons across the tube.

By similar reasoning, the plate-current path for tube V2 is from cathode to plate, upward through the lower half *only* of the primary winding of transformer T2 to the positive terminal of the power supply, through the power supply to ground, and back to the cathode.

Grid-leakage electrons will flow out of the tubes via the control grids whenever a control-grid voltage is more positive than the cathode voltage. In tube V1 this will occur near the end of the first half-cycle, and in V2, near the end of the second half-cycle.

Grid-leakage electrons from both tubes must flow back to their respective cathodes *through* half of the secondary winding of T1 and through resistor R1. Prior to entering R1, the electrons will accumulate on the top plate of capacitor C1 and form a reservoir of negative voltage (known as the grid-leak bias voltage). This voltage biases the control grids of both tubes to the same negative value.

Fig. 4-5 gives the time relationship between the voltages at the two control grids, the two plate currents, and the output current flowing in the secondary winding of transformer T2 in graphical form.

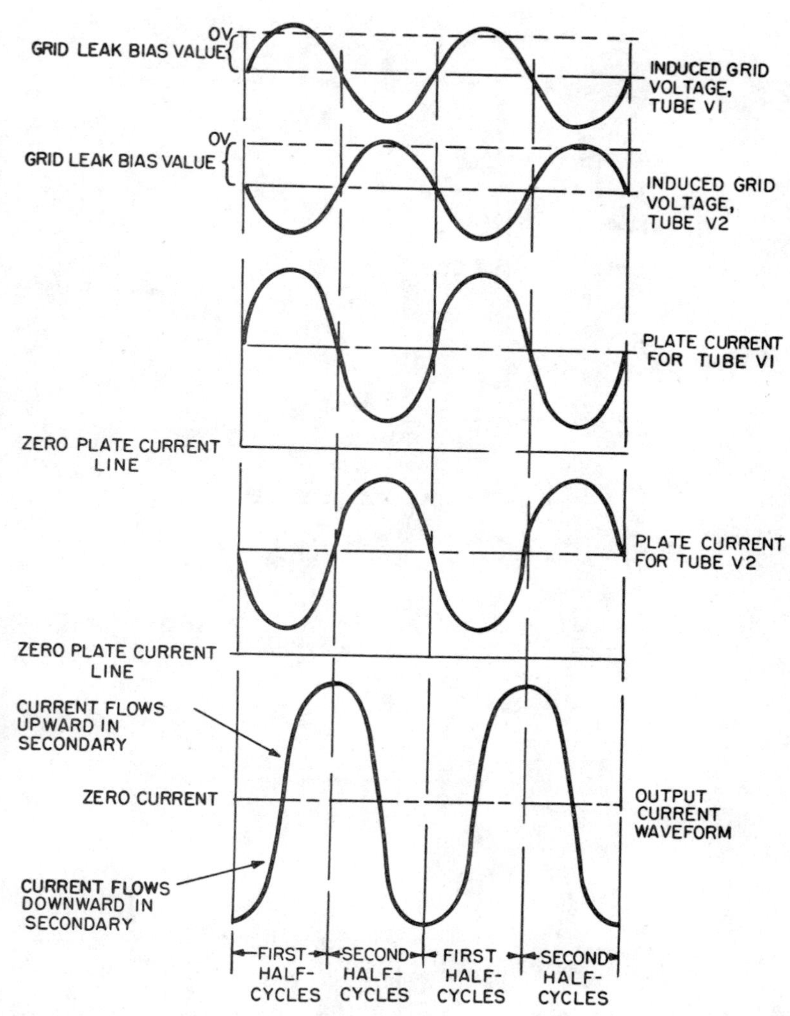

Fig. 4-5. Voltage and current waveforms in the amplifier of Figs. 4-3 and 4-4.

PUSH-PULL AMPLIFIER USING CATHODE BIAS

Another common type of push-pull power amplifier is given in Figs. 4-6 and 4-7. Notice that here two signals, each 180° out of phase, drive the grids of the push-pull amplifier tubes. (How these two 180° out of phase signals are obtained is explained in

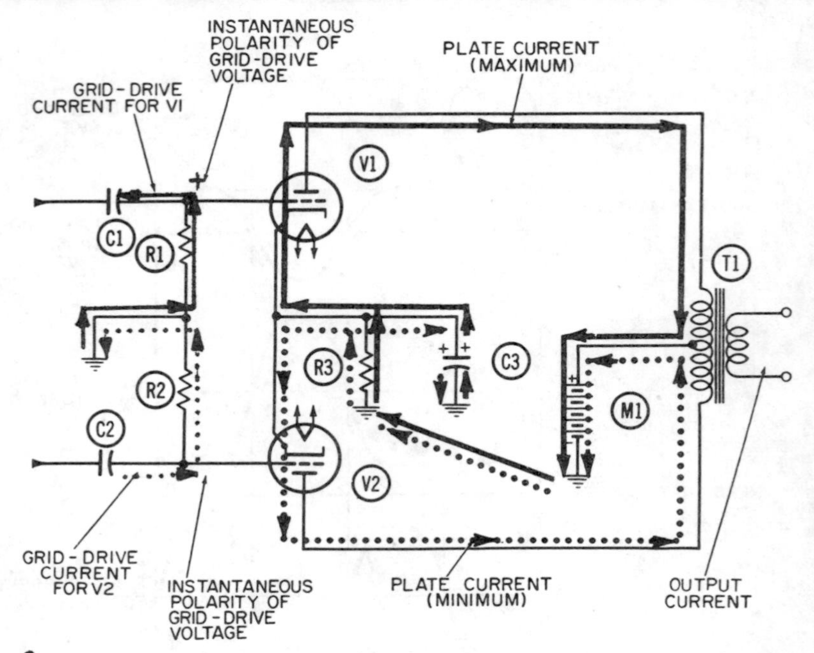

Fig. 4-6. Operation of a push-pull AF power amplifier using cathode bias—first half-cycle.

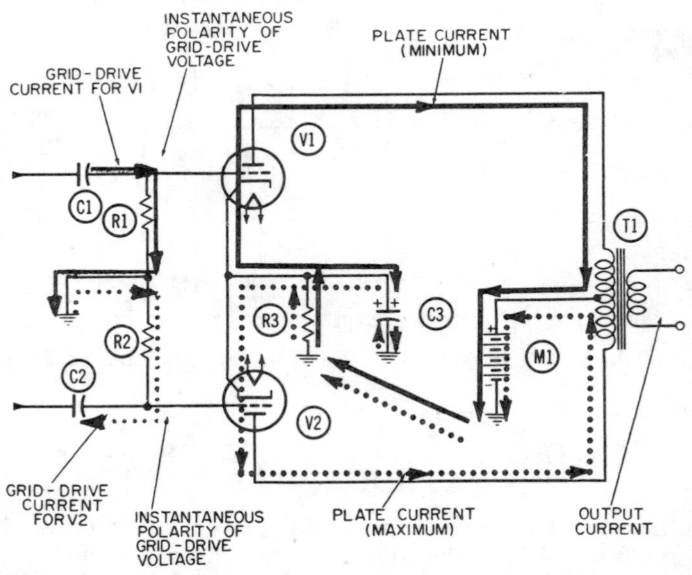

Fig. 4-7. Operation of a push-pull AF power amplifier using cathode bias—second half-cycle.

the next chapter. For now, just assume that two signals of opposite polarity are applied to the two inputs.)

The components of the circuit are:

R1—Grid-driving and grid-return resistor for V1.
R2—Grid-driving and grid-return resistor for V2.
R3—Cathode biasing resistor for both tubes.
C1—Grid-coupling capacitor to V1.
C2—Grid-coupling capacitor to V2.
C3—Cathode filter capacitor.
T1—Output transformer.
V1 and V2—Triode power-amplifier tubes.
M1—Common power supply for both tubes.

There are five currents at work in this circuit which must be identified and analyzed to understand the operation of the circuit. The currents representing them in Figs. 4-6 and 4-7 are:

1. Grid-drive current for V1.
2. Grid-drive current for V2.
3. Plate current for V1.
4. Plate current for V2.
5. Output current to speaker.

Circuit Operation

During the first half-cycle (Fig. 4-6) electrons are being drawn upward through R1 to the right plate of C1 as shown a thin line. This electron flow indicates that the polarity of the signal voltage applied to the left plate of C1 is becoming more positive. At the same time, electrons (shown in small dots) are flowing upward through R2, indicating that the signal voltage applied to the left plate of C2 is becoming less positive (more negative). The polarities of the grid-driving voltages resulting from the electron flows through R1 and R2 are indicated by the thin plus sign at the grid of V1 and the thin minus sign at the grid of V2. During the second half-cycle (Fig. 4-7), the conditions are reversed, the grid-driving currents are reversed and the grid of V1 is driven negative and the grid of V2 positive. Thus, the conditions are essentially the same as in Figs. 4-3 and 4-4. That is, when the grid of one tube is driven positive, the other is driven negative.

Operation of the two push-pull tubes is practically identical to that of the push-pull amplifier with transformer input discussed previously. The plate current for each push-pull tube must flow out of ground and through cathode biasing resistor R3, so that both plate currents will contribute to the total bias voltage developed across R3.

The path for the plate current in tube V1 includes a journey through R3, then upward through the tube from cathode to plate, and downward through the upper half of the output-transformer primary winding. The current exits from the transformer at the center top and is drawn onto the positive terminal of the power supply for delivery back to the common ground. Through the upper portion of the transformer primary, the V1 plate current will flow in pulses, reaching its maximum during the half-cycle shown in Fig. 4-6 and minimum during the half-cycle in Fig. 4-7. The flow of pulsating DC through the primary winding of the transformer will produce an alternating current (a solid line) in the secondary winding. This action is identical to the transformer action of the push-pull amplifier discussed previously.

Whenever the plate current in tube V1 is *increasing* (Fig. 4-6), the secondary current (solid) will flow *upward*. By the same token, it will flow *downward* whenever the plate current is *decreasing,* as shown by the second half-cycle of Fig. 4-7.

The plate current for the lower push-pull tube, V2, is shown as large dots. It originates at the cathode, flows to the plate and upward through the lower half of the primary winding, then goes through the power supply to ground. From ground the current flows up through resistor R3. This completes its journey back to the cathode. V2's current, which is also a pulsating DC, is 180° out of phase with the pulsations going through the upper tube. Thus, the resulting transformer action between primary and secondary will drive additional secondary current in the transformer, the two components of which will be in phase with each other.

As with the previous push-pull amplifier circuit, the secondary current will be twice as heavy as for a single tube. Because the secondary winding has fewer turns of wire than either half of the primary, the secondary current is increased. This step-up action is desirable, because the secondary current normally drives some type of transducer—such as a speaker—which requires a low voltage but a high current.

Cathode Filtering

Filter capacitor C5 performs two filtering actions simultaneously. If they could be looked at separately, each would appear as shown in Figs. 4-6 and 4-7. For instance, when excess electrons flows *onto* the top plate of C3 in Fig. 4-6 (because tube V2 refuses to accept them), an equal number would flow off the bottom plate. Both components of filter current will reverse their direction when V2 conducts heavily (see Fig. 4-7).

The filtering current associated with the fluctuations of plate current through the upper tube, V1, is also shown. At all times it is flowing 180° out of phase with the filtering current associated with tube V2. During *ideal* operating conditions, these two filter currents will at all times be equal in amount but opposite in flow direction. They would cancel each other out completely and the filter capacitor could be dispensed with entirely, with no fear of degeneration.

In actuality such an ideal condition can never be achieved. For one thing, the grid driving voltage developed for each tube will vary slightly. Further, it is highly unlikely the emission characteristics of two identical tubes would remain the same throughout the operating life of the tubes—or even after their first day in service. Consequently, if both grid-driving voltages are equal, and both push-pull tubes have exactly the same emission characteristics, the current shown flowing between the bottom plate of C3 and ground will not exist. However, in any practical circuit a certain amount of current will flow here. The important idea is to visualize the two filtering actions separately. Once this is done, the exact degree to which they may be canceling each other out is of small concern.

CLASS-B PUSH-PULL AMPLIFIER

Often, the bias and plate voltages are selected so that, when no driving signal is present, each of the push-pull tubes will be exactly at the point of conduction. By proper selection of components, the bias voltage is made to exactly equal the grid cutoff value, as determined by the characteristic curves of the tube. Thus, each tube will conduct from cathode to plate whenever an applied signal voltage makes its grid voltage more positive than its cutoff value as shown in Fig. 4-8. Obviously, each tube will not conduct electrons whenever an applied signal voltage makes its grid voltage more negative than its cutoff value. This is illustrated by the plate-current curves in Fig. 4-8.

The net result is that when two equal amplitude–opposite-phase sinusoidal voltages drive the two push-pull tubes, each tube will conduct electrons during half of a cycle but will be cut off during the other half. This is what is known as Class-B operation of the tubes.

The output, or speaker current, which flows in the secondary winding of the output transformer is alternately driven by each plate current as it flows through its half of the transformer primary winding. This is the same as for the two previous push-pull

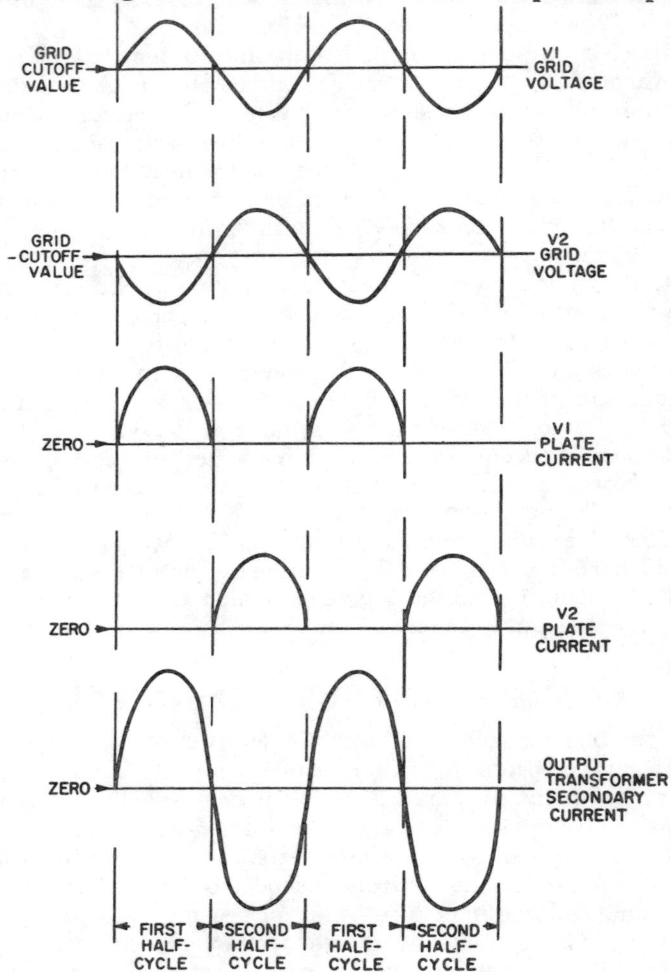

Fig. 4-8. Voltage and current waveforms in a Class-B power amplifier.

amplifier examples, except that current is flowing through only half of the output transformer at a time.

Class-B amplifiers have a much higher efficiency than Class-A amplifiers. Therefore, even though only half of the output transformer has current flowing through it at a given instant, approximately *twelve* times as much audio power can be delivered by two tubes operating Class-B than by their Class-A counter-parts.

Chapter 5

AUDIO PHASE-INVERSION CIRCUITS

In the previous chapter, the need for two signals equal in amplitude but opposite in phase to drive the push-pull audio amplifiers was discussed. One method of obtaining the required signal (using a center-tapped transformer) was given in Figs. 4-3 and 4-4. In this chapter two methods of obtaining the required signal using R-C coupling will be given.

SINGLE-TRIODE PHASE INVERTER

Figs. 5-1 and 5-2 show common phase inversion circuit. This circuit provides the two 180° out of phase output signals necessary to drive the control grids of the two push-pull power amplifier tubes.

The components of this circuit are:

C1—Input capacitor from previous voltage-amplifier stage.
C2—Cathode bypass or filter capacitor.
C3—Plate coupling capacitor.
C4—Cathode coupling capacitor.
R1—Grid driving and grid return resistor.
R2—Cathode biasing resistor.
R3—Cathode "following" resistor.
R4—Plate load resistor.
R5—Grid driving and grid return resistor for V2.
R6—Grid driving and grid return resistor for V3.
V1—Triode tube used for phase-inversion purposes.
V2 and V3—Power-amplifier tubes.
M1—Power supply.

The currents in this circuit are:

1. Grid-driving current for V1.
2. Plate current.
3. Grid-driving current for V2.
4. Grid-driving current for V3.
5. Cathode filter current.

In addition to these currents, it is always possible, and sometimes inevitable, that grid leakage current will flow from any vacuum tube. Because the grid-leakage phenomenon is not used in this particular circuit, grid-leak currents have not been shown in the diagrams of Figs. 5-1 and 5-2. However, all tube circuits invariably take care of any grid-leakage electrons by having *a closed path back to ground and the cathode* from the grid. This path may consist of a single resistor or any number of such components in combination, as long as grid-leakage electrons have a path available through which they can return to the cathode of the tube from which they were emitted. In Figs. 5-1 and 5-2, this path is provided by R1 between the grid and ground, then through R3 and R2 to the cathode.

Also not shown, nor seldom referred to in discussions of circuit operation, are the cathode *heating* currents. These current,

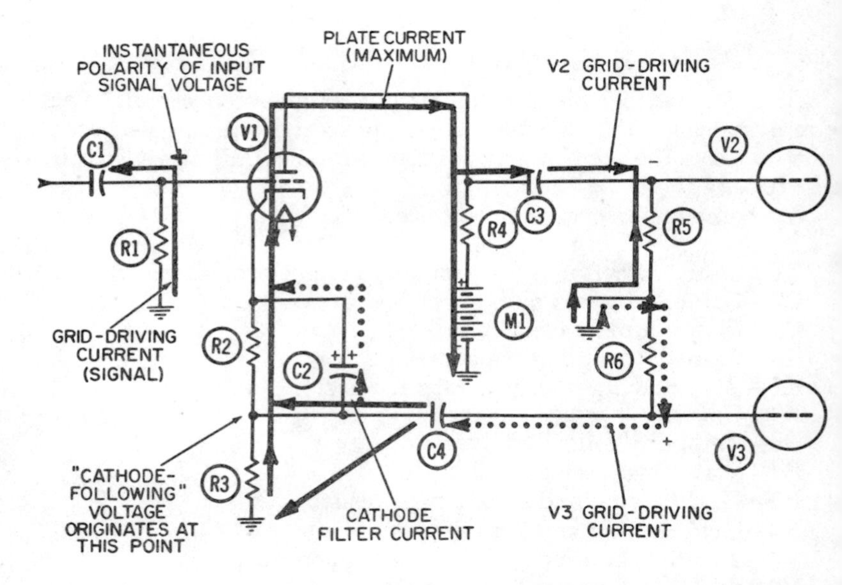

Fig. 5-1. Operation of a cathode-follower phase-inverter circuit—first half-cycle.

which frequently go by the name of filament (or filament heating) currents, were discussed more fully in the opening chapter. They perform the essential function of heating the tube cathodes so that electron emission can occur.

The remaining currents, which might be termed "operating" ones, are shown in Figs. 5-1 and 5-2. The initial, or input, function in this circuit is that of driving the grid of tube V1. This action is provided by the electron current (thin line) as it flows upward through grid resistor R1 during the first half-cycle depicted by

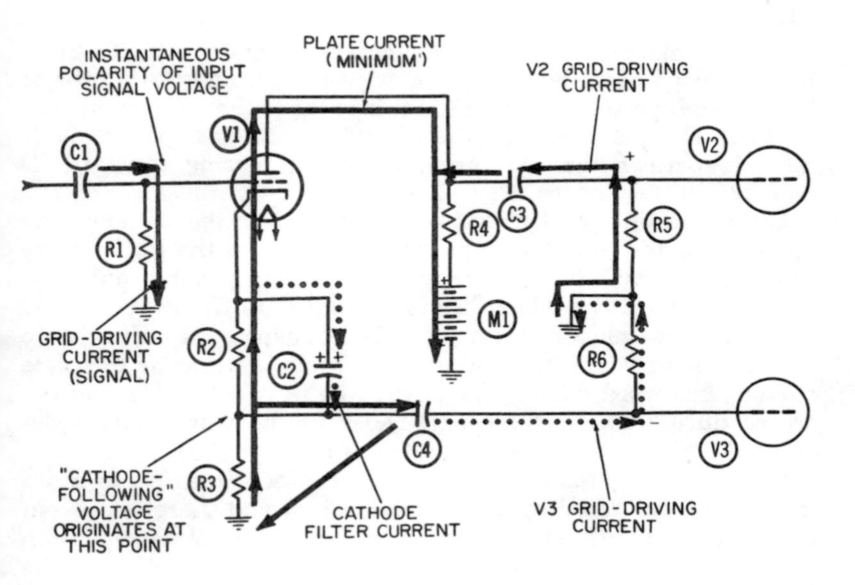

Fig. 5-2. Operation of a cathode-follower phase-inverter circuit—second half-cycle.

Fig. 5-1. The current here is, in all probability, being driven by the plate-current and -voltage combination of a preceding voltage-amplifier stage.

When current is being drawn upward and onto the right plate of capacitor C1, this indicates a positive voltage at the top of resistor R1. The voltage, indicated by a plus sign, is also applied to the control grid of tube V1, releasing a large pulsation of electron current into the plate circuit of the tube. This plate current (solid line) flows onto the left plate of coupling capacitor C3, driving another current *downward* through grid resistor R5. This is the grid driving current for tube V2, and is shown a solid line . Its complete path lies between coupling capacitor C3 and the ground connection at the bottom of R5. Its downward direction during this first half-cycle tells us the voltage at the top of resistor R5 has to be negative. This negative voltage is indicated by a minus sign, to show the instantaneous polarity of the grid voltage.

Since the polarity at the grid of tube V1 is *positive* during this first half-cycle, we see that the signal has experienced the normal phase reversal in its passage from the first to the second tube.

The Cathode-Follower Principle

The complete path of the plate current, shown by the solid lines, is from cathode to plate, downward through load resistor R4, through power supply M1 and into the common ground connection. The plate current then has free access to the cathode by being drawn *upward* through cathode biasing resistors R3 and R2.

By virtue of the electrons which have left the cathode and gone across the tube to the plate, the voltage at the top of these two resistors will be positive with respect to ground, and it is this positive voltage which draws electrons up from ground and completes the plate-current path. If no bypass capacitor were present, this positive cathode voltage would increase as the plate current increased during the positive half-cycles, and would decrease during the negative half-cycles as the plate current decreased. This phenomenon is known as degeneration.

Capacitor C2 is used for filtering purposes across R2 in order to keep the voltage drop across R2 constant and thereby prevent degeneration across this resistor. A small amount of degeneration will occur across remaining cathode resistor R3, however, for the following reason. As the plate current through V1 increases, the positive voltage at the top of resistor R3 will rise slightly, because of the Ohm's law relationship between the value of the resistor, the amount of current, and the resulting voltage across the resistor.

Capacitor C4 couples this small rise in positive voltage to the control grid of the lower push-pull tube, V3. This action can be visualized by looking at the currents flowing in and out of C4. As the positive voltage at the top of R3 *increases,* it will tend to draw electrons *away* from the left plate of C4. Their departure draws an equal number of electrons onto the right plate of C4. This electron flow, shown in large dots, constitutes the grid driving current for the lower push-pull tube, V3. Their downward flow through grid resistor R6 indicates that the voltage at the bottom of R6 will have a positive polarity (indicated by the heavy plus sign) since electrons will always flow away from an area of negative voltage and toward an area of positive voltage. This is the positive grid voltage for tube V3.

When an alternating voltage is taken, or "coupled," from it, the cathode will have the same polarity as the input or driving voltage. In the example of Fig. 5-1, the moment the grid voltage reaches its most positive value, the voltage across the lower cathode resistor, R3, will also reach its most positive value and be coupled to the grid of the next tube, making the grid positive. This function is known as cathode following which means the voltage coupled out of the cathode circuit has the same polarity as the voltage at the grid of the same tube. In that sense the cathode voltage "follows" the input voltage.

In the second half-cycle, depicted by Fig. 5-2, most of these conditions are reversed. The grid-drive current flows *downward* through resistor R1, making the control grid of tube V1 negative. As a result, the plate current through tube V1 is reduced and its plate voltage rises. This rise in plate voltage draws electron away from the left plate of C3 and causes an equal number to flow upward through R5 to the right plate of C3. The resulting voltage drop across R5 makes the grid of V2 positive as indicated by the plus sign.

This reduction in the current flowing upward through R3 and into the tube causes a *smaller* voltage drop across this resistor than occurred during the first half-cycle. This means the positive voltage at the top of resistor R3 will fall and thus permit electrons to return toward the left plate of capacitor C4. This return flow will be accompanied by (1) a reversal in the direction of the grid driving current flowing through resistor R6, and (2) a change in polarity, of the grid drive voltage for tube V3, from plus to minus.

Hence, it is evident that the phase of the driving signal has now been inverted—when one push-pull tube is driven positive, the other is driven negative and vice versa.

Cathode Filtering

Capacitor C2 is bridged across resistor R2 to bypass the plate-current pulses around R2. It does this by alternately *receiving* some electrons onto its top plate during negative half-cycles such as the one in Fig. 5-2 and *relinquishing* them into the tube during positive half-cycles (Fig. 5-1). When electrons are given up into the tube, an equal number will be drawn onto the lower plate of the capacitor; this current is shown in large dots. Fig. 5-2 shows what happens when the tube refuses to take the electrons coming up through the cathode resistors. The excess electrons flow onto the upper plate of capacitor C2 and drive an equal number away from the lower plate.

The small current flowing in and out of the capacitor is the cathode filter current. Its action is somewhat analogous to that

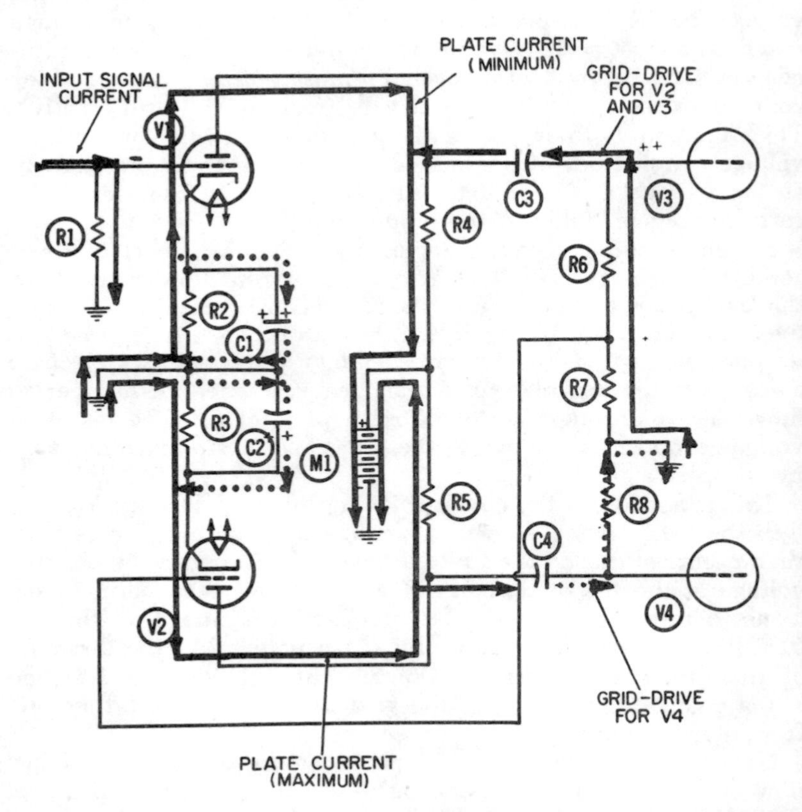

Fig. 5-3. Operation of a two-tube phase-inverter circuit—first half-cycle.

of a shock absorber on an automobile—it keeps the voltage at the cathode from fluctuating with the current going into the tube.

TWO-TRIODE PHASE INVERTER

Figs. 5-3 and 5-4 depict another common type of circuit arrangement for achieving push-pull amplification at audio frequencies. An additional driving tube is utilized, its purpose being to invert the phase of the input signal.

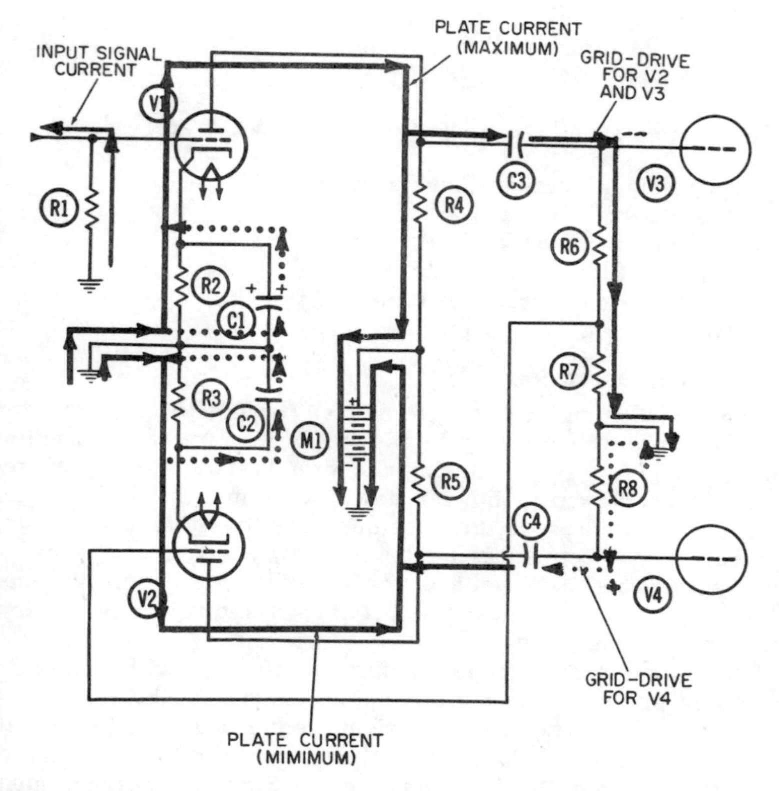

Fig. 5-4. Operation of a two-tube phase-inverter circuit—second half-cycle.

The components which make up this circuit are:

R1—Input grid drive resistor.
R2—Cathode bias resistor for V1.
R3—Cathode bias resistor for V2.
R4—Plate load or coupling resistor for V1.
R5—Plate load or coupling resistor for V2.
R6—Grid drive resistor for V3.
R7—Grid drive resistor for V2 (the phase-inverting resistor).
R8—Grid drive resistor for V4.
C1—Cathode bypass capacitor for V1.
C2—Cathode bypass capacitor for V2.
C3—Coupling and blocking capacitor between V1 and V3.
C4—Coupling and blocking capacitor between V2 and V4.
V1—Input driving tube.

V2—Input phase-inverting tube.
V3 and V4—Push-pull amplifier tubes.
M1—Power supply.

The currents, and the colors they are shown in are:

1. Input-signal current flowing through R1.
2. Plate current for V1.
3. Plate current for V2.
4. Cathode filter currents for V1 and V2.
5. Grid driving current for tubes V2 and V3.
6. Grid driving current for tube V4.

Analysis of Operation

The grid driving current (thin line) flowing through resistor R1 would in all probability be the output current of a preceding voltage-amplifier stage. During the first half-cycle depicted by Fig. 5-3, this current is flowing downward through R1. This tells us the top of the resistor must be more negative than the bottom, since electrons (being negative themselves) will *always* flow away from a more negative voltage and toward a more positive one. This negative grid voltage is depicted by a minus sign at the top of the resistor in Fig. 5-3.

Throughout the entire first quarter-cycle, the grid voltage is becoming progressively more negative. Likewise, the plate current through V1 is being progressively reduced. The result is that the plate current (a solid line) will have dropped to its minimum value by the *end* of the first quarter-cycle. Since this current must flow downward through plate load and coupling resistor R4, the voltage drop across R4 will be smaller and the plate voltage higher than at any other moment during the entire cycle.

Ordinarily, any fluctuations in plate voltage will be coupled, across capacitor C3, to the grid driving resistor for the next stage. The change in plate voltage just described, which occurs during the first half-cycle, can only come about if more electrons are taken away from the plate than are delivered onto it. This means the positive terminal of the power supply draws more electrons downwards through plate load resistor R4 than the plate current of tube V1 is delivering. These extra electrons can only come from the left plate of capacitor C3. An equal number must then be drawn onto the right plate of the same capacitor, to compensate for the loss.

The electrons being drawn onto the right plate of C3 can only come upward through resistors R6 and R7 from the common ground connection at the bottom of R7. In flowing through R6 and

R7, these electrons comprise the grid driving current for both tubes V3 and V2 and the voltage at the top of R6 becomes progressively more positive throughout the first quarter-cycle.

Recall that the grid voltage for tube V1 became progressively more *negative* during this same quarter-cycle. So it should be obvious the conventional phase shift or inversion from grid to plate of tube V1 has occurred. A *small portion* of this V3 grid driving voltage is now used to drive the grid of V2. This was the purpose in putting low-resistance R7 in series with high-resistance R6. By selecting the proper values for R6 and R7, the voltage developed across R7 by the grid driving current can be made *equal in value* but *opposite in polarity* to the voltage being developed across resistor R1 by the original input-signal current. Thus, if the voltage amplification achieved by V1 has a value of 100 (not unusual for voltage amplifiers), resistor R7 would need to be only 1/100th the value of resistor R6.

Since the control grid of V2 is connected directly to the junction of R6 and R7, the voltage at this point becomes the grid drive voltage for the tube. Thus, during the first half-cycle shown in Fig. 5-3, the control grid of V2 becomes progressively more positive throughout the half-cycle, causing a steady increase in the current through the tube.

We associate an increase in plate current with a *decrease* in the positive plate voltage applied to the plate. This decrease in voltage can only come about if more electrons are delivered into the plate area *at any instant during the entire half-cycle* than the power supply is drawing away. If a temporary surplus of electrons is created on the plate during this half-cycle, the only place they can accumulate is on the left plate of capacitor C4. Moreover. they cannot flow onto the left plate of C4 unless an equal number are permitted to flow away from the right plate. The electrons which flow away (actually, they are "pushed" away) from the right plate of C4 have only one path available to them. and that is upward through resistor R8 to the common ground connection.

These electrons become the grid driving current for push-pull tube V4. Their amount increases steadily, or sinusoidally (since all grid driving voltages in this circuit have the same sinewave shape as the input-signal voltage), throughout the entire first quarter-cycle of Fig. 5-3. Consequently the grid voltage for tube V4 increases sinusoidally in the *negative direction*. causing the plate current through tube V4 to decrease from its maximum to its minimum value.

During the second half-cycle, the grid-driving currents and the resulting voltages will be reversed, as shown in Fig. 5-4. Current will flow upward through R1, driving the grid of V1 progressively more positive during the third quarter cycle. This increase in plate current results in a greater voltage drop across R4; hence the plate voltage for V1 will be reduced. Electrons flow onto the left plate of C3 driving an equal number off the right plate of C3 and down through R6 and R7 as shown by the solid line.

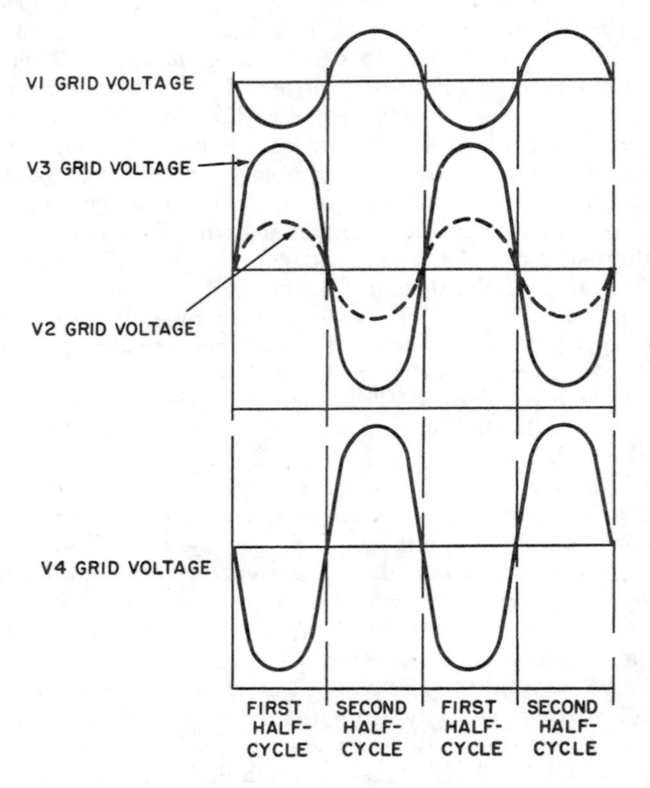

Fig. 5-5. Grid-voltage waveforms for the circuit in Figs. 5-3 and 5-4.

This electron flow drives the grid of V3 negative and the voltage at the junction of R6 and R7 will be a negative polarity but equal in value to the voltage at the grid of V1.

The negative voltage at the grid of V2 reduces the plate current through this tube; hence, its plate voltage increases. This increase in the V2 plate voltage draws electrons away from the left plate of C4 and an equal number of electrons (indicated by the dotted deep line) flow from ground, through R8, and onto the right plate of the capacitor. The flow of electrons onto the right plate of C4 becomes the grid-driving current for V4. The grid-driving voltage produced by this current through R8 is indicated by the plus sign at the grid of V4.

Thus, we have established that two sinusoidal voltages, both equal in amplitude, but 180° out of phase, are driving the two push-pull tubes. Fig. 5-5 shows the relationship between the four grid voltages.

The cathode filter current, indicated by the dotted lines in Figs. 5-3 and 5-4, smooth out the fluctuations in the cathode voltages that would normally exist because of the pulsations in current through the cathode resistors. This is the same as was explained for Figs. 5-1 and 5-2, except that now we have two capacitors, and two filter currents.

Chapter 6

IMPEDANCE-COUPLED RF VOLTAGE AMPLIFIER

The prime function of the radio-frequency (RF) voltage amplifier is to amplify the radio-frequency voltage which is already in existence in the grid portion of the circuit—in other words, to get a higher voltage at the same radio frequency in the plate circuit.

The radio-frequency band includes those frequencies lying within the broadcast band which extends from 540 to 1600 kilocycles per second. In addition, frequencies up to several hundred megacycles per second are also considered radio frequencies.

Pentode tubes are usually employed for RF amplifiers operating in the broadcast band and the simplest type of coupling between stages is impedance coupling. That is, a single coil is used, instead of a transformer between stages.

Sometimes the coupling circuit is made adjustable and tuned to an exact frequency. Then, signals at the desired frequency will receive most of the amplification and all other frequencies will be rejected. In other words, the RF amplifier *selects* a certain frequency for amplification. The degree to which this is accomplished is termed the *selectivity* of the circuit.

Often, it is more desirable to amplify all RF signals present at the grid. The gain and selectivity will not be as great, but the *bandwidth,* or range of frequencies amplified, will be greater in the untuned circuit used in this application.

UNTUNED IMPEDANCE-COUPLED AMPLIFIER

Figs. 6-1 and 6-2 show two half-cycles in the operation of a radio-frequency voltage amplifier using untuned impedance coupling.

The load circuit consists of radio-frequency choke coil L1. The various components of this circuit are:

R1—Grid driving and grid leakage resistor.
R2—Cathode biasing resistor.
R3—Screen-grid dropping resistor.
R4—Grid driving resistor for next stage.
C1—Cathode filter capacitor.
C2—Screen-grid filter capacitor.
C3—Coupling and blocking capacitor.
L1—Radio-frequency choke coil.
V1—Pentode amplifier tube.
V2—Next amplifier tube.
M1—Power supply.

The following currents are at work in this circuit:

1. Grid driving current .
2. Plate current .
3. Screen-grid current.
4. Screen-grid filtering current.
5. Cathode filtering current.
6. Amplified grid driving current for next stage .

Basically, this circuit and the ones discussed previously are alike in principle and function. As before, the grid is driven by causing an alternating current to flow up and down through the grid resistor. This grid driving current, usually called the signal current, flows at a radio-frequency rate in this example.

At an assumed frequency of 1,000 kilocycles per second, this grid driving current would make one million complete journeys down and up through the grid resistor, causing the grid voltage to change from negative to positive and back to negative this many times each second.

This varying control-grid voltage turns the plate-current stream through the tube down and up a million times a second. The resulting plate current flowing into radio-frequency choke L1 is a pulsating DC. The impedance of this choke coil is high enough that the *pulsations* of plate current cannot pass through the coil. Instead most of them flow onto the left plate of coupling capacitor C3, driving electrons away from its right plate and downward through resistor R4. This situation, depicted in Fig. 6-2, is called the second half-cycle of operation.

During those periods *between* the pulses of plate current, this grid drive current for the next stage flows back upward through R4. The purpose of the tube stage is to *amplify* the input-signal voltage applied across resistor R1. This amplification cannot be

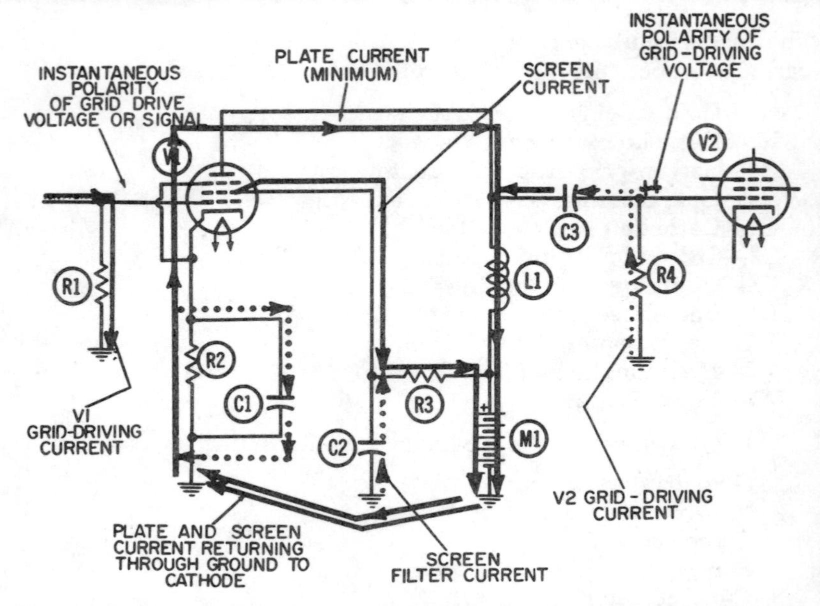

Fig. 6-1. Operation of the untuned impedance-coupled amplifier—first half-cycle.

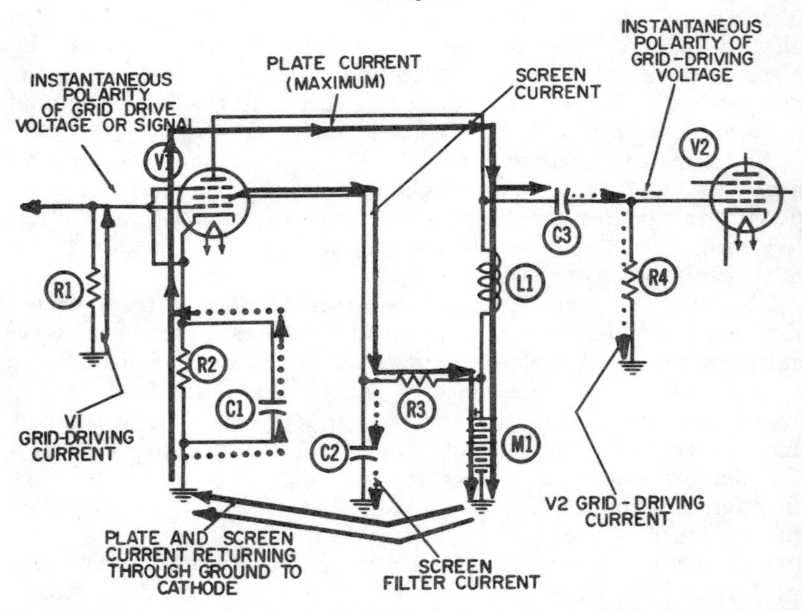

Fig. 6-2. Operation of the untuned impedance-coupled amplifier—second half-cycle.

accomplished unless the alternating voltage developed across R4 is significantly larger than the voltage across R1. If the voltage amplification (gain) achieved by this amplifier is equal to 50, then 50 times more alternating voltage will be developed across R4 than across R1. Should the two resistors have the same ohmic value, then the current flowing through R4 would also be 50 times greater than the current through R1.

The filtering of the pulses out of the plate current at the cathode, and out of the screen current by means of capacitor C2, are routine actions. Cathode filtering has already been adequately described. Screen-grid filtering is of course required only with tetrodes or pentodes, which have the extra grid. On the second half-cycle (Fig. 6-2), when the number of electrons in the plate-current stream increases, the number exiting from the tube via the screen grid will increase accordingly. If filter capacitor C2 were missing, this increase in screen current would cause a corresponding increase in the voltage drop across screen resistor R3. As a result, the voltage at the screen would be lowered during these positive half-cycles.

During the negative half-cycles (Fig. 6-1), when the negative control grid reduces the number of electrons in the plate-current stream, the number exiting from the tube at the screen grid will be reduced accordingly. Without filter capacitor C2, this reduction in screen-grid current would cause a corresponding *decrease* in the voltage drop across resistor R3, and a resultant *rise* in the positive voltage applied to the screen grid.

These voltage changes at the screen grid are undesirable because they constitute a form of degeneration which reduces the available amplification from the circuit as a whole. The reason is that the screen-grid voltage has a definite effect on the amount of plate current through the tube. With this type of degeneration present, the voltages at the screen and control grids will be working at cross-purposes: As the control grid tries to *increase* the plate current, the screen grid will act to reduce it, and vice versa.

The resulting loss in amplification can be avoided by the addition of C2. This filter capacitor absorbs the pulses of screen current during the positive half-cycles. The filter current (shown in large dots is now driven from the lower plate of C2 to ground and permitted to return upward on the negative half-cycles.

With the audio circuits discussed in earlier chapters, filter connections were usually made directly to ground. When radio-frequency currents are being handled, cathode and screen filter capacitors should normally be connected to the lower end of the cathode resistor rather than directly to ground. Once the radio-frequency currents are permitted to enter the common ground (usually the chassis), they could be coupled through it to some other point and introduce undesirable feedback there.

As is true with all vacuum-tube circuits, the plate and screen currents must eventually return to the cathode after passing through the load and power supply M1. This return path has been indicated beneath the diagrams.

As mentioned previously, a circuit using an untuned impedance as a load lacks the high selectivity of circuits with tuned loads (discussed later). It is this feature which gives rise to the use of this circuit, for there are applications where radio-frequency amplifiers in series can become *too* selective and pass too narrow a band of frequencies to carry the desired intelligence, or modulation. By using an occasional untuned impedance load like this one, the selectivity is broadened somewhat, but at the expense of gain.

TUNED IMPEDANCE-COUPLED AMPLIFIER

This pentode radio-frequency amplifier has a tuned circuit connected to the grid and another one in the plate circuit. These circuits are normally tuned to be resonant at the same radio frequency.

Figs. 6-3 and 6-4 depict the two half-cycles of the tuned impedance-coupled RF voltage amplifier. The various functions of this circuit, and the manner in which they are accomplished, will be discussed in detail in this section. The circuit components include:

R1—Grid resistor.
R2—Cathode biasing resistor.
R3—Screen-grid dropping resistor.
R4—Grid resistor for next tube.
C1—Grid tank capacitor.
C2—Grid coupling capacitor.
C3—Cathode filter capacitor.
C4—Screen-grid filter capacitor.
C5—Plate tank capacitor.
C6—Grid coupling capacitor for next tube.
L1—Grid tank coil.
L2—Plate tank coil.
V1—Pentode tube.
M1—Power supply.

There are three different groups, or "families," of currents at work in this circuit during normal operation. The first of these includes the radio-frequency alternating currents in the two tank circuits, together with their output currents and voltages. These have been shown a thin line in the grid circuit, and as solid in plate circuit. The second family is made up of the unidirectional currents, shown in solid red, which flow primarily in one direction only; their paths are through the vacuum tube, and also through the power supply.

The third family consists of filtering currents (both are shown in figures as dotted lines) flowing into and out of capacitors C3 and C4.

All these currents will have radio-frequency components—it should not be inferred that the ones labeled "radio-frequency alternating currents" are the only currents with RF components.

Fig. 6-3 shows the current flow and voltage polarities during the first half-cycle. It is labeled a "negative half-cycle" because the grid driving voltage becomes progressively more negative during this half-cycle, until at the end it reaches its most negative value.

Fig. 6-4 shows the current flow and voltage polarities during the second, "positive," half-cycle. During this half-cycle, the grid-driving voltage becomes progressively more positive and at the end reaches its most positive value.

Fig. 6-5 shows the significant current and voltage waveforms which occur during circuit operation. Vertical lines have been drawn on the diagram to separate the half-cycles.

RADIO-FREQUENCY ALTERNATING CURRENTS

Fig. 6-3, which depicts current conditions during an entire half-cycle of radio frequency and the voltage conditions at the *end* of the half-cycle, shows electron current (in thin line) flowing upward through grid tank inductor L1 and charging the upper plate of grid tank capacitor C1 with electrons. The negative voltage created at this point causes electrons to flow onto one plate of coupling capacitor C2. An equal number are driven off the opposite plate and flow downward through grid resistor R1. This electron flow is also shown as a thin line.

The grid will have its maximum negative voltage at the same time this current is flowing downward through the grid resistor at its maximum rate. It will always help, in determining whether a grid is negative or positive, to remember that electrons always flow *from* negative *to* positive. Since they are flowing downward through the resistor, toward ground, this tells us the ground voltage is more positive than the voltage at the top of the resistor, which is connected to the grid.

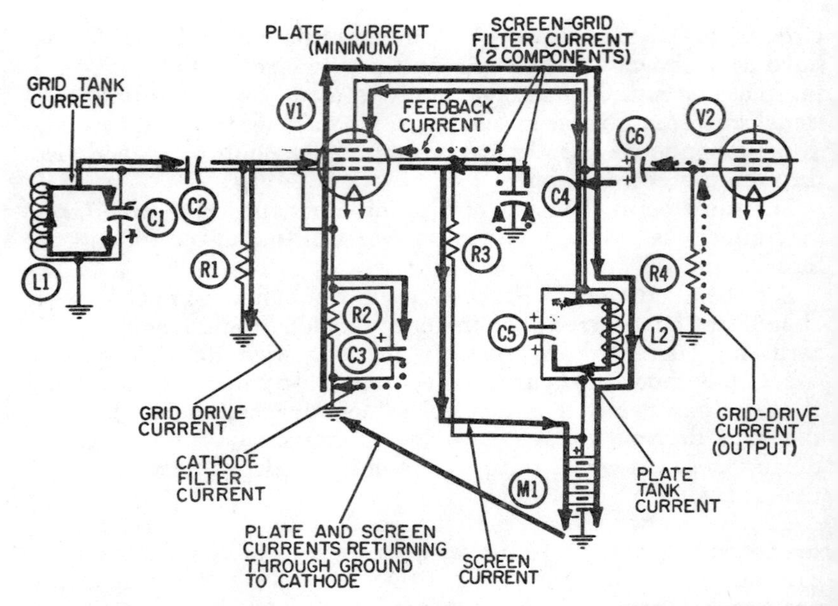

Fig. 6-3. Operation of the tuned impedance-coupled amplifier—first half-cycle.

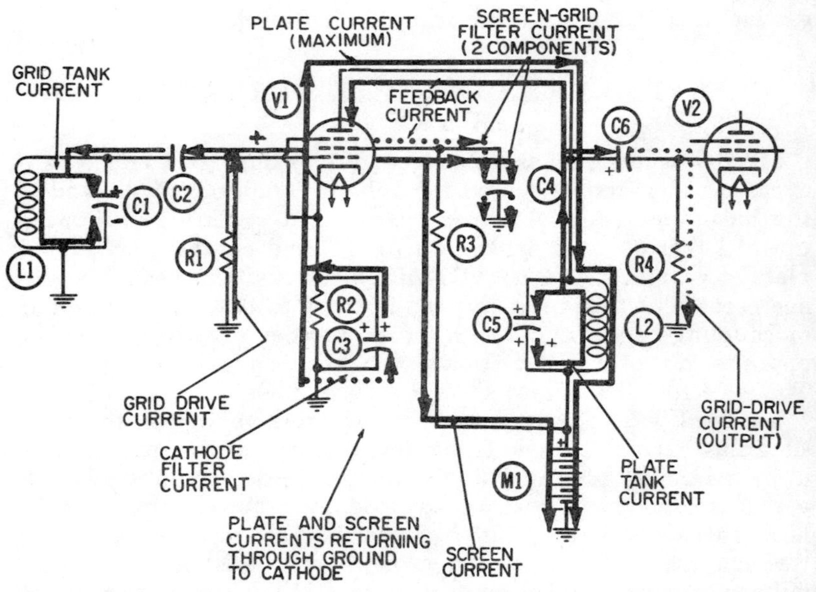

Fig. 6-4. Operation of the tuned impedance-coupled amplifier—second half-cycle.

Special attention should now be given to the waveforms for this circuit, in order that phase relationships between the various currents and voltages can be understood.

Grid tank current flows from the bottom plate to the top plate of capacitor C1 during the entire negative half-cycle and stops at the end of the half-cycle. In the waveform representing grid tank current in Fig. 6-5, that portion of the waveform *below* the center, or reference, line represents electrons flowing *upward* through tank inductor L1. Conversely, that portion *above* the reference line represents electron current flowing *downward* through L1.

At those points where the current waveform actually *crosses* the reference line—at the beginning and end of each half-cycle—*zero* current in flowing. That is, electrons have stopped flowing in one direction, and have not yet started in the opposite direction.

In Fig. 6-5 we see that at the end of the negative half-cycle, when the upward electron flow through the tank inductor has been completed, the grid tank voltage shown in line 2 has reached its *negative* peak value. The reason is that the tank current has delivered the maximum concentration of electrons to the upper plate of capacitor C1.

Line 3 of Fig. 6-5 shows the direction and amount of the grid drive current, which is caused to flow by the tank voltage of line 2 and is "in phase" with this voltage. The grid drive current, in Fig. 6-3, flows downward through grid resistor R1. This current is also known as the tank circuit's "external current." As long as there is a negative charge, meaning negative voltage, on the upper plate of C1, this electron current will flow *away* from the negative charge (or downward through the grid resistor). When there is a positive charge, meaning positive voltage, on the upper plate of the tank capacitor, these electrons will be drawn *toward* this positive voltage and consequently, *upward* through the grid resistor.

The voltage developed *across* any resistor is always in phase with and proportional to the current *through* it.

Thus, the voltage developed across resistor R1 (the grid driving voltage) will have its maximum negative value at the end of the negative half-cycle, when the tank's external current reaches its maximum downward flow. At the end of the positive half-cycle, when the external current achieves its maximum upward flow, the grid driving voltage will attain its maximum positive value. Thus, the tank voltage and the grid driving voltage will be in phase with each other at all times.

V1 is being operated under what are called Class-A conditions. That is, the tube conducts electrons during the entire cycle, as shown by the plate current waveform (line 4) of Fig. 6-5.

Another RF alternating current is shown in the plate tank circuit—this time a solid line. This plate tank current (which is oscillating continuously between capacitor C5 and inductor L2) consists of considerably *more* electrons than does the grid tank current. The plate tank current is actually an "amplified" version of the grid tank current.

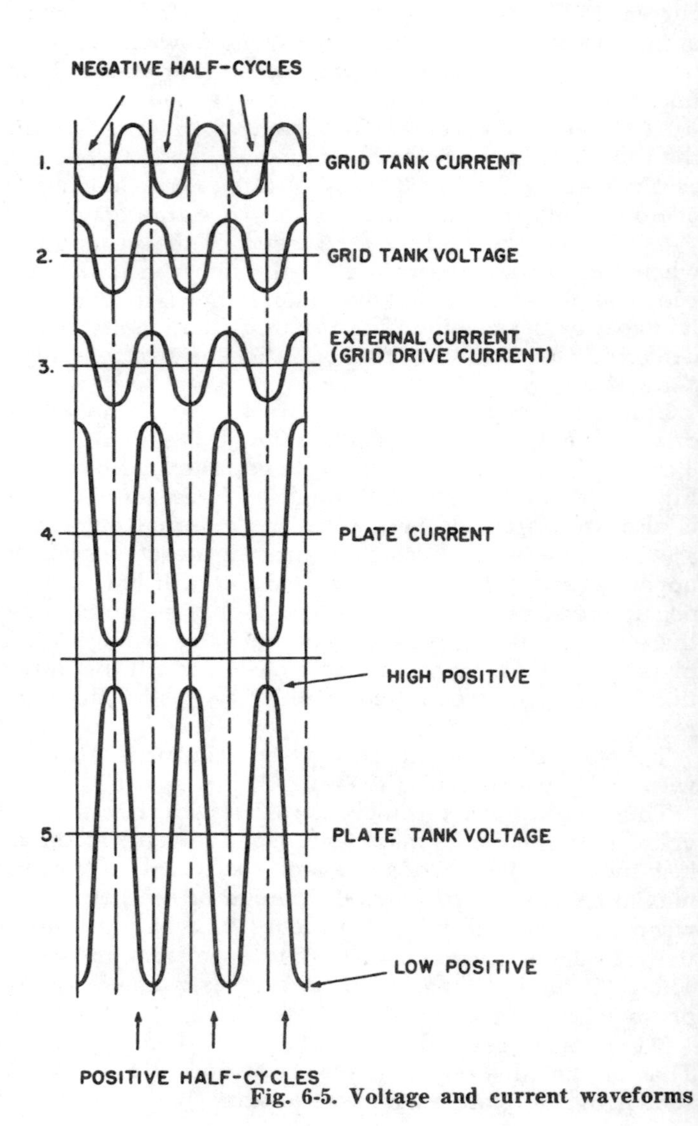

Fig. 6-5. Voltage and current waveforms

The plate tank voltage, which is measured at the top of coil L2 and capacitor C5, is also an amplified version of the grid tank voltage.

This plate tank current supports two other currents, both shown as a solid line. One is the output current flowing in and out of capacitor C6. The other is the feedback current, which flows

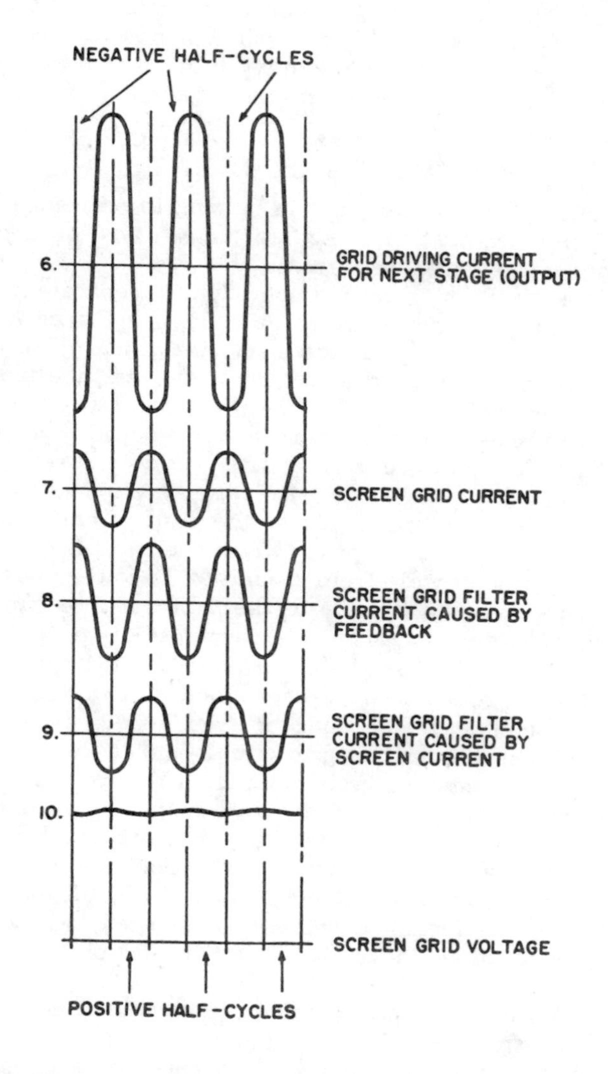

in the tuned impedance-coupled amplifier.

between the tank and the plate of the tube. Both act as "loads" on the electrons which are oscillating in the tank. During the last portion of the negative half-cycle and the first portion of the positive half-cycle, when the plate tank voltage is most positive, electrons are being drawn from the left plate of C6 and from the plate of the vacuum tube. This action draws electrons upward through grid-resistor R4 of the next tube and onto the right plate of C6. This flow of electrons (shown in small dots), is the grid-driving current for the next tube. It will produce a positive voltage, with respect to ground, at the top of R4.

FILTERING CURRENTS

The function of the screen-grid, in helping to prevent a circuit from going into self-sustained oscillations, can now be explained by reference to the feedback current.

The feedback current, shown a solid line, is separate from the unidirectional plate current, shown a solid line During the last portion of the first half-cycle, the electrons which make up the feedback current are being drawn *away* from the tube plate and *toward* tank circuit inductor L2. Their movement away from the tube plate draws electrons *toward* the screen grid from its external circuit. A filtering current (large dots) therefore flows in the screen grid circuit. During the latter portion of the negative half-cycle and the first portion of the positive half-cycle, this filtering current is flowing upward from the top plate of C4 to the screen grid, and upward from ground to the lower plate of C4.

The screen grid and the plate of the tube act like two plates of a capacitor. The feedback current, which is driven by the oscillating plate tank voltage, in turn drives the filtering current along its path to ground through capacitor C4. If this screen grid were not in the tube, the feedback current would drive an unwanted current in the control grid circuit and the entire circuit could go into self-sustained oscillation. Both the control grid and the screen grid have interelectrode capacitance with respect to the plate of the tube. However, the screen grid is much closer to the plate, and therefore has much more capacitance to the plate, than does the control grid. Consequently, most of the energy represented by the feedback current is diverted harmlessly to ground, through filter capacitor C4 by means of the filtering current.

Additionally the wire mesh, which makes up the screen grid, shields the control-grid wires from the plate of the tube and thus greatly reduces the amount of interelectrode capacitance between control grid and plate.

C4 must do an additional filtering job—it must bypass to ground the fluctuations in the unidirectional screen grid current. The electron stream which flows through the vacuum tube consists of the part which goes to the plate and its external circuit, and the part which exists at the screen grid. The number of electrons in this stream is being continually varied by the changing voltage on the control grid. During the positive half-cycle (represented by Fig. 6-4), the control grid is made positive and a heavy surge of current goes through the tube. That portion which exists at the screen grid finds two alternate paths—it can flow directly into the screen-grid resistor, which is a comparatively high impedance; or it can flow momentarily onto the upper plate of capacitor C4. A size of capacitor is chosen that will act as a comparatively low reactance (impedance) at the particular frequency. Consequently, the surge of electrons in the screen-grid current during the positive half-cycle flows into C4, driving an equal number of electrons from the lower plate of C4 as shown by the dotted lines. A half-cycle later, during the conditions depicted by Fig. 6-3, the screen-grid current is reduced considerably, and the excess electrons resting on C4 from the previous half-cycle will now be drawn off through the screen-grid resistor R3 and to the power supply. At the same time, electrons will flow upward from ground to the lower plate of C4. These electrons, shown in large dots are the filtering component for the unidirectional screen-grid current. They of course are independent of the filtering current for the feedback energy previously described, even though they flow along the same path.

Resistor R2 and capacitor C3 function as an RC filter in much the same fashion as R3 and C4. All the unidirectional current which passes through V1 must first flow upward through R2 to reach the cathode. Since this current is constantly fluctuating (because of the voltage variations at the control grid), the amount of current through the resistor would also fluctuate if it were not for filter capacitor C3. If the current were allowed to fluctuate the voltage across R2 would also fluctuate, causing a normally undesirable condition known as *degeneration.*

During the period depicted by Fig. 6-3, the control-grid voltage has been made negative and the flow of electrons through the tube is thus restricted. The cathode is being held at a positive voltage by the deficiency of electrons (indicated by the plus sign) on the upper plate of C3. This positive voltage will continue to attract electrons upward through resistor R2, and if the negative control grid prevents them from entering the tube, they will flow directly onto the upper plate of C3, as indicated by the arrow. This action would *tend* to reduce the positive voltage in storage on the capacitor, and such reduction would be equivalent to a half-cycle of degeneration.

Fig. 6-4 indicates that the positive control grid increases the amount of electron current through the tube. This increased demand for electrons to leave the cathode may be supplied either through resistor R2 or capacitor C3. A size of capacitor is chosen that will have a reactance of a very few ohms at the operating frequency. Consequently, a sudden surge of electrons will be drawn into the tube from the upper plate of C3. This action will *tend* to increase the positive voltage on C3 (and at the cathode of the tube) and thereby constitute another half-cycle of degeneration.

The size of a capacitor (meaning its amount of capacitance) is paramount in determining whether such degeneration will occur, or whether it will be held within acceptable limits if it does. As long as the "pool" of positive ions on the upper plate of capacitor C3 is overwhelmingly large, in comparison with the number of electrons coming into or going out of capacitor C3 on successive half-cycles, then the "pool" voltage will not change significantly from one half-cycle to another and degeneration will not occur.

The cathode filtering current is shown in large dots between the lower plate of C3 and ground. It is moving downward during the first half-cycle, being driven by the influx of electrons onto the top plate of C3. During the second half-cycle, as electrons surge away from the upper plate and into the tube, the filtering current is drawn upward from ground and toward the lower plate. The free movement of this filtering current is essential to the filter function. If something prevented this current from flowing (say, a broken connection), then the fluctuations of current onto and away from the top plate of C3 (shown a solid line) cannot occur. The electron flow through resistor R2 will then vary in unison with the variations in electron flow through the tube, and the resulting voltage across the resistor will also vary. In other words, we would have degeneration.

The most obvious effect of degeneration is a decrease in the over-all amplification achieved by the circuit. This amplification is a comparison between the input and output voltages. The amount of output voltage in this figure is proportional to the number of electrons (shown a solid line) oscillating in the plate tank circuit. The size of any plate tank oscillation will depend on the size, or strength, of the current or voltage which replenishes it each cycle. This oscillation is replenished by the *variations in amount* of unidirectional current which reaches the tank from the tube.

Fig. 6-3 shows a negative voltage on the control grid, with the resulting *decrease* in the number of electrons through the tube. If degeneration were occurring, this decrease in tube current through cathode resistor R2 would also decrease the positive voltage at the cathode. The *difference* in voltage between cathode and grid would not then be as great as desired, and the *reduction* in tube current from its normal value would not be as great as desired.

Fig. 6-4 shows a positive voltage on the control grid, with the resulting *increase* in the number of electrons going through the tube. If degeneration were occurring, this increase in tube current through the cathode resistor would increase the positive voltage at the cathode. The difference in voltage between control grid and cathode again would not be as great as desired, and the *increase* in tube current flowing would not be as great as desired.

The plate tank oscillation is sustained by the surges in the unidirectional plate current. The size, or strength, of each surge is determined by the difference in the number of electrons flowing through the tube at these two moments (namely, at the grid-voltage negative and positive peaks). Larger surges of electrons will sustain larger oscillations. As the surges become smaller, so will the oscillations. When degeneration is occurring, the minimum current through the tube will not be low enough and the maximum current will not be great enough. As a result, their *differences* will be reduced and the plate tank oscillation will be smaller accordingly. The output voltage, which is directly proportional to the size of this oscillation, will also be reduced accordingly.

The total path of the unidirectional plate current (solid line) is a closed circuit which can be considered to originate and terminate at the ground connection below the cathode resistor. Normal vacuum-tube action includes a heated cathode which will emit electrons into the tube, and a positive voltage on the plate (anode) to attract the electrons across the tube. Once across the tube, they continue flowing toward the point of highest positive voltage —the point known as B+, or the power supply. Plate current flows through the power supply, entering at its positive terminal and exiting at its negative terminal, or ground. Once this current is returned to a common ground, the condition of having a closed path for the plate current has been met.

When a varying grid voltage imposes fluctuations in the quantity of the current, as is done in this example, these fluctuations are available for sustaining the oscillation in the plate tank circuit.

The size of the current fluctuation determines the size of the tank oscillation which will be supported, each individual fluctuation replenishing and thus reinforcing an individual cycle of oscillation.

The ability of a vacuum tube to act as an amplifier stems from the fact that small changes in control-grid voltage will cause much larger changes in the amount of current which flows through the tube (plate current). These large fluctuations in plate current can then give the oscillations much greater voltage swings than they had in the grid circuit.

The second member of the "unidirectional tube-current family" is the current which exits from the tube via the screen-grid electrode (also shown a solid line). Going by the name of *screen-grid current,* it flows (in unison with the unidirectional plate current) upward through the cathode resistor and into the tube via the cathode. Because the screen grid is maintained at a high positive voltage, a substantial portion of the electrons passing through the tube are attracted to it. After striking the screen-grid wires, these electrons exit from the tube and are drawn through screen-grid resistor R3, toward the point of highest positive voltage in the circuit (the entrance to the power supply). Their closed path is through the power supply and back to common ground.

If there were no fluctuations in the amount of over-all tube current, there would be no significant fluctuations in the amount of screen-grid current. (Such a condition is referred to as static operation of the tube.) The plate tank oscillation would not be sustained under these conditions. However, when the signal voltage is impressed at the grid circuit, the electron stream through the tube does fluctuate in quantity and the tube is said to be operating under dynamic conditions. The control-grid wires, through which this stream must pass, act like a control valve on this stream. A positive voltage at the grid turns the stream "up," by increasing the number of electrons that flow. This condition is depicted by Fig. 6-4, where plate current is maximum. A less positive (more negative) voltage at the control grid turns this stream "down," by reducing the number of electrons passing through the tube. This condition is depicted by Fig. 6-3, where plate current is minimum.

The manner in which capacitor C4 filters out these fluctuations in screen-grid current was described in the discussion on filtering currents, and will not be repeated here. The RC combination of R3 and C4 must be a "long time-constant" combination to the radio frequency in question, and the combination of R3 and C4 operates in much the same fashion as R2 and C3 in the cathode circuit. The filtering action assures that a steady rather than a fluctuating flow of electrons occurs in screen-grid resistor R3. If this current were allowed to fluctuate, then the voltage across R3 would also fluctuate and vary the voltage at the screen grid accordingly, whereas the latter voltage should be constant. In other words, the screen-grid voltage would decrease whenever the tube current increased and vice versa. Thus, degeneration would be occurring, and the end result would be the same as when degeneration occurs in the cathode circuit—namely, a decrease in amplification.

This circuit is called a *voltage amplifier*. We are seeking to amplify the driving voltage applied to the control grid of the amplifier tube. This is the voltage developed across grid resistor R1 by the flow of grid driving current through R1. This grid drive voltage has the same value and phase as the grid tank voltage shown in line 2 of Fig. 6-5.

The plate tank voltage, represented by the plus signs on the upper plate of capacitor C5 in Fig. 6-3 and 6-4, alternates at the same frequency as the grid tank voltage, but its amplitude is much larger. The voltage value is, of course, directly related to the number of electrons in oscillation in the plate tank circuit. For these reasons, it is an "amplified" version of the grid tank and grid driving voltages.

Chapter 7

TRANSFORMER-COUPLED RF VOLTAGE AMPLIFIERS

There are three methods of using transformer coupling to amplify voltages at radio frequencies. They are:

Untuned primary—tuned secondary.
Tuned primary—untuned secondary.
Tuned primary—tuned secondary (frequently called bandpass amplification).

In the discussions that follow, the operation of these circuits will be explained in detail.

UNTUNED-PRIMARY–TUNED-SECONDARY COUPLING

Between the first two variations given in the foregoing, the untuned primary—tuned secondary offers more advantages and hence enjoys wider usage. Figs. 7-1 and 7-2 show two successive half-cycles of such a circuit. Any tuned transformer coupling will introduce the problem of *coupled impedance*. The principles of coupled impedance apply equally to the first two examples given previously. In other words, it matters not which winding is tuned and which is not, as far as explaining the principles is concerned.

The elements for the circuit in Figs. 7-1 and 7-2 include:

R1—Grid driving and grid return resistor for V1.
R2—Cathode biasing resistor for V1.
R3—Cathode biasing resistor for V2.
C1—Cathode filter capacitor for V1.
C2—Tuning capacitor for secondary winding.
C3—Cathode filter capacitor for V2.
T1—Radio-frequency transformer.

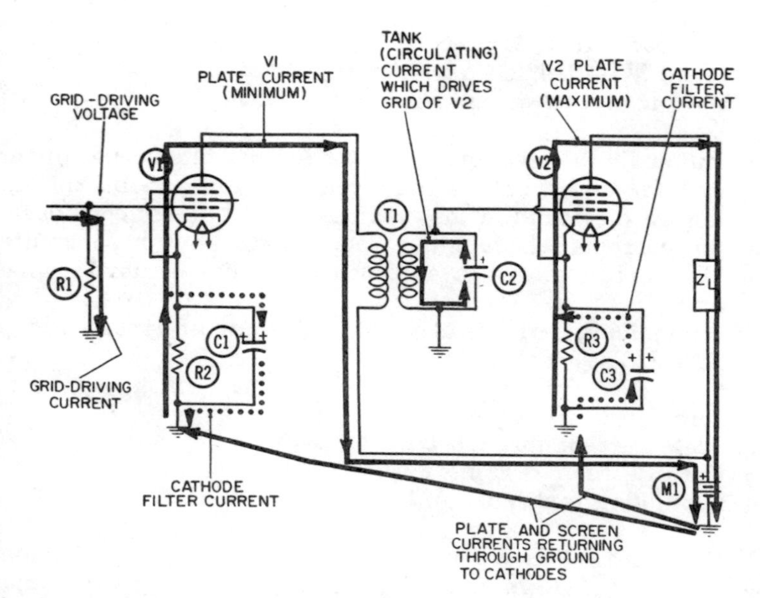

Fig. 7-1. Operation of the untuned-primary—tuned-secondary RF amplifier—first half-cycle.

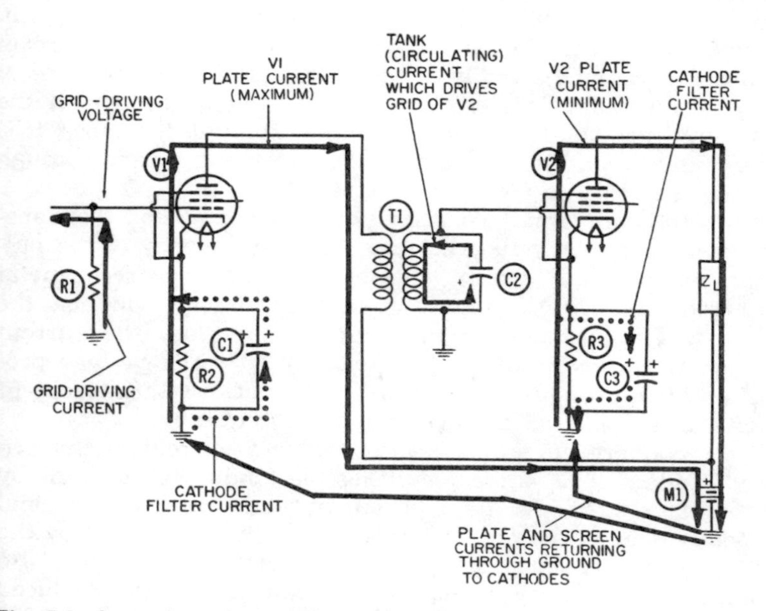

Fig. 7-2. Operation of the untuned-primary—tuned-secondary RF amplifier—second half-cycle.

Z_L—Load circuit for second stage.
V1 and V2—Pentode amplifier tubes.
M1—Common power supply for both stages.

Additionally, a screen-grid resistor and its filter capacitor are required for each pentode tube. Since screen-grid biasing and filtering were covered in the previous chapter, these components and the currents which flow through them have been omitted from the circuit diagrams and also from the discussion which follows.

The remaining currents at work in this circuit appear in Figs. 7-1 and 7-2. They include:

1. Initial grid driving current or signal.
2. Plate current through both tubes.
3. Cathode filter currents for both tubes.
4. Oscillating tank current.

Notice, both tubes for this circuit are served by a common power supply, M1. In actual practice, a single power supply normally supplies the B+ voltage to all tubes in the radio or television set. A B+ line is run through the set and every plate and screen grid requiring high positive voltage is tapped into it. If we could look inside the B+ line, we would see a vast stream of electrons moving steadily toward the high-voltage terminal of the power supply. This electron stream is composed of all the plate currents and all the screen currents from all the tubes. It is drawn into the positive terminal of the power supply and through the power supply to common ground.

When the tank circuit, consisting of capacitor C2 and the transformer secondary winding, is tuned to the frequency of the pulsations in the plate current, the tank is said to be resonant at that frequency. In flowing through the primary winding, the pulses will induce a secondary current that lags the current through the primary winding by exactly 90°. The voltage produced by the secondary current will be in phase with the current when the tank is resonant at the applied frequency.

The secondary current will also induce a current in the primary winding. This current will lag the "inducing" current by another 90°; so it will be 180° out of phase with the original applied voltage in the primary. Also, this current induced by the secondary into the primary will have a companion voltage, also induced in the primary winding. These two, the primary induced current and voltage, will be in phase with each other but 180° out of phase with the original applied voltage and current.

Currents which are out of phase *oppose* each other and subtract from each other. Voltages do likewise.

In the example of an untuned primary winding and a tuned secondary winding that is exactly resonant at the frequency of applied current and voltage, the coupling between the two windings *reduces* the strength of the original applied current and voltage. In turn, the induced secondary current and voltage suffer the same fate.

Mathematically, the reduction in applied current and voltage in the primary is the result of *coupled impedance*—in other words, the impedance from the secondary is coupled back to the primary.

This sequence can be understood by relating the two currents and voltages to the tank current *circulating* in the tuned tank, and also to the tank voltage associated with this tank current. We are in reality talking about *four* currents and four voltages on the two sides of the transformer.

Fig. 7-3 shows four successive quarter-cycles in this type of coupling circuit. The four currents are identified as follows:

1. Primary plate current.
2. Secondary induced current.
3. Secondary tank current
4. Primary induced current.

The four voltage polarities are:

"Applied" voltage in primary.
Induced voltage in secondary.
Secondary tank voltage.
Primary induced voltage (also known as counter emf).

By inspection of the four parts of Fig. 7-3, we see that at the end of each quarter-cycle, the applied plate current in the primary winding goes through its characteristic motions as follows:

1. It is *decreasing* at its maximum rate (Fig. 7-3A).
2. It has its *minimum* value (Fig. 7-3B).
3. It is increasing at its maximum rate (Fig. 7-3C).
4. It achieves its maximum value (Fig. 7-3D).

We know that whenever pulsating DC (like this one) flows through one winding of a transformer, a current will be induced in the secondary winding, and this current will lag the primary current by 90°. We can observe this phenomenon in Fig. 4-11. The primary current (plate current) reaches its minimum value at the end of the second quarter-cycle, but the induced voltage does not reach its maximum value until the end of the third quarter-cycle.

In Fig. 7-3D the plate current has reached its maximum value, whereas the voltage induced in the secondary winding does not reach its peak until 90° later, or at the end of the first quarter-cycle shown in Fig. 7-3A.

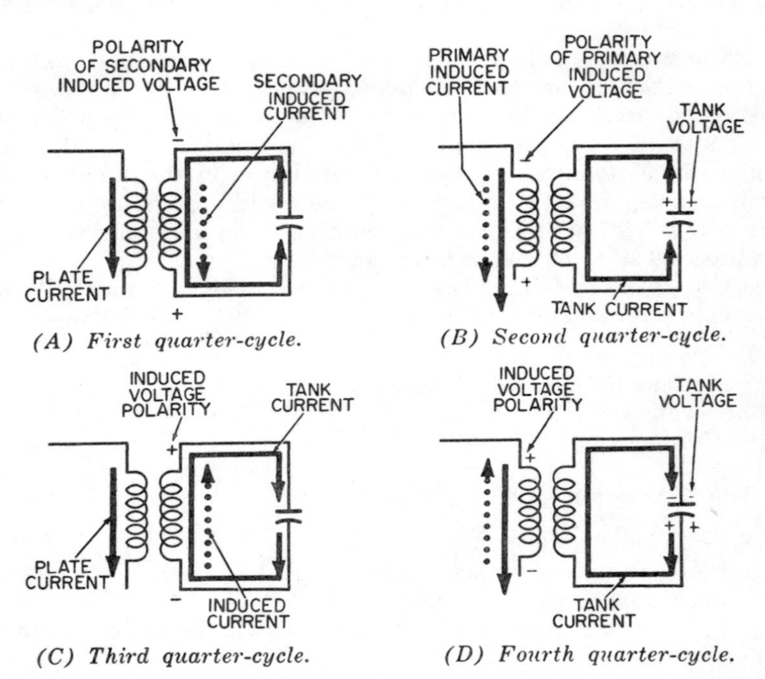

(A) First quarter-cycle. *(B) Second quarter-cycle.*

(C) Third quarter-cycle. *(D) Fourth quarter-cycle.*

Fig. 7-3. The four quarter-cycles of operation of the tuned secondary circuit at resonance.

On the other hand, the induced secondary voltage and current are in phase with each other. This means they achieve their peak values at the same instant as indicated in Figs. 7-3A and C.

It is worth taking the time to understand *why* the induced voltage and current within the secondary winding are in phase with each other. One of our initial assumptions was that the tank circuit was tuned so it would be resonant at the frequency of the applied current. Part of the definition of resonance includes the stipulation that the inductive reactance and the capacitive reactance must be equal in value but opposite in polarity, so that they cancel each other. This means they "add up to zero" and the only remaining impedance is the insignificant resistance in the wire of the coil. Consequently, the induced current 'sees" only a low-resistance path within the tank and the current and voltage will be in phase.

Although the tank circuit, consisting of capacitor C2 and the secondary winding of transformer T1 in Fig. 7-1 and 7-2, is an example of *parallel* resonance as far as delivering an output voltage to the grid is concerned, the capacitor and secondary winding are in *series* to the induced current. This anomaly occurs frequently and can easily confuse you. However, the true criterion in determining whether components are in series or in parallel with each other is not the manner in which they are connected, but the path they present to certain currents. The tank circuit in this example is parallel-resonant to the external circuit to which it is connected, and series-resonant to currents flowing within the tank (the induced and circulating currents.) This point will be discussed further in the later example involving both a tuned primary and secondary winding.

We see from inspection of Fig. 7-3 that the induced current in the secondary winding is flowing in phase with the circulating current. The circulating current (also called the tank current) is in reality an *oscillation* of electrons sustained by the induced current. With a tuned circuit having a reasonably high value of Q, a fairly substantial oscillation can be built up from a fairly small amount of sustaining energy.

The tank voltage and current are of course 90° out of phase with each other. A *downward* current peak is achieved at the end of the first quarter-cycle in Fig. 7-3A, but another quarter-cycle is required before this current flow can deliver enough electrons to the lower plate of capacitor C2 to create a voltage peak. There will be a negative polarity on the bottom plate of C2, and a positive polarity (deficiency of electrons) on the upper plate. By this time the induced current has come to a halt. Likewise, an *upward* peak of electron current is achieved through the secondary winding at the end of the third quarter-cycle; but another quarter-cycle is required before this current flow can deliver enough electrons to the upper plate of C2 to create a negative voltage peak here.

It is a characteristic of a coupling transformer that its coupling works both ways—the primary current will induce a current and a companion voltage in the secondary winding; at the same time, the induced secondary current will itself induce a second current and companion voltage in the primary winding. In our example (Fig. 7-1 and 7-2) the pentode has a "plate resistance" of several hundred thousand ohms, which is very high. The total impedance of the primary circuit is essentially resistive, so the current and voltage induced in the primary winding will be i phase with each other. This means they will reach their pea maximum and minimum values at the same moment. These in duced current and voltage peaks occur at the end of the secon and fourth quarter-cycles. This happens to be 90°, or a quarte of a cycle, later than the peak values of the *inducing* curren (the tank current). Thus, the voltage reinduced in the primar winding is essentially 180°, or half a cycle, out of phase with th original (applied) voltage and thus reduces the effectiveness o the applied voltage. Accordingly, less current and voltage wi be induced in the secondary winding than would be possible the reinduced voltage did not exist.

Fig. 7-4 graphically illustrates the currents and voltages o Fig. 7-3. The currents shown above their reference lines ar flowing upward through the coil, except for the plate curren which always flows downward as a pulsating DC.

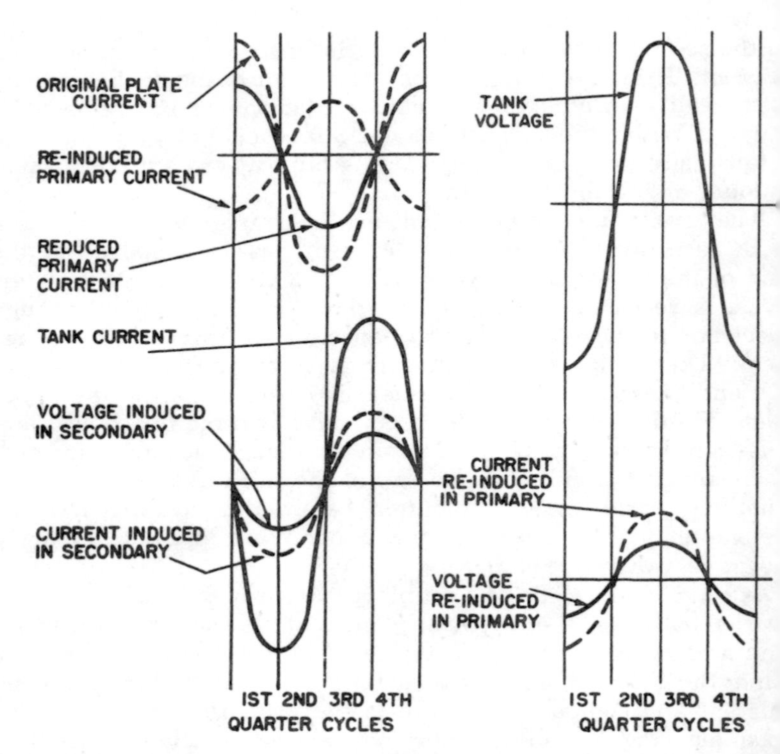

Fig. 7-4. Voltage and current waveforms in the tuned secondary circu at resonance.

The total current flowing in the primary winding is of course the sum of the primary plate current, plus the current induced in this winding by the circulating tank current. Since they are out of phase, the primary plate current is reduced by the amount of the induced current. It is convenient to think of this reduction in primary current as being caused by an additional impedance "coupled" into the primary winding from the secondary circuit, because impedances always change the amount of current flowing in a circuit (usually they *reduce* it—but not always, as we shall see in the following example).

The amount of impedance coupled from one winding to the other, and the amounts of current and voltage induced in one winding by current in the other one, can be precisely calculated. However, these calculations will not be elaborated upon here. Instead, we will take the opportunity to grasp some of the important terminology of coupled circuits—the degree of coupling, tight or loose coupling, coefficient of coupling, mutual coupling, and coupled impedance.

In an iron-core transformer like the one in audio-output or low-frequency power circuits, practically all the magnetic lines of force set up by a current through one winding will flow through the iron core and encircle or "link" the other winding. Under these circumstances, the degree, or *coefficient* of coupling between the two windings is practically unity.

Higher frequencies force us to use very small values of inductance, because inductive reactance increases as the frequency does. A coil designed for radio frequencies will consist of only a few turns of wire wound around an air core, and its inductance will be 100 to 200 microhenrys. Two such coils constitute a radio-frequency transformer, and the coefficient of coupling between coils (represented by the symbol k) will be extremely low—perhaps one or two per cent. With an air-core transformer, it is almost impossible to avoid a great amount of leakage flux (the term applied to magnetic lines of force which are set up by a current through one winding and which do not link the other winding).

The mutual inductance between two coils, L1 and L2, and their coefficient of coupling, k, are related to each other by the following formula:

$$M = k\sqrt{L1 \times L2}$$

where,

M is the mutual inductance in henrys,
k is the coefficient of coupling,
L1 is the primary inductance in henrys,
L2 is the secondary inductance in henrys.

As an example, if we assumed a perfect iron-core transformer, the coefficient of coupling would be unity, or 1. If we assumed also that L1 is equal to L2, then the mutual inductance, M, would be found by:

$$M = 1 \times \sqrt{L1 \times L1}$$
$$= L1.$$

In other words, whenever two windings have the same inductance, and the coefficient of coupling is unity, the mutual inductance between them is equal to the inductance of either winding.

It is re-emphasized that this is an unusual example and that two assumptions had to be made:

1. The tank circuit must be tuned to the frequency of the applied current, so that the induced secondary voltage and current will be in phase with each other, and
2. The total resistance of the primary winding and its circuit must be much greater than the inductive reactance of the primary, so that the current and voltage induced in the primary will be in phase with each other, and also that the applied primary plate current and applied primary voltage will be in phase with each other.

Either of these assumptions is easily invalidated by changing components—and hence, conditions—in the circuit. A slight change in frequency of the applied current would have the same result. Or the use of a triode tube (which has a lower plate resistance than a pentode does) would invalidate the second assumption.

TUNED-PRIMARY–TUNED-SECONDARY COUPLING

Figs. 7-5 and 7-6 show two successive half-cycles in the operation of a radio-frequency voltage amplifier that has a tuned primary circuit on the plate (input) side and a tuned secondary circuit on the output side leading to the next grid. This is the second of the two most widely-used adaptations of transformer coupling at radio frequencies, the first being the untuned-primary—tuned-secondary one just considered.

A pentode tube again is used. It is not impossible to use specially-designed triodes at radio frequencies, but the pentode has several natural characteristics which make it much more suitable at radio frequencies.

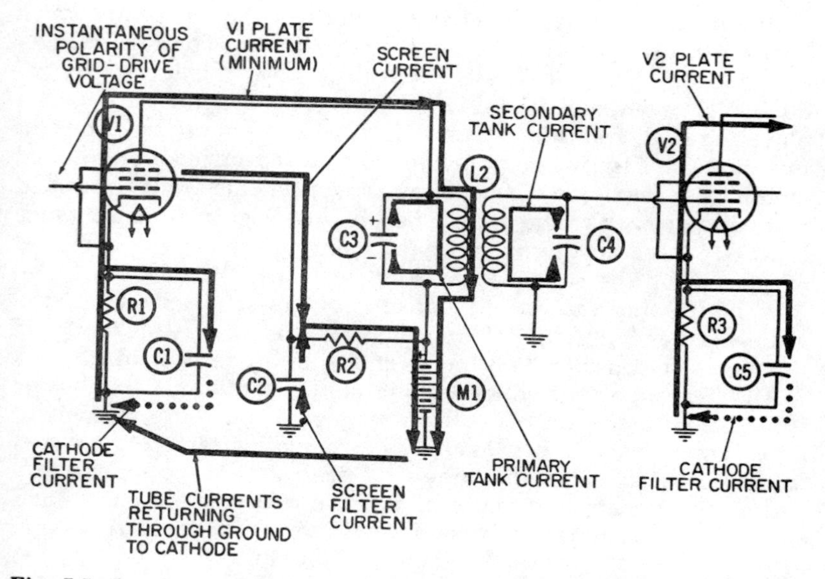

Fig. 7-5. Operation of the tuned-primary—tuned-secondary RF amplifier —first half-cycle at resonance.

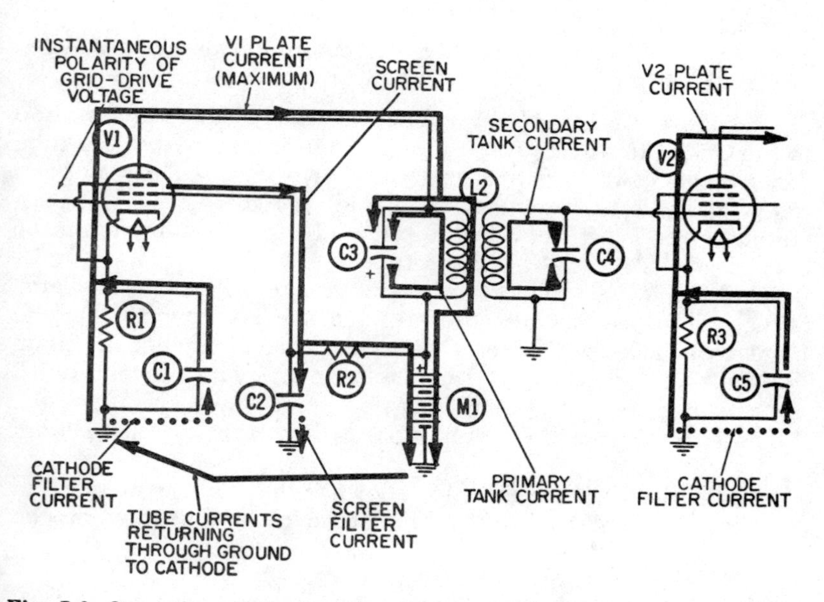

Fig. 7-6. Operation of the tuned-primary—tuned-secondary RF amplifier —second half-cycle at resonance.

One of these important characteristics is the *shielding* which the screen grid provides between the plate and control grid. As explained in Chapter 1, such shielding does much toward reducing or even eliminating interelectrode capacitance between plate and grid, and thus prevents feedback of energy from output to input. In Fig. 6-3 and 6-4 for the tuned-impedance-coupled amplifier, a feedback current was shown flowing from the tuned plate circuit back to the plate. Then by interelectrode capacitance it was coupled to the screen grid, from where it was bypassed to ground by the screen-grid filter capacitor.

Although this undesired feedback current is not shown in the radio-frequency circuits of this chapter, it is always present, and it is a major consideration in the choice of a pentode tube.

Another important characteristic of the pentode is its shorter *transit time*. An electron will always require a finite amount of time, however small, to cross over to the plate after being emitted from the cathode. To the tube, this transit time becomes more and more of a limiting factor as the applied frequency rises and each cycle becomes correspondingly shorter. Here is a simple formula which relates the frequency to the duration of each cycle:

$$f = \frac{1}{T}$$

where,

f is the frequency of applied current in cycles per second,
T is the duration of an individual cycle in seconds.

If a half-cycle of applied voltage is nearly equal in duration to the transit time, electrons will still be in transit within the tube (as a result of the positive grid voltage) after the advent of the next half-cycle has now made the grid voltage negative. The resultant phase shift between output and input usually limits or even destroys the utility of the tube.

A third and very important characteristic of the pentode tube is its high plate resistance. Recall from the first chapter that the plate resistance is the *ratio* between a small change in plate voltage and the resultant small change in plate current when the grid voltage is maintained constant.

The several electron currents at work in this circuit include:

1. Grid driving current (not shown), the voltage polarity of which is negative on the first half-cycle and positive on the second.
2. Plate current through both pentodes.
3. Screen-grid current.
4. Cathode and screen-grid filter currents.

5. Plate tank current.
6. Grid tank current for next stage.

Both of the amplifier stages indicated in Figs. 7-5 and 7-6 are using cathode biasing. Since the existence of a cathode biasing voltage depends on the continuing flow of plate current through the tube, we have come to identify cathode biasing with Class-A operation of the tube—meaning that some plate current flows at all times throughout the entire cycle.

The details of grid driving, cathode biasing and filtering, screen-grid filtering, etc., have been covered in many prior examples and will not be rediscussed here. The following discussion will be oriented around the phase relationships between the plate current, plate tank current, and grid tank current for the next stage. Once these currents and their phase relationships can be visualized for the condition of resonance and critical coupling, it becomes relatively simple to understand the small but important variations from this special case. These variations arise first from small changes in the signal or carrier frequency; as a result, the tuned circuits operate either above or below the natural frequency of the circuit. The second important variation arises from the degree of coupling.

The current diagrams of Fig. 7-3 can be adapted to the two tuned circuits (plate and grid tanks) of Figs. 7-5 and 7-6, but only where the latter circuits are being operated *exactly* at resonance. The five individual currents which should be identified separately would include:

1. Plate current.
2. Circulating current in plate tank. This is the primary *inducing* current, which function was performed by the primary plate current.
3. Secondary induced current.
4. Circulating current in grid tank.
5. Current reinduced in the primary by the circulating current in the grid tank.

The circulating current in the plate tank and the secondary induced currents are identical, as they were in the untuned example, and for the same reason. The circulating, or oscillating, current is in reality sustained or built up by this continual inducing of a current in the secondary winding.

The circulating current in the grid tank and the current reinduced in the primary by this current are exactly *out of phase* with each other. As a result, the reinduced current opposes the

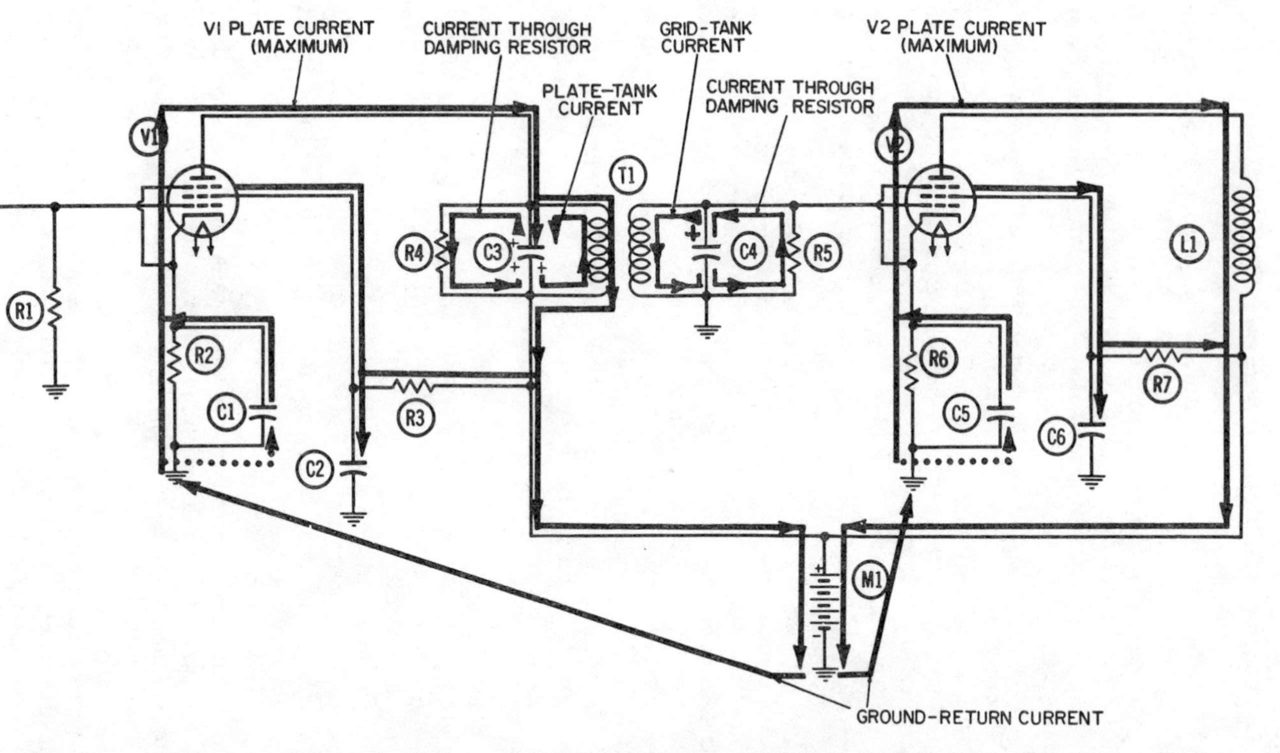

Fig. 7-7. Operation of the bandpass RF amplifier—first half-cycle.

primary tank current and thereby reduces the amount of current circulating there. This is similar to the untuned example discussed previously, where the current reinduced in the primary winding was opposite in phase to the applied plate current through this winding.

In this tuned example, the plate current supports or replenishes the oscillation of electrons in the plate tank, whereas the re-induced current opposes this oscillation. Obviously, the plate current is strong enough to overcome this opposition and support the circulating currents in the two tuned tanks.

As in the untuned example, the reduction in primary tank current is due to the impedance coupled back from the secondary to the primary winding. When the opposing current is out of phase with the primary circulating current (as in this resonant example), the coupled impedance acts like a pure resistance inserted within the tuned circuit with a size that would reduce the primary current the same amount as it is in fact reduced by the out-of-phase induced current. If the flow directions of the tuned tank currents and induced currents are related to the polarities of the tank voltages and induced voltages, we will find that at resonance the two tank votlages will be 90° out of phase with each other.

THE BANDPASS AMPLIFIER

Figs. 7-7 and 7-8 show two half-cycles in the operation of a radio-frequency voltage amplifier being used as a bandpass amplifier. This type of circuit differs from the preceding example of a tuned-primary—tuned-secondary circuit only in the addition of the two resistors across the two tuned tank circuits. Their purpose is to *reduce* the response of the particular tank circuit at resonance. Fig. 7-9 shows a typical response curve for a highly selective, or high-*Q*, resonant circuit. It is obvious that the closer the current being amplified is to the resonant frequency of the tuned circuit, the larger the gain (response).

Should we desire to amplify only a single frequency, the amplifier circuit could be tuned to that frequency and maximum amplification or gain would be realized. More frequently than not, however, we face the problem of amplifying not a single frequency, but a *band* of frequencies. The reason being that radio frequencies are usually carriers of intelligence superimposed at the transmitter by one of several processes of modulation. Whenever a carrier signal is modulated by an audio signal, new radio frequencies—known as side frequencies or *sidebands*—are introduced.

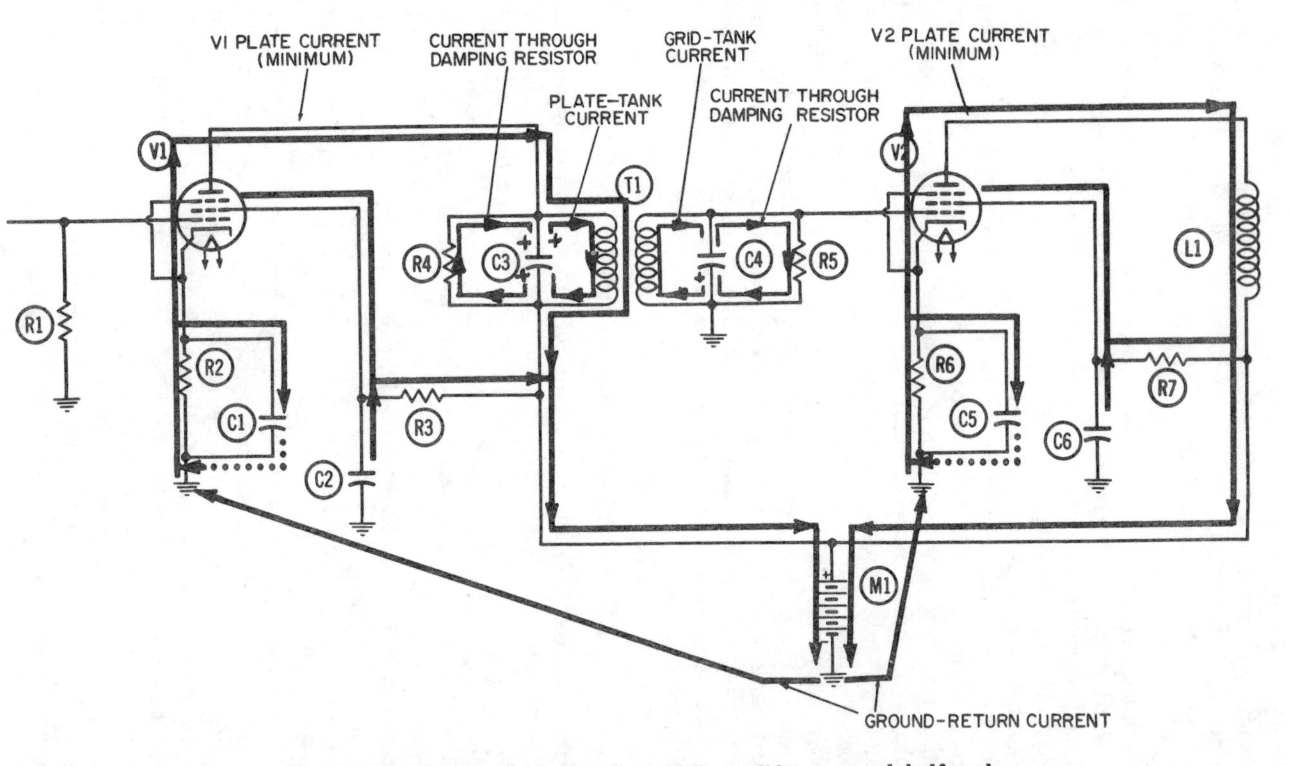

Fig. 7-8. Operation of the bandpass RF amplifier—second half-cycle.

As a simple example of what this means, assume we desire to amplitude-modulate a 1,000-kilocycle carrier signal with a 5,000-cycle audio signal. The radio-frequency signal would then consist of these three components:

1. The original carrier of 1,000 kc.
2. The carrier *plus* the modulating frequency, or 1,005 kc.
3. The carrier *minus* the modulating frequency, or 995 kc.

The 1,005-kilocycle signal is known as the *upper* sideband, and the 995-kilocycle signal is the *lower* sideband. To be acceptable, an amplifier must provide fairly equal amplification of all three. It must "pass" a band of frequencies 10 kilocycles wide and centered at the 1,000-kilocycle point. An "ideal" response curve would look like a rectangle standing on end as shown by the dashed line in Fig. 7-9. Its response would be equal or flat over the 10-kilocycle band, and it would not respond whatsoever to any frequencies above or below this band. A more practical and realizable response curve is shown as a broader curve superimposed on the ideal curve. It is characterized by a rounded rather than a flat top, indicating reasonably uniform application over the passband. It also has sloping rather than vertical sides, indicating an acceptable (if not complete) attenuation of any frequencies above or below the passband.

Use of Damping Resistor

If a resistor is placed across a tuned tank, the tank voltage will be prevented from building up as high as it would otherwise. The resistor acts as a "load" on the oscillation of electrons—any unbalance in voltage across the tank capacitor will begin redistributing itself through the resistor. From Fig. 7-7, we see electron current flowing from the upper plate of tank capacitor C3, to the lower plate, through resistor R4. This direction of flow will persist as long as the positive voltage on the lower plate of C3 is higher than the positive voltage on the upper plate. Also, in Fig. 7-7 we see electron current flowing out of the bottom plate of grid tank capacitor C4 and upward through resistor R5 to the top plate. This flow will persist as long as the voltage is more negative on the lower than on the upper plate.

In both tanks in Fig. 7-7, the tank currents flow *onto* the respective plate of each capacitor, trying to build up a high peak voltage; at the same time, electrons are being drained off through the resistors. It is this continual drainage that prevents the achievement of high peak voltages and sharp response curves at resonance. A resistor of high ohmic value drains off only a small amount of current and so constitutes a "light" load on the tank circuit. Conversely, a low-value resistor will drain off much current and constitute a "heavy" load.

In Fig. 7-8, the polarities of the two tank voltages have been reversed, and so have the directions of electron flow through the two bandpass resistors. The tank-voltage polarities, tank-current directions, and flow directions of the resistor currents will be discussed later for each quarter-cycle.

Critical Coupling

Two tuned circuits are said to be "critically" coupled whenever they give a single response peak at their resonant frequency. Either one of the response curves in Fig. 7-9 could result from critically coupled tuned circuits. However, a damping resistor such as R4 or R5 (Figs. 7-7 and 7-8) is not normally used with critical coupling. When it is desired to pass a band of frequencies, the tuned circuits would be "overcoupled." Fig. 7-10 shows three response curves for three degrees of overcoupling. Contributing to the width of these double-humped response curves are the size of the resistors across the tuned circuits, their *Q* (quality), and finally the degree of overcoupling between them. Overcoupling is the opposite of "loose" coupling.

The value of critical coupling can be determined by the formula:

$$K = \frac{1}{\sqrt{Q_1 \times Q_2}}$$

where,

K is the value of critical coupling,
Q_1 is the *Q* of the plate tank circuit,
Q_2 is the *Q* of the grid-tank circuit.

As an example, if each tuned circuit has a *Q* of 100 (which is not unreasonable), the value of critical coupling, K, would be .01 and a single-peaked response curve similar to those in Fig. 7-9 could be achieved.

Overcoupling

If the two inductor windings are wound closer together, or if the coupling between them is increased by adjustment of a variable core, the coefficient of coupling will be greater than .01 and the circuits will be overcoupled—the objective in the design of bandpass amplifiers. The double humps in the response curves of Fig. 7-10 are the result of resonances occurring at frequencies on either side of the natural frequency. Fig. 7-11 has been drawn to represent four successive quarter-cycles in the operation of the two tank circuits of Fig. 7-7, *when these circuits are being operated above the resonant frequency.*

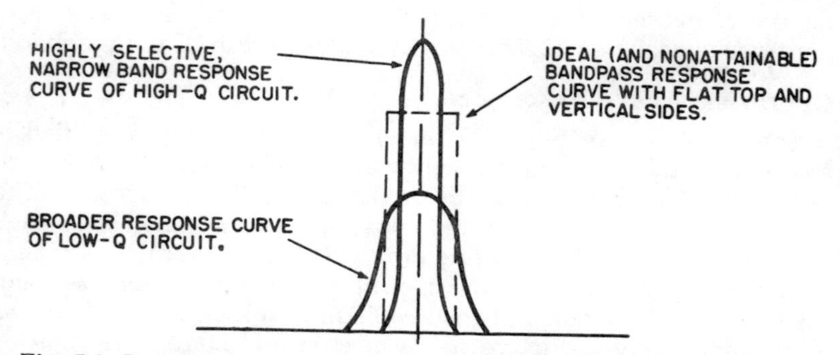

Fig. 7-9. Response curves of high- and low-Q circuits at resonance.

It was previously stated that when two tuned circuits are coupled together at their resonant frequency, the two resulting tank voltages are 90°, or a quarter of a cycle, out of phase with each other. However, as soon as the frequency being amplified deviates even slightly from the resonant frequency of the tanks, the phase difference will increase to a maximum of 180°, or half a cycle. With very high-Q circuits, this 180° phase difference between the two tank voltages will occur fairly close to the resonant frequency. The inside curve of Fig. 7-10, with its two resonant peaks fairly close together, represents the response of a high-Q circuit and the outer curve, with its resonant peaks far apart, the response of a lower-Q circuit.

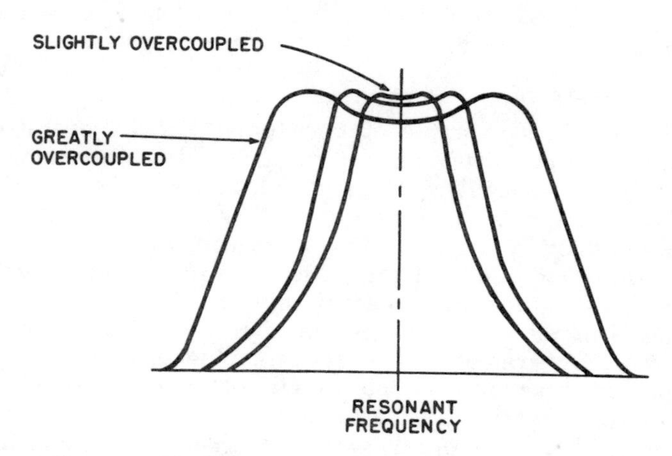

Fig. 7-10. Response curves of over-coupled tuned circuits.

Thus we see that either the Q of the tank circuits, or their degree of overcoupling, can have an effect on where these resonant peaks occur and consequently on the width of the passband.

Operating Above Resonance

Let us assume in Fig. 7-11 that the current being amplified is *higher* than the resonant frequency of the tank circuits. Then each tank circuit will appear to be *inductive* to these currents. The reason is that the inductor and capacitor are in series with each other (as far as the circulating current is concerned), and whenever the frequency rises, the inductive reactance *increases* and the capacitive reactance *decreases*. Either tank circuit will appear to be completely inductive to its circulating current as soon as the internal resistance of the tank becomes negligible (compared with the inductive reactance). Since capacitors and connecting wires have minute amounts of resistance, most of the internal resistance of a tank is due to the resistance of the wire composing the coil.

In a high-Q coil this resistance is of course very small, as illustrated by the formula for the Q of a coil:

$$Q = \frac{\omega L}{R}$$

where,

ω equals 2π times the frequency of operation (called omega),
L is the inductance of the coil in henrys,
R is the resistance of the coil in ohms.

The quantity, ωL, is also the inductive reactance of the coil at the frequency of operation.

Phase Relationships

In the previous example of two tuned circuits operated at resonance, the primary tank current induced a secondary tank current. Along with this secondary tank current was an induced secondary voltage, and the two were *in phase* with each other, since the inductive and capacitive reactances had equal but opposite values and canceled each other. So the series impedance of the secondary tank circuit was purely resistive.

In the example we are considering here, when the secondary tank circuit is operated at some frequency higher than resonance, the capacitive reactance is opposite but *not equal* to the inductive reactance. Only part of the latter is canceled, and the circuit is inductive rather than purely resistive. Therefore, the primary tank current still induces a secondary tank voltage which lags the primary tank current by 90°. However, instead of being in phase with the induced secondary voltage, the secondary induced current *lags it by another 90°*, because in an inductive circuit the current always lags the voltage by 90°.

During the first quarter-cycle, in Fig. 7-11A, the primary tank current is flowing *upward* through its inductor at its maximum rate. However, the induced secondary current is flowing *downward* through *its* inductor at its maximum rate. This denotes that the two are 180°, or half a cycle, out of phase. Both currents, the inducing and the induced, are shown in blue so they can be more easily related to each other.

The secondary tank current, shown in green, is an oscillation of electrons supported by the secondary induced current (in fact, these two currents are normally indistinguishable). Like all electron oscillations in tuned L-C circuits, energy is stored alternately in magnetic and in electric fields. The magnetic field is maximum when the tank current reaches maximum, at the end of the first and third quarter-cycles. The electric field is maximum when the voltage across the tank capacitor is maximum, at the end of the second and fourth quarter-cycles.

The secondary tank current also reinduces a new voltage in the primary circuit, which lags the secondary current by 90°. The primary circuit is inductive for the same reason the secondary is. When the operating frequency is higher than the natural frequency of the circuit, all capacitive reactance is canceled but some inductive reactance is left over. When this inductive reactance becomes significantly larger than the winding resistance the primary tank circuit will appear as a pure inductance to the induced primary current. This current will thus lag the induced primary voltage by another 90°, and consequently be 180° (or half a cycle) behind the phase of the original tank current in the primary circuit.

Two events which are a whole cycle out of phase with each other are actually "in phase," so that the primary induced current adds to and thus strengthens the primary tank current. Thus, the response curve is peaked at a frequency higher than the resonant frequency and the primary tank appears to be operating at resonance, when we know this cannot be true.

In each of the four quarter-cycles shown in Fig. 7-11 we see an induced primary current flowing in phase with the primary tank current. There is also an induced secondary current flowing in phase with the secondary tank current.

Operating Below Resonance

The diagrams of Fig. 7-11 cannot be applied equally well to two tuned circuits operating *below* their resonant frequency.

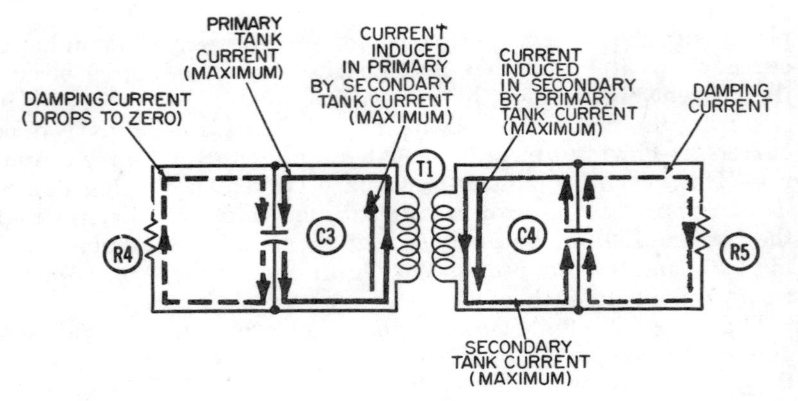

(A) First quarter-cycle.

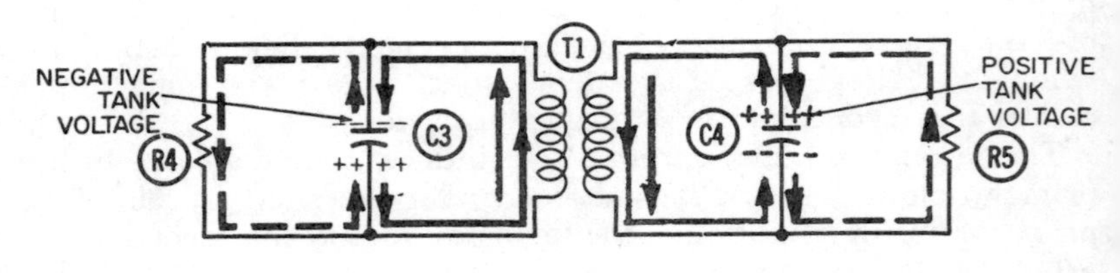

(B) Second quarter-cycle.

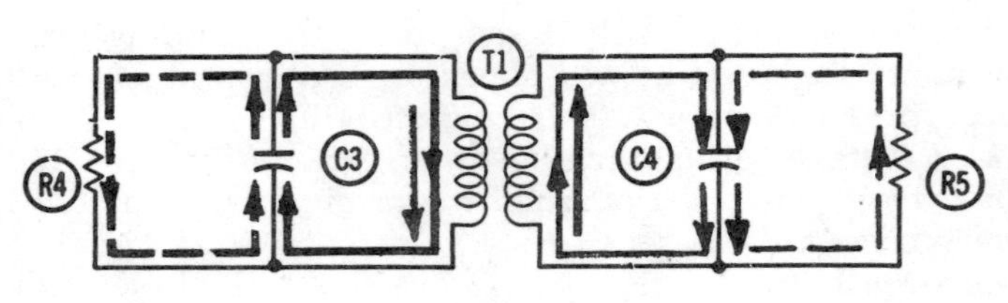

(C) Third quarter-cycle.

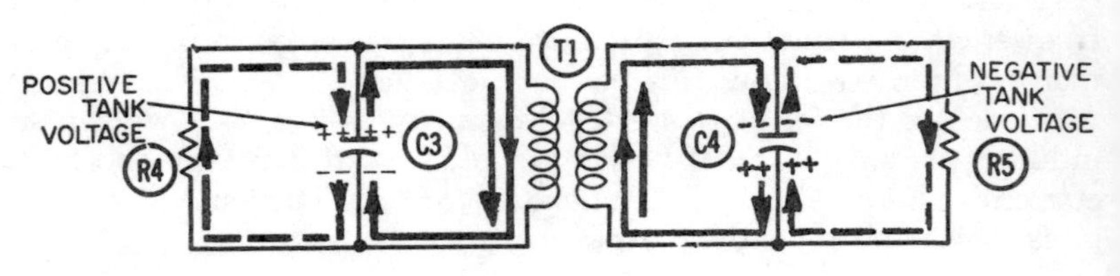

(D) Fourth quarter-cycle.

Fig. 7-11. The four quarter-cycles of operation of the bandpass tank circuit at a frequency above resonance.

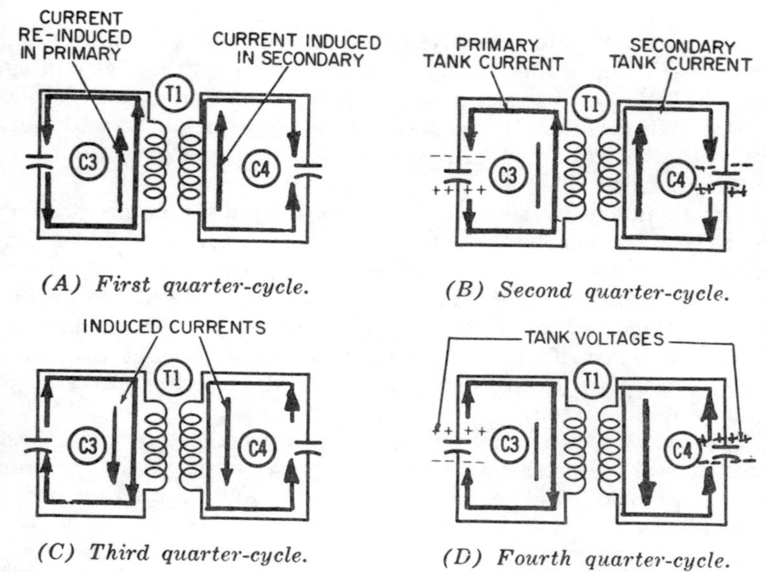

(A) First quarter-cycle.

(B) Second quarter-cycle.

(C) Third quarter-cycle.

(D) Fourth quarter-cycle.

Fig. 7-12. The four quarter-cycles of operation of the bandpass tank circuit at a frequency below resonance.

When this happens, each of the inductive reactances of the two windings will decrease and the capacitive reactances of the two tank capacitors will increase, in accordance with the standard reactance formulas. To a current circulating within either tank circuit, the over-all impedance of its series path will be capacitive, since all inductive reactance in each path will be canceled but some capacitive reactance will be left over. Now it becomes interesting to analyze the phase relationships between the two tank currents and their voltages, and between the two induced voltages and their currents.

In Fig. 7-12A, the primary tank current is flowing *upward* through the primary winding and again induces a voltage in the secondary winding which lags the primary current by 90°. Therefore, the secondary voltage is zero at the end of the first quarter-cycle. This induced voltage will have an induced current flowing with it; but since this current is flowing in a circuit that is capacitive, it will *lead the voltage.* This means maximum current occurs 90°, or a quarter of cycle, in advance of the maximum voltage or exactly in phase with the inducing current. In the four quarter-cycles of Fig. 7-12 we see the induced secondary current, shown above , flowing in phase with the inducing primary current, also shown above . As in the previous example, the induced secondary current supports the oscillation of electrons in the second tank and these two currents are indistinguishable. Both the primary and the secondary tank currents cause tank-voltage peaks at the end of the second and fourth quarter-cycles. Unlike the previous example, these tank voltages are in phase with each other—that is, their positive and negative peaks each occur simultaneously.

The secondary current also induces a voltage in the primary which lags the secondary current by a quarter of a cycle. The induced primary current associated with this induced voltage *leads* the voltage by a quarter of a cycle, because of the capacitive reactance exhibited by the primary circuit when operated below resonance. Thus, the induced primary current turns out to be exactly in phase with the original inducing current, or primary tank current.

Consequently, when operated below resonance, both induced currents will reinforce their respective oscillating tank currents, causing them to build up in a manner similar to a current build-up at resonance.

The damping resistors of Fig. 7-11 have been omitted from Fig. 7-12 for clarity. However, they would be required in the normal bandpass amplifier, whether it is being operated above or below resonance. During the first and third quarter-cycles, electrons flow through these resistors in unison with the electrons which flow through the respective inductors. During the entire first quarter-cycle, some excess electrons are amassed on the lower plate of capacitor C3, and they flow from this plate upward through the inductor and resistor. During the entire third quarter-cycle, some electrons are amassed on the upper plate of capacitor C3, and they flow from this plate downward through the resistor and inductor.

During the second and fourth quarter-cycles, the currents through the resistor and inductor do not appear to be flowing in unison with each other. During the second quarter-cycle, while the oscillating tank current is delivering electrons to the upper plate of C3 and trying to build up a tank-voltage peak, the resistor is letting electrons drain off. During the fourth quarter-cycle, a similar situation is occurring at the bottom plate of C3.

Resistor R5, bridged across the secondary tank circuit, acts in a similar manner to prevent the attainment of a high peak voltage across the secondary tank.

SERIES *VERSUS* PARALLEL IMPEDANCE

One of the hallmarks of a *parallel*-resonant tank circuit is the high impedance it presents to an external circuit. At the same time, the tank circuit presents an extremely low impedance to a current circulating within it. The reason is that the inductive and capacitive components are in *series resonance* to the circulating or oscillating current, so their respective reactances cancel each other.

When we say that a parallel-resonant tank circuit offers high impedance to an external circuit, we are referring to its opposition to the passage of an external current. A tuned tank in the plate circuit of an amplifier tube, as discussed earlier, offers a good example from which to explain the meaning of the term, *high parallel impedance*. Whenever any of these tuned plate tanks is operated at resonance, the *ratio* between tank voltage and line current (plate current) will reach its highest value. This ratio, called the tank impedance, is derived from the fundamental Ohm's-law relationship which states that the voltage *across* a resistor (or network) is proportional to the current through it.

In a pentode tube, as we tune a tank towards resonance the tank voltage rises, meaning a large number of electrons are in oscillation, whereas the plate current is virtually unchanged. This gives a high ratio of tank voltage to line current, and consequently a high resonant impedance. When the circuit is operating off-resonance, the amount of circulating or oscillating current is greatly decreased. This leads to a lowered tank voltage and tank impedance.

In a triode amplifier using a tuned plate tank, the amount of plate current will be greatly reduced when the tank is at resonance, because the negative (or low-positive) voltage peak at the plate occurs simultanously with the positive grid-voltage peak. The latter is trying to release maximum current through the tube, whereas the low plate voltage restricts the plate current. A simple operating test for resonance is to tune the tank slowly near the resonant frequency and observe the amount of plate or line current with an ammeter. The point where minimum plate current is achieved indicates the tank is resonant and its tank voltage is maximum. Obviously, the ratio between maximum tank voltage and minimum line current indicates the highest attainable value of tank impedance, or resonant impedance.

SECTION 3

RECTIFIER CIRCUITS

Chapter 1

INTRODUCTION

It is frequently said of rectifier power supplies and detector circuits that their functions are based on the peculiar ability of the vacuum tube to permit "unidirectional current flow"—in other words, the flow of electron current through a vacuum tube in only one direction. While this statement is intrinsically true, it requires considerable qualification, *since all other vacuum-tube circuits* also depend directly on this same tube characteristic. For instance, all circuits in the other volumes of this series also depend for their operation on the fact that the vacuum tubes used will pass electron current in only one direction.

RECTIFICATION

The simplest application of the fundamental principle of unidirectional current flow through a vacuum tube is illustrated in the rectifier power supply. For this reason, the first portion of this volume is devoted to such circuits. In order for any type of circuit to find broad usage, there must of course be a need for the function it performs. Rectifier power supplies are needed to convert alternating current to direct current because of two basic facts:

1. Most of the electric power available throughout the world today, both for home consumption and for industrial and commercial use, is in the form of alternating current.
2. The vast majority of vacuum-tube applications require that the tube be supplied with one or more sources of relatively pure DC (meaning direct current and hence a steady, or direct, voltage).

Alternating current is brought to practically every wall plug in every home, laboratory, and factory in the country. Consequently, every piece of electronic equipment requiring steady, or "DC," voltages for its operation must carry within itself the circuitry required to convert alternating current to direct current, and then be able to adequately filter, or smooth, the output so that it becomes a relatively "pure" DC voltage instead of a pulsating one. Many such circuits will consist of a single vacuum tube with an appropriate combination of the three fundamental components—resistors, capacitors, and inductors—to provide the required filtering action.

The block diagram in Fig. 1-1 shows the basic action of the rectifier circuit, if we can assume the rectifying and filtering functions as separate operations. Of course, in practice both occur together. In fact (as we shall see in later chapters), without filter capacitors, some rectifier circuits would not operate at all. In Fig. 1-1, the alternating AC voltage is applied to the rectifier. This is the 60-cycle AC from the wall plug. Notice that the voltage waveform extends an equal amount above and below the zero line.

The waveform at the rectifier output represents the voltage after rectification but before filtering. Here, the waveform extends only in the positive direction from the zero reference line. It still varies in amplitude, but always above—never below—the line. Likewise, the electron flow through the circuit will always be in the same direction but in varying amounts. Hence, the current and its associated voltage at the rectifier output are known as *pulsating DC*.

The waveform in Fig. 1-1 is for a full-wave rectifier. A half-wave rectifier will not conduct during half-cycles 2, 4, and 6, when the input waveform is negative. Hence, these half-cycles will not be present in the waveform at the rectifier output, and the voltage will be zero during these periods.

The filter circuit follows the rectifier in Fig. 1-1. As we said before, filtering and rectifying occur simultaneously. However, the two are shown in this order so the basic reason for each can be better understood. At the filter output, the waveform shows that the filter circuit has removed all pulsations in the applied

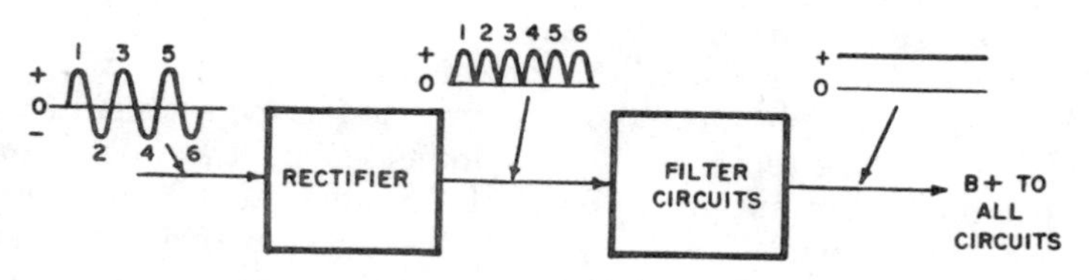

Fig. 1-1. Basic action of a rectifier circuit.

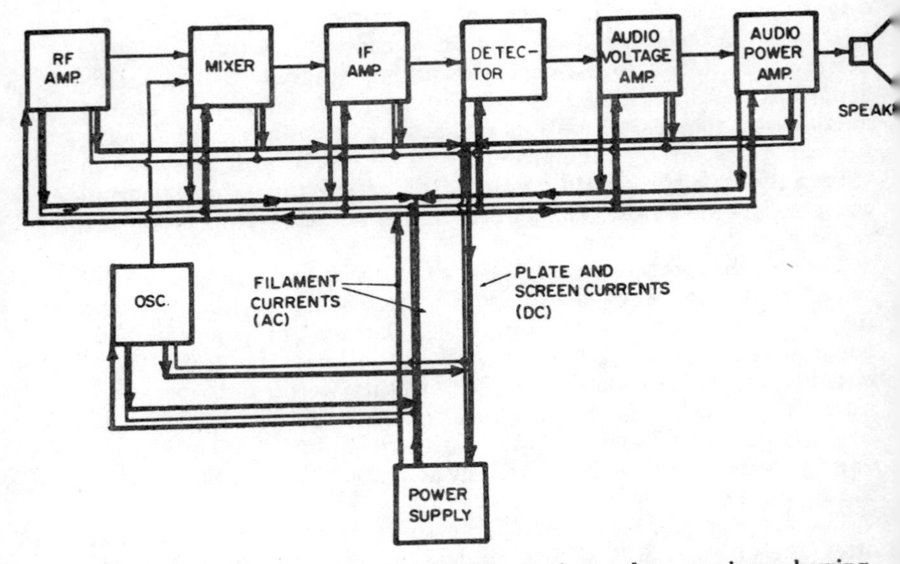

Fig. 1-2. Block diagram of a typical AM superheterodyne receiver, showing plate and filament currents.

waveform. This can be construed as a voltage of a constant positive amplitude, or as a direct current of constant magnitude.

Some ripple will always be present in the output waveform of a rectifier. For all practical purposes, however, the waveform can be considered a steady one, as shown in Fig. 1-1.

Fig. 1-2 shows a block diagram of a superheterodyne AM radio receiver. Each functional block designates a particular type of circuit necessary for complete operation of the receiver. The blocks labeled "detector" and "power supply" are the subject of this book. Because of certain operating similarities, these two circuits can be studied together. As indicated by Fig. 1-2, however, they perform quite different functions.

A detector, for instance, receives an input signal from the final RF or IF amplifier and delivers an output signal to the audio amplifiers. In doing so, the detector has detected (demodulated) the signal carrier. More about the detector circuit later; now we will limit our discussion to the power supply.

The power supply provides an essential service function to *all* circuits in the receiver. This is indicated by the lines which connect the power supply to each of the other fuctional lines in Fig. 1-2. As a general rule, every oscillator, converter or mixer, and amplifier circuit will require at least two separate services from the power-supply circuit, as follows:

1. A high positive voltage for application to the plates and screen grids of the amplifier and oscillator tubes.
2. A low alternating voltage for heating the filaments and/or cathodes, a process which is mandatory in order for electron emission to occur.

A detector circuit does not require the first of these two voltages, but does require the filament heating voltage.

Fig1-2 has been designed to show the flow of currents between these functional blocks and the power supply. The deep lines might be called the B+ current, which flows *from* the plates and screen grids of all amplifier, converter, and oscillator tubes *toward* the high positive-voltage source in the power supply. The various rectifier circuits discussed later make clear why these B+ currents, being made up of electrons, must always flow toward the power supply—never away from it.

The thin lines represent the filament heating currents required by the vast majority of vacuum tubes. These are alternating currents, as indicated by the fact that the arrows go in both directions. Sometimes these currents are obtained from a separate winding, of relatively few turns, on the power transformer. Often referred to as the tertiary winding, it supplies the necessary low voltage (usually 6.3 or 12.6 volts) to the filaments. The circuit diagrams for the full-wave rectifier power supply, and the associated text material (Chapter 3), constitute a full qualitative discussion of the operation of a typical low-voltage filament winding. In the vast majority of radio receivers and in many other pieces of electronic equipment, however, the teritary winding is not used. Instead, all filaments are connected in series; that is, the filament current must flow through each tube in succession before reaching the common ground point. Filament values are selected so that the voltage developed across all tubes in series is equal to the voltage of the power line. The tubes are connected directly across the power line. Of course, if the filament of one of the tubes in the series string should open, no filament current will flow through any of the tubes.

Since the plate and filament voltages are applied to the various stages through completely separate wiring, Fig. 1-2 indicates separate paths for the two types of currents associated with the two voltages. It has been emphasized repeatedly throughout this series that voltages do not *flow* between any two points. When it is desired to apply any particular voltage to a remote point, the natural action or mechanism by which this is accomplished is for an appropriate electron current to flow between the two points.

DETECTION

The term "detection" comes from the early days of radio, when circuits were devised to detect, or discover, the presence of a radio signal from a distant transmitting antenna. From these humble beginnings, the usage of the term "detector" has been broadened so that it is used almost interchangeably with "demodulator." Today the function of detection is almost indistinguishable from the function of demodulating a modulated signal.

The *need* for circuits to detect or demodulate the signal can be seen from Fig. 1-3. Two signals are shown at the transmitter input. The RF carrier is a signal of constant amplitude and frequency. (For the broadcast band, the frequency will be from 540 to 1,600 kilocycles per second.) Before this signal can be used, it must be modulated with the particular intelligence it is to carry. The term "modulate" in this sense means to change some characteristic of the RF carrier so that it will convey the intelligence (music, speech, etc.) to be broadcast. In Fig. 1-3, this intelligence is the audio signal shown entering the transmitter.

Fig. 1-3 shows the carrier being amplitude-modulated and the signal leaving the transmitter antenna. Notice that this signal still varies at the same rate as the RF carrier, but that the amplitude of each cycle varies. If the audio signal at the input were superimposed on either the upper or the lower tips of the mod-

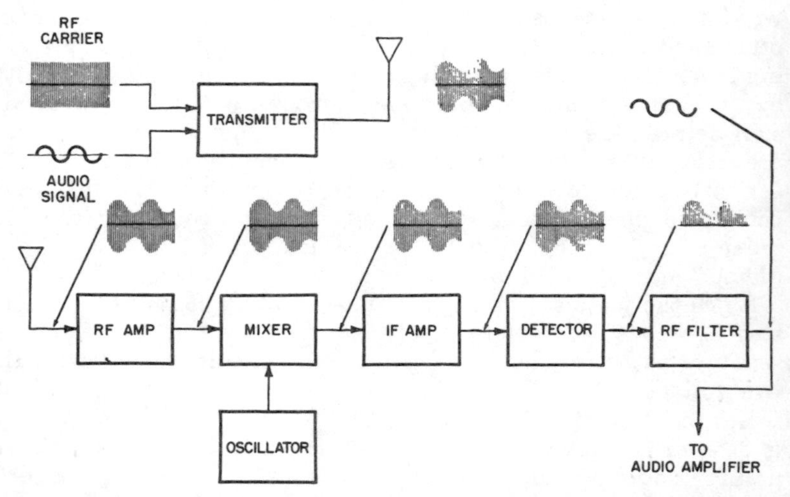

Fig. 1-3. The signal, from modulation at the transmitter to demodulation at the receiver.

ulated carrier, the variations in the modulated carrier would exactly match those of the audio signal. That is, whenever the audio signal went positive, the amplitude of the modulated signal would increase. Likewise, the amplitude would decrease whenever the audio signals were negative.

The modulated RF carrier is picked up by the receiver antenna and amplified by the RF amplifier. Next it is normally converted to a lower frequency (the IF) by the mixer-oscillator circuit. The IF signal, usually 455 kilocycles, is then further amplified by the IF amplifier. Except for its frequency and strength, the signal at the IF-amplifier output is the same as the transmitted signal—the modulations have been faithfully retained.

Now the signal is applied to the detector, which "separates" the audio signal from the RF (IF) carrier. Again, the detecting (rectifying) and filtering actions are shown separately, even though they occur together. The detector performs a function similar to that of the rectifier discussed previously. Notice that the signal at the detector input is still varying at the RF rate, but that only the positive half-cycles of voltage are shown. Thus, this signal is the same as the one applied to the detector, except that the bottom (negative) half has been removed. In other words we have a pulsating DC voltage like we had at the rectifier output. Only now, the pulsations are occurring at the IF rate, and their amplitude varies at the same rate as the original audio signal instead of remaining constant as in the rectifier.

The filter circuit removes the IF pulses, leaving a DC voltage the amplitude of which varies in accordance with the original audio signal. As we shall see in later chapters, this DC subsequently changed to AC before it is further amplified by the receiver.

The foregoing discussion assumed the RF carrier was amplitude-modulated. In addition to amplitude modulation, detectors for demodulating frequency-modulated (FM) signals are analyzed in this volume. With FM modulation, the amplitude of the transmitted RF carrier is held constant and the frequency is varied. This necessitates an entirely different method of detection than described in the foregoing. (FM detection is discussed in Chapters 7 and 8.) The basic purpose of the detector—that of extracting the audio signal from the carrier—remains the same, however—whether the modulation is AM or FM.

There are many additional methods of modulating a carrier wave, such as phase modulation, pulse modulation, frequency-shift keying, etc. In an elementary work like this, however, attention must be directed only to those circuits which have the widest application. Besides, the various types of modulation which have important special application in advanced circuitry will all prove to be *special cases* of either amplitude or frequency modulation. Consequently, a working knowledge of the detectors presented in this volume lay valuable groundwork for anyone pursuing advanced applications of modulation and demodulation.

Almost all detectors and rectifier circuits in this volume depend for proper operation on the principle of the "long time-constant" combination of a resistor and a capacitor. This is undoubtedly the most widely used combination of circuit components in the electronics field. For this reason, it is advisable to understand thoroughly how such a combination works, and how the particular values of components chosen are related to the frequency of the current under consideration. The physical actions in the various R-C combinations of this volume are fully explained as they are encountered in each chapter. Consequently, they will not be reviewed here.

Chapter 2

HALF-WAVE RECTIFIER CIRCUITS

Half-wave rectifier circuits are very common in power supplies of electronic equipment. This is especially true where low cost and high output voltage are essential. In this chapter, the following circuits will be discussed:

Half-wave transformerless rectifier.
Half-wave voltage-doubler rectifier.
Half-wave rectifier with transformer.

TRANSFORMERLESS HALF-WAVE RECTIFIER

Fig. 2-1 and 2-2 show two successive half-cycles in the operation of a transformerless half-wave rectifier power supply. This type of power supply may be plugged directly into either an alternating- or direct-current power line. Because of this feature, the power supply is known as an AC-DC supply. (Obviously, when the unit is operated from a DC line, there will be no such thing as an alternating half-cycle.)

The essential components of this power supply are:

R1—Filter resistor.
R2—Voltage-divider resistor for taking ouput voltages.
C1—First filter capacitor.
C2—Second filter capacitor.
C3—Isolating capacitor.
V1—Half-wave rectifier tube.

Only two main types of currents are at work in this circuit. They are:

1. Main rectifier current.
2. Filter currents.

When the applied power-line voltage makes the plate of the rectifier tube more positive than the cathode, the tube will conduct electrons from cathode to plate. Their path starts at the other side of the power line. These electrons flow upward through voltage-divider resistor R2, and then to the left through filter resistor R1, enroute to the cathode.

Resistor R1 and capacitor C1 form a long time-constant filter of the type discussed in the introductory chapter to the oscillator volume in this series. Resistor R2 and capacitor C2 form another long time-constant filter combination. These two RC combinations provide adequate filtering and consequently adequate reduction in ripple voltage.

When the circuit is turned off and is at rest, zero output voltage will be measured at either output point. However, as soon as power is applied, electrons will leave the cathode during each positive half-cycle and cross to the plate of the rectifier tube. This action leaves the cathode with a *deficiency* of electrons and consequently with a positive voltage. The upper plate of capacitor C1 immediately assumes this value of positive voltage.

The only way this positive voltage can be reduced to zero is by drawing electrons up from the other side of the line, through the two resistors. (This is a solid line.) Because these are long time-constant filters, the positive cathode voltage undergoes only insignificant reduction before the next positive half-cycle of power-line voltage is applied. Each positive half-cycle draws more electrons across the tube. During negative half-cycles like those represented in Fig. 2-2, no electrons cross the tube. However, the upward flow of electrons through the resistor path continues throughout the entire cycle.

Ohm's law tells us that the voltage across a resistor is proportionate to the current through it. This is written symbolically as:

$$E = I \times R$$

where,

E is the voltage in volts,
I is the current in amperes,
R is the resistance in ohms.

Coulomb's law tells us that the amount of voltage across a charged capacitor is proportionate to the quantity of electrons (or positive ions) stored there. This is written symbolically as:

$$Q = C \times E$$

where,

Q is the number of coulombs of charge stored (1 coulomb $= 6.25 \times 10^{18}$ electrons),
C is the size of the capacitor in farads,
E is the voltage across the capacitor in volts.

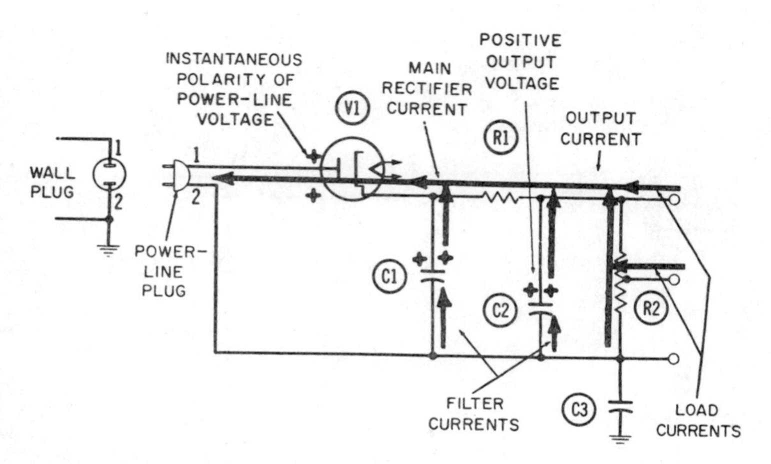

Fig. 2-1. Operation of the transformerless half-wave rectifier—conducting half-cycle.

This formula can be rewritten as:

$$E = \frac{Q}{C}$$

Since the voltage across capacitor C2 is identical to the output voltage across R2, we can make these voltage equations equal to each other and say:

$$\frac{Q}{C2} = I \times R2$$

where,

Q is the quanity of positive ions stored on the upper plate of C2,

I is the main rectifier current (solid) flowing upward through R2.

The positive voltage on the upper plate of C1 draws electrons to the left, through resistor R1. The deficiency of electrons created on the upper plate of C2 becomes the output voltage. It is this positive voltage that draws electron current upward through voltage-divider resistor R2. Any fraction of the total output voltage can be obtained by placing an appropriate tap on the voltage divider.

Filter Currents

The RC filtering used in this power supply is identical in principle to the cathode filtering discussed in the oscillator and am-

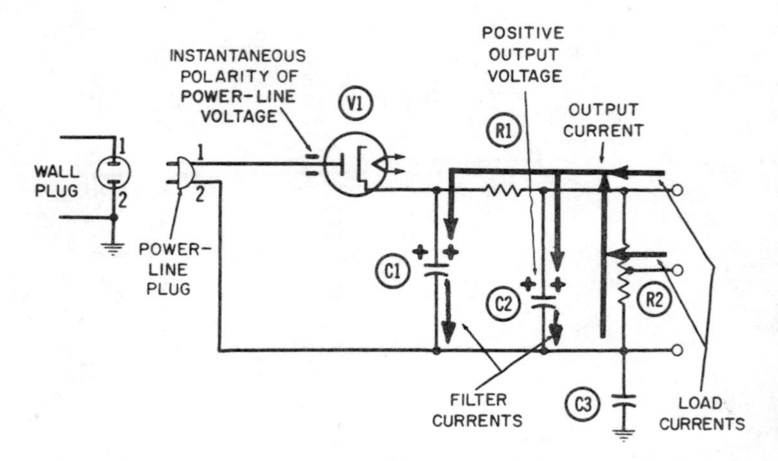

Fig. 2-2. Operation of the transformerless half-wave rectifier—nonconducting half-cycle.

plifier volumes of this series. As electrons are drawn into the rectifier tube, they can come from one of two places—either directly through the high-impedance path composed of R1 and R2, or from the upper plate of C1 (which offers a much lower impedance). Naturally, most of the tube-current electrons come from the top of capacitor C1. In order for this to happen, an equal number must be able to flow *onto* the capacitor's lower plate. These are the electrons, thin line, flowing upward into C1 during the first half-cycle (Fig. 2-1).

On the negative half-cycle such as are shown in Fig. 2-2, these electrons flow back down to the other side of the line. These two movements constitute the filter action of the capacitor.

A similar filter current flows between the lower plate of C2 and the neutral side of the line. This filter current is much smaller because C1 has already filtered out most of the fluctuation in the main rectifier current.

Capacitor C3 isolates the equipment from ground to eliminate the danger of shock. As shown in Figs. 2-1 and 2-2, the two leads on the left connect to the two prongs of an ordinary electric plug. One of the two wires leading to the wall outlet in the home is normally grounded and the other is "hot." Capacitor C3 could be omitted and the lower end of R2 grounded directly, provided one could always be sure prong 2 would be plugged into the ground side of the power line and prong number 1 into the "hot" side. However, since the plug can be inserted either way, the user runs the risk of connecting the "hot" side of the line directly to the receiver ground, which is usually the chassis. This could result in a dangerous shock, so C3 is added between the line and the receiver chassis as a safeguard.

TRANSFORMER INPUT HALF-WAVE RECTIFIER

Figs. 2-3 and 2-4 show two successive half-cycles in the operation of a half-wave rectifier power supply using a step-up input power transformer. The filter circuit used for converting the resultant pulsating DC into pure DC is similar to the previous circuit used with the transformerless power supply. Both use a resistor (R1) instead of the bulkier and more costly filter choke. Another disadvantage of a filter choke is that the continual passage of current in a single direction through the winding establishes a permanent magnetic field. This feature can be seen more clearly in the next chapter, where the full-wave réctifier is discussed. However, using a resistor in lieu of a filter choke is not without its disadvantage. There is a definite voltage drop and power loss, occasioned by the flow of all the rectified current through the resistor. This loss in voltage and power is in addition to the power consumed by the rectifier current flowing upward through output resistor R2.

Circuit Components

This circuit is made up of the following components:

R1—Filter resistor.
R2—Output resistor, suitably tapped to provide any desired partial voltages.
C1—Input filter capacitor.
C2—Output filter capacitor.
T1—Power transformer.
V1—Diode tube.

Identification of Currents

The following currents are at work in this type of circuit:

1. Power-transformer primary current.
2. Power-transformer secondary current.
3. Diode tube current.
4. Cathode heating current.
5. 60-cycle filter currents.

Fig. 2-3 shows the half-cycle when the power transformer makes the diode plate positive, so that electrons can cross the tube. An attempt has been made to depict this transformer action just as it actually occurs in these figures. The current in the primary, which is the driving current, can be considered to flow downward through the primary in the first half-cycle, in response to an applied negative voltage. This applied polarity has been indicated by the minus sign at the top of the primary winding.

This primary current *induces* a voltage in the secondary winding which is in opposition to the applied voltage. The polarity of this "back emf" has been indicated by the plus sign at the top of the secondary current associated with this induced voltage, and its flow direction is indicated as upward during the first half-cycle.

This induced voltage and current will have their greatest values twice during each cycle, when the primary inducing current is *changing* at the maximum rate. While the primary inducing current has its greatest value, the secondary induced current and voltage will be zero. These moments also occur twice during each cycle—midway in each half-cycle.

Heating Current

The power transformer used here and in most power-supply applications has a third, or tertiary, winding which delivers a large flow of current to heat the filament or cathode, in order that emission may occur. This current, shown as **large dots** , follows the closed path through the winding and through the filament or cathode. The example used here is a directly heated cathode which requires no additional filament.. This heating current flows at the same frequency as the power current, which is 60 cycles per second in most parts of the United States. It is shown flowing clockwise during the first half-cycle and counterclockwise during the second. The wire connecting the filter network to the cathode is usually brought to the center tap of this winding. In this way, a minimum of 60-cycle ripple (due to the heating operation) will be coupled into the filter system.

During that portion of the first half-cycle while tube current is flowing, it enters the cathode network through this center tap and flows through *both* sides of the line to the cathode before entering the tube. Neither of these two currents in the tertiary winding and in the cathode "disturbs" the other one in any sense.

The tertiary winding constitutes a voltage step-down and current step-up device. Tube cathodes and heating filaments are designed for a certain applied voltage, which varies in size depending on the application. Common values are 5, 6.3, and 12.6 volts. However, 25- and even 35-volt heaters are used.

The *amount* of current is of course regulated by the amount of resistance in the winding and that portion of the cathode or fil-

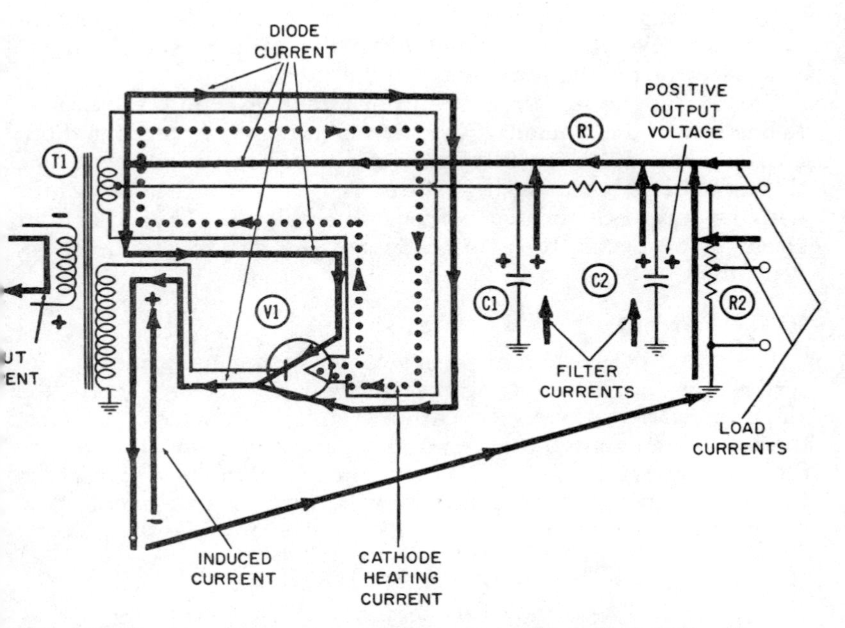

Fig. 2-3. The transformer input half-wave rectifier—first half-cycle.

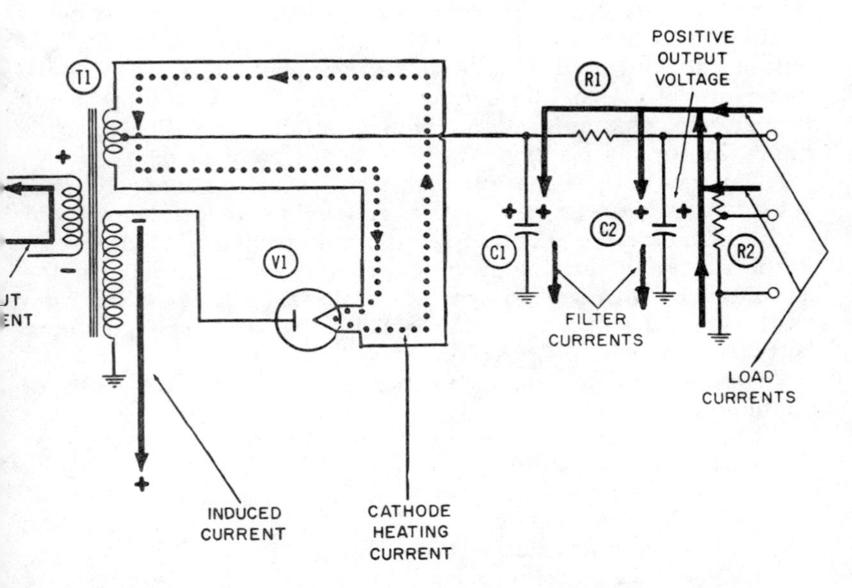

Fig. 2-4. The transformer input half-wave rectifier—second half-cycle.

ament through which the heating current must flow. The tube is designed so that its current heats the cathode sufficiently to emit electrons without burning up the tube.

Sometimes the lead from the filter circuit is connected directly to one side of the filament. Operation is the same, except the diode current flows from the filter circuit directly to the filament, and through the tube to the transformer. In other versions, a tube with an indirectly heated cathode is employed. Then the filter circuit is connected to the cathode and the heater is entirely independent.

Filter Currents

The filter network here operates on the same principle described for the preceding transformerless power supply. As the electron current leaves the cathode and flows across the tube, these filter currents (thin line) flow upward from ground into filter capacitors C1 and C2. When the electron flow across the tube stops during the second half-cycle, these filter currents flow downward to ground.

VOLTAGE DOUBLER

Figs. 2-5 and 2-6 show two successive half-cycles in the operation of a dual-diode rectifier system being used as a voltage doubler. This particular circuit includes a step-up input transformer. However, it is even more common for the voltage-doubling principle to be used with a transformerless input. The output voltage would then be limited to about 2.8 times the *peak* value of the alternating supply voltage. With a step-up transformer, the maximum output voltage attainable will equal this figure, multiplied by the turns ratio of the transformer.

As an example, if the input voltage is the standard 110 volts (effective value), its peak value will be 1.4 times as great, or about 155 volts. When doubled, this would come to 310 volts; this is the maximum attainable in a transformerless circuit. In the circuit shown in Figs. 2-5 and 2-6, if the transformer turns ratio were 2-to-1, then the maximum attainable output voltage would be twice as much, or 620 volts.

The following circuit components make up this voltage-doubling circuit:

R2—Output resistor.
C1—Output capacitor.
C2—Output capacitor.
T1—Power transformer.
V1 and V2—Diode tubes.

Identification of Currents

The following electron currents are at work in a circuit of this type:

1. Input power current .
2. Current through V1 .
3. Current through V2 .

Details of Operation

The plate of diode V1 and the cathode of diode V2 are connected directly together and also to the top of the transformer secondary winding. In the first half-cycle shown in Fig. 2-5, transformer action makes the top of this winding positive. V1 conducts from cathode to plate. V2 is of course unable to conduct, since its cathode is more positive than its plate.

The flow of electrons from cathode to plate of V1 charges the upper plate of capacitor C1 to a positive voltage which is almost equal to the peak transformer secondary voltage. The reason is that the capacitor gives up electrons to provide the necessary tube current; the resultant electron deficiency constitutes a positive voltage. As a result of this current flow through the diode and the positively charged capacitor, electron current is drawn upward through resistor R2. This is part of the output current of the rectifier, and it has been shown in solid lines. The voltage it will develop across R2 will always be proportionate to the amount of current flowing, in accordance with Ohm's law, and will always be *exactly equal* to the voltage across capacitor C1.

After passing from cathode to plate, these electrons head downward through the transformer secondary winding and enter the neutral, or ground, point between the two capacitors.

During the second half-cycle, shown in Fig. 2-6, transformer action causes the top of the secondary winding to assume a negative voltage. This negative voltage appears at the plate of V1 and prevents any electron curent from flowing through this tube. However, the same negative voltage at the cathode of V2 causes a full flow of electrons across this tube. These electrons originate at the ground point between the two capacitors and flow upward through the secondary winding, then downward to the cathode of V2 and through the tube to the lower plate of capacitor C2. Here they accumulate, forming a negative voltage almost equal in value to the negative peak value of the induced voltage across the secondary winding. This negative charge on the lower plate of C2 drives a continuous electron current upward through resistor R2. This current, shown in dotted lines, is also part of the rectifier output current. It develops a voltage across R2 that will at all times be exactly equal to the negative voltage existing across capacitor C2.

Consequently, these two currents flowing upward simultaneously through R2 will develop a voltage across it which is approximately twice as great as either current could develop alone. The total voltage across R2 is of course equal to the sum of the two capacitor voltages, and almost equal to twice the peak secondary voltage. Because of this feature, the circuit is referred to as a voltage-doubler circuit.

ADVANTAGE OF HALF-WAVE OPERATION

The principal advantage of using a half- rather than full-wave power supply (discussed in the next chapter) is that a power transformer is unnecessary. However, when a power transformer is used, a higher output voltage can be obtained for a given *size and weight of transformer.* A higher voltage can be obtained with half-wave operation because the entire amount of secondary voltage is applied to the plate of the rectifier tube. The DC voltage developed at the cathode can never exceed this peak secondary voltage, although the two will usually be very close in value.

If the secondary of this transformer were center tapped, with the two ends of the winding going to the plates of two rectifier tubes or a twin diode tube (as it is in the full-wave rectifier), then only half the available voltage swing would be applied to each plate, and only half as much output voltage would be available across the output resistor. This point can be clarified by comparing the circuit of Figs. 2-3 and 2-4 with those of the next chapter.

Let us assume the transformers used in two circuits are identical in construction, turns ratio, size, weight, etc., except that one has a center-tapped secondary to allow for full-wave operation. Let us also assume that turns ratio is 1-to-3 and that the *peak* voltage swing of the primary power is about 150 volts, which is reasonably close to home supply conditions (110 volts effective value equals 155 volts peak). Under these conditions, the peak secondary voltage will be 450 volts. In the half-wave rectifier of this chapter, all this voltage will be applied to the diode plate. The resultant cathode voltage will be close to this value, perhaps 420 volts. (The output voltage across R2 will be somewhat less because of the voltage dropped or lost as the rectifier current passes through R1.)

In the full-wave rectifier, with the center-tap winding connected to ground each plate would be driven alternately between

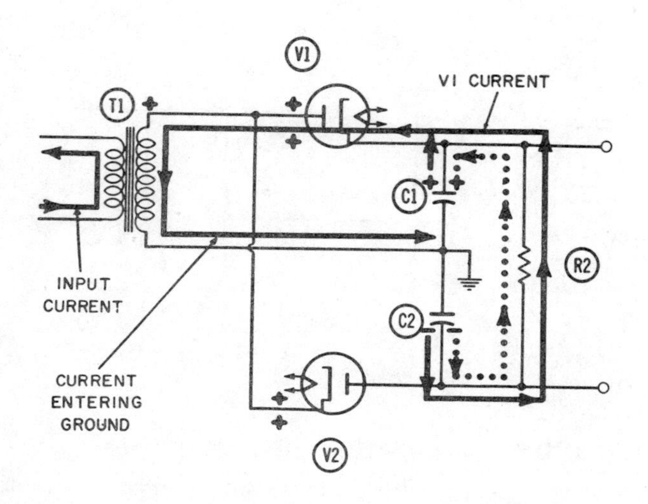

Fig. 2-5. The voltage-doubler rectifier—positive half-cycle.

plus and minus 225 volts. The maximum attainable output voltage would thus be limited to 200 volts or so.

Since a transformer is bulky and heavy, there are many applications where its elimination is desirable. For example, most table-model radios use the circuit of Figs. 2-1 and 2-2 and many television receivers employ the voltage-doubler circuit in Figs. 2-5 and 2-6. At other times the circuit of Figs. 2-3 and 2-4 is selected because the advantage of higher output voltage without transformer size or bulk has overriding attractions. An obvious disadvantage which must be accepted in choosing half-wave over full-wave operation is the probability of somewhat poorer *voltage regulation.*

VOLTAGE REGULATION

Voltage regulation is defined as the percentage variation in output voltage of a power supply in going from no load to full load. It may be expressed by this formula:

$$\text{Voltage regulation} = \frac{E_2 - E_1}{E_1} \times 100$$

where,

E2 is the output voltage at no load,
E1 is the output voltage at full load.

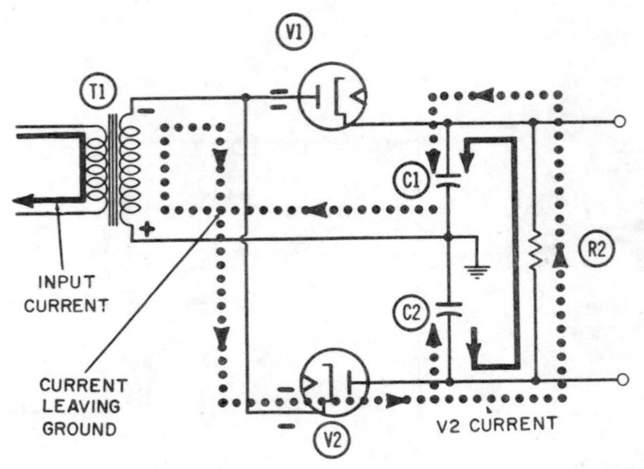

Fig. 2-6. The voltage-doubler rectifier—negative half-cycle.

The no-load condition with a power supply means it is not delivering current to any external circuits. Figs. 2-1 through 2-4 show some load current being drawn toward the rectifier tube from the right side of the diagrams. The electrons which make up this load current are being drawn from all tube circuits supported by this power supply. The power supply delivers current to these load circuits by providing a high positive voltage which draws electrons through them. These electrons are mostly in the form of plate and screen-grid currents of vacuum tubes.

The *total current* a rectifier tube will draw is determined by its construction and by the applied voltage from the transformer secondary. When little or no load current is flowing, the rectifier will draw all its required current up from ground through resistor R2. The output voltage is then equal to the voltage developed across R2 by this current, in accordance with Ohm's law.

When substantial load current is drawn, however, the amount of current drawn through R2 will be decreased equally, and the voltage developed across R2 will also be decreased proportionately. This amounts to a reduction in the output voltage under full-load conditions.

Obviously, a voltage regulation of zero per cent is unattainable in power supplies of this type. In order for the regulation to equal this ideal, the output voltage under full-load conditions would have to *equal* the output voltage under no-load conditions—an impossible situation.

In this type of circuit and using RC filters as described, fairly high values of regulation (meaning *poor* voltage regulation) will be obtained. Values of 30% or 40% are not uncommon.

Chapter 3

FULL-WAVE RECTIFIER CIRCUITS

In this chapter, three versions of the full-wave rectifier circuit will be discussed. Half-wave rectifiers, discussed in the previous chapter, utilize only one half-cycle of the applied alternating voltage. Full-wave rectifiers, on the other hand, use both half-cycles. This results in a ripple frequency twice that of the half-wave rectifier. In other words, using half-wave rectification the rectifier will conduct for a portion of one half cycle; then, during the next half-cycle, the tube does not conduct. This results in a pulse, then a long period of no voltage, followed by another pulse. Of course, the filter circuit "fills in" between pulses so we have a DC voltage at the output. With the full-wave rectifier, conduction occurs on both half-cycles; hence the output contains twice as many pulses of DC, with no long periods between them. All that is required of the filter circuit is to smooth out the output. Thus, smaller filter capacitors can be employed.

CONVENTIONAL FULL-WAVE RECTIFIER

The first full-wave rectifier circuit to be discussed might be called the "conventional" circuit, which is widely used in power supplies for radios, television sets, etc. It employs a center-tapped transformer, a tube with two diode plates, and a common directly-heated cathode. Figs. 3-1 through 3-4 show the four quarter-cycles of operation for the circuit. The components, and their functional titles are:

R1—Load resistor.
C1—First filter capacitor.
C2—Second filter capacitor.
L4—Filter choke.
T1—Power transformer.
V1—Full-wave rectifier tube.

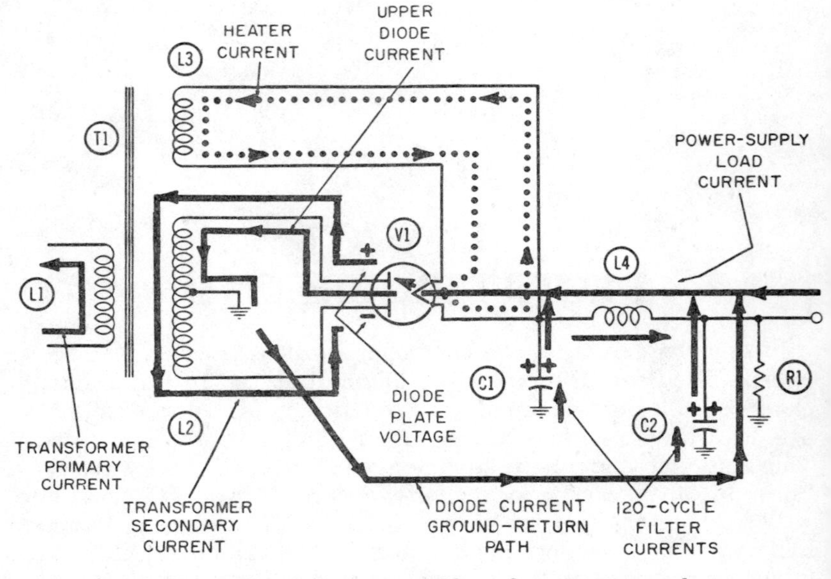

Fig. 3-1. The full-wave rectifier—first quarter-cycle.

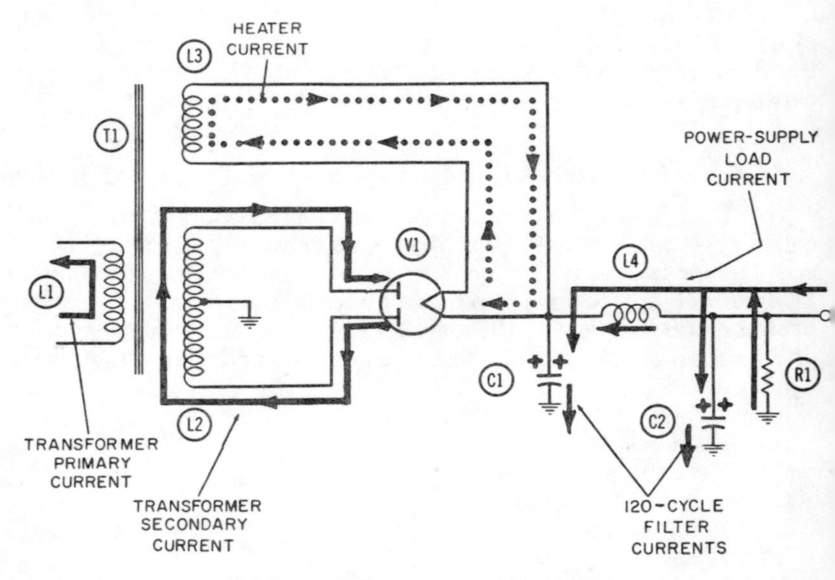

Fig. 3-2. The full-wave rectifier—second quarter-cycle.

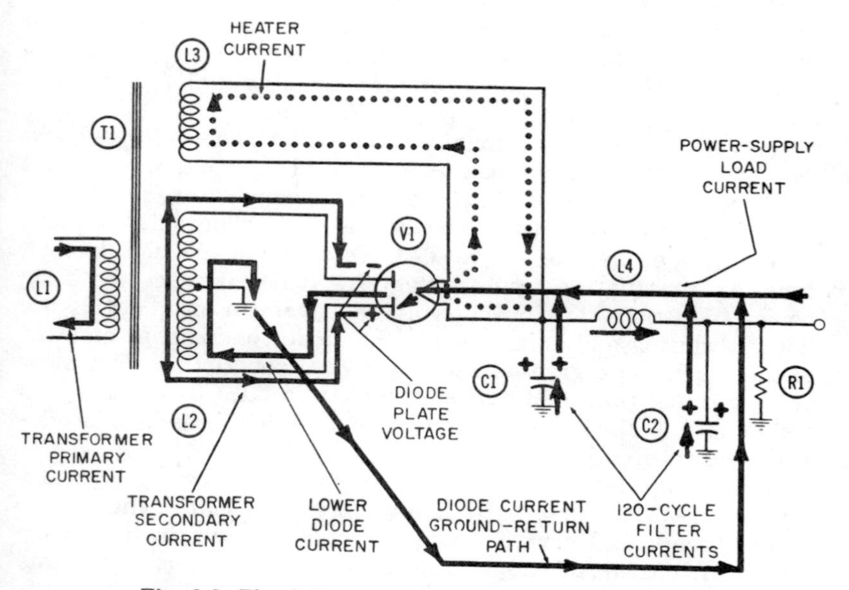

Fig. 3-3. The full-wave rectifier—third quarter-cycle.

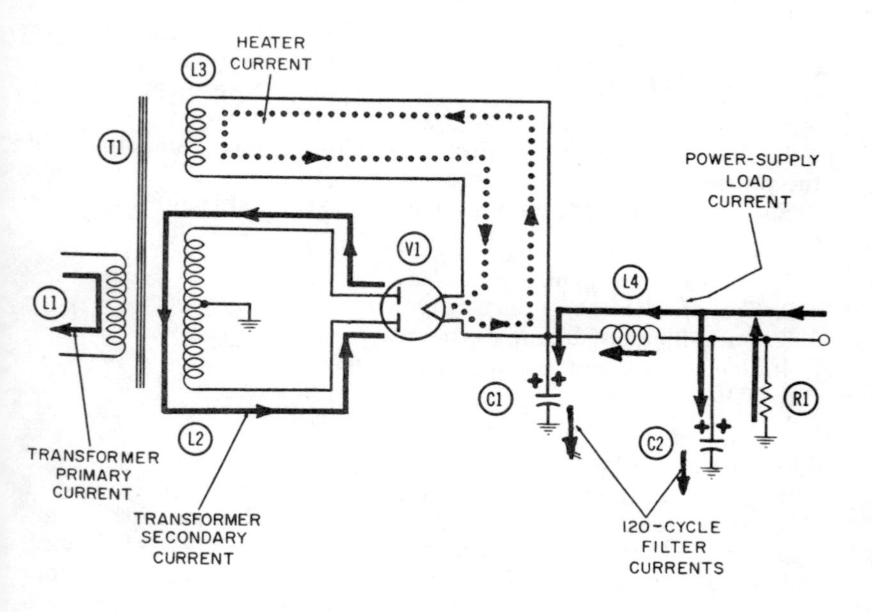

Fig. 3-4. The full-wave rectifier—fourth quarter-cycle.

Fig. 3-5 shows the phase relationships between the transformer secondary current, the upper-diode plate voltage, the upper- and lower-diode currents respectively, and finally the output-voltage waveform. (The ripple component of this output voltage has been exaggerated somewhat for easy identification.)

No waveform is shown for the lower-diode plate voltage. However, such a waveform would be exactly 180°, or half a cycle, out of phase with the upper-diode plate voltage shown in line 2.

It can be seen from these waveforms that each diode conducts once during each cycle and raises the output voltage slightly each time. As a result, the ripple-voltage frequency is twice the basic power frequency. The output-voltage waveform also reveals the fairly abrupt rise in voltage when the diode conducts, as opposed to the much more gradual decay between conduction. This slow decay results from the filtering provided by C1, C2, and L4. This action is explained more fully later in the discussions of the circuit-operation diagrams.

Three separate "families" of currents are shown in Figs. 3-1 through 3-4. The first such family might be called the *power* group, because all currents are driven from the primary power source such as an AC supply line. There are three currents in this group—transformer primary, transformer secondary, and filament heating.

The second family, or group, of currents is the "rectified," or unidirectional, currents; after being filtered, they enable this circuit to provide the stable DC voltage output necessary for the operation of many electronic systems.

The third family is the *filtering* current which "smooths" out the pulsations in the rectified current.

Each of these currents is shown in Figs. 3-1 through 3-4:

1. Transformer primary current.
2. Transformer secondary current.
3. Tube-filament heating current.
4. Rectified or unidirectional current.
5. Filter current.

The Power Currents

The first current of this group flows in the primary winding of the power transformer and is shown as a solid line. In the average household this is a 60-cycle current, and its associated driving voltage is the standard 110 volts. The first quarter-cycle diagram (Fig. 3-1) shows a single quarter-cycle of this driving current as it flows upward through the primary winding. The essential property of a transformer (or any other inductance) is that it will resist any *change* in the amount of current flowing at any given moment. During an entire half-cycle (such as the one represented by Figs. 3-1 and 3-2), this primary-winding current undergoes an entire series of changes. Starting from zero, it rises to maximum and falls to zero again, always in the upward direction.

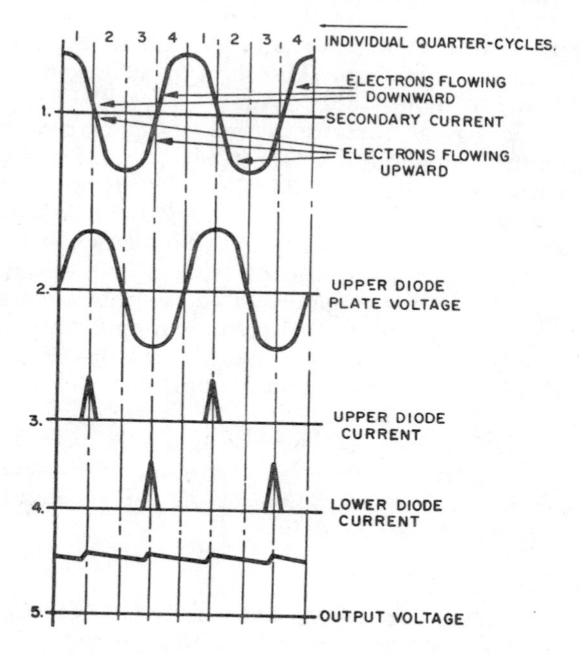

Fig. 3-5. Full-wave rectifier waveforms.

Any changes in direction or amount of this current will cause current movements in other parts of the transformer. The large secondary winding carries a current, also shown a solid line, whose movements are associated with the voltage changes on the plates of the twin-diode rectifier tube. All currents are electrons in motion. Hence, an electron current flowing in one direction will ultimately deliver a certain quantity of electrons to one end of the current path and will leave a scarcity of them at the other end. A concentration of free electrons is by definition a negative voltage. By the same token, a scarcity (deficiency) of electrons results in a concentration of positive ions, which is by definition a positive voltage.

During the entire first quarter-cycle, this current moves *downward* in the transformer secondary. At the end of the quarter-cycle, the lower end of the transformer has achieved its maximum negative voltage, as depicted by the deep minus sign at the lower plate of the rectifier tube. At the same time, a deficiency of electrons will exist at the upper end of the transformer secondary. This is shown by the deep plus sign at the plate of the upper rectifier tube.

The upper diode will conduct the maximum amount of rectified (solid) current at the same instant this diode plate reaches its maximum positive voltage. The first quarter-cycle diagrams (Figs. 3-1 and 3-5) depict this instant.

The third and final current in this group of power currents flows in the tertiary winding of the transformer. Its purpose is to heat the cathode of the twin-diode rectifier tube. This winding consists of a relatively few turns of wire. Thus it is a low-voltage winding which delivers high current. The current in the primary winding drives this cathode heating current at the same 60-cycle frequency. To clarify the transformer action, this current is shown flowing in opposite phase to its driving current—in other words, downward during the first quarter-cycle and upward during the second and third. During the fourth quarter-cycle, it again flows downward.

This current follows the closed path which includes only the tertiary winding and the direct-heated cathode. In the four diagrams depicting circuit operation (Figs. 3-1 through 3-4), this heating current is shown as dotted .

The Rectified Currents

The rectified currents shown eventually flow through the rectifier tubes, from the common cathode to either plate. At the end of the first quarter-cycle, as shown in Figs. 3-1 and 3-5, this current flows from the cathode to the upper plate of the two diodes. (The current-voltage action in the transformer secondary makes the upper-diode plate positive. This accounts for the electron flow across the tube to the upper-diode plate. Whenever either diode plate is more positive than the cathode, electrons will leave the cathode and cross the open space to that plate.)

Any electrons leaving the cathode must be provided with a return path to the cathode. This is a fundamental requirement of vacuum-tube operation. The return path in the circuit of Fig. 3-1 is from the upper plate of the diode to the top of the transformer secondary, then through its upper half to the center tap. Here the current enters a common ground and returns to the bottom of R1, which serves as both a bleeder and a load.

The current then continues upward through R1, then to the left through the filtering inductance, to the cathode.

The diagrams for the third quarter-cycle (Figs. 3-3 and 3-5) show electron current flowing from the cathode to the *lower* plate of the diode. Current flows in this direction through the tube because the secondary current in the transformer has made the lower-diode plate positive. This is indicated by the deep plus sign on the lower plate of the twin-diode tube.

After crossing the lower diode, the rectified current (still shown solid) goes to ground through the lower half of the transformer secondary, then through ground and back to the load resistor. It returns to the cathode by flowing upward through this resistor, and to the left through the filtering inductance L4.

Another component of rectified current is shown entering the filtering inductance from the right. This portion does not pass through the load resistor. Instead, it comes from those vacuum-tube circuits which receive their DC power from this rectifier circuit. Once it enters the rectifier, this current follows the same paths as the other through the diodes and transformer secondary to the common ground. From here, it then has a free return path to the cathodes of the other tubes.

An important measure of any rectifier's performance is its capability for voltage regulation—that is, the stability of its output voltage under varying loads. The output voltage of this rectifier is the positive voltage on the upper plate of capacitor C2, and consequently at the output terminal. This voltage, represented by deep plus signs, should be as steady as possible under all circumstances from no load to full load—in other words, it should be "regulated," or have "regulation."

In any charged capacitor (such as C2), there is a definite relationship between the amount of charge (electrons or positive ions) stored there and the amount of negative or positive voltage produced. Coulomb's law states this relationship as a formula:

$$Q = C \times E$$

where,

Q is the quantity of charge in coulombs,
C is the size of the capacitor in farads,
E is the voltage in volts.

Since one coulomb of charge is always equal to a precise number of negative electrons or positive ions (6.25×10^{18}, to be exact), the voltage across a charged capacitor is always *directly proportionate* to the quantity of electrons or ions stored in the capacitor.

The Filtering Currents

In addition to good regulation under varying load conditions, it is desirable that the output voltage on the charged capacitor, C2, have a minimum ripple component. Inadequate filtering of the power supply causes this ripple voltage. Fig. 3-1 shows the rectified currents flowing through the upper diode during that portion of the first quarter-cycle when the transformer current-voltage actions (shown a solid line) makes the upper-diode plate more positive than the cathode. As electrons leave the cathode, others are drawn to it from the filter of the power supply. Thus, current flows *upward* from both capacitor plates at this instant.

A component of filtering current is a thin line beneath each capacitor. As electrons move *away* from the top plates of each capacitor and head toward the cathode, the same amount of filtering current (number of electrons) flows upward from ground. It is the essence of capacitor action that the same amount of current flows onto (or away from) one plate as flows away from (or onto) the other.

Figs. 3-2 depicts the operation during the second quarter-cycle. It is inserted here to show the completion of one cycle of filtering. The filtering currents, a thin line, are flowing in the opposite direction, back into ground. The electrons in these currents were drawn upward by the other electrons leaving the upper plates. However, with nothing to hold them on the bottom plate, they quickly fall back to a point of neutral voltage (in other words, ground). At the same time, the rectified currents are shown *entering* rather than leaving the upper plates of these two filter capacitors.

Thus, during a single half-cycle of the primary power current, one complete filtering cycle has occurred. Also, during the same period, rectifier current flows off the upper plates of the filter capacitors, and then back onto them. When the primary power frequency is 60 cycles per second (the standard household supply throughout most of the United States), this filtering action will occur twice as fast, or 120 times per second.

Two things determine the amount of voltage across filter capacitor C2—its size, and the number of positive ions stored on its upper plate. This voltage is also the output of the whole rectifier circuit. Near the end of the first quarter-cycle, as shown by Fig. 3-5, electrons are leaving this plate and passing through the tube. Hence the output voltage, being already positive, will tend to become even more so. The *inflow* of electrons a moment later, when the plate current stops flowing, will tend to reduce this output voltage. These undesirable fluctuations are termed the ripple voltage. They will occur at twice the input frequency, since two complete fluctuations will occur for each cycle of input frequency.

One obvious means of reducing the ripple voltage is to use a larger output capacitor C2. Then, a greater *quantity* of positive ions, in comparison with the number of electrons coming in or going out during each quarter-cycle of the filtering action, will be stored on the capacitor. If the capacitor is made large enough, the number of electrons becomes insignificantly small in relation to the number of ions, and the output voltage may be considered steady.

The first and third quarter-cycle diagrams show the components of rectifier current flowing *away* from the upper capacitor plates. In turn, compensating filtering currents flow upward from ground, toward each of the lower capacitor plates. The second and fourth quarter-cycles show currents flowing *onto* the upper capacitor plates, with compensating currents flowing back into ground from the lower plates.

Capacitor action can be clarified by remembering that the quantity of current (or electrons) flowing into one side of a capacitor is always equal to the quantity flowing away from the other side. If the flow of either current is impeded, its counter-part on the other side will also be reduced and by the same amount. Thus, if the connection between either lower plate and ground becomes broken, or "open," the component of filtering current shown will be unable to flow. The capacitor will then cease to act as a filter, or "shock absorber." Now, when rectified current passes through either diode, the surge of electron current which *would have been* drawn from the upper capacitor plate must be drawn from other parts of the circuit.

Specifically, if this trouble occurred to capacitor C2, the surge of electrons being drawn into the filter choke from C2 would have to come instead from the external circuit or load resistor, or both. The momentary increase in electron current through the resistor would momentarily raise the voltage at the output point. This voltage rise would be repeated each time electrons flow through either diode, or 120 times a second. The resultant fluctuation in output voltage would constitute an unacceptable component of ripple voltage.

The action of choke coil L4 in contributing to the total filtering function is rather crudely portrayed by green current lines flowing back and forth in the coil. At any given instant, these currents flow in such a direction that they oppose any *change* in the amount of rectified current being drawn through the coil. In the first quarter-cycle, the upper diode requires a surge of electrons. They are drawn from both the upper plate of capacitor C1 and the left end of the choke coil. Any inductance will oppose any *change* in the amount of current flowing through it. This opposition takes the form of a voltage opposite in polarity to the imposed voltage and is called the "counter emf" or "back emf." As the imposed voltage draws the excessive current through the coil to the cathode, the counter emf generates a "counter current," a solid line. The latter flows in the opposite direction to the normal current, and the two currents therefore will tend to cancel each other. Practically speaking, the "counter current" can never be as large as the normal current. Also, since the counter current depends on changes in the normal current, it dies out quickly as soon as the normal current ceases to change and approaches a steady value.

The second quarter-cycle shows the filter current as having reversed itself so that it is now flowing *toward* the cathode. The rectified current through the diode is decreasing at this time; thus, the filter current tends to keep the total current through the coil from changing.

Fig. 3-3 represents the third quarter-cycle of operation. Here current is flowing through the lower diode. This demand for additional electrons must again be supplied from the upper plate of capacitor C1 and from the rectified current flowing through choke coil L4. As before, the existence of inductor (choke) action is shown by the momentary current, thin line flowing in the opposite direction through the coil. This current momentarily opposes the change in total current through the coil.

The fourth quarter-cycle (Fig. 3-4) shows this hypothetical current flowing *toward* the cathode again. This time it opposes the decay (decrease) in rectified current which will occur as the lower diode ceases to conduct and before the upper diode starts. Each diode conducts for only a relatively small portion of a quarter-cycle. This occurs when the current-voltage relationship in the transformer secondary winding makes the plate of each diode more positive than the voltage on the upper plate of capacitor C1.

Likewise, the filtering currents in the inductor coil do not flow continuously for each full quarter-cycle before reversing direction. Their basic frequency will of course be 120 cycles per second.

Because of this additional filtering in the choke coil, the output voltage at the right end of the coil will be steadier than the voltage at the left end. In other words, the ripple voltage has been reduced by this additional filtering. As more and more coils and capacitors are added to the filter, the output voltage will become freer and freer from ripple.

If C1 and C2 have the same capacitance, then the positive voltages on their two upper plates will represent equal amounts of positive ions in storage, since the voltages are assumed to be equal. However, since one of these two voltages fluctuates more widely, different amounts of electron current must flow in and out during each cycle. The voltage on C2 is steadier than on C1. Consequently, *fewer* electrons *leave* the upper plate of C2 during each conduction period than leave C1.

Also, the filtering current (thin) flowing between C2 and ground will consist of fewer electrons than are in the filtering current between C1 and ground. Each current is driven by the arrival and departure of electrons at or from the upper plates of the respective capacitors. To repeat, the *same quantity of electrons* must always be in movement on both sides of any capacitor.

FULL-WAVE BRIDGE RECTIFIER CIRCUIT

Figs. 3-6 and 3-7 show two successive half-cycles in the operation of a full-wave bridge rectifier circuit. The symbols used here are for solid-state rectifiers such as selenium, silicon, germanium, etc. However, the circuit would function equally well using conventional tubes. A solid-state rectifier functions like a diode—electron current flows essentially in only one direction through it. Electron flow through a solid-state rectifier is in the direction opposite the arrowhead indication. When the flat end of the rectifier (we call this end the cathode) is made more negative than the triangular end, electron current will be drawn across the junction with relative ease. However, when the rectifier anode (triangular end) is made more negative, practically no electron current flows in the reverse direction.

The components of this circuit include:

R1—Output resistor.
C1—Output filter capacitor.
T1—Input power transformer.
M1 through M4—Solid-state rectifiers.

Identification of Currents

Four separate electron currents perform key functions in this circuit. They are:

1. Transformer primary current.
2. Secondary induced current.
3. Current through rectifiers M2 and M3.
4. Current through rectifiers M1 and M4.

Details of Operation

The phase and amplitude relationships between the primary and secondary currents and their associated voltages were discussed earlier. Hence, they will not be repeated here. Input transformer T1 will probably be of the voltage step-up variety. Thus, the amplitude of the voltage induced across the secondary winding will be N times the amplitude of the primary input voltage (where N is the ratio between the number of turns of wire in the primary and in the secondary).

In the first half-cycle, represented by Fig. 3-6, the polarity of this secondary induced voltage has been indicated by a deep plus sign at the top of the winding and a deep minus sign at the bottom. Since this positive polarity is applied to the junction of rectifiers M1 and M2, it will draw electron current through M2. Simultaneously, application of a negative voltage to the junction of rectifiers M3 and M4 will in effect "push" electrons through rectifier M3.

The paths for both electron currents are identical—in fact, they are the same current. Shown a solid line, it flows only during this half-cycle, through rectifier M2, downward through the transformer secondary, through rectifier M3, and to the common-ground return line. From here it enters output resistor R1 and flows upward through it to re-enter rectifier M2.

During the second half-cycle, in Fig. 3-7, the polarities of the induced voltages are reversed. Now the top of the secondary winding is negative and the bottom is positive. The negative polarity at the junction of rectifiers M1 and M2 in effect "pushes" electrons through rectifier M1, in the direction shown. Simultaneously, the positive voltage at the junction of rectifiers M3 and M4 draws electrons through M4. These two currents also have identical paths—*upward* through the transformer secondary winding and then through rectifier M1 to ground. From ground, the path is through output resistor R1 and rectifier M4 to the bottom of the transformer secondary.

The presence of capacitor C1 serves as a symbol of the filtering action needed to convert the pulsating voltage output across R1 to a pure DC voltage or one having a low ripple content. Without adequate filtering, the current through R1 would consist of a series of half sine waves. Moreover, this current would momentarily cease to flow twice during each cycle, as the current switches from one set of diodes to the other.

Such a single R-C combination would of course be entirely inadequate for filtering the output of a power supply. A much more elaborate system, like the one discussed in connection with the power supply in Figs. 3-1 through 3-4, would be required instead. The capacitor was used here for one reason only—to demonstrate the principle of preserving a steady DC voltage at the output point, even while the current is switching from one set of diodes to the other end and, consequently, not flowing through either set. The presence of this output voltage is indicated by the plus signs on the upper plate of C1.

Since the anodes of M1 and M3 are connected to a common ground point, the two rectifiers cannot conduct electrons unless their cathodes are *more negative* than the ground voltage. Likewise, the cathodes of rectifiers M2 and M4 are connected to the output point, which is maintained at a fairly high positive voltage. Hence, they cannot conduct electrons either, unless their anodes are more positive than this output voltage.

FULL-WAVE VOLTAGE DOUBLER

Figs. 3-8 and 3-9 show two alternate half-cycles in the operation of a full-wave voltage-doubler power supply with transformer input. The transformer used here might be identical to the one shown in Figs. 3-1 through 3-4. Because of the different circuit construction, however, the output voltage developed across output resistors R1 and R2 will be twice the output voltage developed across R1 in Figs. 3-1 through 3-4. There are two important differences in the construction of the voltage-doubling circuit. First, the lower end of the transformer secondary is connected to ground rather than to the plate of the lower diode, V2. Also, the position of diode V2 is reversed from the previous example; here its cathode is connected directly to the plate of diode V1. In addition the junction of output resistors R1 and R2 is connected to ground; now this circuit will deliver both a negative and a positive output voltage.

Although two separate tubes are shown in Figs. 3-8 and 3-9, two halves of a single tube (with separate cathodes) are often employed. At other times, solid-state rectifiers such as used in Figs. 3-6 and 3-7 are utilized.

The necessary components of this circuit correspond as closely as possible to components doing similar jobs in Fig. 3-1. They include:

R1 and R2—Output resistors.
C1—Output filter capacitor.
C2—Output filter capacitor.
T1—Power transformer.
V1—Upper-diode rectifier tube.
V2—Lower-diode rectifier tube.

For simplicity, the choke coil in Fig. 3-1 has been omitted from this circuit, and also the tertiary winding for heating the filaments. Likewise, the filament heating current, shown in large dots in Fig. 3-1, does not appear in Figs. 3-9 and 3-10.

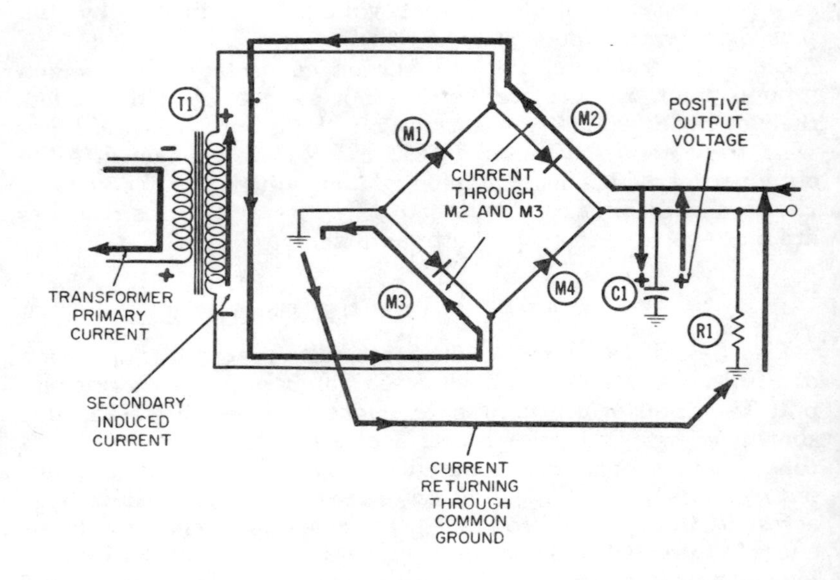

Fig. 3-6. The full-wave bridge rectifier—first half-cycle.

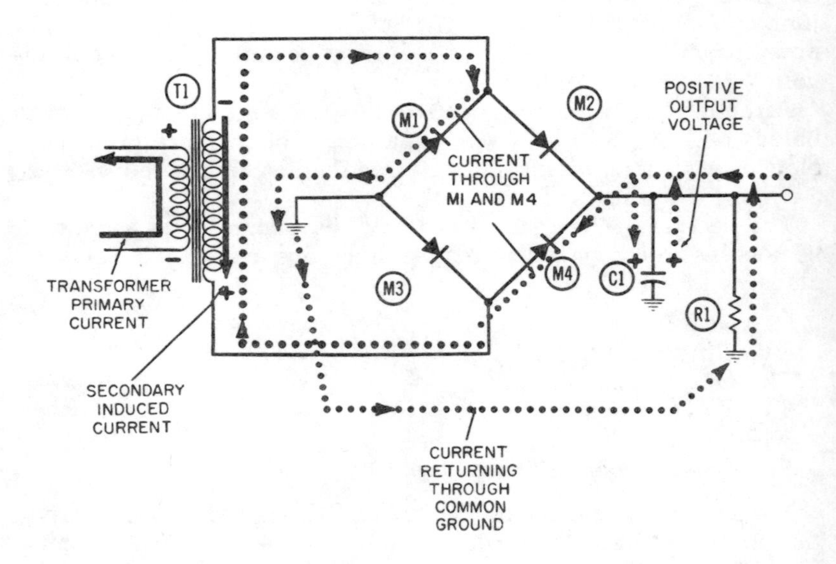

Fig. 3-7. The full-wave bridge rectifier—second half-cycle.

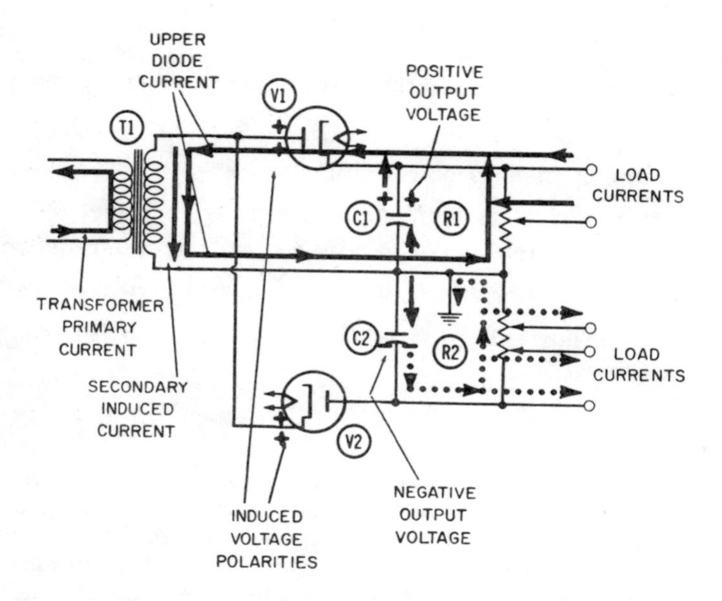

Fig. 3-8. The full-wave doubler rectifier—positive half-cycle.

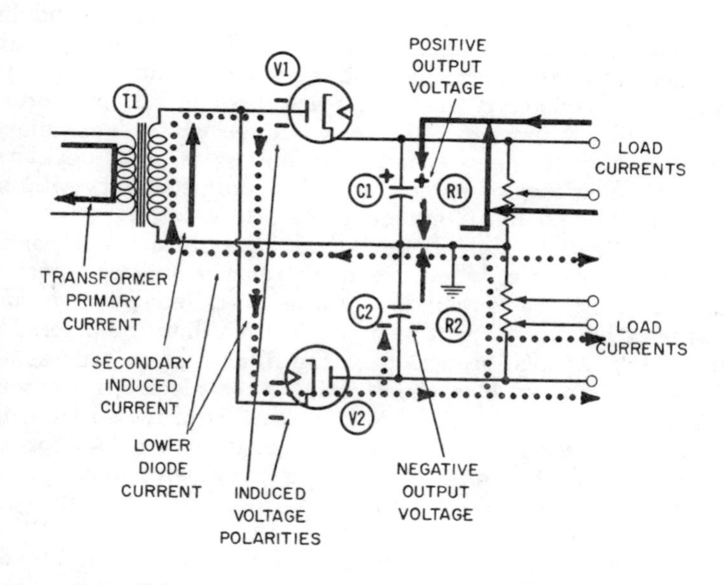

Fig. 3-9. The full-wave doubler rectifier—negative half-cycle.

Identification of Currents

Six principal electron currents are at work in this circuit, as follows:

1. Primary current in power transformer.
2. Induced secondary current in power transformer.
3. Current through upper diode V1 and its filter network.
4. Current through lower diode V2 and its filter network.
5. Two filter currents between C1 and ground and between C2 and ground.

Details of Operation

Fig. 3-8 has been designated the positive half-cycle of operation, because at this time the inductive action between the primary and secondary winding drives the upper end of the secondary winding positive. This positive voltage is indicated in Fig. 3-8 by the deep plus signs at the plate of diode V1 and also at the cathode of diode V2.

Diode V1 will conduct electrons, now that its plate is more positive than its cathode. This is the current shown a solid line and its complete path is from cathode to plate within the tube, then downward through the secondary winding and to the right. Here it enters resistor R1, flows upward through R1, and returns to the cathode. Almost the full peak voltage which is developed across the transformer secondary winding is also developed across R1, since the lower end of it and the secondary winding are both connected to a common ground.

Electrons flow through diode V1 during only a small portion of the first half-cycle, when the plate reaches its peak positive voltage. During the remainder of this so-called positive half-cycle, the cathode of V1 will be more positive than the plate. This is due to the integrating action of the filter system represented by R1 and C1. During that short period when electrons are being drawn across diode V1, this excess demand is provided from the upper plate of capacitor C1. This accounts in Fig. 3-8 for the upward flow of electrons leaving this plate, and also for the continuing positive voltage on it, as indicated by the plus signs in Figs. 3-8 and 3-9.

During the nonconductive portion of the positive half-cycle, and during the entire negative half-cycle, electrons flow *downward* onto the upper plate of capacitor C1. This direction of flow is indicated in Fig. 3-9 only. The filter current (a thin line in Fig. 3-8) between the lower plate of C1 and ground flows upward when electrons are drawn away from the upper plate of C1 for passage through the tube. During the remainder of each cycle the current flows downward, as indicated in Fig. 3-9.

No current can flow through diode V2 during the first half-cycle of Fig. 3-8 since its cathode is positive and its plate is negative (for reasons we shall discuss later).

The positive voltage created at the upper plate of capacitor C1 is one of the output voltages of the power supply. It serves to draw electron current toward it from any load or loads to which it is connected. Two such load currents are shown in Figs. 3-8 and 3-9. One enters the filter at the main output tap, and the other through a potentiometer tap. Such a potentiometer could be made to provide a positive output voltage with any value between zero (ground) and the full output value.

During the second, or negative, half-cycle depicted by Fig. 3-9, the inductive action across the transformer makes the upper end of the secondary winding negative. This temporary negative voltage is indicated by the minus signs at the plate of V1 and the cathode of V2. At the negative peak of this half-cycle, the cathode of V2 will become more negative than the plate. As a result, this diode will conduct. The complete path of this current (shown in large dots) is from cathode to plate within the tube, upward through R2, turning left to the lower end of the secondary winding. Then it flows upward through this winding, and downward to the cathode again.

R2 prevents each pulsation of plate current from flowing immediately to ground as it passes through the lower diode. Consequently, these electrons flow onto the lower plate of capacitor C2. A pool of negative electrons is soon created, and consequently a source of negative voltage. This pool of electrons persists during the entire cycle, and it drives a continuous electron current *upward* through resistor R2, toward the ground connection. This current can be seen in Figs. 3-8 and 3-9.

The filter current which flows between capacitor C2 and ground changes its direction in accordance with the main electron current through diode V2. When V2 is conducting, electrons flow onto the lower plate of C2. This action forces the filter current to flow upward from C2, toward ground. When V2 is not conducting, the filter current flows back onto the upper plate of C2 from ground. The latter condition is depicted most clearly in Fig. 3-8.

The three output points on resistor R2 constitute sources of negative rather than positive voltage. Negative voltages are required for various reasons in some types of equipment. A negative voltage source *delivers* electrons *to* a load, rather than drawing them away as a positive voltage source does. For this reason, load currents are shown flowing *away* from the various taps on resistor R2.

Since the full transformer-secondary voltage appears across each resistor separately, and since the two resistors are in series, the input voltage has in effect been doubled. As an example, if the peak primary voltage were 150 volts and the transformer turns ratio were 2-to-1, then the peak secondary voltage would be 300 volts. Thus, the positive output voltage on capacitor C1—measured between the upper plate of C1 and ground—would approach 300 volts, and so would the negative voltage output on capacitor C2, measured between the lower plate of C2 and ground.

A voltage measurement across the entire output-resistor path, consisting of R1 and R2 in series, would indicate a difference of 600 volts. This is twice the peak input voltage across the transformer secondary.

Chapter 4

DIODE-DETECTOR CIRCUITS

The diode detector is one of the simplest and most widely used circuits for detection of a carrier wave. The term "detection" is frequently used synonymously with "demodulation"—particularly whenever some modulation is imposed on the radio-frequency carrier wave at the transmitting source. The detector circuit produces a steady, or DC, voltage which will at all times be proportionate to the strength of the carrier wave. In fact, this voltage depends on the carrier wave for its existence. Consequently, if the carrier signal is turned on and off by keying the transmitter, the detector circuit will provide a means of knowing this at the receiver, and therefore a means of "reading" the keying.

CIRCUIT DESCRIPTION

Figs. 4-1 and 4-2 show the two half-cycles of operation of the diode detector. The circuit components and their functional titles are:

R1—Diode load resistor.
C1—RF tank capacitor.
C2—Coupling capacitor.
T1—Output IF transformer.
V1—Diode tube.

There are only two main families of currents at work in this circuit:

1. Radio-frequency alternating currents .
2. Unidirectional tube current .

The radio-frequency tank consists of transformer T1 and capacitor C1. This circuit is resonant at the particular radio frequency being received. An oscillation of electrons will be built up in this tank and be sustained by another RF current which exists at the same frequency in the primary winding of the transformer. Both currents are shown.

RF TANK CURRENT

Fig. 4-3 shows a series of sine waves representing the radio-frequency voltage and current in the tank. Unless otherwise stated, such a voltage sine wave normally applies to the voltage at the *top* of the tank; this is the point of coupling to the plate of the diode detector. The first half-cycle is centered at the instant of maximum negative voltage, as indicated in Fig. 4-1 by the deep minus signs on the top of plate tank capacitor C1. The solid line indicating electron current is shown during this half-cycle as flowing upward through the transformer secondary, thus delivering electrons into C1.

The tank voltage is represented by the plus and minus signs on the upper plate of C1 in Figs. 4-1 and 4-2. It drives an external current (a straight line) up and down through diode load resistor R1. When this tank voltage is at its maximum negative value midway in the negative half-cycle, the electrons in this external current will be flowing *downward* through R1 at their maximum rate.

By the same token, when the tank voltage is at its maximum positive value midway in the positive half-cycle, the external-current electrons will be flowing *upward* through R1 at their maximum rate. This will cause the top of the resistor to reach its maximum positive voltage, which is just high enough to overcome the negative voltage caused by the accumulation of electrons on the right plate of C2. As a result, the diode can again conduct.

The normal path for this external current driven by the tank current is up and down through diode load resistor R1. There is a simple rule of thumb for determining the instantaneous polarity of a voltage associated with any current flow. Recall that current is nothing more than electrons in motion; also, that all electrons have a negative charge and are therefore repelled by each other and by negative voltages in general. So, in flowing through a conductor, electrons always head *toward* a point of more positive (less negative) voltage than the point *from* which they are leaving. Since the lower end of the resistor is tied to ground, it will always be at zero voltage. Therefore, when electrons flow downward through R1, as they do in Fig. 4-1, the driving voltage at the top of the resistor must be negative. Conversely, when they flow upward as in Fig. 4-2, this driving voltage must be positive.

UNIDIRECTIONAL CURRENT

Diode tubes have one important characteristic—when their plate voltage is more positive than their cathode voltage, electrons will cross the evacuated space from cathode to plate. This condition is fulfilled during a portion of the second half-cycle (but not during all of it), and electron current is a solid line crossing the tube. The complete path for this current is from cathode to plate and downward through diode load resistor R1 to ground, then through ground and back to the cathode. During its journey through the tube and to the top of the resistor, this DC current is pulsating at the basic radio frequency. Because of the delay in entering and passing through the resistor, the electrons from the detector tube accumulate on the right plate of capacitor C2. Here they form a constant negative voltage, as indicated by the minus signs on the capacitor. Some electrons flow into this pool during each pulsation from the detector, and an equal number flows out during the entire cycle. The latter flow is pure rather than pulsating DC.

CIRCUIT OPERATION

It is important to distinguish between the two different currents which flow through the diode load resistor. The current just discussed, along with its associated negative voltage on the right plate of capacitor C2, applies a fixed negative voltage to the plate of the diode. Before the diode can conduct, the other current (a solid line) must first overcome the fixed negative voltage. This will occur only during that portion of the positive half-cycle when this current is flowing upward at or near its maximum rate. During these moments, it will exceed the amount of fixed current which regularly flows downward through the resistor each time the diode conducts. The net result of the two opposing voltages is a small, momentary positive voltage at the diode plate, permitting more tube current to flow. These moments occur midway in each positive half-cycle. Fig. 4-3 shows a small amount of plate current flowing whenever the tank and plate voltages reach their positive peaks. It can be seen that the plate voltage in line 3 of Fig. 4-3 is negative most of the time. This is due to the accumulation of electrons on the right plate of C2 as a result of diode conduction, followed by the filtering action of C2 and R1.

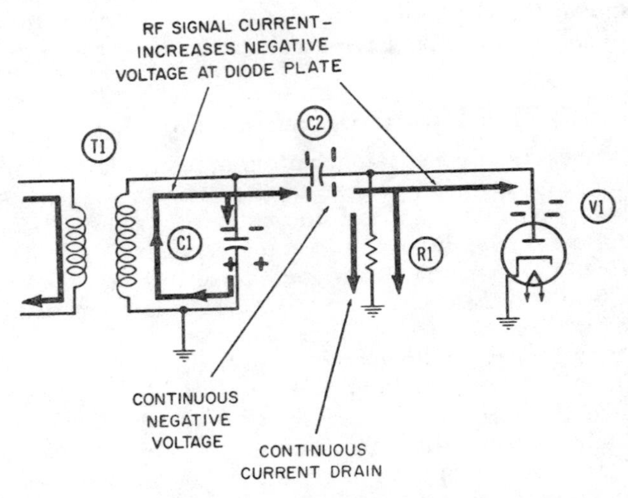

Fig. 4-1. The diode-detector—negative half-cycle.

The action of the tank voltage alternately increases and decreases this negative voltage, so that the plate voltage at any instant is the sum of the two voltages. Only when the diode plate is more positive than the cathode will plate current flow.

Let us look more closely at line 3 of Fig. 4-3, which represents the voltage at the diode plate. This voltage is mostly negative and is caused by the accumulation of electrons on the right plate

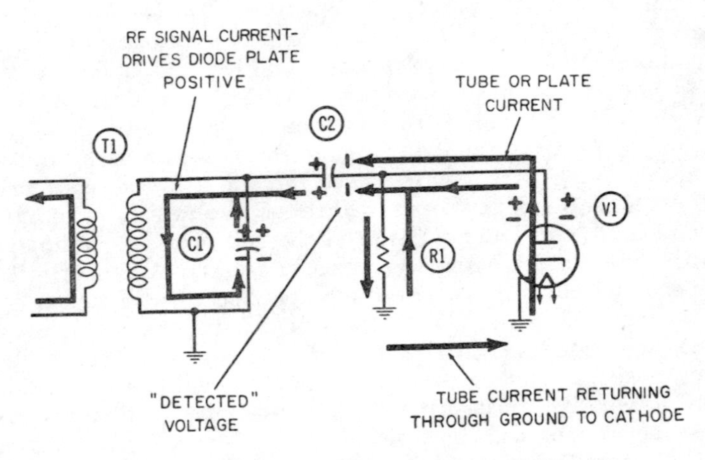

Fig. 4-2. The diode-detector—positive half-cycle.

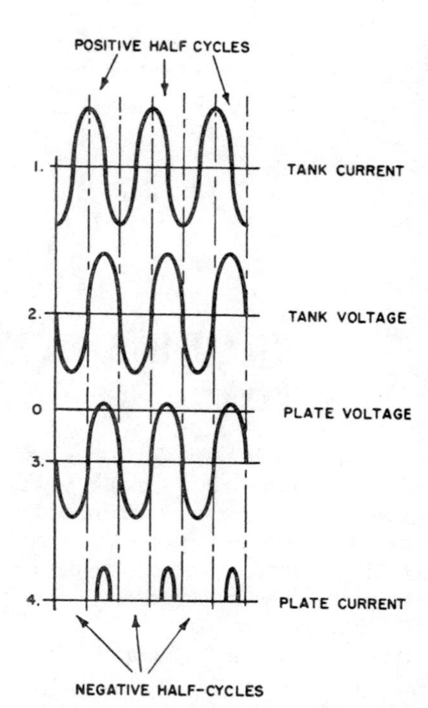

Fig. 4-3. Tank- and plate-current and -voltage waveforms in the diode detector circuit.

of C2. The *size* of this negative voltage depends on the strength of the radio-frequency current in the tank; and the latter, of course, depends in part on the strength of the received signal. If the carrier signal is cut off at the transmitter, as is done during keyed CW communications, the tank current will die out in a relatively few cycles. The diode will then stop conducting, and the negative voltage on the right plate of capacitor C2 will drop to zero as the electrons stored there drain downward through the diode load resistor to ground.

When the carrier signal is resumed at the transmitter, the tank current comes back into being, the diode conducts, and a negative voltage again builds up at the top of the diode load resistor. Thus, with this circuit, what amounts to a "DC voltage" has been produced; its existence depends on the existence of a carrier signal, and the strength of this voltage will vary in accordance with the amplitude variations (modulation) of the carrier. Hence, we can say the carrier signal has been "demodulated," or "detected."

Chapter 5

DIODE DETECTOR WITH AUTOMATIC VOLUME CONTROL

Most diode detectors also include an automatic volume control (AVC) circuit, along with the basic detector discussed in the previous chapter. Automatic volume control is a fairly elaborate function within radio receivers. Its purpose is to maintain a constant audio output, or volume, in the face of possible changes in signal strength due to abnormal propagation conditions. Signal fades and build-ups are the obvious results of these irregularities in propagation. It would be most disconcerting for the listener to have to continually adjust the volume on his receiver in order to compensate for a program becoming suddenly too loud or faint. Therefore, automatic volume control is added to virtually all radio receivers.

SUMMARY OF ILLUSTRATIONS

A total of eight circuit diagrams and seven waveform diagrams are used to illustrate the various actions in the operation of a diode detector with AVC.

The first two circuit diagrams (Figs. 5-1 and 5-2) show successive negative and positive half-cycles of unmodulated radio-frequency signal current being received and detected by diode detector V1 (this tube is an integral part of the total AVC circuit). Certain DC voltages come into existence when this detection occurs. These are shown as the three detected voltages on the upper plates of capacitors C2, C3, and C4, and also as the AVC voltage on the upper plate of capacitor C5. The electrons which constitute these detected voltages come originally from the plate current of the tube. Consequently they, too, are shown as lines. Fig. 5-3 shows negative and positive half-cycles of tank voltage and current, and also of detector plate current and AVC voltage.

The plate current leaves the tube in the form of pulsations at the signal frequency. Filter capacitor C2 bypasses these pulsations to ground. This bypassed, or filtered, RF current is shown.

The next two circuit diagrams (Figs. 5-4 and 5-5) show the actions occurring in the audio half-cycles which precede a modulation trough and peak. Also, Fig. 5-6 shows a modulated waveform for each condition. It also portrays the pulsations in plate current, the negative audio and AVC voltages, and finally, a sine-wave representation of the current which must flow through an AVC resistor (such as R4) when RF signals are being demodulated.

The next two circuit diagrams (Figs. 5-7 and 5-8) show the voltage-current conditions while a carrier signal is fading. The waveform diagrams in Figs. 5-9 and 5-10 show the changes occurring in the carrier signal, the resultant audio voltage, and finally the change in AVC voltage during a carrier fade.

The last two circuit diagrams (Figs. 5-11 and 5-12 show the voltage-current conditions while a carrier signal is building up. The associated waveform diagrams of Figs. 5-13 and 5-14 illustrate the changes occurring in the carrier signal, the audio voltage, and the AVC voltage during a carrier build-up. Fig. 5-15 is a graphical representation of the audio voltage developed across output resistor R2 during a period of carrier fade and a period of carrier build-up.

Some of the important truths the reader should expect to learn from these diagrams and the text include:

1. The changes in audio voltage on capacitors C2, C3, and C4 from modulation peak to trough, and what causes them.
2. Why the AVC voltage on capacitor C5 does not change between modulation troughs and peaks.
3. What causes an audio current to flow up and down in output resistor R2.
4. What causes the AVC current to flow back and forth in AVC resistor R4 at the audio frequency, and why this current flow does not change the amount of AVC voltage stored on capacitor C5.
5. What causes the negative AVC voltage to decrease during a carrier fade and to increase during a build-up.

CIRCUIT DESCRIPTION

The components which make up this circuit, and their functional titles are:

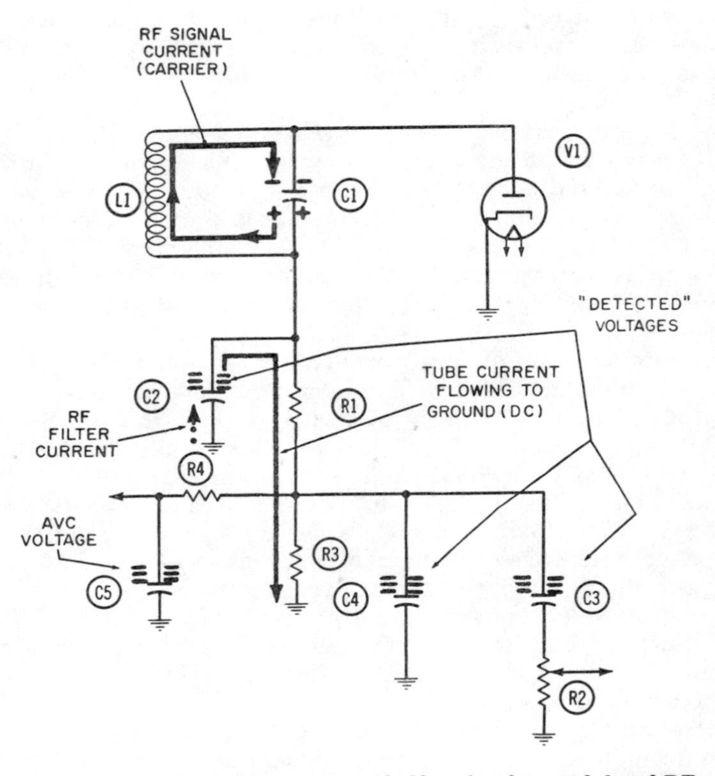

Fig. 5-1. The AVC circuit—negative half-cycle of unmodulated RF.

R1—Filter resistor (works in conjunction with C2).
R2—Audio-output resistor.
R3—Primary load resistor.
R4—AVC resistor.
C1—RF tank capacitor.
C2—RF filter capacitor.
C3—Output coupling capacitor.
C4—Additional RF filter, or bypass, capacitor.
C5—AVC storage capacitor.
L1—RF tank inductor.
V1—Diode tube.

Five electron currents are at work in this over-all circuit. They are:

1. Radio or intermediate-frequency tank current.

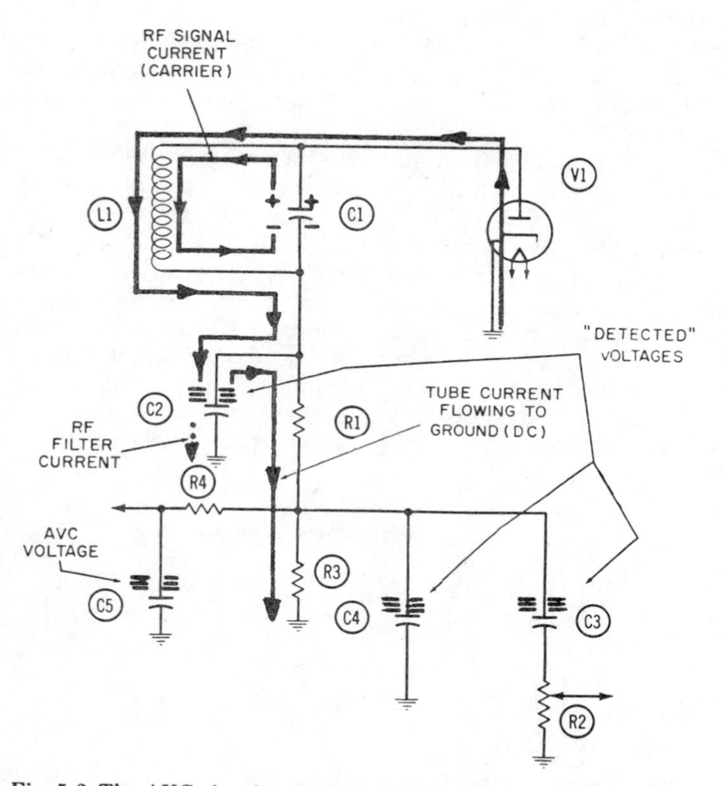

Fig. 5-2. The AVC circuit—positive half-cycle of unmodulated RF.

2. Pulsating radio or intermediate-frequency DC flowing through the detector.
3. AVC filter current.
4. Modulation signal, usually called audio.
5. Radio-frequency filter currents.

Voltage levels, both for audio and for automatic volume control, are shown by thin minus signs. There are three kinds of time periods, which might be referred to as cycles, that are important in understanding this circuit. These periods are the short time required for one RF or IF cycle, the somewhat longer time for one modulating cycle, and the much, much longer time required for signal fades or build-ups. A typical intermediate frequency (IF) in broadcast receivers is 455 kilocycles per second. Thus the time required for one IF cycle is 1/455,000 second, or slightly more than two microseconds.

If the audio tone being demodulated is middle C, then its frequency will be 256 cycles per second. Thus the time for one such audio cycle is 1/256 second, or about four milliseconds. This is 4,000 microseconds—about *two thousand* times longer than the time of one IF cycle.

Even this is short compared with the time required for a signal to completely fade or build up. The latter may take several seconds or even minutes. Thus, many hundreds or even thousands of audio cycles may occur during one fade or build-up.

These relative time values are important to understanding demodulation and the automatic control of volume.

In order to depict one complete audio cycle in the accompanying diagrams, it would be necessary to show the 2,000 IF cycles occurring. This would hardly be practical. Likewise, to show a complete signal fade followed by a build-up, we would have to show several thousand audio cycles—and consequently many millions of IF cycles.

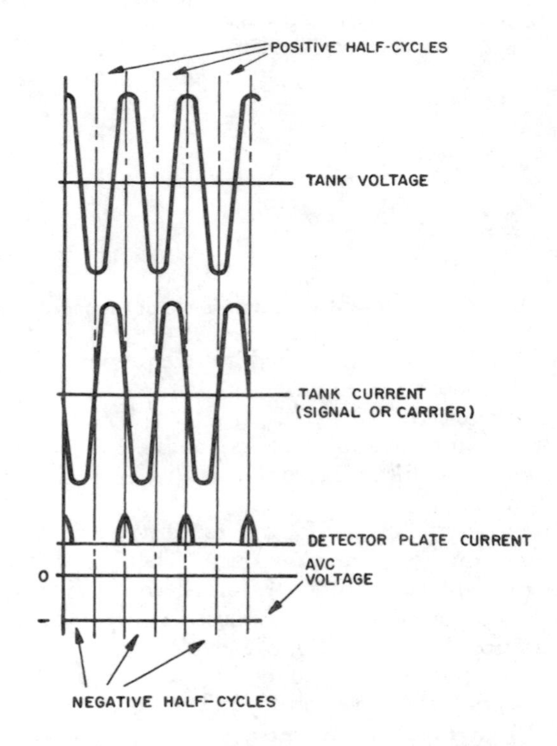

Fig. 5-3. AVC circuit waveforms when unmodulated RF is being received.

As is true with any vacuum tube, a closed path must be available between plate and cathode of the detector diode in order for detector current to flow. This path includes coil L1 and resistors R1 and R3. The thin line has been chosen to represent detector current; it completes a path from plate to ground by flowing through L1, R1, and R3.

R1 and C2 together act as a radio-frequency filter. Resistor R3 and capacitor C4 act as an additional RF filter. Resistor R2 and capacitor C3, in conjunction with R3 and C4, develop an audio-voltage output from the detector or demodulator stage.

Resistor R4 and capacitor C5 are the two components which provide the necessary compensation for varying signal strengths. Without them, the remainder of the circuit would continue to function as a fairly elaborate diode demodulator. In order to understand the AVC function, it is necessary to begin with the diode detector and to consider how it operates to detect, or demodulate, an incoming radio-frequency signal.

CIRCUIT OPERATION

In Fig. 5-4 when a modulation trough is occurring, we see two thin minus signs on the upper plate of filter capacitor C2. Between the lower plate of this capacitor and ground, the dotted arrows indicate current flow in both directions. This current, which is flowing at the basic radio frequency, represents the RF pulsations being filtered out of the detector current before it continues its journey back to the cathode.

Capacitor C2 is of such a size that its reactance at this particular radio frequency will be substantially less than the resistance of the alternate path available to the detector current. This path is through the series combination of R1 and R3 to ground.

Figs. 5-4 and 5-5 each cover a period which includes *many* RF cycles. The reader should recognize that this filtering current flows *downward* when a pulsation of detector current flows onto the top plate of C2, and that it flows *upward* in the interim before the next pulsation occurs.

When a detector pulsation of current is small, as it is near a modulation trough, only a small filtering current must flow. On the other hand, when a pulsation is large, as it is near a modulation peak (Fig. 5-5), a large amount must flow.

Although this filtering current is driven by the pulsations of detector current (a thin line), the filtering current is shown in large dots in order to help clarify the actual closed path of detector current from plate to ground. While the energy in the

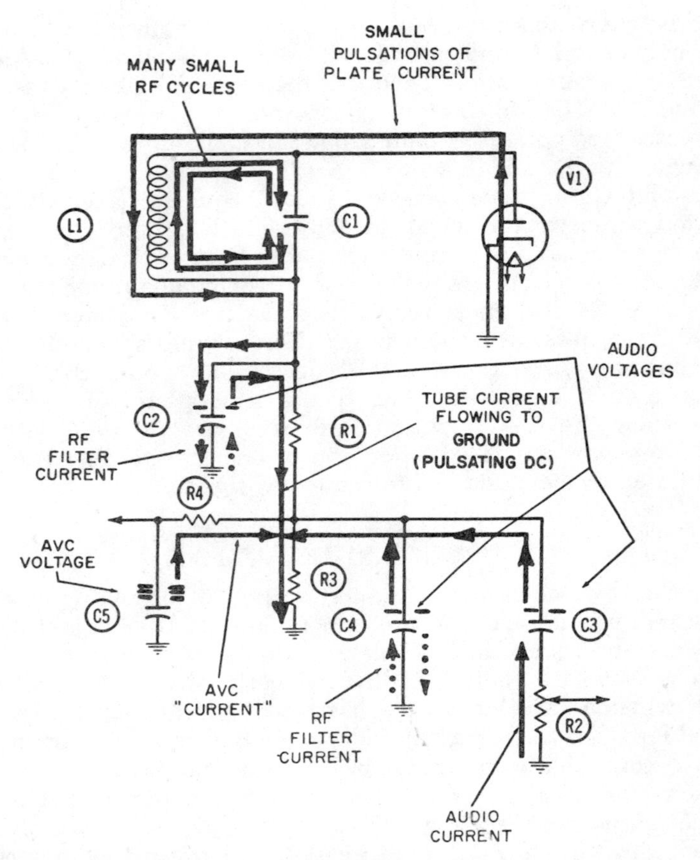

Fig. 5-4. Operation of the AVC circuit—audio half-cycle preceding a modulation trough.

individual pulsations is passed across the capacitor plates, the electrons which compose the detector current form a pool of negative voltage on the upper plate of C2 until they can escape downward through R1 and R3 to ground and the cathode.

This pool of electrons on the upper plate of C2 constitutes the first appearance of the audio voltage in a receiver system. Receivers are customarily analyzed in the order in which a signal progresses through from antenna to speaker. Until this detector stage is reached, no audio voltage has been generated in the system.

In Fig. 5-4, the amount of negative voltage on capacitor C2 is indicated by the two minus signs. In Fig. 5-5 this negative voltage has been greatly increased, as indicated by a total of ten such

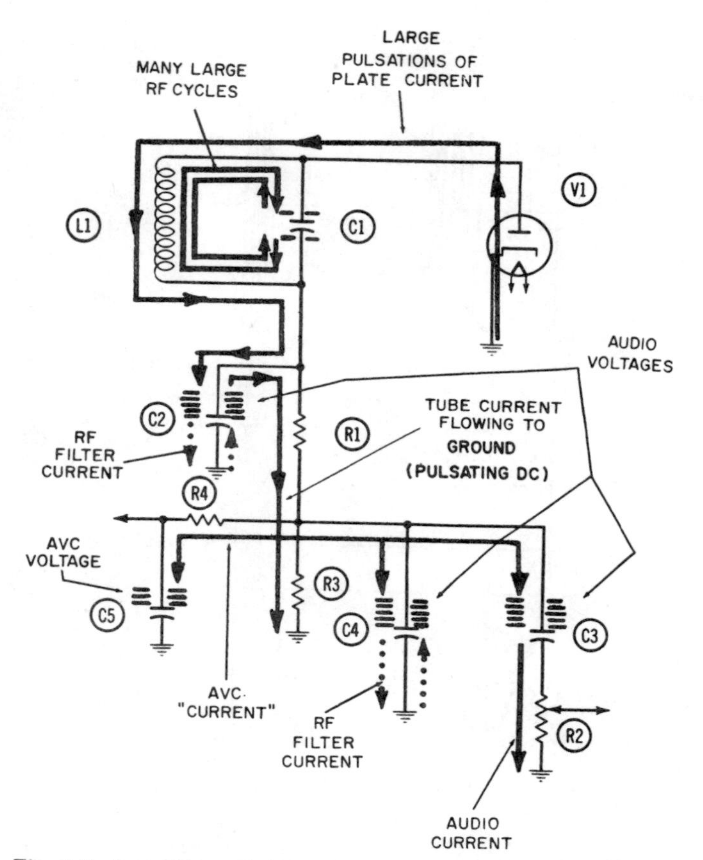

Fig. 5-5. Operation of the AVC circuit—audio half-cycle preceding a modulation peak.

signs. These two figures constitute one whole cycle of the audio, or modulating, voltage. The voltage on this capacitor indicates the rise and fall of the audio voltage in accordance with the peaks and troughs of the modulating envelope. During each half-cycle of the audio voltage, many hundreds or even thousands of pulsations of detector current will be coming into the capacitor.

The waveforms of Fig. 5-6 should be studied in conjunction with Figs. 5-4 and 5-5. Note that the sine wave of audio voltage in Fig. 5-6 is below the reference line of zero voltage at all times. In other words, the audio voltages which appear simultaneously on capacitors C2, C3, and C4 are negative at all times; and they fluctuate from their low negative value at the modulation troughs, to a high negative value at the modulation peaks.

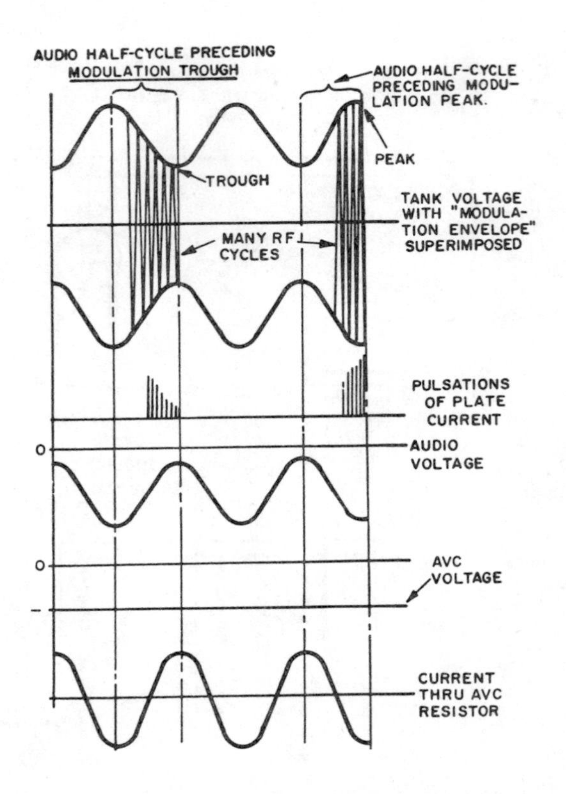

Fig. 5-6. AVC circuit waveforms when modulated RF is being received.

The electron current which flows downward through R1 and R3 is a pulsating unidirectional current, but the pulsations are occurring at the audio frequency. The least current flows when the voltage on C2 is lowest, during the modulation trough. Conversely, the most current flows when the voltage on C2 has its highest value, during the modulation peak.

Since an audio current flows through resistors R1 and R3, an audio voltage is developed across them. This is in accordance with Ohm's law. It is desirable for R3 to be considerably larger than R1, so that *most* of this audio voltage will be developed across R3. The reason for this will be apparent from analysis of the function of resistor R2 and capacitor C3. These two are in series with each other, and the series combination is in parallel with R3. R2 is a variable resistor across which the *output* audio voltage is developed. This potentiometer is commonly used as the volume control in receiver systems.

Capacitor C4 serves primarily as an additional radio-frequency filter, on the likely assumption that the combination of R1 and C2 will not do a thorough job of filtering the RF pulsations out of the detector current and into ground.

The combination of C3 and R2 is critical because the two develop the output audio voltage for the next amplifier stage. The product of these two components must give a short time-constant when compared with the lowest audio frequency being demodulated. When this short time-constant condition has been met, an alternating current at the audio frequencies will flow up and down through resistor R2. This current will be drawn *upward* as electrons flow away from the upper plate of capacitor C3 during the modulation trough, as indicated in Fig. 5-4. Conversely, the current will be driven *downward* through R2 as electrons flow onto the upper plate of C3 during the modulation peak, as indicated in Fig. 5-5.

This two-way current through R2 is the first appearance in the receiver of *alternating* current at the audio frequencies. The audio current which flows through R1 and R3 is unidirectional. In other words, it is DC which pulsates at the audio frequencies. Frequently this current is called pure DC with an alternating component superimposed on it. Both concepts are correct, and the reader should use the one that is clearer to him.

The alternating current through resistor R2 develops an audio voltage across it, in accordance with Ohm's law. This voltage then becomes the output of the detector stage. By moving the arm of the potentiometer to the various points on the resistor, it is possible to apply any given percentage of this output audio voltage to the grid of the first amplifier stage. By this means, the receiver volume can be adjusted to any desired level.

Although the voltage levels on C3 and C4 follow the changes occurring on C2, the two are slightly lower than the voltage on C2. The amount of reduction in audio voltage is an accurate reflection of the relative voltage drops occurring across R1 and R3, respectively. The voltage drop across R1 is a loss, which is made acceptable by the filtering achieved from adding R1 and C2 to this circuit.

Resistor R4 and capacitor C5 form the AVC filter combination. In order to understand how these components perform, it is desirable to visualize the current through R4, the voltage on the upper plate of C5, and their interrelationship with the audio voltage on C4 and C3. Since the voltage level on these capacitors is rising and falling in accordance with the modulation, we can visualize an *average* voltage value for a whole modulating cycle. The detected voltages shown in Figs. 5-1 and 5-2, when the car-

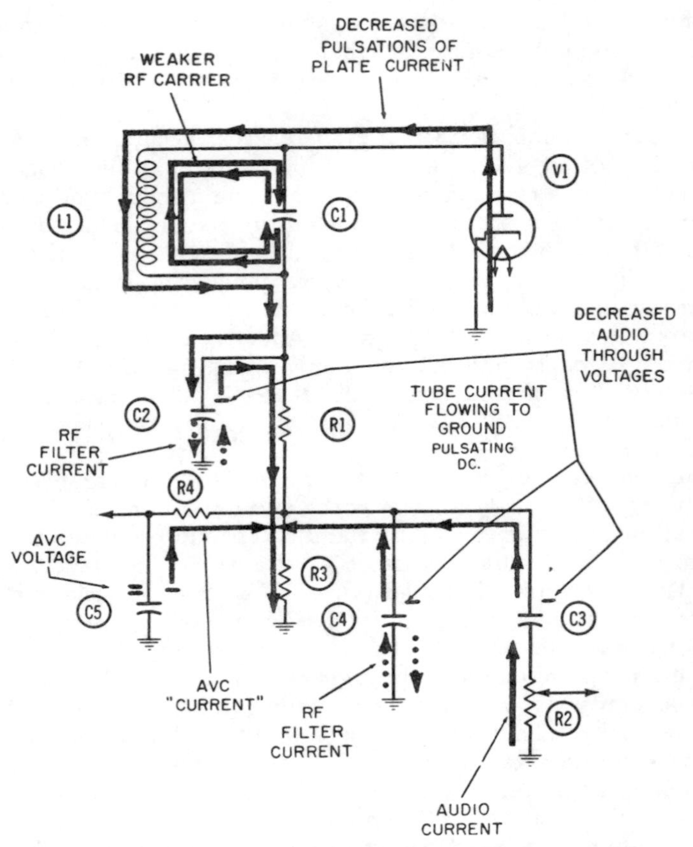

Fig. 5-7. Operation of the AVC circuit—modulation trough during a carrier fade.

rier is unmodulated, are a fair approximation of what this average value would be. During the modulation trough of Fig. 5-4, the voltage level on these capacitors is *below* this average level. Conversely, it is *above* the average value during a modulation peak in Fig. 5-5.

Even with changes in percentage of modulation, this average voltage level on the capacitors will not change, although the amounts of voltage above and below the average will. For instance, with a very low percentage of modulation impressed on the carrier wave, the capacitor voltage during the modulation trough will be only slightly below average. On the other hand, during a modulation peak it will be an equal percentage above the average level.

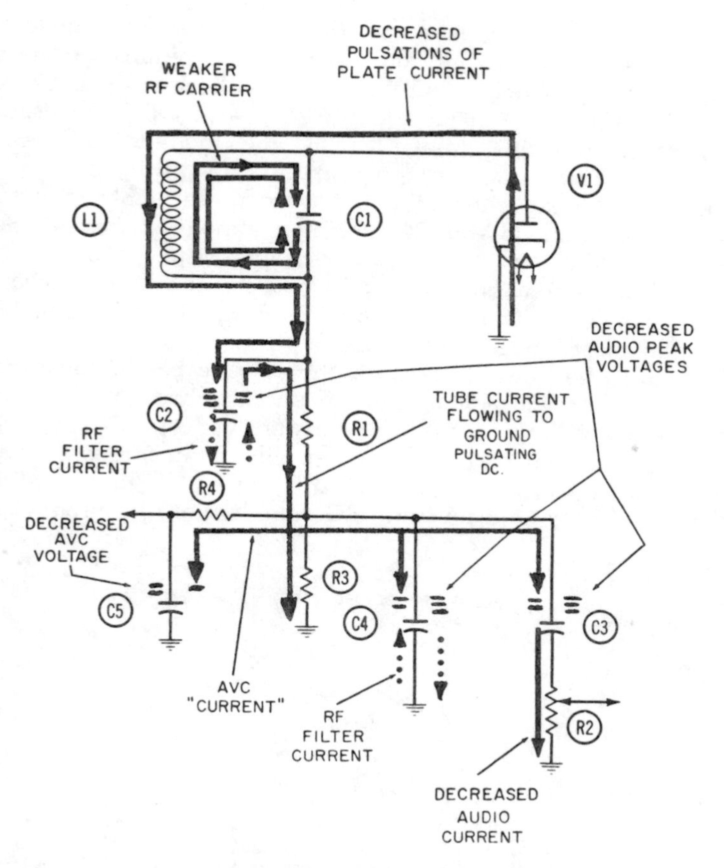

Fig. 5-8. Operation of the AVC circuit—modulation peak during a carrier fade.

Let us consider how capacitor C5 acquires its charge of negative voltage, as indicated by the series of thin minus signs. If there were no modulation at all on the carrier signal, then the average voltage levels on C3 and C4 would be quickly achieved and maintained because of continued detection of the unmodulated carrier. Under this condition, electron current would flow from the upper plates of C3 and C4, through resistor R4, to the upper plate of C5. Eventually C5 would become charged to the same average value of negative voltage. This leads us to an important point in the operation of AVC circuits(and in understanding them)—namely, *the AVC capacitor will charge to the average value of the audio voltage.*

Once the reader visualizes that this can and does happen to an unmodulated carrier, only one more step is required to demonstrate that this principle still holds where there is some modulating voltage on the carrier signal. Figs. 5-4 and 5-5 are useful for demonstrating this point. During the modulation trough depicted by Fig. 5-4, electron current (a thin line) will flow *out* of AVC capacitor C5 and toward those points of lesser negative voltage on C3 and C4. During the modulation peak depicted by Fig. 5-5, a greater negative voltage exists on C3 and C4. Now electron current flows *away* from these capacitors, toward the AVC capacitor. The voltage on this capacitor has remained substantially unchanged during the entire cycle.

The reason the voltage level on the AVC capacitor does not change during an individual cycle of audio voltage can best be understood by calling on the time-constant concept. When a resistance in ohms is multiplied by a capacitance in farads, the resultant product has the dimensions of time and is measured in seconds. In a long time-constant RC combination, its product is

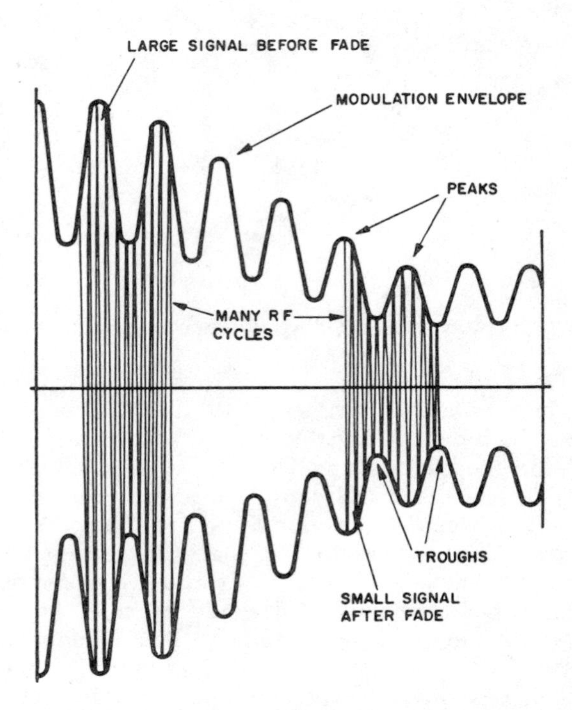

Fig. 5-9. The modulated carrier during a signal fade.

A word of caution: this circuit arrangement, including component values and the applied voltage value of 4 volts, has been chosen for one reason only—to clarify some of the important fundamentals of transistor operation in a sample circuit. They might not prove to be the best component values for low-frequency audio amplification in a particular circuit. Also, even though the 2N105 transistor works equally well in a higher voltage range, that portion of its characteristic curve which applies in this range has been omitted from Fig. 1-11.

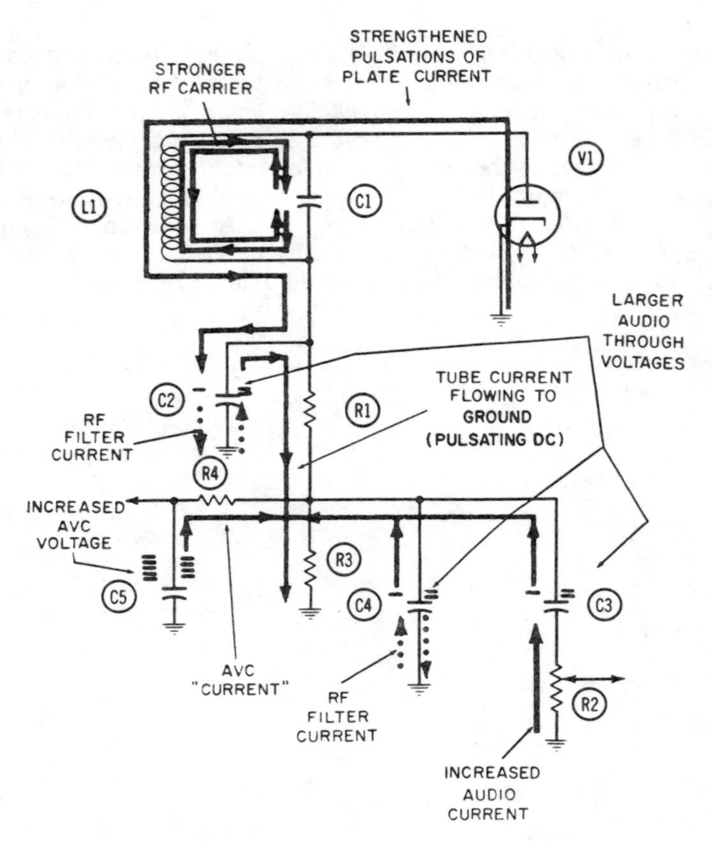

Fig. 5-11. Operation of the AVC circuit—modulation trough during a carrier build-up.

through it as the voltage levels on C3 and C4 change in accordance with the modulation peaks and troughs. This small amount of electron traffic in and out of C5 will be insignificant *in comparison with the electrons stored there.* Consequently, the AVC voltage—which is the negative voltage stored on the upper plate of AVC capacitor C5—*does not change with the audio or modulating voltage.* This is an important point in understanding automatic volume control.

The AVC line connected to the junction of R4 and C5 leads directly to the control grids of one or more RF amplifier tubes. The carrier signal must be processed through these tubes before it reaches tuned tank L1-C1. Hence, this increased negative voltage shows up as a larger negative bias on the grids of these tubes

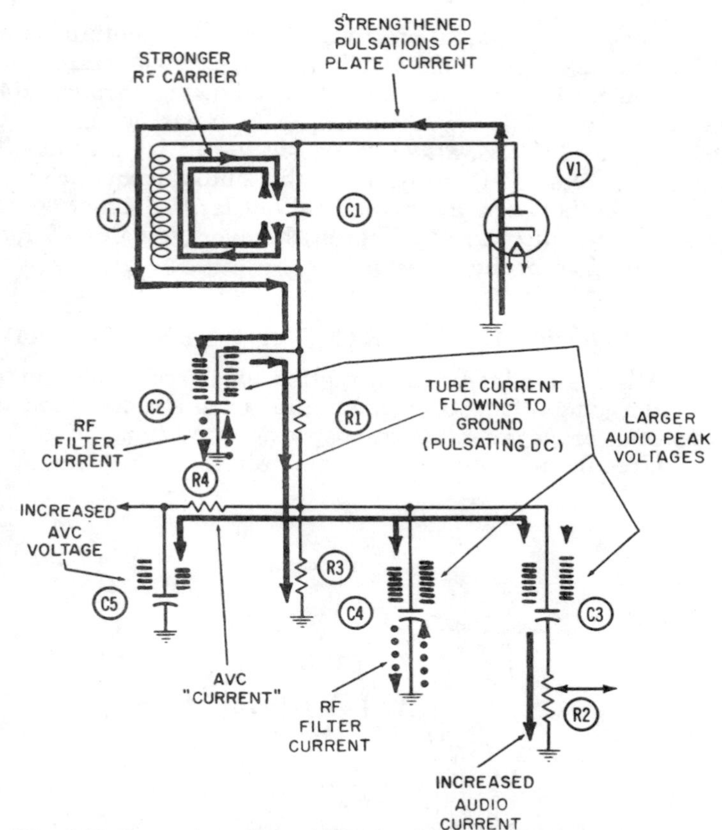

Fig. 5-12. Operation of the AVC circuit—modulation peak during a carrier build-up.

and reduces their amplification, to compensate for the stronger signal.

A signal fade or build-up may take place over a period of several seconds. During this time, many hundreds of cycles of audio voltage will probably occur. The audio current (a solid line) which flows back and forth through R2 will of course continue to do so during this time. However, the voltage on C5 can only be changed by the current through resistor R4. This resistor, along with R3, provides the only path to ground for the electrons which make up the AVC voltage on C5. The only way this voltage can be decreased or increased is for electrons to empty out through R4, or for more electrons to come in through that path.

In the example shown, when a signal build-up occurs more electrons must flow into C5 during the modulation peaks than flow out during the modulation troughs. This flow continues until the AVC voltage has become stabilized at the new average audio voltage. During this interim, the current flowing through R4—to the left in Fig. 5-12—will exceed the current flowing to the right in Fig. 5-11. Thus, although two-way current continues to flow through the AVC resistor at the audio frequencies, its amount varies between individual half-cycles. In this way, the AVC voltage is automatically kept at the average value of audio voltage coming from the detector.

CIRCUIT ACTIONS DURING A CARRIER FADE

Fig. 5-9 is a simplified representation of a modulated carrier signal undergoing a fade, or reduction in signal strength, because of abnormal propagation or atmospheric conditions. Fig. 5-10 shows the resultant reduction in audio voltage after detection or

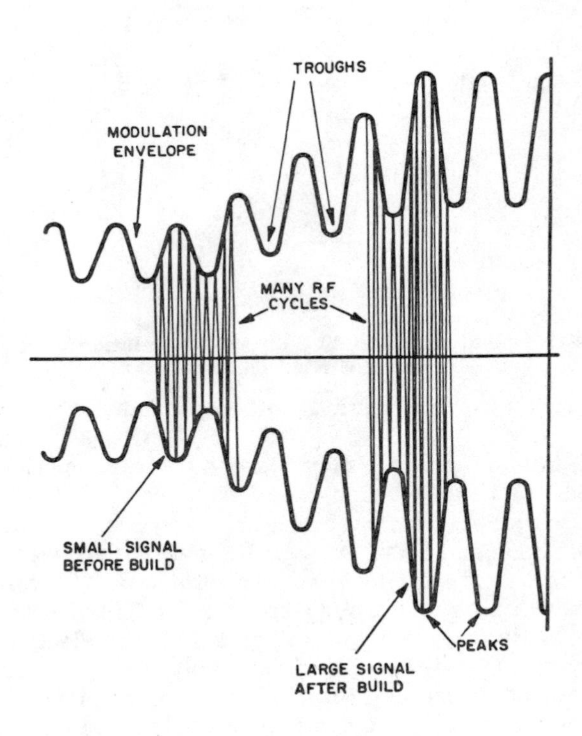

Fig. 5-13. The modulated carrier during a period of signal build-up.

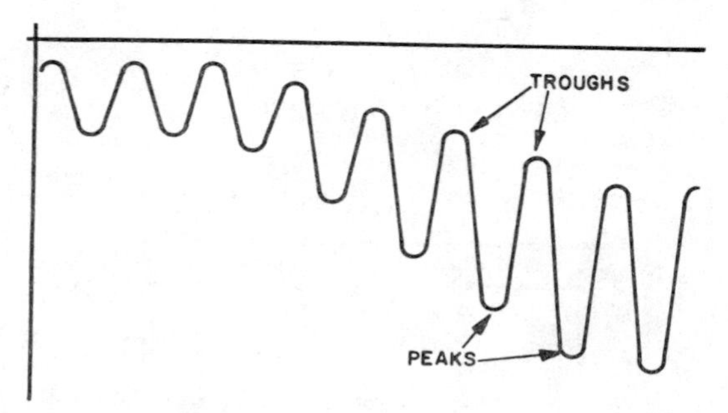

(*A*) *Audio voltage after demodulation.*

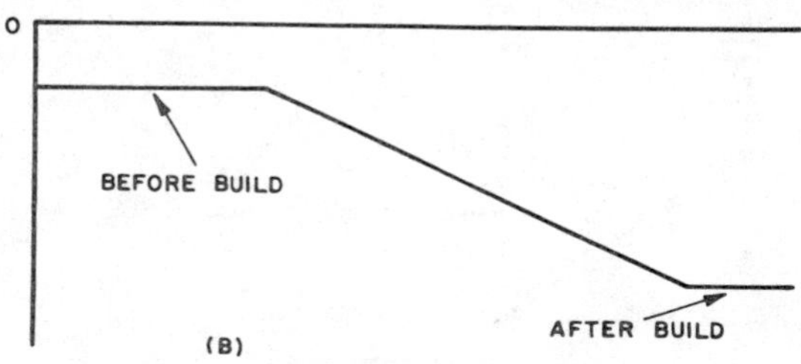

(*B*) *AVC voltage.*

Fig. 5-14. The audio and AVC changes occurring during a period of signal build-up.

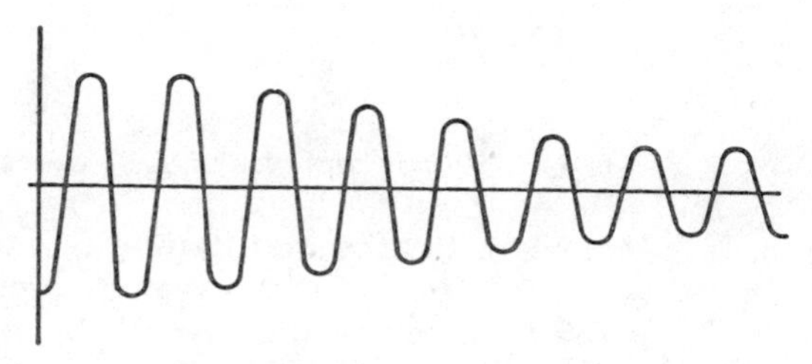

(*A*) *During signal fade.*

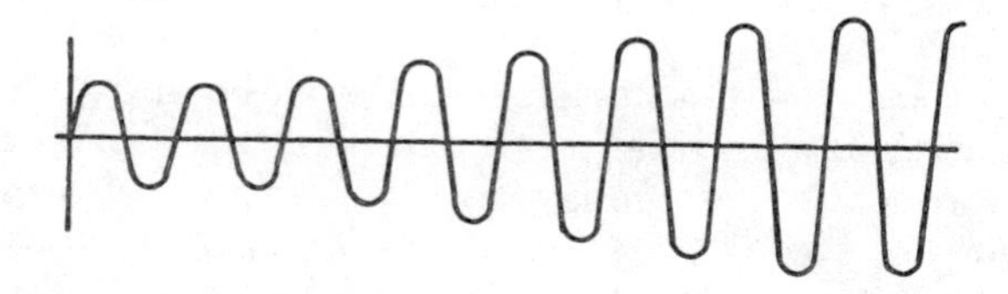

(*B*) *During signal build-up.*

Fig. 5-15. The audio output voltage during a signal fade and a signal build-up.

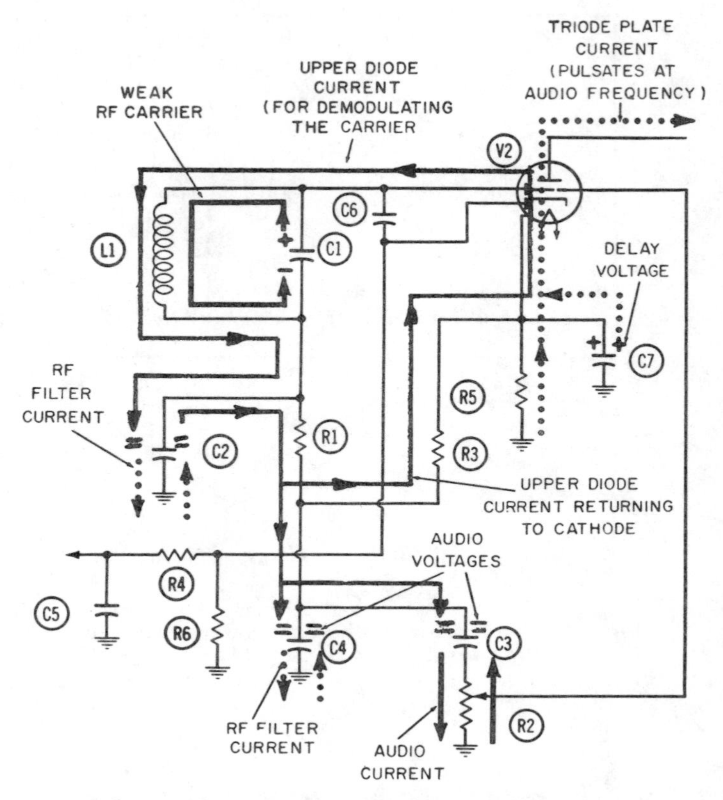

Fig. 5-16. The delayed AVC circuit—weak-signal operation.

demodulation. It is important to note that individual audio cycles are larger before the fade. Also note the more obvious fact that the *average value* of the audio voltage decreases from a high to a low negative value during the fade. The AVC voltage line in this figure represents the negative voltage stored on capacitor C5 in Figs. 5-7 and 5-8. This line faithfully reflects the average value of audio voltage.

Fig. 5-7 shows the relative sizes of the audio and AVC voltages at the moment of a modulation trough, after the carrier has faded. Fig. 5-8 shows the same voltages at the moment of a modulation peak. Compare these voltages with those that existed at the same points before the carrier faded (Figs. 5-4 and 5-5). You will see that the audio and AVC voltages are both lower. However, the amounts of AVC voltages indicated in Figs. 5-7 and 5-8 are intended to be equal, because the AVC voltage does not change

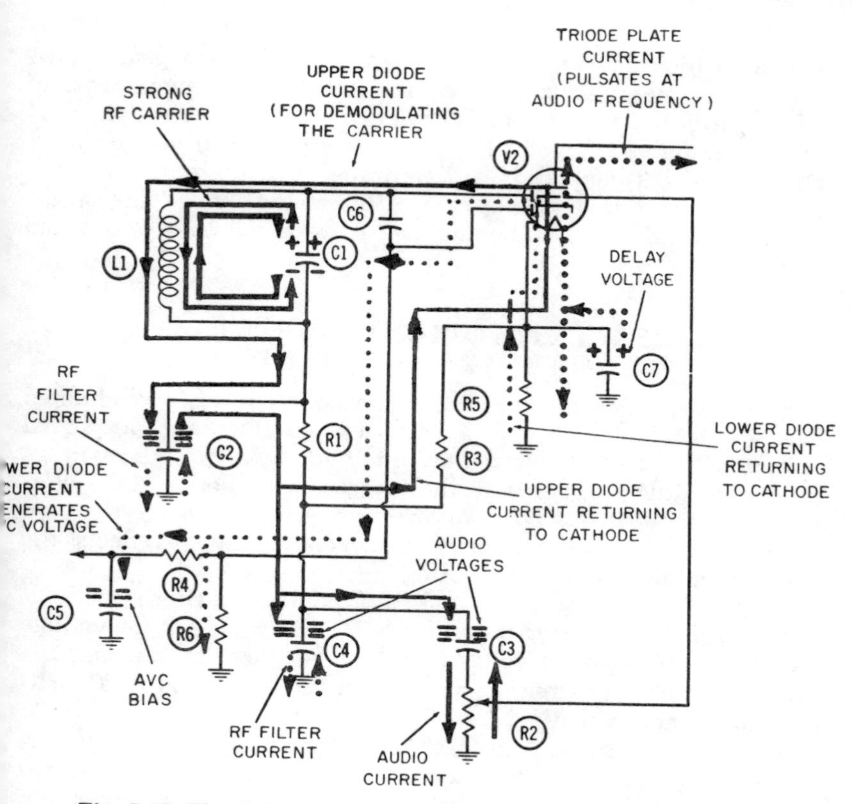

Fig. 5-17. The delayed AVC circuit—strong-signal operation.

from modulation trough to peak. This is due to the long time-constant nature of the R4-C5 filter combination, as previously discussed.

During a carrier fade, the AVC current which flows through resistor R4 reduces the number of electrons stored on capacitor C5. This is normally a minute two-way current that flows at the audio frequency. During a modulation trough this current flows to the right, as shown in Figs. 5-4, 5-7, and 5-11, because the momentary audio voltage at the right end of the resistor is more negative than the stored AVC voltage at the left. Electrons will always flow from an area of higher to one of lower negative voltage as long as a conducting path is provided. During a modulation peak this current flows to the left, as shown in Figs. 5-5, 5-8, and 5-12.

Under normal conditions, the amount of AVC current which flows to the right during a modulation trough will just equal the amount which flows to the left during a modulation peak. However, when a carrier fade is occurring, these two components of current are no longer equal. More current flows to the right during the trough half-cycle than flows to the left during the peak half-cycle. This inequality will continue until the fade has been completed, and in this way the AVC voltage will "discharge" electrons until it again equals the average value of the audio voltage.

CIRCUIT ACTIONS DURING A CARRIER BUILD-UP

Fig. 5-13 shows a simplified representation of a modulated carrier signal undergoing a build-up, or increase, in signal strength. Fig. 5-14 shows the resultant increase in audio voltage after demodulation, and also the increase in negative AVC voltage as a result of this build-up in signal strength. As in the case of the fading signal, the AVC voltage accurately reflects the *average* value of the audio voltage.

Fig. 5-11 shows the audio and AVC voltage during a modulation trough, after a carrier build-up. By comparing these voltage pictures with the same voltages in Fig. 5-4, you can see that the build-up in carrier strength has caused the audio and AVC voltages to go more negative.

Fig. 5-12 shows these voltages to be larger than their counterparts in Fig. 5-5 before the carrier build-up. Between the modulation trough of Fig. 5-11 and the peak of Fig. 5-12, however, the AVC voltage itself does not change. The AVC current flowing through resistor R4 again provides the mechanism by which the AVC voltage becomes stabilized at the average value of audio voltage.

AUDIO-OUTPUT VOLTAGE

Fig. 5-15 shows the audio-output voltage developed across potentiometer R2 as the audio current (a solid line) is driven up and down through R2. As indicated in Figs. 5-4, 5-7, and 5-11, this current flows upward during modulation troughs. During modulation peaks, it goes downward, as shown in Figs. 5-5, 5-8, and 5-12. Excess electrons from the detector tube flow onto the upper plate of capacitor C3 during modulation peaks, driving an equal number away from the lower plate and down through the resistor. (This is the basic capacitor action.) As these excess electrons flow away from the upper plate of C3 during a modulation trough and discharge to ground through R3, an equal number will be drawn upward through R2 to the lower plate of C3.

During a carrier fade, less audio current will flow and therefore less output voltage will be developed across resistor R2, as indicated in Fig. 5-15A. The resultant reduction in volume would be undesirable to the listener. However, if we connect the grid of a previous RF amplifier tube to the AVC voltage point on capacitor C5, the lower AVC voltage will *reduce* the negative grid bias on the RF amplifier tube. This will increase the amplification of the signal and thus largely nullify the effects of the fading signal.

Likewise, during a carrier build-up the spread between peak and trough voltages on capacitor C3 is greater and more audio current will flow in output resistor R2. According to Ohm's law, the voltage developed *across* a resistor is proportionate at all times to the current flowing through the resistor. Thus, the audio-output voltage increases as the carrier strength builds up. This higher voltage, shown by Fig. 5-15B, may increase the loudness to the point where it is most unpleasant to the listener.

When the higher negative AVC voltage on capacitor C5 is connected to the control-grid circuit of a previous amplifier tube, the negative grid bias of the tube will be increased and its amplifying properties reduced. This reduces the total amplification of the signal and largely nullifies the undesirable effects which would otherwise result from the stronger signal.

DELAYED AVC

Figs 5-16 and 5-17 show two modes in the operation of a circuit designed to provide delayed automatic volume control. There are considerable similarities between this circuit and the one shown previously. To accentuate these similarities, individual components having identical functions in the two circuits have been assigned the same names. The component names and functions are:

R1—Filter resistor.
R2—Audio-output resistor.
R3—Primary load resistor.
R4—AVC resistor.
C1—RF tank capacitor.
C2—RF filter capacitor.
C3—Output coupling capacitor.
C4—Additional RF filter capacitor.
C5—AVC storage capacitor.
L1—RF tank inductor.

Additional components in the delayed AVC circuit, which were not present in the previous AVC circuit, are:

R5—Cathode filter capacitor.
R6—AVC return resistor.
C6—Isolating and coupling capacitor between the two diode plates.
C7—Cathode filter capacitor.
V2—Dual-diode/triode tube.

Principles of Operation

Delayed AVC allows us to eliminate the AVC on weak signals, but to enjoy its benefits when the incoming signal level reaches a predetermined strength. These objectives are accomplished by using separate diodes for signal demodulation and for automatic volume control. Recall that the electron current through the single diode in the preceding example was eventually used to provide an audio-output voltage as well as an AVC bias voltage.

Each diode plate in tube V2 operates independently from the other, and also independently from the triode operation within the same envelope. The fundamental principle of diode operation prevails in either case—that whenever a diode plate is more positive than the cathode, electrons will be drawn across the open space between the two electrodes.

Both diode plates are driven by the same alternating tank voltage. The tank current associated with this tank voltage is shown circulating between the inductance and capacitor within the tank. The half-cycle in Fig. 5-16 represents operation during a weak signal. At the moment shown, the tank current is flowing downward through L1 and thus delivering electrons to the lower plate of capacitor C1, making it negative and the upper plate positive.

This positive voltage is coupled directly to the upper diode plate and capacitively (by means of C6) to the lower diode plate. Even with a weak RF or IF signal applied from the tank, the upper diode plate will draw some electron current on the positive half-cycles. This diode current, shown by the solid thin lines, follows a similar path as the diode current in the previous example, with one important difference. After flowing through resistor R3, it goes directly to the cathode rather than ground.

This manner of connecting R3 to the cathode assures that the signal will be detected, even with weak signals. Since the multiple tube has a common cathode and since cathode biasing is used in connection with the amplifying function of the tube, the cathode is necessarily placed at a positive voltage. This voltage persists as long as electron current flows through the tube.

Depending on the requirements of the particular circuit, this positive cathode voltage might have any reasonable value—but for convenience, let us assume it is +3 volts. *This becomes the amount of delayed AVC the circuit is using.* The lower diode plate is returned to ground through resistor R6. So this diode plate will assume a reference voltage of zero. The alternating voltage resulting from the IF oscillations in the tuned tank is coupled to this diode plate by capacitor C6. As a result, the diode plate voltage will vary in either direction from this zero reference value.

In the weak-signal example shown in Fig. 5-16, the IF oscillations are considered small enough that they never raise the lower diode plate any higher than +3 volts. Consequently, no electron current can flow between the cathode and this plate, and no automatic-volume-control voltage will be developed.

In the strong-signal example of Fig. 5-17, the IF oscillations are large enough that they raise the lower diode plate well above +3 volts on each positive peak. Fig. 5-17 depicts one such positive peak occurring. As a result, the lower diode plate is now more positive than the cathode, and conduction occurs. The complete path of this electron current, shown in small dots , is from cathode to lower diode plate within the tube, then downward to the junction of the two AVC resistors R4 and R6. The ultimate destination is of course back to the chassis and then to the cathode. However, resistor R6 is made large enough to impede this flow and divert some of the electrons into resistor R4. Eventually a pool of electrons collects on the upper plate of capacitor C5. These accumulated electrons become the AVC voltage, which is applied to one or more control grids of the preceding RF (or IF) amplifier tubes.

This negative AVC voltage increases as the signal becomes stronger and vice versa, in the same manner shown in earlier figures of this chapter. Because of the similarity in function between these two circuits, a detailed analysis of AVC will not be repeated here.

Chapter 6

GRID-LEAK DETECTOR CIRCUITS

Grid-leak detection combines the functions of detection and amplification into one triode vacuum tube. As we said before, the primary purpose of detection is to demodulate a carrier current, or carrier wave, which has been modulated (i.e., changed) by the audio-frequency current representing the intelligence being conveyed. Like the diode detector discussed in the two previous chapters, the grid-leak detector is useful only for amplitude-modulated carrier currents.

CIRCUIT DESCRIPTION

Figs 6-1 through 6-4 represent successive half-cycles of operation during a modulation trough and a modulation peak. Since the peak is separated from the trough by many hundreds or even thousands of radio-frequency cycles, the operating conditions depicted by Fig. 6-3 do not immediately follow those of Fig. 6-2.

The circuit components in Figs. 6-1, 6-2, 6-3, and 6-4 and their functional titles are as follows:

R1—Grid biasing resistor.
R2—Load resistor.
C1—RF tank capacitor.
C2—Grid leak capacitor.
C3—RF filter capacitor.
C4—Coupling and DC blocking capacitor to the headphones.
L1—RF choke.
T1—RF transformer.
V1—Triode tube.

The tank (consisting of C1 and the secondary winding of T1) is adjusted, or tuned, to resonance at the basic carrier frequency being transmitted. After being received and amplified by the receiver, the amplified signal will then be delivered to the grid-leak detector circuit for demodulation.

A total of four different groups, or families, of currents are at work in this circuit. These currents are as follows:

1. Radio-frequency alternating currents.
2. Radio-frequency unidirectional currents.
3. Audio-frequency unidirectional currents.
4. Audio-frequency alternating currents.

The order in which these currents are listed follows closely the one in which they are normally considered in a circuit. This conforms with the idea that the "input"—i.e., control-grid portion—of a circuit must normally be understood before the "output," or plate portion, is discussed.

The four lines chosen do not follow exactly this breakdown of currents into four families. However, this was not done to confuse the reader! One of the most important points to be understood about any current is its complete path through a circuit. This essential information can frequently be made clearer in this series by using different lines than by some other means. Consequently, the reader should not try to match lines faithfully with these current families.

RADIO-FREQUENCY ALTERNATING CURRENTS

The current shown solid in the grid tank circuit is the final amplified version of the modulated carrier wave. This oscillatory current is sustained by a similar current (also solid) which flows in the transformer primary. In Fig. 6-1, the tank current is shown moving *upward* through the transformer secondary. Since a current is made up of electrons in motion, electrons are withdrawn from the lower plate of tank capacitor C1, making the plate positive as indicated by the deep plus sign. These electrons are then delivered to the upper plate of C1, making the plate negative as indicated by the deep minus sign. This negative voltage in turn delivers electrons to the left plate of grid capacitor C2, but they cannot flow onto the plate unless an equal number flow away from the right plate. The latter component of current, also a solid line, leaves the grid capacitor and moves toward the grid, making it momentarily negative and thus restricting the flow of electron current through the tube (but not cutting it off entirely). The solid line which follows the con-

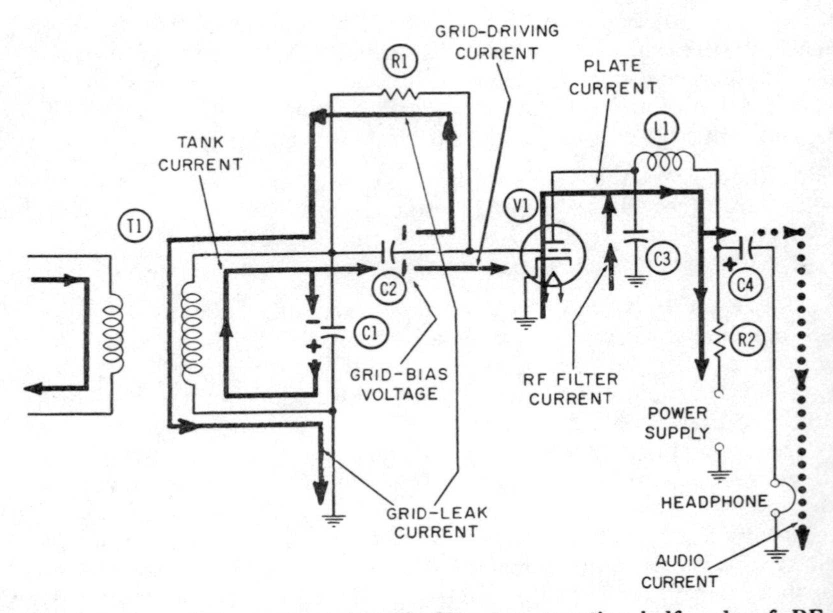

Fig. 6-1. Operation of the grid-leak detector—negative half-cycle of RF during modulation trough.

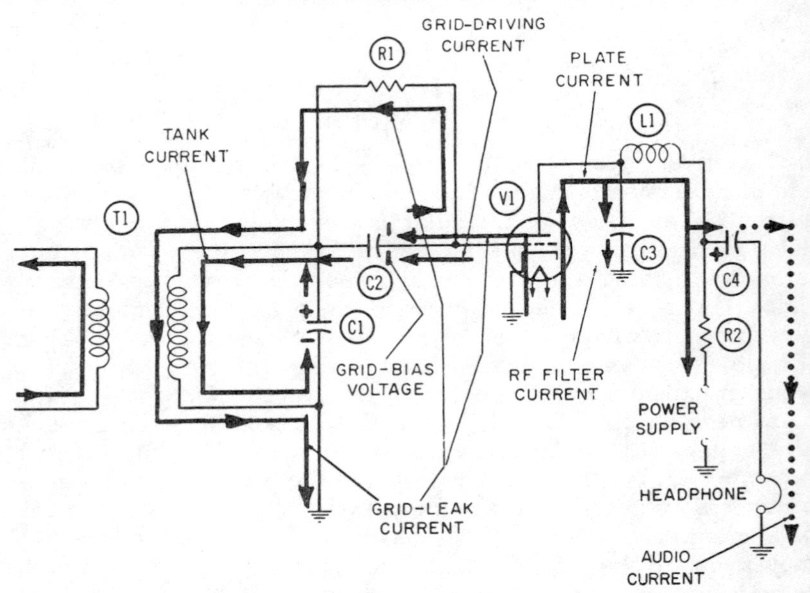

Fig. 6-2. Operation of the grid-leak detector—positive half-cycle of RF during modulation trough.

ventional plate-current path indicates that some plate current is going through the tube.

When the operating conditions of a tube are such that the plate current is never completely cut off—even when the grid reaches its most negative voltage—the tube is said to be operating "Class A." That is true of the tube in this example, and it accounts for the flow of current through the tube when the grid reaches its most negative value.

Figs 6-1 and 6-2 depict the approximate conditions during a modulation trough, when the carrier wave is weakest. Fig. 6-2 shows the tank current flowing *downward* through the secondary winding of the transformer. Electrons are delivered onto the lower plate of tank capacitor C1, making its voltage negative as indicated by the deep minus sign. At the same time, electrons are withdrawn from the upper plate of tank capacitor C1 and from the left plate of grid coupling capacitor C2, making both points positive as indicated by the deep plus signs on the upper plate of C1.

Normal capacitor action will simultaneously draw electrons onto the right plate of grid coupling capacitor C2. This makes the grid voltage momentarily positive and thereby permits more plate current to flow. (The plate current will be discussed more fully under "Radio-Frequency Unidirectional Currents.")

Figs. 6-3 and 6-4 represent conditions during the modulation peak. The only essential difference between the oscillating grid current from trough to peak is its size—meaning the quantity of electrons that make up the current. In Fig. 6-3 the tank current flows in the same direction as in Fig. 6-1 and achieves the same results, except for being larger. This is indicated by an additional solid line following the path of the tank current.

A larger tank current results in a larger tank voltage. This is indicated by the additional minus and plus signs on the upper plate of the tank capacitor in Figs. 6-3 and 6-4. Since the amount of conduction by the tube depends on the grid voltage, conduction will be greatest when the grid is most positive (Fig. 6-4).

One more radio-frequency alternating current is at work in this circuit—the filter current which flows between ground and the lower plate of filter capacitor C3. This current, a solid line, will be discussed in the next section, after the tube currents.

RADIO-FREQUENCY UNIDIRECTIONAL CURRENTS

Tube currents—whether plate, screen-grid, or grid-leak—are essentially unidirectional. In other words, they flow in one direc-

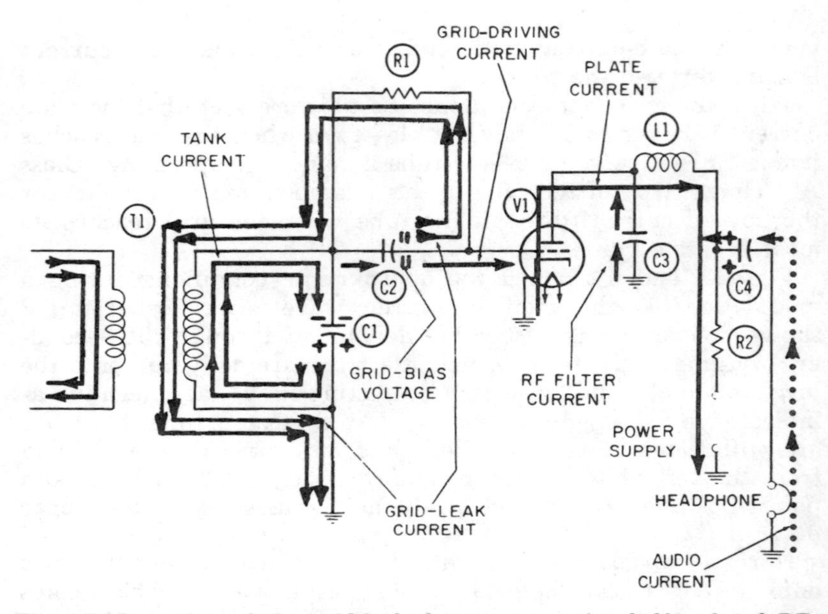

Fig. 6-3. Operation of the grid-leak detector—negative half-cycle of RF during modulation peak.

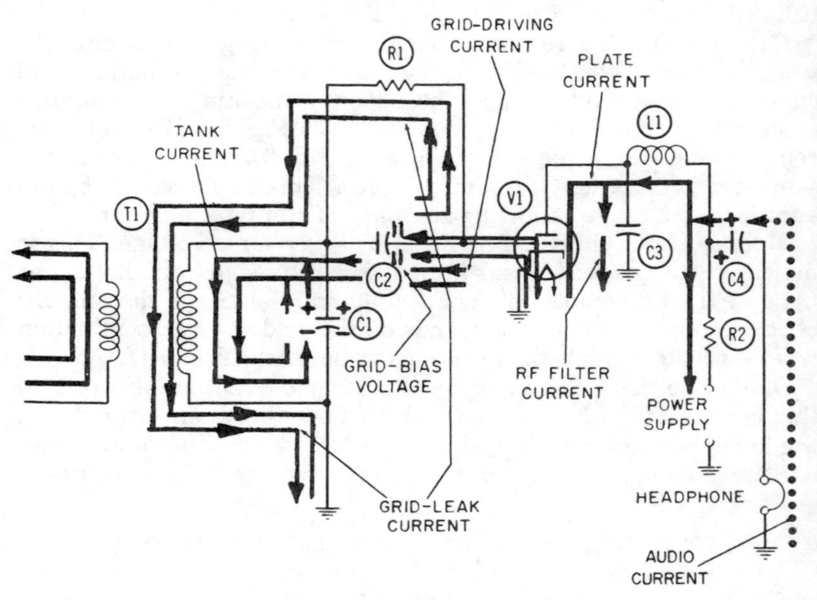

Fig. 6-4. Operation of the grid-leak detector—positive half-cycle of RF during modulation peak.

tion only, and never reverse their direction. This characteristic distinguishes them from alternating currents, which reverse directions at regular intervals. In this chapter, the plate current is shown solid , and the grid-leak current as thin.

Referring to Fig. 6-1, note that the alternating tank voltage has made the control grid negative. However, the tube has not been cut off entirely—a small amount of plate current flows through it, but no grid-leak current does during this half-cycle.

In the second diagram (the next half-cycle), the control grid has been made positive and more plate current is released through the tube. Since these grid-voltage changes are occurring at the basic radio frequency being demodulated, the plate current will be turned "down" and "up" in this fashion once each cycle. Therefore, even though it is a direct current, the plate current pulsates at the applied radio frequency. Hence, the plate current is said to be DC with a radio-frequency alternating component superimposed on it.

Fig. 6-2 reveals a small amount of grid-leak current (thin line) coming out of the tube at the control grid. These electrons accumulate on the right plate of grid capacitor C2, until they can leak out through grid resistor R1 and the secondary of T1 to ground, then through ground to the cathode. The grid-leak current also flows in one direction only. Since it does not flow from the tube unless the grid is positive, it is obviously an intermittent rather than continuous current. Because of the combined action of the capacitor (which stores the electrons) and the resistor (which passes them), the grid-leak current is smoothed out and becomes a continuous current for the rest of its journey. This resistor-capacitor action is referred to as integration, a term derived from its mathematical background. The RC action is also referred to as a "long time-constant," another term derived from the underlying mathematics. In other words, the electrons come out of the tube on each positive half-cycle and accumulate on the capacitor, but are unable to discharge completely to ground before the next positive half-cycle comes along. A continuous discharge of electrons therefore takes place as the number leaking out during an entire cycle is equaled by the number entering during the brief period when the control grid is positive.

Fig. 6-5 shows a typical transfer characteristic curve for an amplifier tube. This curve relates an instantaneous grid voltage to the resultant instantaneous plate current, using assumed or average values of power-supply voltage, plate-load resistance, etc.

Graphical representations of two currents and two voltages appear in Fig. 6-5. Each current is shown in the same color as

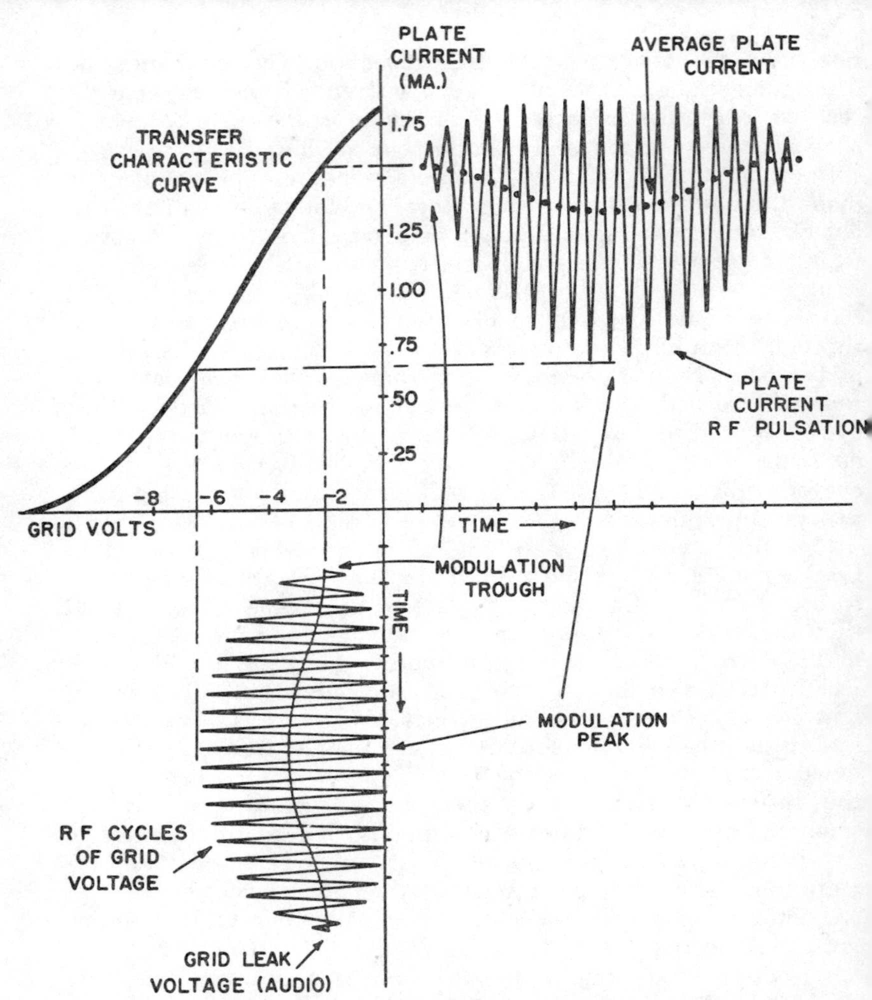

Fig. 6-5. Graphical representation of the grid-voltage, plate-current characteristics of the grid-leak detector.

its related voltage or current in Figs. 6-1 through 6-4. Thus, the radio-frequency grid voltage is shown solid because it is directly related to the radio-frequency tank current.

The thin line down the middle of the RF grid-voltage sine wave represents the grid-leak bias voltage. This voltage is indicated in the circuit diagrams by the thin minus signs (representing negative electrons) on the right plate of capacitor C2. From the grid-voltage scale in Fig. 6-5, note that the grid-leak voltage goes from −2 volts during a modulation trough to approximately

−4 volts during a peak. In other words, the strength of these voltage swings rises and falls in step with the modulation imposed at the transmitter.

The grid-leak voltage in this type of detector circuit marks the first appearance of the desired audio voltage in the receiver system. This voltage will more or less faithfully reproduce the variations in frequency and amplitude of the audio voltage used at the transmitter to modulate the transmitted radio-frequency carrier signal.

As previously discussed, the strength of this grid-leak voltage depends on the amount of grid-leak current drawn from the tube during each half-cycle of radio-frequency voltage. By projecting upward from any point on the grid-voltage sine wave to the characteristic curve and thence horizontally, it is thus possible to construct a curve of instantaneous plate currents. This curve is a solid line because it is most directly related to the plate current, also solid. Reference to the scale in the middle of the diagram reveals that the plate current varies between a low and a high positive value—in the example shown, from 0.6 milliamp to about 1.8 milliamps.

The dotted line which appears to bisect the plate-current RF sine wave is the average plate current. Reference to the plate-current scale shows that the average current varies roughly between 1.35 and 1.6 milliamperes—or 0.25 milliampere peak-to-peak.

In the circuit diagrams, the audio voltage on the left plate of capacitor C4 is shown rising to its highest value (two plus signs) during a modulation peak, and falling to its lowest value (one plus sign) during a trough.

Because the characteristic curve is not a straight line, some distortion of the audio voltage will occur during amplification. (This is known as "square-law" detection. The derivation of this term is beyond the scope of this book.)

On the graphical portion of the diagram, one cycle or pulsation of plate current can be seen occurring for each cycle of radio-frequency current. Since the largest pulsations occur during modulation peaks and the smallest during modulation troughs, it is possible to filter out the radio-frequency characteristics of the plate current and retain only the trough-to-peak variations. Fig. 6-5 shows these variations as "average plate current." They constitute the audio intelligence which the carrier wave has carried from transmitter to receiver.

Capacitor C3 provides the necessary RF filtering in the plate circuit. Its reactance is normally only a few ohms at the carrier frequency. Reactance—in simplest terms, opposition to electron flow—is inversely proportional to both the frequency and the capacitor size, in accordance with the formula:

$$X_C = \frac{1}{2\pi fC}$$

where,

X_C is the reactance in ohms,
f is the current frequency in cycles per second,
C is the capacitance in farads.

The radio-frequency pulsations of plate current which reach the external plate circuit find three alternate paths available. They can flow directly into the filter capacitor, or else they can flow through filter choke L1 and into either coupling-blocking capacitor C4 or directly into resistor R2. The strength of these pulsations will divide betwen the three paths in inverse proportion to the opposition (resistance and/or reactance) offered by each path.

When one or both types of reactances (capacitive and inductive) are combined with resistance, the common name of impedance represents the sum total of this opposition to the passage of electron current. The filter choke in this circuit is designed to have a high inductive reactance at the radio frequencies in use here. Inductive reactance is directly proportional to both the inductance and the frequency, in accordance with the standard formula:

$$X_L = 2\pi f L$$

where,

X_L is the inductive reactance in ohms,
f is the frequency of the current being passed or impeded in cycles per second,
L is the inductance of the coil in henrys.

Since filter choke L1 in effect is in series with load resistor R2 and blocking-coupling capacitor C4 as far as plate-current pulsations are concerned, the impedance of each alternate path can be added to the filter-choke impedance to determine the total impedance of either path. It is hardly necessary to state that most of the plate current (which is pulsating at the basic radio frequency) will choose the low-impedance path to ground offered by filter capacitor C3, rather than the two high-impedance paths through L1 and R2 or C4.

Remember that reactive ohms and resistive ohms cannot be added directly by simple arithmetic; they can only be added vectorially. This is a more complex mathematical process than simple arithmetic and is beyond the scope of this book.

The filter combination of C3 and L1 could be made more elaborate by adding another capacitor leading from the right terminal of choke L1 to ground. It could also be less elaborate, consisting of capacitor C3 without the filter choke. Even so, the impedance of each alternate path for the RF pulsations in the plate current would be great enough that C3 would bypass most of the energy to ground. The load resistor, for example, will normally be several thousand ohms, as opposed to the few ohms offered by C3 to radio frequencies.

The reason why the other alternate path through C4 will also reject radio-frequency currents requires more explanation. Capacitor C4, which must couple the audio or modulating voltage to the headphones (or speaker), must have a low reactance to current flow at audio frequencies. Since audio frequencies are much lower than radio frequencies, C4 must be much larger in value than C3. Therefore, C4 will offer only negligible reactance to the radio-frequency pulsations in the plate-current stream.

At first glance it appears that the energy of these pulsations will flow into C4 and be coupled to the headphones (rather than the desired objective of flowing into capacitor C3 and being filtered back to ground). However, this does not occur because of the resistance of the transducer (headphones) beyond the coupling capacitor. (A transducer is a device for converting energy from one form to another. A headphone or speaker, for example, converts electrical energy to sound energy.) A headphone has an internal resistance of several hundred or even several thousand ohms. The RF pulsations will not flow onto the left plate of C4 unless pulsations of equal size are permitted to flow off the lower plate. Since the headphone resistance is in series with the capacitor, the RF pulsations would also have to flow through the phones. Consequently, the total impedance—or opposition to electron flow—at radio frequencies is much greater than the impedance offered by filter capacitor C3. As a result, C3 filters almost the entire radio-frequency component of the plate current to ground. In other words, the RF pulsations in the plate current flow onto the upper plate of C3. Each pulsation drives an equal number of electrons off the lower plate, toward ground. These directions of flow exist only on the alternate half-cycles when the grid is positive (Figs. 6-2 and 6-4).

On the alternate half-cycles when the grid voltage is negative and less plate current flows, these directions are reversed, as depicted in Figs. 6-1 and 6-3. During these half-cycles, the electrons which came out of the tube during the preceding pulsation will complete the path followed by all plate current and move downward through load resistor R2, into the power supply.

The current which flows between the lower plate of capacitor C3 and ground is thus one of the radio-frequency alternating currents. Named the "RF filtering current," it will faithfully reflect the frequency of the carrier current. Also, each of its cycles will always equal the quantity of electrons involved in the accompanying RF pulsation of plate current.

AUDIO-FREQUENCY UNIDIRECTIONAL CURRENTS

The first appearance of an audio-frequency voltage and current in this circuit is at the grid, where grid-leak detection has occurred. This is depicted by the increased number of electrons stored on the right plate of grid capacitor C2 during the modulation peak (Figs. 6-3 and 6-4). Another thin line is shown flowing through the grid resistor and transformer secondary to ground and back to the cathode. This higher grid-leak voltage and its associated current are due to the increasing strength of the tank voltage as a modulation peak is approached. A close scrutiny of the grid-voltage representation in Fig. 6-5 will assist the reader in understanding the several, intricately related actions occurring with the circuit.

Fig. 6-5 graphically reveals a series of radio-frequency cycles of grid voltage. These begin small during a modulation trough, build up during the modulation peak, and then decrease again as the next modulation trough approaches. They are of course driven by the oscillating voltage in the grid tank circuit. Observe that the grid voltage cannot be raised appreciably above zero. Thus, as the grid-voltage swing increases, the grid merely becomes more negative during negative half-cycles. A solid thin line represents the biasing voltage, which is built up by the grid-leak current from the tube. This changing bias voltage is indicated by the collection of electrons, thin line, on the right plate of grid capacitor C2. During the first whole cycle (Figs. 6-1 and 6-2), this collection of electrons consists of one line, and two lines during the second whole cycle (Figs. 6-3 and 6-4).

The solid thin line in Fig. 6-5 is not actually a smooth curve as shown. Rather, it is a series of short zigzags which are repeated once during each cycle of RF. During positive (second and fourth) half-cycles, grid-leak current will flow and tend to increase the negative bias. The solid bias line "zigs" to the left at the end of each positive half-cycle, since the higher negative voltages are to the left in this diagram.

During any negative half-cycle (represented by Figs. 6-1 and 6-3), the solid line "zigs" back to the right toward the region of lower negative voltages, indicating that the bias voltage is discharging toward zero. This discharge is shown by the thin lines going from the grid-storage area (right plate of the grid capacitor), through the grid resistor and transformer secondary, to ground.

In this type of circuit, there can be a few cycles, close to each modulation trough, when the grid voltage is not driven to the zero-voltage line. Consequently, no grid-leak current can begin to flow and no electrons are added to those already stored on the grid capacitor. As a result, the bias voltage discharges continuously toward ground and zero volts during these few cycles. The normal condition is for some grid-leak current to flow on each positive half-cycle, so that the biasing voltage becomes a fair approximation of the carrier wave's modulation envelope and therefore of the audio voltage "carried" by the wave. The difference between an approximation and an exact reproduction of input waveshape is known as distortion, and this circuit is characterized by high distortion.

Fig. 6-5 is a conventional waveform diagram for this type of circuit. It was made using the transfer characteristic curve of the tube. The reader is cautioned that the radio-frequency cycles (a straight line) are grid-*voltage* cycles, whereas those a straight line are plate-*current* cycles; hence, they are pulsating direct current.

Likewise, the audio cycle, a thin line, is an audio-*voltage* cycle representing the accumulation of grid-leak electrons on the right plate of grid capacitor C2, whereas the audio cycle shown in large dots is a current cycle and represents the approximate average amount of plate current flowing. The latter cycle is shown in large dots to relate it to the audio current which flows through the headphone, or load, between the right plate of capacitor C4 and the common ground.

Figs. 6-1 and 6-2 are devoted to a modulation trough. In both diagrams the audio current flows *downward* through the headphones, because of the high *average* plate current and hence the somewhat lower plate voltage. The amount of plate voltage is represented by the number of plus signs on the left plate of capacitor C4. During a modulation trough, the plate voltage has a low positive value.

The only way this voltage can be altered is by adding or withdrawing electrons. The positive voltage stored on the left plate of capacitor C4 can be thought of as a "pool" of positive ions. There is a continuous inflow of electrons into this pool from the plate current, and a continuous drain of electrons out of this pool and through R2 to the higher positive voltage of the power supply.

AUDIO-FREQUENCY ALTERNATING CURRENT

During a period of higher average plate current, electrons are added to this positive-voltage pool faster than they can be drained away through load resistor R2. Consequently, this positive voltage decreases. The normal capacitor action of C4 is such that when additional electrons flow onto the left plate, an equal number are driven off the right plate. This accounts for the downward direction of the audio current through the headphones in Figs. 6-1 and 6-2.

During a period of *lower* average plate current, the power supply (because of its high positive voltage) withdraws electrons from the positive-voltage pool, making it even more positive. Again, the normal capacitor action of C4 is such that an equal number of electrons are drawn onto the right plate of C4. This accounts for the upward direction of the audio current through the headphones in Figs. 6-3 and 6-4.

Chapter 7

DISCRIMINATOR CIRCUITS

In Chapters 4, 5, and 6, the signals to be detected were amplitude-modulated (AM). In this and the next chapter, methods of recovering the modulating signal from a frequency-modulated (FM) carrier will be discussed. Since the FM signal differs from the AM signal in the previous chapters, a brief discussion of the signals will be given before operation of the discriminator circuit is explained.

THE FM SIGNAL

Fig. 7-1 shows a series of waveforms from which it is possible to derive some essential definitions associated with frequency modulation. Figs. 7-1A and 7-1C are audio-frequency sine waves of voltage which have been used to modulate the carrier waves of Figs. 7-1B and 7-1D. Fig. 7-1A represents an audio voltage of low volume of amplitude (meaning loudness), and Fig. 7-1C, an audio voltage of high volume or amplitude. For simplicity we might assume that the audio frequency in each case is 1,000 cps. Hence, an individual audio cycle will be 1/1,000th of a second.

The frequency of the carrier wave may be anywhere in the assigned band of frequencies allocated for FM broadcasting. The latter is from 88 to 108 megacycles. This is considerably above the 550-1600–kc band assigned for amplitude-modulation broadcasting.

In the waveforms of Fig. 7-1, the carrier varies both above and below an assigned frequency called the *center frequency.* When no sound and consequently no modulation are imposed on the carrier at the transmitting end, the carrier frequency will be stabilized at its assigned center frequency. Let us assume this center frequency to be 100 megacycles.

When some audio-modulating voltage is applied to the carrier, its frequency will vary on either side of the center frequency, *but will return to the center frequency twice during each audio cycle.* This occurs as the audio voltage passes through its own reference line and has a value of zero. These are the moments when the audio voltage is changing its polarity.

The number of cycles or kilocycles by which the carrier deviates from its center frequency is called the *frequency deviation.* It is measured from the center frequency to either the highest or the lowest frequency attained by the carrier. The maximum deviation from the center frequency prescribed by regulation for commercial FM broadcasting is 75 kilocycles. Thus, a carrier with an assigned center frequency of 100 megacycles may legally vary, or "deviate," between 99.925 and 100.075 megacycles.

Sound has two distinguishing features—frequency (pitch) and volume (loudness). To be successful, any radio transmitting system must be capable of modulating the radio-frequency wave in such a manner that these two important sound characteristics of pitch and loudness are faithfully reproduced. Fig. 7-1 provides some idea of how this is done in a frequency-modulated carrier.

Audio Amplitude

The amount in kilocycles by which the carrier deviates from its center frequency is determined by the *amplitude* of the modulating audio voltage—in other words, by its volume, or loudness. Two successive whole audio cycles, or four half-cycles, are indicated in Fig. 7-1. At the start of each half-cycle, the audio voltage is neither positive nor negative. Rather, it is passing through zero and the carrier is at its center frequency. In the middle of the first half-cycle, the modulating audio voltage reaches a positive peak. Accordingly, the carrier frequency will have deviated to a somewhat higher frequency—perhaps 25 kilocycles above the center frequency, or 100.025 megacycles.

Midway in the second half-cycle, the modulating audio voltage has reached its negative peak and the carrier frequency accordingly will have deviated to a somewhat *lower* frequency. Only this time the deviation is 25 kilocycles *below* the center frequency, or 99.975 megacycles.

As stated before, Fig. 7-1A represents the waveform of an audio voltage of *low volume,* or amplitude. Fig. 7-1C represents the waveform of an audio voltage of the same frequency but with a *high volume,* or amplitude. Because of this higher volume, the frequency deviations of the carrier will be greater. Midway in the first half-cycle in Fig. 7-1C and 7-1D, the audio voltage reaches its positive peak. At this time the carrier frequency will have deviated 50 kilocycles *above* the center frequency, to a new frequency of 100.050 megacycles. Midway in the second half of the audio cycle in Figs. 7-1C and 7-1D, the modulating audio voltages reaches its negative peak. Now the carrier will have deviated the same amount of 50 kilocycles *below* the center frequency, to a new value of 99.950 megacycles.

In a properly adjusted modulating system, the frequency deviation should at all times be proportionate to the amplitude of the modulating voltage. Consequently, for a frequency deviation of 50 kilocycles to be achieved as in Fig. 7-1D, *versus* the deviation

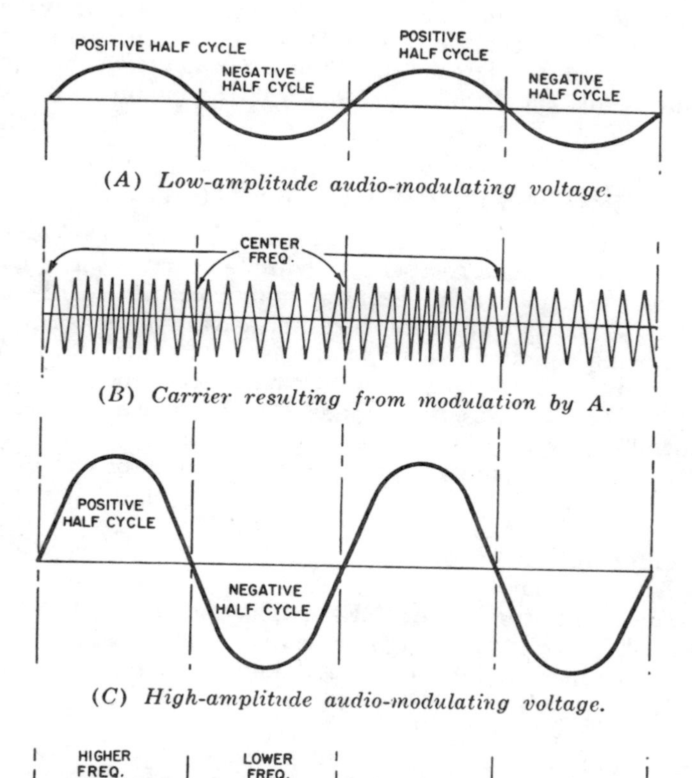

(*A*) *Low-amplitude audio-modulating voltage.*

(*B*) *Carrier resulting from modulation by A.*

(*C*) *High-amplitude audio-modulating voltage.*

(*D*) *Carrier resulting from modulation by C.*

Fig. 7-1. The relationships between the amplitude and frequency of the audio signal with the carrier.

of 25 kilocycles achieved in Fig. 7-1B, the modulating audio voltage in Fig. 7-1C would have to have twice the amplitude of the voltage in Fig. 7-1A.

Audio Frequency

The rate at which the carrier deviates in frequency between its two extremes is determined by the frequency of the modulating audio voltage. This should be fairly evident from Fig. 7-1. During the two audio cycles shown in this figure, the carrier completes two cycles of deviation. From center to maximum frequency, to center and to minimum, and back to the center frequency constitutes one complete cycle. The rate at which the carrier frequency deviates is sometimes called the *excursion* frequency, and it will always be equal to the frequency of the modulating audio voltage.

THE DETUNED DISCRIMINATOR

Figs. 7-2, and 7-4 show a fairly standard discriminator circuit used for demodulating a frequency-modulated carrier wave. Each of these figures depicts a different operating condition for the circuit. Fig. 7-2 shows significant currents and voltages when the circuit is being operated exactly at the center frequency. Fig. 7-3 shows currents and voltages when the carrier frequency has deviated *above* the center frequency, and Fig. 7-4, when the carrier frequency has deviated *below* the center frequency.

The circuit components of this type of discriminator circuit are as follows:

R1—Filter resistor for V2.
R2—Filter resistor for V3.
C1—Tank capacitor for final RF- or IF-amplifier stage.
C2—Tank capacitor for "high" deviation tank circuit.
C3—Tank capacitor for "low" deviation tank circuit.
C4—Filter capacitor for V2.
C5—Filter capacitor for V3.
L1—Tank inductor for final amplifier stage.
L2—Tank inductor for "high" deviation tank circuit.
L3—Tank inductor for "low" deviation tank circuit.
V1—Pentode RF- or IF-amplifier tube.
V2 and V3—Diode detector tubes.

Identification of Currents

Operation of this type of circuit can be understood from a discussion of the following electron currents:

1. Plate current through tube V1.
2. Oscillating current in final amplifier tank circuit.
3. Current through first rectifier diode, V2.
4. Oscillating current in upper tuned tank circuit.
5. Current through second diode rectifier tube, V3.
6. Oscillating current in lower tuned tank circuit.
7. Radio-frequency filter currents in two filter capacitors, C4 and C5.

Details of Operation

The three tank circuits are each tuned to different frequencies. The first circuit, consisting of L1 and C1, will be tuned exactly to the center frequency. Since frequency conversion undoubtedly has already taken place, the center frequency in the demodulation process will be much lower than the center frequency that exists during transmission and the initial reception. When the carrier initially was received, its center frequency may have been 100 megacycles and the frequency deviations due to modulation may have caused it to vary 75 kilocycles (the maximum allowable) on either side. However, after frequency conversion (called heterodyning), the new center frequency may be 10 megacycles. The same frequency deviations will have been preserved, of course. In other words, if the original carrier deviated the full 75 kilocycles—from 100 to 100.075 megacycles—the new IF carrier will also deviate 75 kilocycles, from 10 to 10.075 megacycles.

Assume that tank L1-C1 is tuned to this new center frequency of 10 megacycles. The tank consisting of inductor L2 and capacitor C2 will be tuned to a higher frequency, and the other tank (L3 and C3) will be tuned to a lower frequency. Hence, these two tank circuits are not tuned to the center frequency to which the first tank is tuned. They are said to be detuned and the circuit is named a detuned discriminator.

Discriminator Response Curve

Fig. 7-5 shows a typical frequency-response curve for the average discriminator circuit. A response curve is defined as a graphical representation of the response, or *output,* of the circuit throughout the entire range of frequencies in which the circuit operates. Fig. 7-5 has three essential parts: The response curve of the *upper*-detector circuit (solid) consists of tank L2-C2, diode V2, load resistor R1, and filter capacitor C4. The response curve

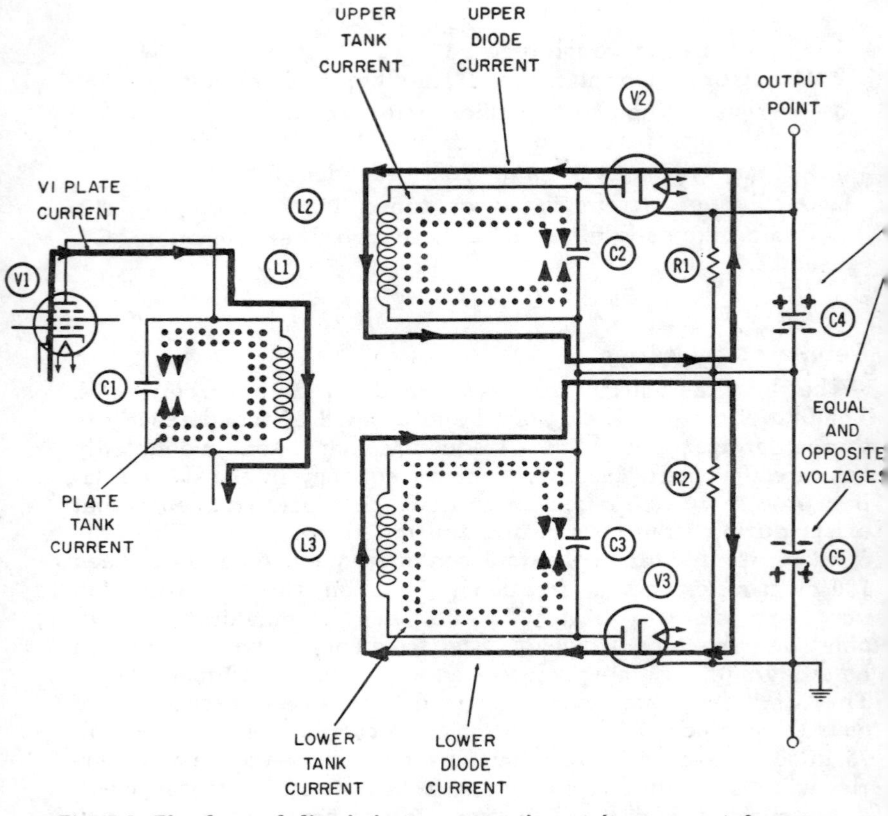

Fig. 7-2. The detuned discriminator—operation at its resonant frequency.

of the *lower*-detector circuit (thin) consists of tank L3-C3, diode V3, load resistor R2, and filter capacitor C5. The response curve of the two detector circuits taken together is shown in black. It is desirable for a discriminator to be operated along the linear portion of its over-all response curve. (See Fig. 7-5.) The response curve of the whole circuit is merely an algebraic addition of the two individual response curves. On either side of the center frequency, the sum of these two curves gives a fairly straight line, as indicated. As the carrier frequency deviates farther from the center frequency, the response of one of the two diode-rectifier circuits dwindles to the vanishing point. Eventually the total response curve has almost the same shape as the response curve of the other diode.

If the linear portion of the response curve extends, say, 75 kilocycles on either side of the center frequency, the two detuned

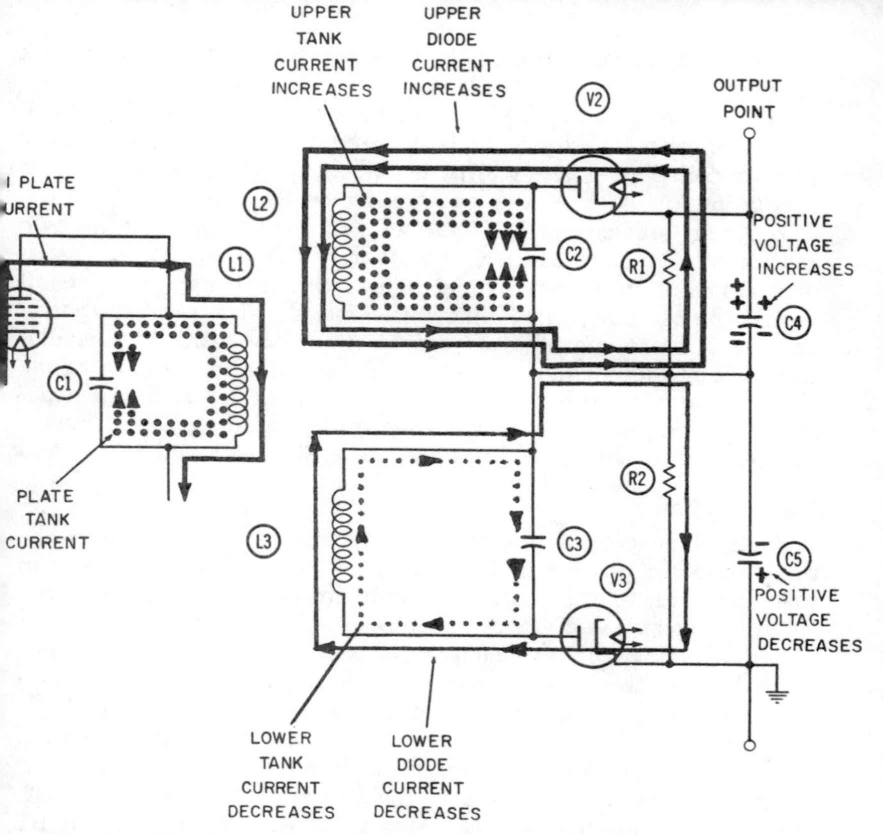

Fig. 7-3. The detuned discriminator—operation above resonance.

peaks must be resonant at frequencies *farther* removed than this amount from the center frequency. As an example, tank L2-C2 might be tuned 150 kilocycles *higher* than 10 megacycles, to a resonant frequency of 10.150 megacycles. Tank L3-C3 would then have to be tuned 150 kilocycles *lower* than the center frequency, or 9.850 megacycles.

Operation at the Center Frequency

Fig. 7-2 shows the significant currents and voltages when this discriminator circuit is being operated at the exact center frequency. The frequency at which this circuit is being operated is of course determined by the carrier frequency. When it is exactly 10 megacycles, ten *million* pulsations of plate current will pass through amplifier tube V1 each second. (This is the current shown by the solid line.) Each such pulsation sustains one cycle of oscillation of the electrons which make up the plate-tank current.

Since both L2 and L3 act as secondary windings in their relationships with L1, the plate tank current will induce a current in each coil. This current in turn will set up and sustain electron oscillations in each of the two tank circuits, L2-C2 and L3-C3. The upper tank circuit (L2-C2) will be operating *below* its own resonant frequency, whereas the other will be operating *above* its own resonant frequency. Nevertheless, the most important result is that the oscillation in each tank is considerably weaker than if each tank were operated at resonance. Further, the two oscillations are reduced equally in strength, so that they remain equal. (This presupposes that the two diode circuits have identical designs and degrees of coupling to primary winding L1. Unfortunately, such design objectives are not easily achieved.)

Rectification

Each diode circuit operates in the manner described in the diode-detector portion of this book, the only difference being in the placement of the RC filter combination. In that chapter, the RC filter was located where it could "bias" the diode plate with a negative voltage. In this chapter, this bias is a positive voltage. In both cases the bias is the detected voltage.

In the upper-diode circuit, each positive half-cycle will drive the plate of diode V1 to a slightly more positive voltage peak than the cathode. This causes electron current (shown a solid line) to cross from cathode to plate, downward through inductor L2, across to the junction with resistor R1, and upward through R1 to the cathode. This is not a continuous current, but an intermittent one that flows only at the peaks of positive half-cycles of tank voltage. However, the portion that flows through resistor R1 is of course a continuous current, since it is drawn upward by the electron deficiency (positive voltage) on the upper plate of capacitor C4.

The output for the upper-diode circuit is this positive voltage stored on the upper plate of C4. Under the conditions described earlier, it might have a value of +10 volts when the carrier is exactly on the center frequency. Since the output voltage for the entire circuit is measured across *both* R1 *and* R2, this portion of the output voltage must be added algebraically to that portion developed independently across R2 by the current through the second diode. The latter will also equal 10 volts, but the *bottom* of R2 will be positive. Let us see why.

During the peaks of the positive half-cycles of tank voltage in the lower tank circuit, diode V3 will conduct from cathode to plate. These electrons, shown by solid thin lines, then flow upward through inductor L3 to the junction with R2, and downward through R2 to the cathode. Like the current through diode V2, this current is also intermittent except through R3, where it flows continuously towards the cathode.

When we say the voltage at the cathode is a positive 10 volts, we mean of course that it is positive with respect to the voltage at the *other end* of resistor R2. Since the cathode of V2 has been grounded in this particular circuit, the voltage at the top of R2 must be −10 volts. The voltage at the cathode of the upper diode is +10 volts with respect to this voltage at the junction of the two resistors, by virtue of the upper-diode current. Hence, the voltage at the cathode of V2, as at the cathode of V3, will have a true value of zero.

Since the voltages at the two diode cathodes both equal zero, the difference between them—which is also the output voltage for the entire circuit—is zero. This is as it should be, and can be confirmed by referring to Fig. 7-1. At those two moments in each cycle when the audio-modulating voltage was passing through the reference line and therefore was equal to zero, the carrier was exactly equal to the center frequency.

Operation Above the Center Frequency

Fig. 7-3 shows significant currents and voltages when the carrier is higher than the center frequency. Now tuned circuit L2-C2 will be operating more nearly at its own resonant frequency. In accordance with the selectivity or response curve of Fig. 7-5, a stronger oscillation of electrons will be built up in the tank, as indicated by the three dotted lines. Each positive voltage peak which is applied to the plate of diode V2 will be proportionately higher, and more current will flow through the diode. This increases the electron deficiency (positive voltage) on the upper plate of capacitor C4, and of course the amount of current flowing upward through R1. The amount of voltage stored in the capacitor is always the same as developed across the resistor.

Since tuned circuit L3-C3 is tuned *lower* than the center frequency, in Fig. 7-3 this circuit will be operating even farther from its resonant frequency than in Fig. 7-2. Its response curve from Fig. 7-5 indicates that a comparatively weak oscillation of electrons (indicated by the single dotted line) will be in motion in the tank. Consequently, diode V3 will conduct fewer electrons than before. Less current will flow through resistor R2, and less voltage will be developed across it.

Since the output voltage of the entire circuit is the algebraic sum of the two separate voltages across R1 and R2, it is obvious

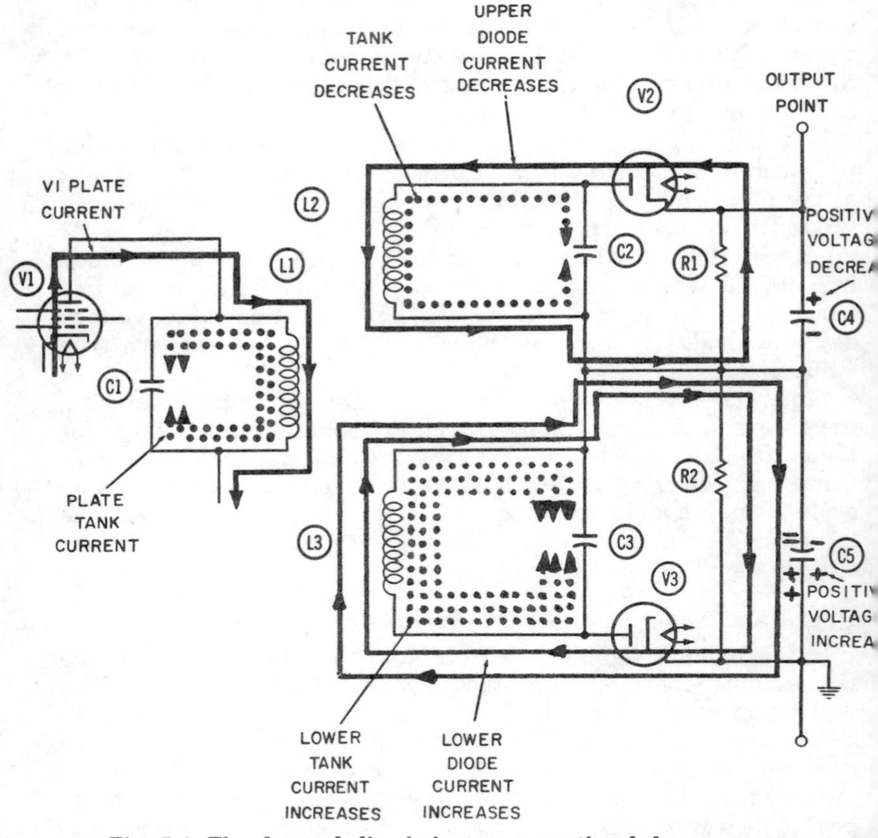

Fig. 7-4. The detuned discriminator—operation below resonance.

that when the circuit is operated above the center frequency, the output voltage at the top of resistor R1 will be positive. The *amount* of this voltage depends on the amounts of conduction in each diode tube. In turn, these amounts depend on the strength of each electron oscillation in the two tuned tanks, L2-C2 and L3-C3. As the operating frequency moves higher and higher toward the resonant frequency of tank L2-C2, the amount of voltage developed across R1 increases steadily, whereas the amount across R2 *decreases*. Consequently, the sum of these two voltages —which is the output of the circuit—is a positive voltage which *increases* as the carrier frequency does. From Fig. 7-1 we deduce that a positive voltage output is associated with a positive peak in the modulating audio voltage. Also, we see that the output-voltage peak is directly proportionate in size to the modulating audio voltage.

Operation Below the Center Frequency

Fig. 7-4 shows the important currents and voltages in this circuit when the carrier is below the center frequency. Tuned circuit L3-C3 will now be operating more nearly at its resonant frequency. The oscillation of electrons in that tank becomes stronger, and more current flows through diode V3 and resistor R2. This results in a much greater voltage drop across R2.

At the same time, the oscillation in upper tank L2-C2 will be drastically weakened because it will be operating far from its resonant frequency. The response curve of Fig. 7-5 which applies to this tank indicates a very low response. So only a small current will flow upward through R1 and develop only a small voltage across this resistor. This voltage will be positive at the top of R1.

The output is the sum of the two voltages across R1 and R2. The junction of these two resistors is at a fairly large negative voltage with respect to ground, by virtue of the larger current flowing through R2. Thus the total output voltage, measured at the top of R1, will also be negative. Again this is consistent with Fig. 7-1, which indicates that a negative audio peak voltage is associated with a lower operating frequency. It is also consistent with Fig. 7-5, which illustrates that a lower operating frequency leads to negative output voltages.

Summary

From the foregoing discussion it is evident that the output voltage varies from positive to negative at an audio rate which de-

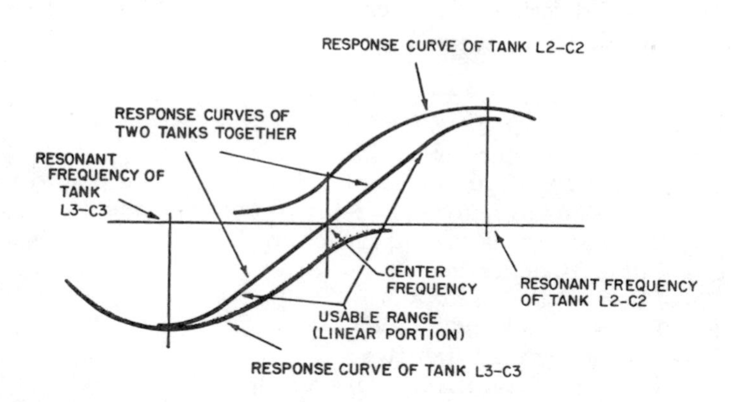

Fig. 7-5. Response curve of the detuned discriminator.

pends on the excursion frequency of the carrier. The latter is of course the same as the frequency of the modulating audio voltage. Thus the audio frequency has been reproduced by this type of demodulation.

The *amplitude* of the audio-output voltage depends on how greatly the carrier deviates from the center frequency. A deviation toward a higher frequency causes a *positive* voltage at the output point, and a deviation toward the lower frequencies causes a *negative* voltage. In either case the size of the deviation determines the size of the output voltage. Since the amount of frequency deviation is controlled by the amplitude of the modulating audio voltage within the transmitter, this discriminator circuit can accurately reproduce the intended audio amplitude as well as frequency.

TUNED DISCRIMINATOR

Figs. 7-6, 7-7, 7-8, and 7-9 represent four quarter-cycles in the operation of a tuned discriminator circuit at the center frequency. As with the detuned discriminator discussed previously, the function of this circuit is to demodulate a frequency-modulated signal. Unlike the detuned discriminator, however, two tuned circuits are used instead of three, and the primary and secondary tuned circuits are coupled inductively *and* capacitively. (The significance of this dual coupling will be discussed later.) The operation of this circuit is somewhat more difficult to visualize than it is for the detuned discriminator.

Identification of Components

The tuned discriminator includes the following necessary circuit components:

RI—Cathode biasing resistor for V1.
R2—Output resistor for V2.
R3—Output resistor for V3.
C1—Primary tank tuning capacitor.
C2—Coupling capacitor between two tank circuits.
C3—Secondary tank tuning capacitor.
C4—Filter capacitor for V2.
C5—Filter capacitor for V3.
L1—Primary tank inductor.
L2—Secondary tank inductor.
V1—Final IF amplifier tube, usually a pentode.
V2—Upper-diode detector tube.
V3—Lower-diode detector tube.

Identification of Currents

The following electron currents are at work in this circuit:

1. Plate tank current.
2. External current driven by plate tank current.
3. Secondary tank current.
4. Upper-diode current.
5. Lower-diode current.

Details of Operation

To understand the operation of this circuit, it is necessary to review its operation on a single-cycle basis at the center frequency, and also above this frequency. Figs. 7-6 through 7-9 show the four successive quarter-cycles of operation at the center frequency, and Fig. 7-10 the significant voltage waveforms during a single cycle.

Operation at the Center Frequency

It is not readily apparent from the circuit diagrams, but inductors L1 and L2 are two halves of a radio-frequency transformer. Hence, any change in the amount of current flowing through L1 will *induce* a voltage and a companion current in secondary winding L2. Also, any changes in the voltage at the top of the primary tank circuit simultaneously will be coupled—via capacitor C2—to both sides of the secondary tank and to both detector diodes.

It is a well-established fact that when two tuned circuits are operated exactly at resonance, the two tank voltages will be exactly 90°, or a quarter of a cycle, out of phase. Usually the secondary tank voltage is said to lag the primary tank voltage by a quarter of a cycle—meaning the voltage at the top of the secondary tank circuit will reach its positive peak a quarter of a cycle *after* the voltage at the top of the primary tank reaches *its* positive peak.

The resonant frequency of these tank circuits is the center frequency. In the amplifier volume of this series, the chapter on radio-frequency amplifiers contains a detailed discussion of tuned circuits and their operation in the vicinity of their resonant frequency. The entire discussion will not be repeated here; any readers wishing to learn more about tuned circuits are referred to this volume.

Looking at Fig. 7-10 in conjunction with the four quarter-cycle diagrams, you will see that the voltage across primary tank circuit L1-C1 is positive during the first two quarter-cycles, reaching its peak at the end of the first quarter-cycle. During the last two

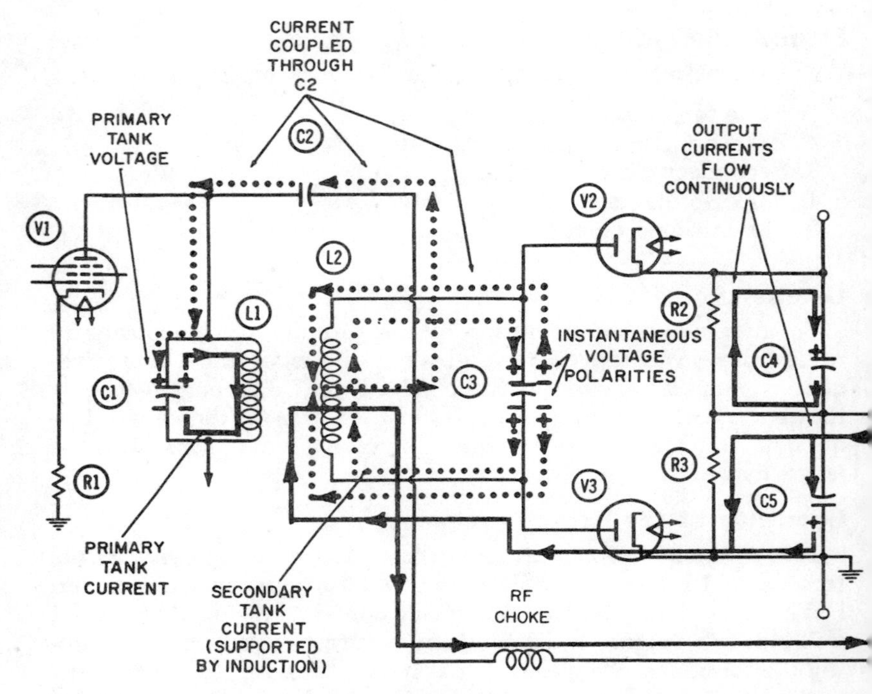

Fig. 7-6. Operation of the tuned discriminator at the center frequency—first quarter-cycle.

quarter-cycles, the tank voltage is negative, reaching its peak at the end of the third quarter-cycle. This oscillatory voltage across the primary tank is accompanied by an oscillatory current *through* it. During the first quarter-cycle (Fig. 7-6), this current flows downward through coil L1 and delivers electrons to the lower plate of tank capacitor C1, leading to a negative peak voltage on the lower plate and a positive one on the upper plate.

During the second quarter-cycle (Fig. 7-7), this tank current reverses and flows upward through the coil. The charge (electrons) stored in the capacitor is redistributed, taking a half-cycle to do so, and the tank voltage reverses polarity. The voltage at the top of the primary tank reaches its negative peak at the end of the third quarter-cycle (Fig. 7-7).

The purpose of capacitor C2 is to couple this tank voltage directly to both sides of the secondary tank circuit. This is accomplished by connecting C2 to a center tap on coil L2. The resultant current flow, shown in dotted lines, flows to the left during the first two quarter-cycles and to the right during the last two.

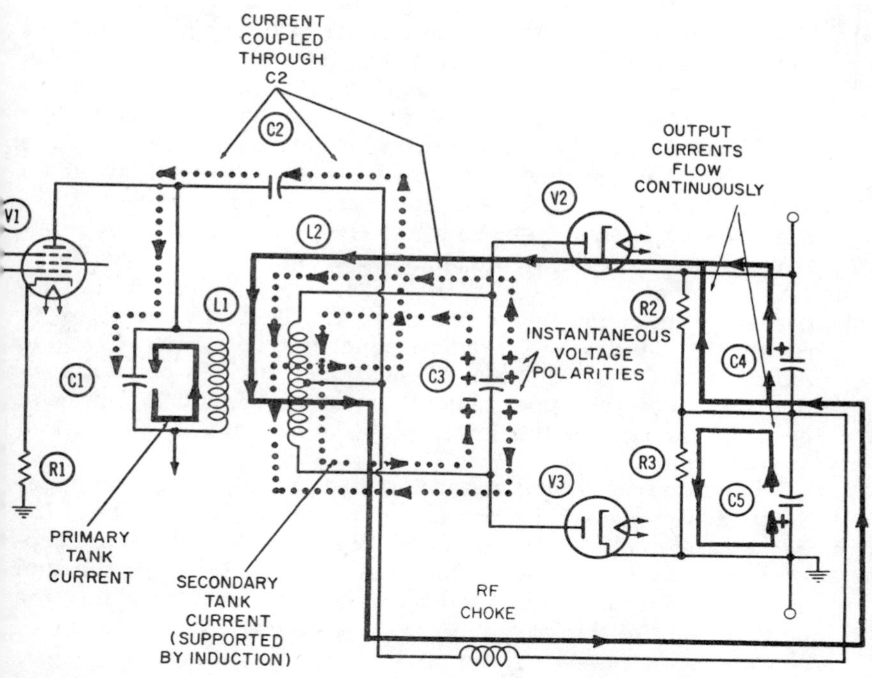

Fig. 7-7. Operation of the tuned discriminator at the center frequency—second quarter-cycle.

Called an "external" current to the oscillation in the primary tank, it constitutes a load on the primary tank oscillation and, like any other loss or load current, should be kept as small as possible.

This current provides the necessary function of *simultaneously* transferring the polarity of the voltage at the top of the primary tank to *both* sides of the secondary tank. This positive polarity—indicated by the deep plus signs on both plates of capacitor C3 during the first two quarter-cycles—results from the fact that the external current from the primary tank is drawing electrons away from both sides of the secondary tank. These electrons, which make up the external current, will continue to be drawn *toward* the primary tank during the first two quarter-cycles as long as the voltage across the primary tank is positive. During the third and fourth quarter-cycles, the tank voltage across the primary tank circuit has a negative polarity at the top and these electrons will be repelled. This accounts for the deep minus signs on both sides of capacitor C3 in Figs. 7-8 and 7-9.

A primary function of coupling capacitor C2 is to isolate—or "block"—the high positive voltage of the power supply connected to the bottom of L1 and C1 from the plates of the two diode detectors. This coupling function can always be performed better by a piece of straight wire than by a capacitor. It should be understood that neither side of the primary tank ever reaches a true negative voltage; instead, the voltage varies between a low and a high positive value. Thus, on alternate half-cycles, each plate of capacitor C1 is shown as being negative, but actually it is negative only *with respect to the other plate.*

The oscillating voltage in the secondary tank circuit is supported by induction between L1 and L2. These two coils are in reality two windings of a radio-frequency transformer. Fig. 7-10 indicates that this oscillation of electrons leads to a peak of positive voltage on the upper plate of capacitor C3, and to a peak of negative voltage on the lower plate, at the end of the second quarter-cycle.

To determine the total voltage applied to the plate of the upper diode at any moment, it is necessary to add the amplitudes of the two separate voltages at that point. Since the amplitudes are represented by the waveforms labeled A and B in Fig. 7-10, the sum of the two can be represented by waveform E. Observe that the positive peak of this voltage sine wave occurs *midway* in the second quarter-cycle—after the peak of coupled voltage but before the peak of oscillatory voltage. This is the moment when upper diode V2 will conduct the most electrons. This diode current (solid line in Fig. 7-7) flows from cathode to plate within the tube, downward through coil L2 to the center tap, and back through the RF choke to the bottom of resistor R2. From here it flows upward through R2 and returns to the cathode. This current causes a positive voltage at the top of R2, and also on the upper plate of capacitor C4. This positive voltage becomes a portion of the discriminator output voltage. The balance of the output voltage is developed across resistor R3 by the other diode current.

The sine wave (labeled waveform C in Fig. 7-10) represents the amplitude of the oscillatory voltage at the *bottom* of tuned secondary tank L2-C3. Obviously, this voltage should always be a half-cycle out of phase with the oscillatory voltage at the top of the tank. From waveform C we can observe that the oscillatory voltage at the bottom of the tank reaches its positive peak at the start of the first quarter-cycle.

The *total* voltage at the bottom of the tank is found by adding waveforms A and C together, giving the waveform labeled E in Fig. 7-10. This voltage reaches its positive peak midway in the first quarter-cycle—after the oscillatory tank-voltage peak but before the voltage peak due to the capacitive coupling from the primary tank circuit. This is the moment when maximum conduction occurs through lower diode V3. The complete path of these electrons is indicated (thin line) in Fig. 7-6 only, since this is probably the only quarter-cycle in which the lower diode conducts at all. The complete path takes the electrons from cathode to plate within the tube, upward through the lower half of coil L2 to the center tap and out, through the common return line to the upper end of R3, then downward through this resistor and back to the cathode. This current flow develops a voltage across R3 which then becomes part of the output voltage. Since the lower end of R3 is connected to ground, the upper end must be more *negative*. The downward flow of electrons toward the more positive lower end verifies this statement.

The detector output voltage is the algebraic sum of the voltages across output resistors R2 and R3. When the discriminator circuit is operated exactly at resonance—which is the center frequency—these two voltages will be equal in value but opposite in sign, so their sum will be zero. Recall that Fig. 7-1 related the modulating audio-voltage waveform to the carrier-frequency waveform. Here you can see that when the modulating audio voltage is crossing its own reference line and is therefore zero, the carrier will be exactly on its center frequency. Thus, when the modulating audio voltage is zero, so is the demodulated (detected) output voltage.

The Filtering Actions

C4 and C5 act as conventional filter capacitors, in conjunction with output resistors R2 and R3, respectively. When upper diode V2 conducts during the second quarter-cycle, it draws electrons from the upper plate of capacitor C4, making this plate positive (electron deficiency). This positive voltage, which persists throughout the entire cycle, accounts for the continuous upward flow of electron current through R2.

By somewhat analogous reasoning, we can show that the upper plate of C5 assumes a negative voltage by virtue of the electrons delivered there during the first quarter-cycle. The negative voltage on the upper plate of C5 will drive electrons downward through resistor R3 during the entire cycle, as depicted in each quarter-cycle diagram.

The fact that the two resultant voltage waveforms of Fig. 7-10 are exactly 90°, or a quarter of a cycle, apart stems directly from the original assumption that the oscillating tank voltage (shown a dotted line) and the capacitively coupled voltage (shown a dotted line) are equal in amplitude. Otherwise, the phase of the resultant voltages would be shifted. As a result, both peaks of

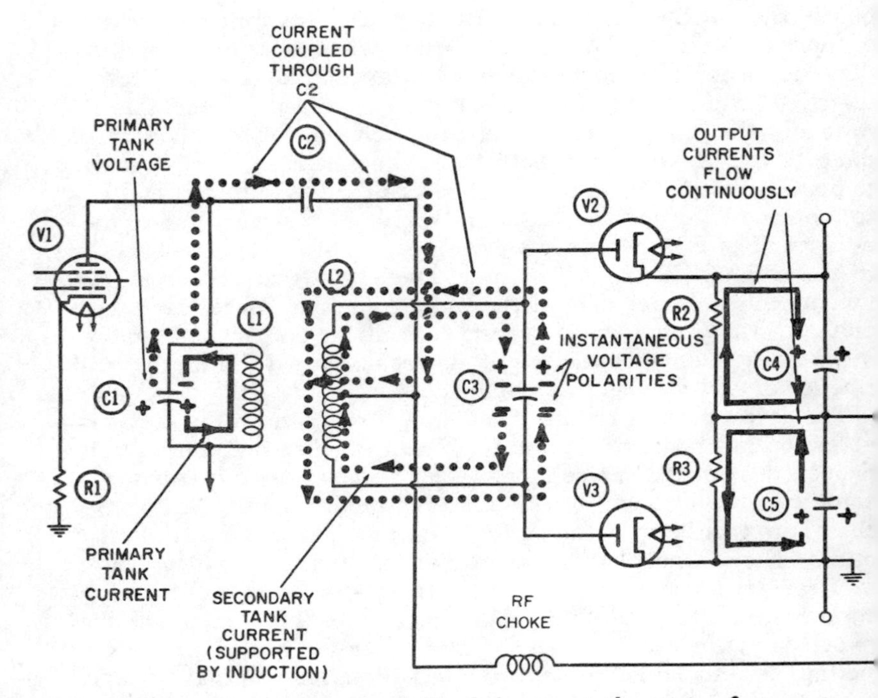

Fig. 7-8. Operation of the tuned discriminator at the center frequency—third quarter-cycle.

positive voltage (waveforms D and E of Fig. 7-10) would shift in phase and occur closer in time to the positive peak voltage of the *larger* driving voltage.

As an extreme example, imagine the inductive coupling between coils L1 and L2 were reduced until the oscillation of electrons in the secondary tank (the current shown a dotted line) could barely be sustained and almost vanished. The amplitude of waveforms B and C of Fig. 7-10 would progressively diminish, too. Should oscillation cease, these two waveforms would be straight horizontal lines coinciding with the reference line. The sum voltage waveform A and each line would give a waveform equal in size and phase to voltage waveform A. Obviously, then, the two diodes would conduct at the same instant. This would occur at the end of the first quarter-cycle, when the primary tank voltage (represented by waveform A) had reached its positive peak.

As another extreme example, consider what would happen if the voltage coupled from the primary to the secondary tank via

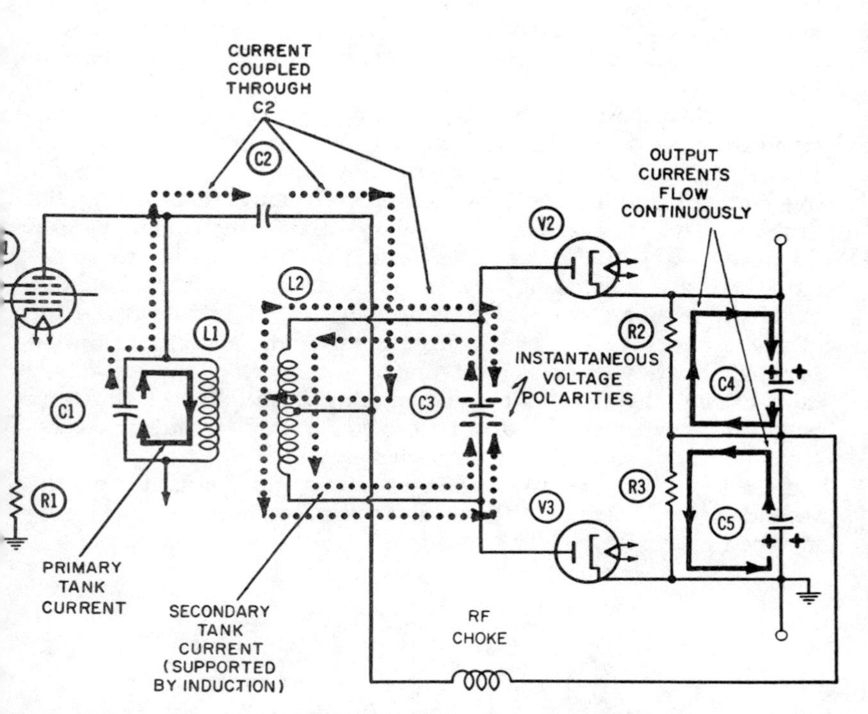

Fig. 7-9. Operation of the tuned discriminator at the center frequency—fourth quarter-cycle.

capacitor C2 were reduced almost to the vanishing point. Voltage waveform A would decrease in amplitude and, if *no* voltage were coupled across C2, would become a straight line coinciding with the reference line. Obviously, if this hypothetical waveform A were added to C, the resultant waveform would be in phase with C and also have the same amplitude. The upper diode would then have maximum conduction at the end of the second quarter-cycle.

By the same token, the algebraic sum of waveform B and a waveform of zero amplitude (a straight line) will give a waveform which is in phase with waveform B and has the same size. The lower diode will thus have its maximum conduction at the end of the fourth quarter-cycle.

Operation Above the Center Frequency

Figs. 7-11 and 7-12 show two successive half-cycles in the operation of the tuned discriminator. Here the operating frequency is *higher* than the center frequency. From Fig. 7-1 we see that the carrier frequency is modulated by the audio voltage, so that frequencies higher than the center frequency represent positive peaks of audio voltage and lower frequencies represent negative troughs.

The voltage polarities applied to the two diode plates are indicated by four deep plus and minus signs on both plates of tank capacitor C3. The signs represent the voltage polarity resulting from the oscillating tank current. Observe that during the first half-cycle of Fig. 7-11, this voltage makes the upper plate of C3 positive and the lower plate negative. But something else occurs during this same half-cycle—*both* plates are made negative by virtue of being coupled capacitively to the voltage at the top of the primary tank. The coupling current, shown in dotted lines, alternately delivers electrons to both sides of the secondary tank during the first half-cycle, making both sides of the tank negative. During the second half-cycle, it then withdraws electrons and both sides are made positive.

As a result of these two voltages acting independently on the secondary tank circuit, neither diode can conduct electrons during the first half-cycle. The reason is that the lower-diode plate is

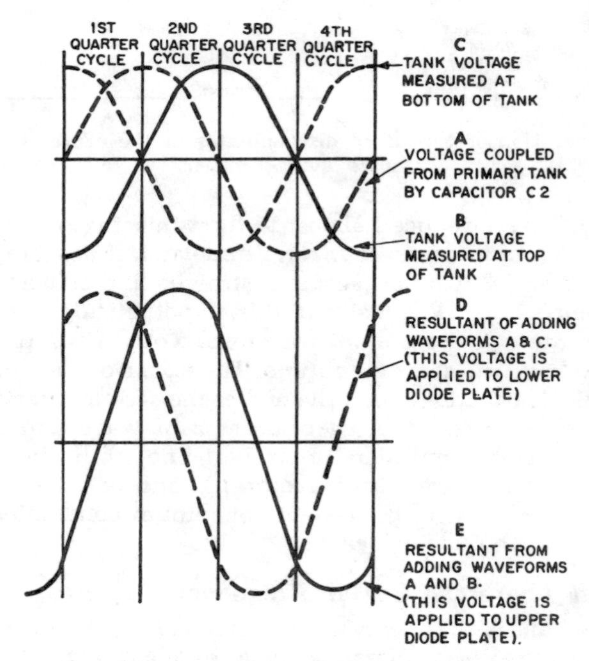

Fig. 7-10. Waveforms in the tuned discriminator circuit when operated at the resonant frequency.

at a high negative voltage, while the upper-diode plate remains at a neutral voltage. (In essence, the positive and negative components cancel each other at the upper-diode plate.) The lower diode conducts electrons during the second half-cycle because both applied voltages are positive. The upper-diode plate is again at a neutral voltage, and the positive and negative applied voltages cancel each other as before. Thus we have a means of assuring that the lower diode will conduct more electrons than the upper diode when the circuit is operated above the center frequency. This fact has a direct bearing on the voltages developed across output resistors R2 and R3—and on the sum of these two voltages, which is the discriminator output voltage.

Under the conditions shown in Figs. 7-11 and 7-12 (the upper diode not conducting), no current will flow through resistor R2 and no voltage will be developed across it. Consequently, the total output voltage across both resistors will equal the voltage drop across R3 resulting from the lower diode current flowing through it. Therefore the output voltage, measured at the top of the resistor combination, will be negative.

The results indicated by this discussion would appear to conflict with the classical response curve of an FM detector in Fig. 7-5. As shown, the voltage is positive above the center frequency and negative below it. This discrepancy is not of great significance. It stems from the assumptions made at the outset, and in particular from the manner in which these assumptions were defined and interpreted. It is accepted, for instance, that when two tank circuits are inductively coupled together and operated at resonance, their oscillatory voltages will be a quarter of a cycle out of phase with each other. However, *it was an act of interpretation* to assume that the points for measuring and comparing these two voltages would be the tops of the two tank circuits, as they are drawn in the various circuit diagrams in this chapter, beginning with Fig. 7-6.

Depending on the manner in which a particular transformer is wound or is connected into the circuit, we might just as naturally be comparing the voltage at the bottom of the secondary tank with the voltage at the top of the primary. If this were done, the output voltage of this tuned discriminator would then be consistent with the classical response curve of Fig. 7-5.

Actually, the human ear cannot differentiate between the negative and positive half-cycles, so it makes no difference whether they are reproduced in accordance with the original modulating signal or not. Also, each stage of amplification following the detector will shift the phase of the signal 180°. Therefore, if it is desired to have the same polarity as the original modulating signal

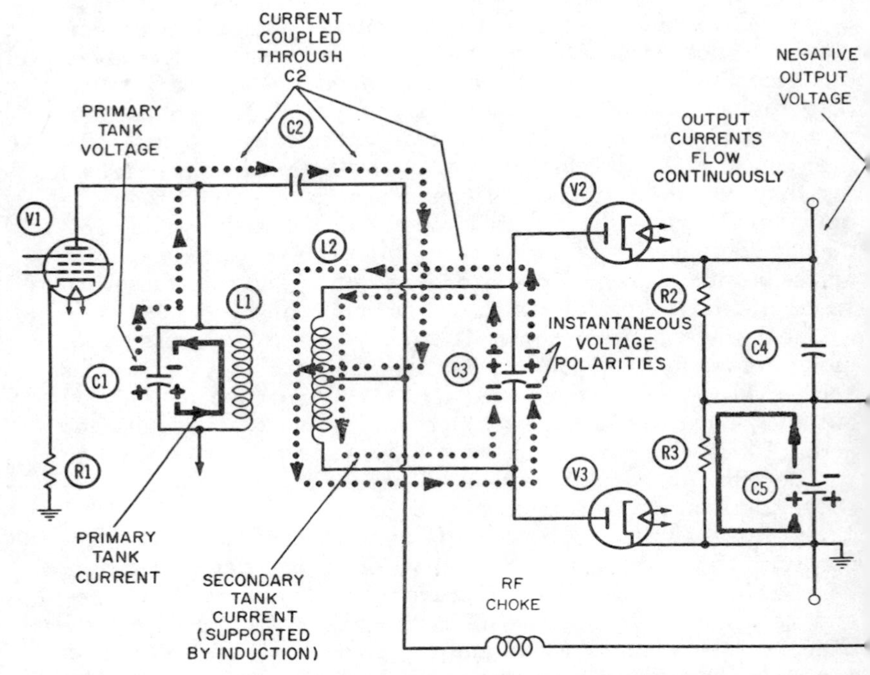

Fig. 7-11. Operation of the tuned discriminator above the center frequency —first half-cycle.

at the output, all that is required is to use an odd number of amplifying stages following the detector.

Simplifying Assumptions

The example in Figs. 7-11 and 7-12 is a special case in which two assumptions have been made to simplify the discussion. These assumptions are that (1) the two voltages applied to the diodes are equal in amplitude or strength, and (2) although exactly in phase on one side of the secondary tank, the two voltages are exactly out of phase on the other side. Fig. 7-13 shows this phase relationship.

Waveform C represents the amplitude of the oscillatory tank voltage *at the lower side* of the tank, and waveform A, the amplitude of the voltage coupled to the tank by capacitor C2. These two waveforms are essentially equal in amplitude and phase, and their algebraic sum is the waveform shown at E. (Note that it has twice the amplitude of either applied voltage.) This voltage is applied to the plate of lower diode V3 during the second half-

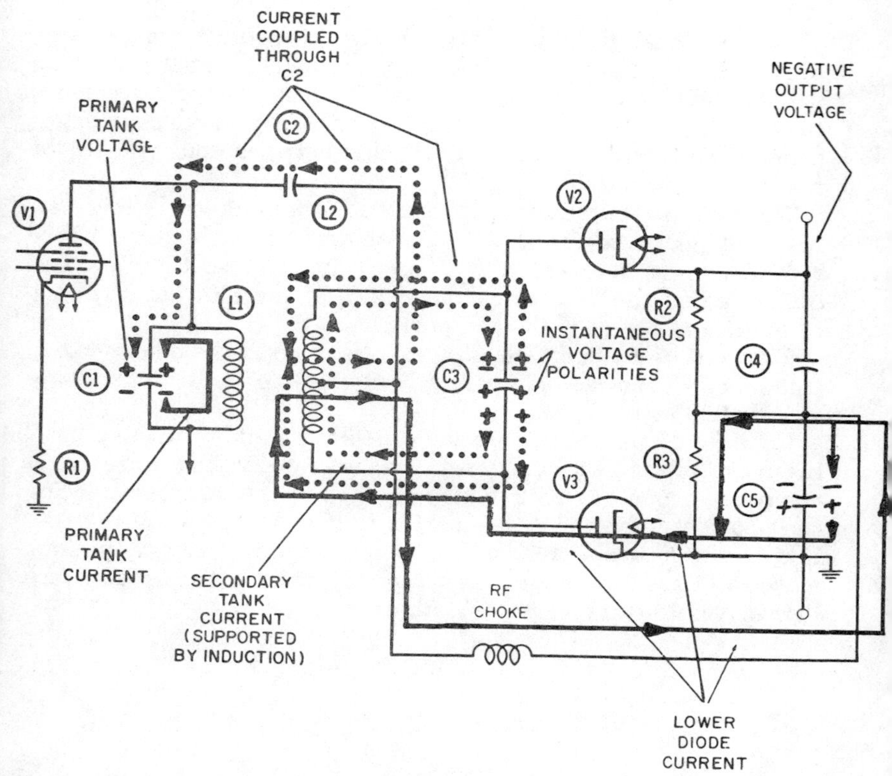

Fig. 7-12. Operation of the tuned discriminator above the center frequency —second half-cycle.

cycle. As a result, the plate is made positive and is able to conduct electrons.

Waveform B of Fig. 7-13 represents the amplitude at any instant when the oscillatory tank voltage is measured at the upper side of the tank. This waveform is assumed to be essentially equal in amplitude to voltage waveform A but 180° (half a cycle) out of phase with it. The algebraic sum of these two waveforms is a sine wave of extremely low amplitude (waveform E). If the two applied voltages were *exactly* equal in amplitude, waveform E would have zero amplitude and thus be a straight line. This sine wave of voltage is applied to the upper-diode plate and prevents this plate from becoming sufficiently positive that the tube will conduct.

There is only one occasion when two inductively coupled tuned circuits like these will have two tank voltages with the precise phase relationship shown. This is when the two coils are slightly overcoupled and when the operating frequency has *one particular value* which is somewhat higher than the true resonant frequency to which the circuits are tuned. What this operating frequency is depends both on the degree of coupling between coils and on the respective circuit Q's.

The response curve for overcoupled circuits indicates that new resonant peaks occur at two particular frequencies, one above resonance and one below. It is precisely at these two new frequencies that the two tank voltages are either 180° out of phase, or exactly in phase, with each other.

Figs. 7-11, 7-12, and 7-13 apply to operation at the *one and only frequency above resonance* at which the two tank voltages are exactly 180° out of phase.

In these examples, the terms "in phase" and "out of phase" mean the phase relationships of the two tank voltages are being compared *as measured at the tops of the two tanks.* In considering the phase of a tank voltage—particularly at each end of a tank—it is always desirable to clarify which voltage is under discussion, since the ones at the top and bottom of any oscillating tank are themselves 180° out of phase with each other.

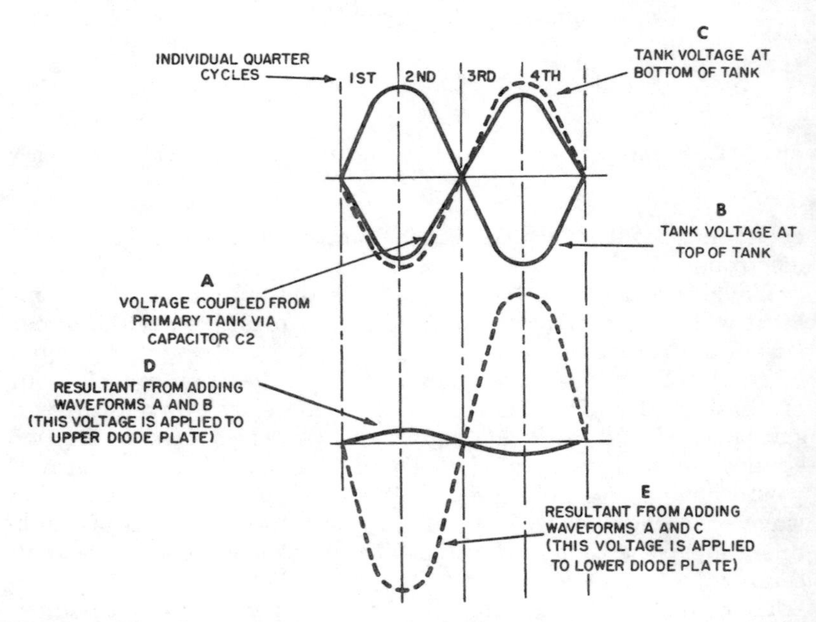

Fig. 7-13. Waveforms in the tuned discriminator circuit when operated above the center frequency.

Operation Below the Center Frequency

Figs. 7-14 and 7-15 show two successive half-cycles in the operation of the tuned discriminator, when the operating frequency is below the center frequency. The same simplifying assumptions have been made as in the previous example. The two tank voltages are now assumed to be exactly in phase, with their positive voltage peaks (measured at the tops of the tanks) occurring at the same instant. This is portrayed in Fig. 7-16, where voltage waveforms A and B are in phase with each other and have essentially the same amplitude. Their algebraic sum is shown by waveform D, which reaches its positive peak during the middle of the first half-cycle. This is the moment when the most electrons are conducted through upper diode V2. Fig. 7-14 shows this tube current a solid line. It follows the expected path from cathode to plate within the tube, then downward through the upper half of winding L2 to the center tap. Here it exits from the coil and returns, via the external line, to the junction of output resistors R2 and R3. This upward flow signifies that the voltage is more positive at the top of R2 than at the bottom.

This positive voltage (indicated in Fig. 7-14 by the deep plus sign on the upper plate of capacitor C3) is preserved throughout the half-cycle of Fig. 7-15, when diode V2 has a negative plate voltage and does not conduct electrons. Fig. 7-15 shows this positive capacitor voltage continuing to draw electron current upward through resistor R2 in an attempt to discharge itself to zero volts.

Waveform C of Fig. 7-16 represents the oscillatory tank voltage as measured at the bottom of the secondary tank circuit. Since it is exactly out of phase with waveform A and essentially of equal amplitude, the algebraic sum of the two is of very low amplitude, as shown by waveform E. In the hypothetical example where the two applied voltages have the same amplitude, their resultant will be a straight line coinciding with the reference line.

Since waveform D achieves no significant amplitude, the plate of diode V3 has a very low positive voltage applied to it during the second half-cycle. Hence, no appreciable current flows through it during the entire cycle. Thus, the output voltage of the discriminator is made up entirely of the negative voltage developed across resistor R2 by virtue of the electron current flowing through upper diode V2.

The examples used here—where upper diode V2 does not conduct when operating *above* the center frequency, and the lower diode does not conduct when operating *below* the center frequency—are extreme cases which might not be desirable or even realized in practice. Nevertheless, the output voltage achieved

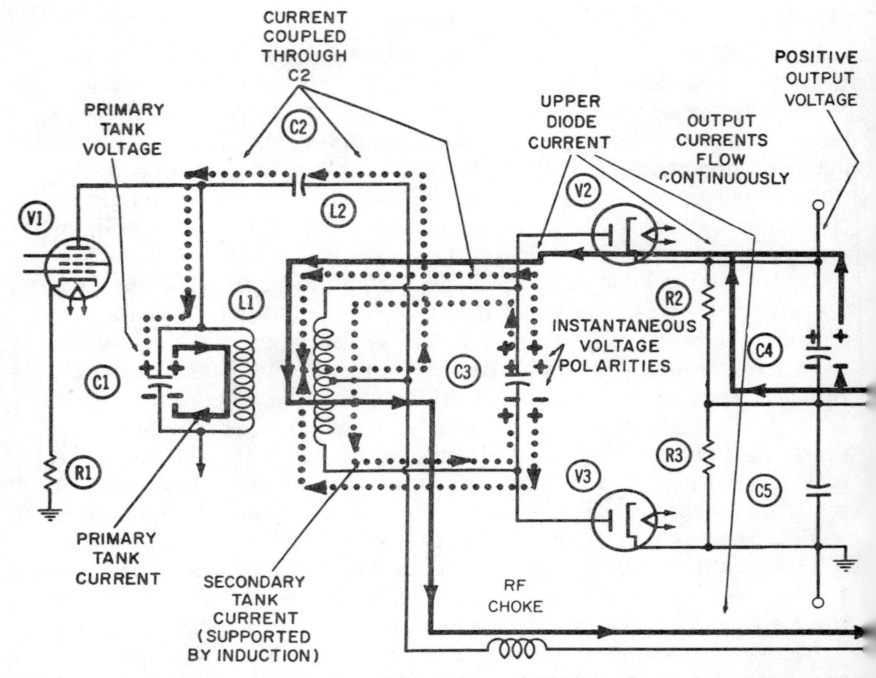

Fig. 7-14. Operation of the tuned discriminator below the center frequency —first half-cycle.

across the two output resistors, when the upper diode does not conduct, would clearly be the maximum attainable negative voltage. Likewise, the output voltage achieved when the lower diode does not conduct would clearly be the maximum attainable positive voltage. The resultant audio voltage, as the output varies between these two positive and negative peaks, would have the maximum amplitude attainable and would correspond to the waveform in Fig. 7-1C.

Audio voltages of lower amplitude, corresponding to the waveform of Fig. 7-1A, would be achieved with smaller deviations of the carrier from the center frequency. In all such cases, the operating conditions shown in Figs. 7-11, 7-12, 7-14, and 7-15 would be modified. The 180° phase relationships between the two applied voltages would of course not exist, and each diode would conduct *some* electrons above and below the center frequency.

Referring to Fig. 7-10, note that the phase relationships between the two applied voltages represent the normal condition when no modulation exists. Consequently the carrier does not

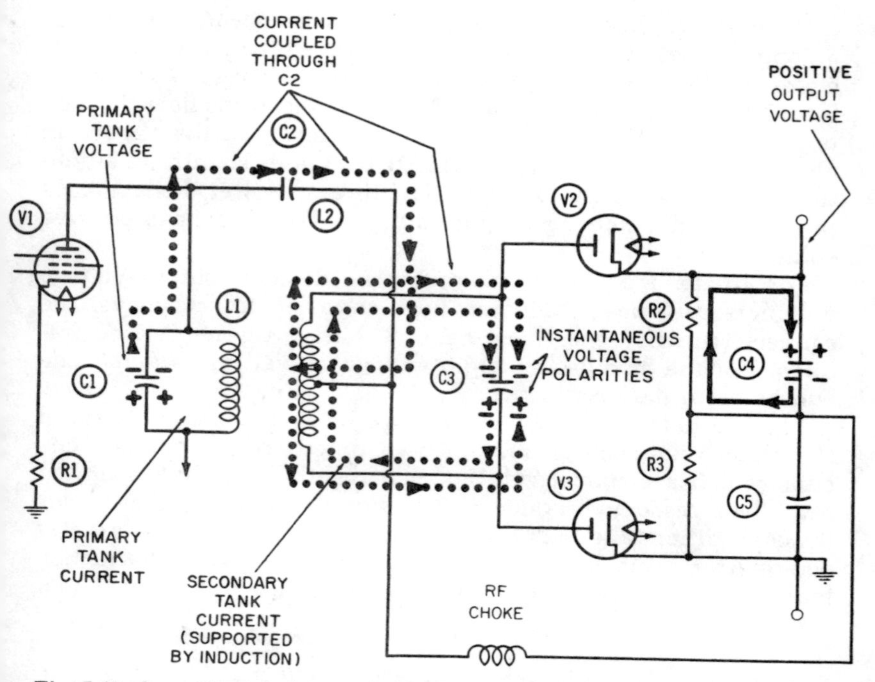

Fig. 7-15. Operation of the tuned discriminator below the center frequency —second half-cycle.

deviate from the center frequency. When two tuned tank circuits are inductively coupled and are operating exactly at resonance, their tank voltages will be 90° out of phase with each other.

When applied to the carrier frequency at the transmitter, an audio voltage of small amplitude causes the carrier frequency to deviate slightly in both directions. At the positive audio peak it goes above the center frequency, and at the negative audio trough it goes below. Any variation from this resonant, or center, frequency destroys the precise 90° phase relationship between the two tank voltages. Inevitably, voltage waveform A is moved closer to waveform C (when above the center frequency) and farther from waveform B.

Under these conditions waveform D would increase slightly in amplitude, causing a slightly greater current flow through the lower diode than existed at the center frequency. At the same time, waveform E would *decrease* slightly in amplitude, and fewer electrons would flow through the upper diode. At this instant, the output voltage has a small negative value.

When the carrier deviated slightly below the center frequency, an opposite set of conditions prevails. Waveforms A and B move closer together in phase. The amplitude of their resultant (waveform E) increases slightly and a few more electrons flow through upper diode V2. Waveforms A and C move farther apart in phase. Their resultant (waveform D) decreases slightly in amplitude, and not as many electrons flow through lower diode V3. At this instant, the detector output voltage will have a small positive value.

For greater frequency deviations, the detector output voltage will increase. Since it goes from positive to negative whenever the carrier frequency goes from below to above resonance, the *frequency* of the modulating audio voltage as well as its *amplitude* will be accurately reproduced.

The key to the operation of this type of circuit is to understand the phase relationships between two tuned radio-frequency circuits which are inductively coupled together and are operated near their resonant frequency. The significant fact is that under these conditions the tank voltages of two circuits will be exactly one-quarter cycle out of phase at resonance. As the frequency increases above resonance, this phase difference will increase to a

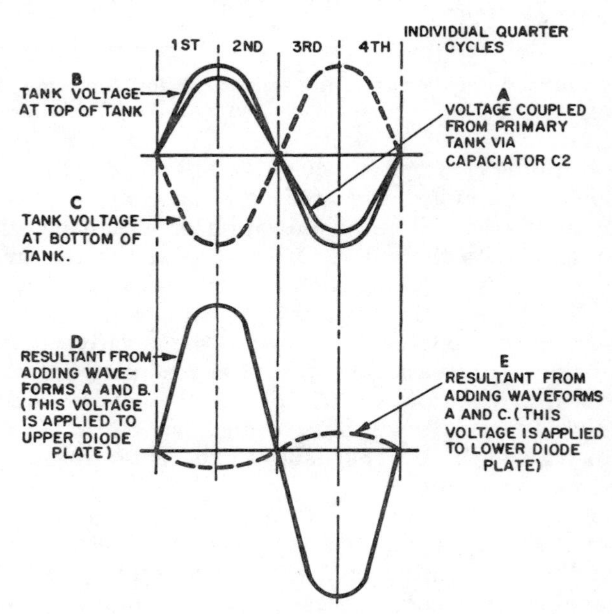

Fig. 7-16. Waveforms in the tuned discriminator circuit when operated below the resonant frequency.

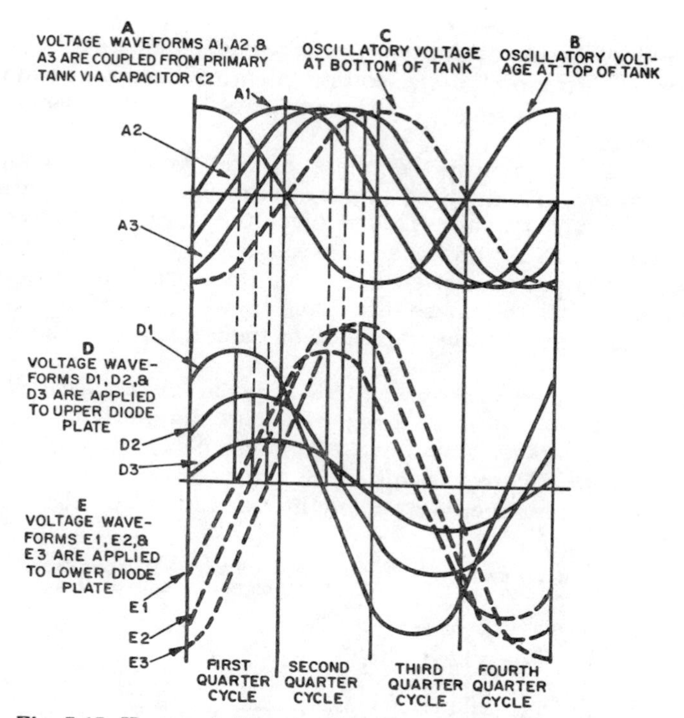

Fig. 7-17. How variations in phase between the two driving voltages will increase the voltage applied to the lower diode and decrease the voltage applied to the upper diode when operating above the resonant, or center, frequency.

maximum of half a cycle, or 180°. Conversely, as the frequency decreases below resonance, this phase difference will decrease from 90° toward zero, meaning the two tank voltages are exactly in phase.

The phase relationships between these two oscillatory tank voltages are important in this circuit only because of what happens when the two voltage waveforms are added algebraically. The use of both capacitive and inductive coupling provides the means for adding these two voltages. Fig. 7-17 shows several sample waveforms resulting when the carrier frequency deviates farther and farther above the resonant, or center, frequency. It can be seen from this figure that the amplitudes of the waveforms at D increase as the frequency excursion does, whereas the amplitude of the waveforms at E decrease in like amounts.

The waveform, shown as curve A1 in Fig. 7-17 and corresponding to waveform A of Fig. 7-10, reaches its positive peak a quarter of a cycle after waveform B (the latter is the oscillatory voltage measured at the top of the tank), and a quarter of a cycle before waveform C (the oscillatory voltage measured at the bottom of the tank). These phase relationships exist only at the resonant, or center, frequency.

Curve A2 shows how the phase of the capacitively-coupled voltage waveform has shifted slightly to the right as the frequency deviates above resonance. This *reduces* the phase difference between waveforms A2 and C and increases the amplitude of their resultant (the curve labeled E2). This phase shift of waveform A2 also *increases* the phase difference between it and waveform B, and thus decreases the amplitude of their resultant (the curve labeled D2).

Waveform A3 represents a further deviation in phase of the primary tank voltage *when compared with the phase of the secondary tank voltage.* Waveform A3 moves closer to coincidence with waveform C. As a result, their sum—shown as curve E3—undergoes another increase in amplitude. Curve A3 also moves

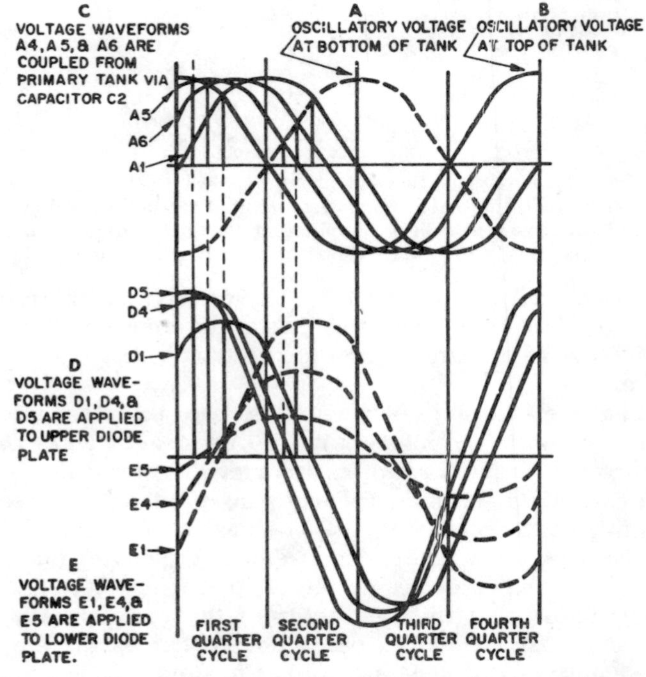

Fig. 7-18. How variations in phase between the two driving voltages will increase the voltage applied to the upper diode and decrease the voltage applied to the lower diode when operating below the resonant, or center, frequency.

farther away from coincidence with waveform B, so that the amplitude of their resultant (labeled curve D3) again decreases.

A similar set of waveforms has been drawn in Fig. 7-18 to illustrate the phase relationships between the two applied voltages when operating below resonance, and the changes in amplitude of the resultant voltages applied to the two diode-detector plates. With such a set of curves, waveform A shifts to the left instead of the right, bringing it toward phase coincidence with waveform B and toward phase opposition with waveform C. Under these conditions, waveform D progressively increases in amplitude as the frequency excursion does, while waveform E progressively decreases in amplitude.

In Fig. 7-18, waveforms A1, B, and C are identical to the waveforms having the same designations in Fig. 7-17. Waveforms A4 and A5, however, represent the progressive shifts in phase of the directly-coupled voltage with respect to the phase of the secondary tank voltage.

Inspection of Fig. 7-18 should reveal that the various waveforms are related as follows:

A1 plus C combine to produce waveform E1.
A4 plus C combine to produce waveform E4.
A5 plus C combine to produce waveform E5.
A1 plus B combine to produce waveform D1.
A4 plus B combine to produce waveform D4.
A5 plus B combine to produce waveform D5.

Waveforms D and E represent the total voltages applied to the plates of the upper and lower diode-detector tubes, respectively. Consequently, their amplitudes will be roughly proportionate to the amount of electron current through each diode, and therefore to the voltages developed across R2 and R3 during any RF cycle.

Because of the opposing polarities of the voltages developed across R2 and R3, the total output voltage across the two resistors will always be negative when operating above resonance and positive when operating below.

Chapter 8

RATIO DETECTOR

Both the detuned and the tuned discriminators discussed in the previous chapter share a disadvantage—the FM signal being demodulated must be of *constant amplitude* when it reaches the discriminator. In addition, the discriminator stage must be preceded by several amplifier stages, in order to build up even the weakest signal enough that the audio output from the discriminator will have sufficient volume. After this elaborate amplification process, the frequency-modulated signal must also be passed through one or more *limiter* stages. Here, any traces of amplitude modulation caused by atmospheric or propagation conditions are removed.

Unless these conditions are met by the two previous discriminators, they will respond to amplitude as well as frequency modulation. The result will then be a distorted audio output. This statement may be clarified with the aid of an example. In the tuned-discriminator operation, the assumption was made that at some frequency above resonance, the secondary tank voltage and the directly coupled voltage (also known as the reference voltage) were exactly in phase with each other. This caused the upper-diode detector to conduct heavily, and prevented the lower diode from conducting at all. The amount of electron current flowing through the upper diode then determined the amount of instantaneous output voltage developed across the appropriate output resistor. This amount of electron current depended of course on the *strength* of the tank and reference voltages.

A little reflection on the nature of these alternating voltages will reveal that both will be large when the received signal is strong, and small when it is weak. Furthermore, these changes in signal strength will be independent of the resonant tank frequency, and also of those frequencies both above and below resonance when the three voltage waveforms are exactly in or out of phase. Consequently, when a strong FM signal is received, the amount of electron current through either diode will be caused partly by the frequency deviations of the carrier (which is good) and partly by the excessively strong amplitude of the carrier (which is bad).

CIRCUIT DESCRIPTION

The ratio detector (Figs. 8-1 through 8-4) is one of the most widely used circuits for demodulating FM signals. One reason is that it responds only to frequency deviations of the signal; it remains unaffected by any amplitude variations caused by unusual propagation phenomena. The fact that the ratio detector is able to do this without additional amplifier or limiter stages makes it possible to build FM receivers which are competitive in price and in size with conventional AM broadcast receivers.

Circuit Components

The components and their functional titles in the ratio-detector circuit include:

R1—Output resistor.
R2—AVC resistor.
C1—Final IF tank capacitor.
C2—Capacitor for coupling the reference voltage to the detector tubes.
C3—Tuned tank capacitor.
C4—Filter capacitor for V1.
C5—Filter capacitor for V2.
C6—Large capacitor for stabilizing the voltage across C4 and C5.
C7—Coupling capacitor for the audio output.
L1—Final IF tank inductor.
L2—Tuned tank inductor.
L3—Radio-frequency choke.
V1 and V2—Diode-detector tubes.

Identification of Currents

The various electron currents at work in this circuit include:

1. Final IF tank current oscillating at the intermediate frequency between L1 and C1.
2. Current which is directly coupled from the top of the primary tank to the center tap of the secondary tank inductor.

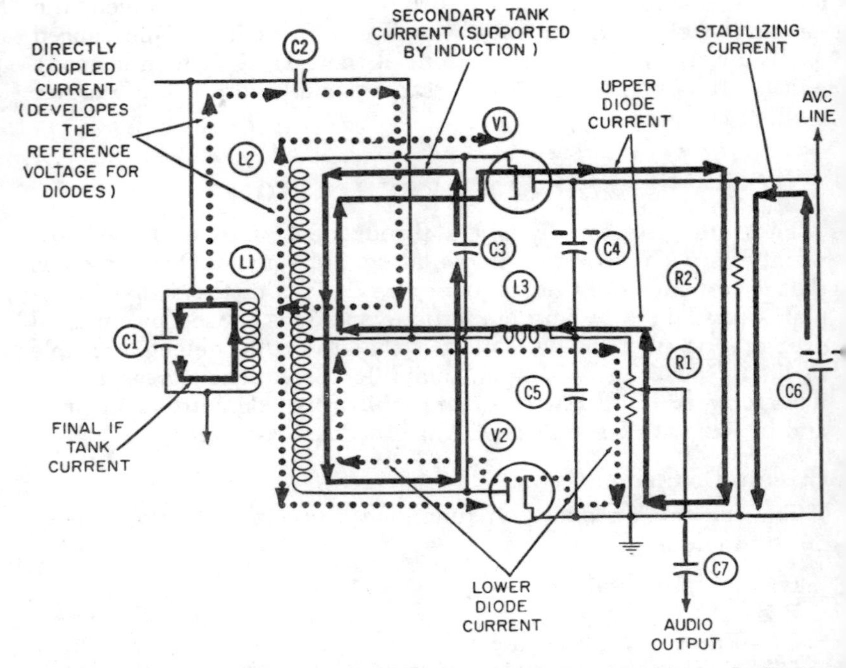

Fig. 8-1. Operation of the ratio detector—at center frequency, during a period of normal signal strength.

3. Secondary tank current which oscillates between L2 and C3 and is supported by induction between L1 and L2 .
4. Current through upper diode V1 .
5. Current through lower diode V2 .
6. Stabilizing current flowing out of C6 and down through R2 .

CIRCUIT OPERATION

This type of circuit differs from the discriminators discussed in the previous chapter in three important particulars. They are:

1. The position of the lower diode, V2, is reversed, so that electron current flows in the opposite direction. This is called a *series-aiding* connection.
2. Both diode currents flow through output resistor R1, whereas in the discriminator circuits they flowed through separate load resistors.

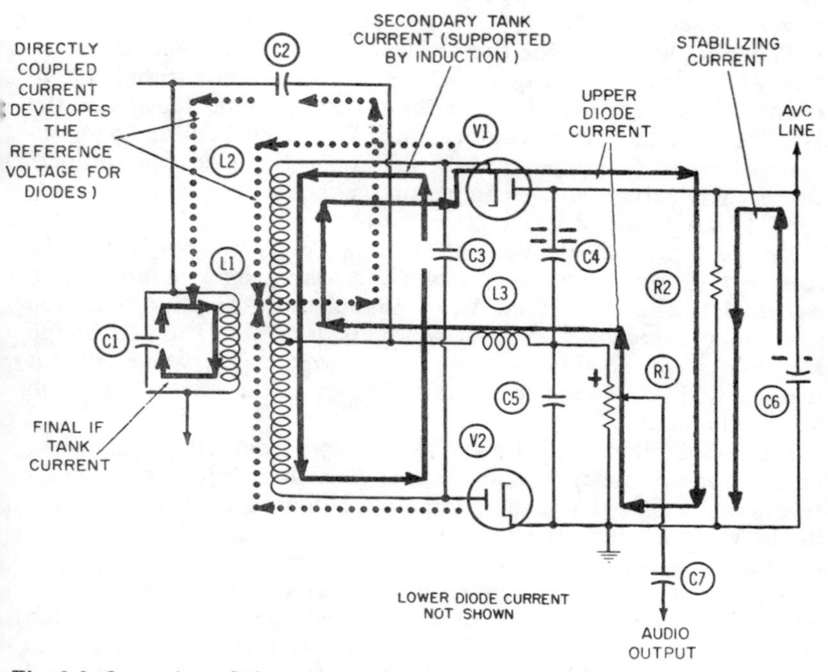

Fig. 8-2. Operation of the ratio detector—above center frequency, during a period of normal signal strength.

3. A long time-constant combination, consisting of R2 and C6, has been added so the circuit will not respond to undesired modulations in signal amplitude.

The same waveform diagrams shown in the preceding chapter on the detuned discriminator also apply to this ratio detector.

The fact that the lower diode, V2, has been reversed means of course that its current flows *clockwise* around the lower portion of the detector circuit. The important difference between the operation of this circuit and the tuned discriminator is that diode V2 conducts during different portions of the whole cycle. However, the quantity of electrons conducted during any individual cycle is the same in each circuit. Because of this, the waveform diagrams of the tuned discriminator will not be repeated here.

Fig. 8-1 depicts the conditions when this circuit is operated at its center frequency. The upper-diode current (solid line) and lower-diode current (small dots) will now flow in equal amounts. Since both diode currents are equal but flow through output resistor R1 in opposite directions, no output voltage is developed across R1.

The complete path for the electron current through upper diode V1 is clockwise through V1, then downward through R2 to ground. From ground, the path is upward through R1 and to the left, through L3, to the center tap of L2. From here it flows upward through the upper half of L2 and returns to the cathode. A steady negative voltage builds up on the upper plate of capacitor C4 as a result of this current flow.

As the frequency deviates above resonance, this negative voltage on the upper plate of C4 will increase proportionately. Fig. 8-2 depicts this condition. V1 conducts more and more as the frequency increases, and V2 conducts less and less. Although Fig. 8-2 does not show any current through V2, some current usually will be flowing through it. However, the current through V1 predominates.

Fig. 8-3 depicts conditions when the ratio detector is operated at any reasonable frequency *lower* than the center or resonant frequency. As the frequency deviation below resonance increases, diode V2 will conduct *more* electron current and diode V1 will conduct progressively less. (As before, the current through V1 is not shown.) The complete path of the electron current through V2 is also *clockwise*, from cathode to plate and upward through the lower half of inductor L2 to the center tap. Here it flows to the right through L3, and down through R1 to the common ground connection, through which it has easy access back to the cathode.

At any frequency below resonance, more current flows through V2 than through V1. The difference between these two currents increases as the frequency deviation does. Because more current flows through V2 than through V1, more current will flow downward through R1 than upward. This establishes the fact that the voltage at any point along R1 is *negative* with respect to ground.

Figs. 8-2 and 8-3 represent positive and negative half-cycles of audio-output voltage, respectively. Capacitor C6 (usually 8 or 10 microfarads) operates in conjunction with resistor R2 to form a long time-constant combination. Its purpose is to restrict the amount of current through the two diodes when the signal is excessively strong, but to encourage their flow when the signal is weak. In this respect, the combination functions somewhat like the AVC combination discussed in a previous chapter.

Fig. 8-4 depicts operating conditions during a signal build-up at a frequency above resonance. Now the received carrier signal is stronger. When this occurs, *both* of the principal currents through the two diodes will be increased proportionately. Since these currents flow in opposite directions through resistor R1,

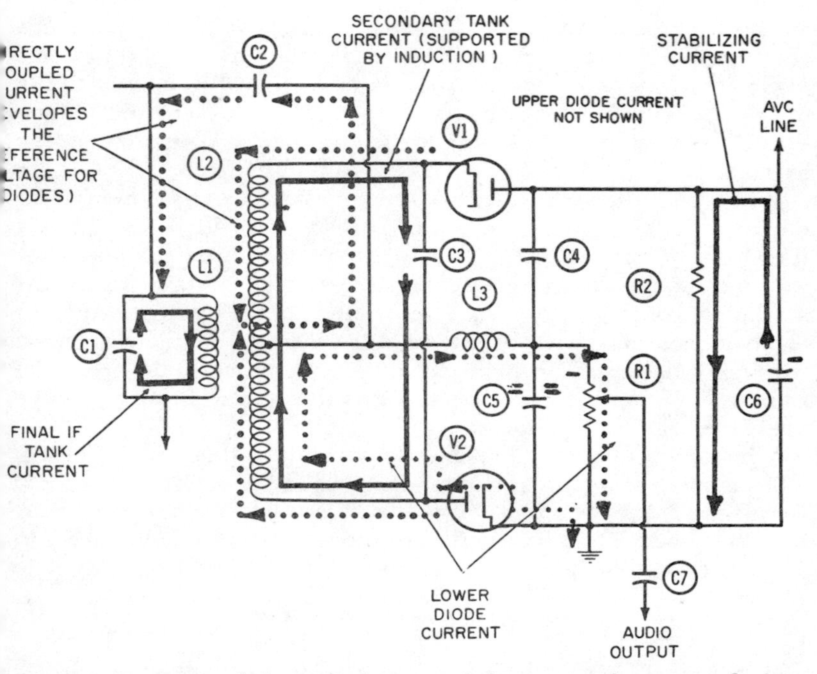

Fig. 8-3. Operation of the ratio detector—below center frequency, during a period of normal signal strength.

the voltage developed across R1 will depend on the *difference* between these two currents, As both currents are increased proportionately, so does the difference between them. Consequently, a signal build-up—one of the commonest forms of undesired amplitude modulation, along with signal fade—will change the instantaneous audio-output voltage across R1. Hence, the circuit will now respond to amplitude modulation.

This undesirable state of affairs is prevented by the voltage accumulated on the upper plate of capacitor C6. As you can see, capacitors C4 and C5 are connected in parallel with C6; therefore, the instantaneous voltages on the upper plates of C4 and C5 must always add up to the voltage on C6, even though these voltages are themselves constantly changing in step with the frequency deviations of the carrier. It is the presence of this fairly stable voltage on capacitor C6 which biases both diodes so they will conduct enough current to produce the two voltages on capacitors C4 and C5.

In summary, the voltage across capacitor C6, and simultaneously across resistor R2, will always be equal to the sum of the voltages across capacitors C4 and C5. This is inevitable, since the tops of C4 and C6 are connected together, as are the bottoms of C6 and C5. The voltages across C4 and C5 always add up to the voltage on C6, but will change individually so that their difference is always proportionate to the frequency deviation of the carrier.

When both diodes in Fig. 8-1 are conducting equal amounts of current, the voltage at the top of resistor R1 will be zero, since its bottom is connected to ground (zero) and equal currents flow in opposite directions through it. A negative voltage will exist at the top of capacitor C4. This is due to the accumulation of electrons on its upper plate as a result of the plate current through diode V1. *The same negative voltage* will exist from top to bottom of capacitor C6 because it is the sum of these twc voltages, one of which is zero.

When operating above resonance, in Fig. 8-2 diode V1 conducts more current and delivers more electrons to the upper plate of C4. As a result, the negative voltage across C4 increases. However, because diode V1 is conducting more electrons than V2, the top of resistor R1 must be more positive than the bottom, which is

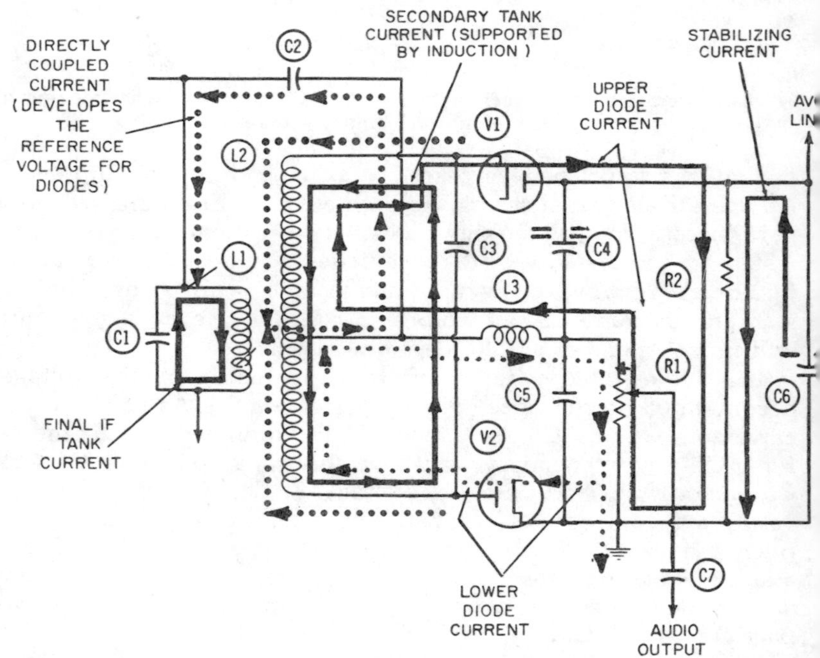

Fig. 8-4. Operation of the ratio detector—above center frequency, during a period of signal build-up.

connected to ground. The reason is that more current is flowing upward through R1 than downward.

Because of the long time-constant of R2 and C6, the voltage across them does not change during the one-quarter of an audio cycle between Figs. 8-1 and 8-2. This voltage is represented by the electrons stored on the upper plate of C6. In Fig. 8-2 this voltage is the sum of the large negative voltage across C4 and the smaller positive voltage across C5.

When operating below resonance, diode V2 in Fig. 8-2 conducts more electrons than V1. Therefore, the top of resistor R1 must be more negative than the bottom. This negative voltage is indicated by the thin minus sign at this point in the figure. The voltage across C6 and R2 is still unchanged and is now equal to the sum of the reduced negative voltage across C4 and the negative voltage which now exists across C5.

SECTION 4

TRANSISTOR CIRCUITS

Chapter 1

TRANSISTOR PHYSICS SIMPLIFIED

Before delving into the action within a semiconductor, and the operation of circuits employing transistors, it will be necessary to review some basic definitions, and the physical principles which support them. The well-known electron-drift process within electrical conductors is a logical starting point.

THE ELECTRON-DRIFT PROCESS

Fig. 1-1 shows a simplified germanium atom. It consists of a positively charged nucleus and four planetary electrons revolving in orbit around it. Each electron carries a negative charge of electricity, so that they together will just neutralize the positive charge of the nucleus, making the entire atom electrically neutral.

(This atomic picture is simplified, in that the germanium atom has a total of 32 planetary electrons in orbit, and the central nucleus has a resulting positive charge of 32 units. Twenty-eight of these electrons are so tightly bound to the nucleus, however, that they are completely unavailable for purposes of being dislodged to form an electron current. These 28 electrons are not shown in Fig. 1-1.)

The electrons shown in the illustration are in the *valence* band, or ring, of the atom. Germanium is normally a good insulator and a poor conductor. This means that only with great difficulty can one of the planetary electrons be dislodged from its orbit by the normal electron drift process. In a good conductor, on the other hand, electron drift occurs quite easily. When an electric field (electric potential, or a voltage) such as from a battery is applied across the terminals of a good conductor, electrons will be driven through the conductor, from the negative to the positive end. This flow of electron is called electric current, and it consists of a long series of "domino" actions between the free electrons in the material (including planetary electrons which can easily be set free). An individual electron, moving into the conductor very quickly, finds itself approaching a head-on collision with another electron, perhaps one in orbit around

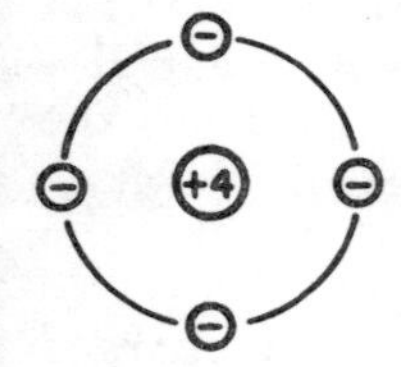

Fig. 1-1. Simplified drawing of a germanium atom, showing the four orbiting electrons.

an atomic nucleus. Since electrons carry a negative charge, they mutually repel each other, and no electron collision occurs. Instead, this orbiting electron will be "knocked out" of its orbit, not by collision, but by the electrical repulsion from the approaching electron. Once set free in this manner, the released electron will assume the velocity and direction of the approaching electron. The approaching electron then finds itself slowed down and in the vicinity of a positively charged atomic nucleus; as a result, it "falls into" the recently vacated orbit. Thus, one complete sequence of electron drift action has taken place—an electron has been repelled from its orbit around a nucleus, and replaced by another electron.

ACCEPTOR ATOMS AND P-TYPE SEMICONDUCTORS

Germanium, the insulator, becomes germanium, the semiconductor, with the addition of selected impurities. Two very common impurities are boron and arsenic. Fig. 1-2 shows the

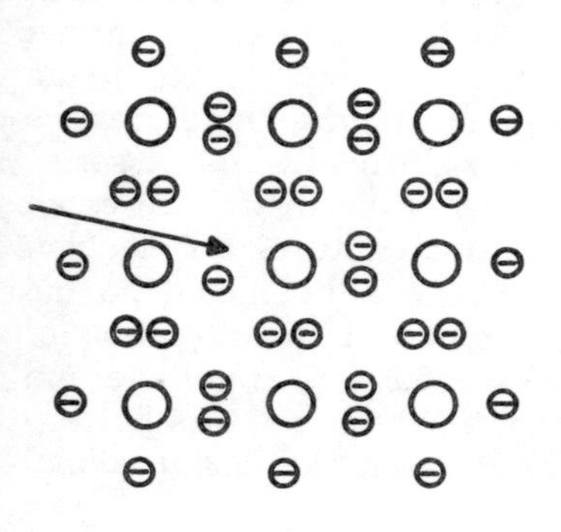

Fig. 1-2. A boron atom tied to four germanium atoms.

atomic realignment that occurs when a small portion of boron is added to germanium. An atom of boron has a total of three planetary electrons in its valence band. One atom of boron will "lock" itself very firmly in place with four adjoining germanium atoms. When this happens, the resulting combination needs or wants one additional electron to complete itself, and will take ("accept") it from a nearby germanium atom. This action gives it the title of an acceptor atom. Before it takes this extra electron from its neighbor, the impurity atom is electrically neutral—neither positive nor negative. After it takes this electron, the acceptor becomes permanently negatively charged.

The Concept of "Positive Holes"

The adjacent germanium atom becomes positively charged after relinquishing one electron. This positively charged atom becomes what is known as a "hole," or the carrier of a positive electrical charge; and the material is called a P-type semiconductor. The hole is indicated by the arrow in Fig. 1-2.

If we consider only this one positive atom and the negatively charged acceptor which "stole" its electron, we might well ask why the stolen electron does not return to its original orbit. How can both positive and negative charges exist side by side without immediately neutralizing themselves? The answer can be found in the physics of atomic structure which is beyond the scope of this book. We must be content with the picture of the acceptor atom, taking and holding tightly the extra electron which gives it an over-all negative charge.

For every atom of impurity, there is eventually created one negatively charged acceptor atom and its positive counterpart, the germanium hole, which is deficient by one electron. Even though the germanium atom by itself will not participate in the electron drift process (it is a poor conductor), the germanium "hole" becomes a fairly good conductor of electricity. It is said that holes move freely through a semiconductor, but this statement requires some elaboration. The positively charged atoms do not move through the semiconductor. By capturing planetary electrons from nearby atomic orbits, the "hole" can be given the appearance of moving fairly rapidly through a series of germanium atoms. In reality, each such atom in turn loses an electron and becomes a hole, then in turn captures an electron from the next atom, causing it to become a hole, and so on.

Thus, the positive charge seems to move in the direction of applied *positive* voltage—meaning *away* from a plus voltage and *toward* a minus voltage such as the negative terminal of a battery This process of hole-current movement resembles conventional electron drift, but with the important difference that the electron drift process presupposes extra electrons traveling through the atomic structure and, in effect, "pushing" planetary electrons out of their orbits. "Hole" current seems to travel in the opposite direction through a semiconductor, by "pulling" planetary electrons out of orbit.

"Mobility" of Charge Carriers

Electron charge is concentrated at a single point, whereas the positive charge of a hole seems to be distributed somewhat—if not over the entire volume of the atom, then at least over a portion of it or a portion of the orbit. This would seem to indicate that the "pulling" of electrons from orbit might be more difficult than the normal process of pushing them from orbit. That this is the case may be verified from the "mobility" of electrons and holes in germanium.

The mobility of an electric charge is a measure of the relative ease or difficulty with which the charge can be moved by an applied electric field (voltage). It is usually expressed as the "drift velocity" of the charge in centimeters per second. The mobility of free electrons in a germanium semiconductor, when the applied electric field has a strength of one volt per centimeter, is about 3,600 centimeters per second.

The mobility of positive "holes" through a germanium semiconductor for the same applied electric field of one volt per centimeter, is about 1,700 centimeters per second. Thus, the drift velocity of free electrons through a germanium semiconductor is more than twice as great as the drift velocity of holes.

DONOR ATOMS AND N-TYPE SEMICONDUCTORS

Fig. 1-3 shows the atomic structure that results when an impurity atom such as arsenic is added to the germanium. Arsenic is chosen as an impurity because it has five planetary electrons orbiting about the nucleus. The arsenic atom will "lock" in place with four germanium atoms; and one of the five valence electrons will be released, or "donated," after this combination occurs. This action gives the resulting atom the name "donor atom." The extra electron is indicated by the arrow in Fig. 1-3. Before giving up one electron, the atom has a neutral electric charge; and afterward, it becomes a positively charged "donor" atom.

Like the negatively charged acceptor atom discussed previously, the donor atom holds on very tightly to its new electrical condition, and does not recapture the electron it released. Consequently, both the positive atom and the negative electron can

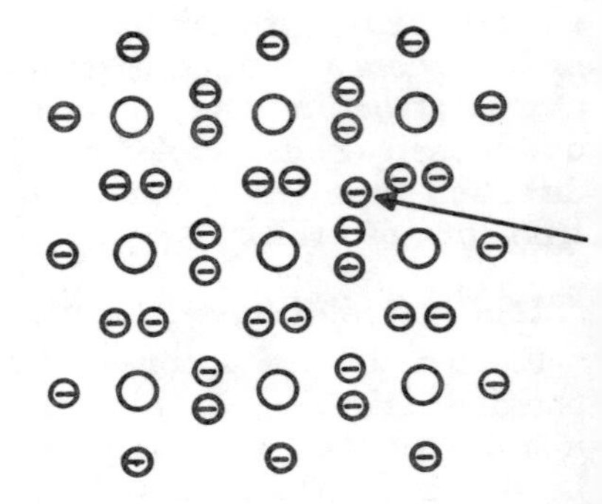

Fig. 1-3. An arsenic atom tied to four germanium atoms.

exist side by side in the same portion of the semiconductor without recombining. Although the total charge of this semiconductor will be zero (since the number of electrons released equals the number of positively charged donor atoms), the semiconductor becomes known as an N-type because of the availability of negative carriers (electrons). The positively charged donor atoms do not act as current carriers, and are not "holes."

We have now created the two basic types of semiconductors. The N-type has an excess of free electrons available for carrying current: whereas the P-type has a series of positively charged germanium atoms, called holes, also available for letting electron current flow through the semiconductor.

IMPURITY CONTENT OF SEMICONDUCTORS

Very small portions of an impurity are required to convert the insulator germanium to a semiconductor germanium—about one part impurity to a million parts germanium. This is due to the conductivity of the material, which varies directly with the density of current carriers (free electrons and holes) contained therein. The number of current carriers in a particular sample will equal the number of impurity atoms.

THE JUNCTION DIODE

When an N-type semiconductor is bonded to a P-type semiconductor, a junction diode is formed. Fig. 1-4 shows the junction diode under three voltage conditions.

The natural tendency for the electrical charges in the junction diode, when no external voltage is applied (Fig. 1-4A), is for excess electrons in the N-material to cross the junction (moving to the left in Fig. 1-4A) and recombine with the positive holes in the P-region. A small amount of such recombining does occur, but it is prevented from happening on a wholesale basis. As soon as a few electrons have left the N-region, it will no longer be electrically neutral, but will have a slightly positive charge. Likewise as soon as the P-region has acquired a few excess electrons, it will no longer be electrically neutral, but will possess a slightly negative charge. This charge redistribution is shown in Fig. 1-4A.

The current carriers in both semiconductors are repelled from the area of the junction. The area close to the junction is variously known as a depletion zone (because it is depleted of current carriers) or as a transition region. The particular charge distribution is referred to as a potential hill across the junction, a potential barrier, a dipole charge layer, a space charge layer, differing energy levels, among others.

In Fig. 1-4B a negative voltage is shown being applied to the P-type material and a positive voltage to the N-type. This condition, known as reverse bias, is not conducive to current flow through the junction. Electrons will try to flow from the P- to the N- material, but will encounter high opposition or resistance in their attempt to flow in the opposite direction. The reverse bias *adds* to the potential barrier set up by the junction when no bias is present.

Fig. 1-4C shows conditions across the junction when forward bias is applied. This term implies that the normal tendency of electron current to flow from N-type to P-type will be aided or encouraged by the applied voltage, and that a substantial electron current will cross the junction and flow through the external circuit.

The foregoing two paragraphs reveal the possibility of using semiconductors as a rectifying device for converting an applied alternating current to a unidirectional current. The rectifying *principle* finds wide application in both rectifier power supplies and detection or demodulation circuits.

THE JUNCTION TRANSISTOR

We are now ready to assemble three semiconductors in such a fashion that they become a transistor. Fig. 1-5 shows a P-type, an N-type, and another P-type semiconductor bonded together to constitute a PNP transistor. The charge distribution, when no external voltage is applied, is as indicated. The so-called charge dipole layer forms at each junction, so that both edges of the center semiconductor are slightly positive. This positive charge is composed of the fixed donor atoms, which do not move and hence do not participate in the electron drift process. The free electrons in the N- material are seen to be bunched near the center. They are held in position by the fixed negative acceptor atoms on the two adjoining faces of the P-type semiconductors, the repelling electric field of which extends through the positive concentrations of donor atoms on either edge of the N- material to the center.

This charge dipole layer at each junction could be shown as being produced by a small simulated battery across each junction, resulting in the existing polarities. It is important to understand that, even under this static or equilibrium condition of no external voltage applied, there actually are small voltages (or electric fields) existing across each junction in a transistor.

A *graphical* representation of this charge distribution is given immediately below the *symbolic* representation in Fig. 1-5. This charge distribution is also called an electric field—or more simply, a voltage. The fixed positive and negative atoms are omitted from this diagram. However, they would be distributed around each junction, just as they are around the single junction of Fig. 1-4A.

The graphical picture (Fig. 1-5B) tells us the same story as the symbolic picture in Fig. 1-5A—namely, that within the base, but adjacent to each junction, there is a positive electric field due to the relative scarcity of free electrons. In the center of the base, however, there is a negative electric field due to the presence of excess free electrons. Within the emitter and collector, but immediately adjacent to each junction, there is a negative electric field due to the relative absence of positively charged "holes." As we move away from each junction, this electric field changes from negative to positive—due now to the presence of an excess of positive holes.

Because the collector is usually manufactured with a much lower conductivity factor than the emitter, considerably *fewer* free charge carrier (positive holes) will be milling around within it. The conductivity of a semiconductor is directly proportional to the density, or concentration, of these charge carriers. Frequently, the emitter material will have a hundred times greater conductivity than the collector material.

In Fig. 1-5A, the lower charge density indicated in the collector reduces the electric field across the base-collector junction, as shown graphically in Fig. 1-5B. And the smaller this electric field or voltage is, the easier will it be to breach the natural tendencies of the junction and get electron current to flow *from the* P-material of the collector into the N-material of the base.

As explained more fully in connection with Figs. 1-8, 1-9, and 1-10, this is accomplished by using a "signal" current (or

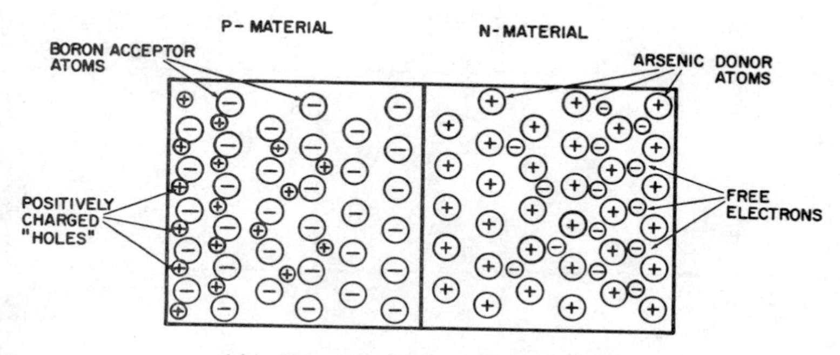

(A) Normal charge distribution.

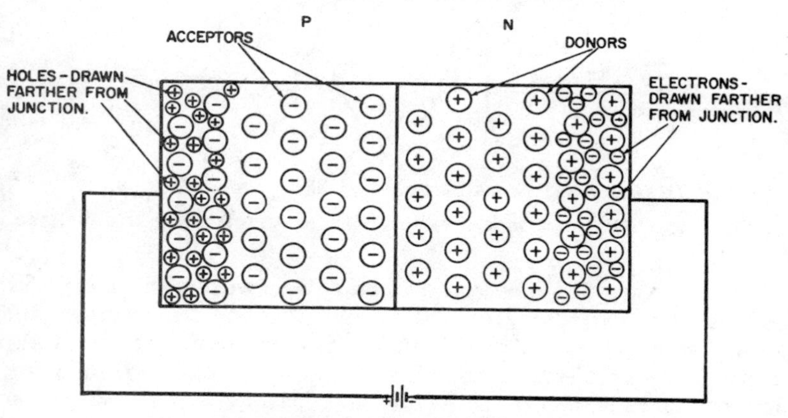

(B) Reverse bias condition.

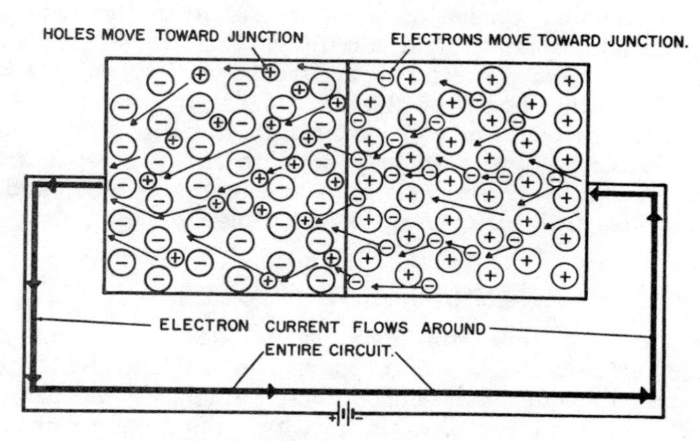

(C) Forward bias condition.

Fig. 1-4. The P-N junction under various bias conditions.

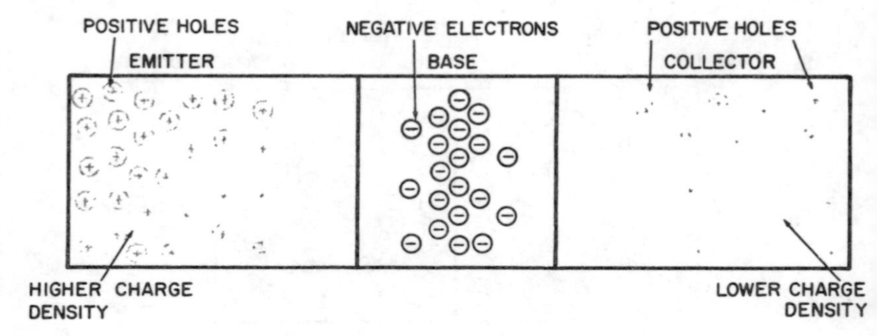

(A) Symbolic representation of charge distribution.

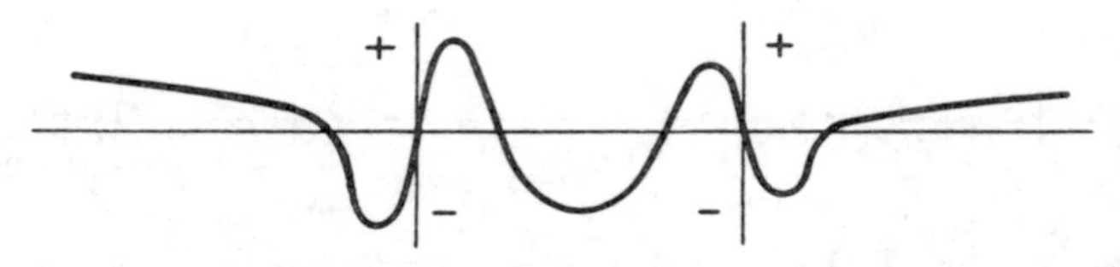

(B) Graphical representation of charge distribution.

Fig. 1-5. The PNP junction transistor with no external applied voltage.

voltage) to vary the electron concentration within the base and thereby vary the strength of the small electric field across the base-collector junction. In this manner, it is possible to regulate the quantity of electrons flowing (as a result of the collector bias voltage) into the collector and through the transistor.

The transistor becomes a useful circuit device when external leads are attached to each semiconductor, and the leads brought to appropriate voltages and circuitry. The transistor can be compared, in some respects, to the triode vacuum tube. As explained in earlier books of this "Basic Electronics" series, the cathode of a triode *emits* electrons into the tube. The control grid, by virtue of the voltage applied to it, then *regulates,* or *controls,* the flow of these electrons through the tube; and the plate *receives* them after they have completed their journey.

TRANSISTOR SYMBOLISM

The *symbolic* representations of the basic types of transistors, PNP and NPN, are shown in Fig. 1-6 and 1-7. In each illustration the left-hand semiconductor is labeled the base, the bottom right-hand one the emitter, and the upper right-hand semiconductor the collector. Since the function of the emitter is to *emit* current through the base to the collector, it is usually compared to the cathode of a tube. By virtue of the biasing conditions (current and voltage) applied between base and emitter, it is possible to control, or regulate, the flow of emitted current to the collector; consequently, the base of a transistor performs the same function as the grid of a tube. The collector *receives,* or *collects,* the current emitted by the emitter and, in that sense, corresponds to the plate of a tube.

ELECTRON-FLOW DIRECTIONS IN TRANSISTOR CIRCUITS

Fig. 1-6 indicates an arrow pointing into the base from the emitter, a symbol which immediately identifies this as a PNP type transistor. The arrow indicates the direction in which the *positive* units of current—in other words, holes—flow. As we shall see in a later example of circuitry, electron flow through a PNP transistor is *into* the collector from the external circuit, then into the base, from there *into* the emitter, and out the emitter to the external circuit again.

Fig. 1-7 indicates an arrow pointing into the emitter from the base, a symbol which immediately identifies any transistor as an NPN type. The arrow again corresponds to the more or less theoretical direction of flow of the *positive* units of current—holes again. In NPN circuitry, it is universally accepted that the *electron* flow is *into* the emitter from the external circuit, then into the base, from there into the collector, and out the collector to the external circuit. Thus, the direction of the electron flow is the same through the NPN transistor and the vacuum tube—*from* emitter (or cathode) *to* collector (or plate). Hence, the analogy between a transistor and a vacuum tube is more precise for an NPN type.

The electron-flow directions through the two types of transistors are indicated in lines in Figs. 1-6 and 1-7. The main electron stream (collector current) is as a line; the base current, which usually carries the applied signal and a biasing current, is a thin line.

Forward Biasing of Emitters

Fig. 1-8 shows the PNP transistor of Fig. 1-5 when biasing voltages are applied to the emitter and collector. The voltage applied to the emitter (left-hand semiconductor) is positive. Its polarity is such that electrons are drawn *from* the base *into* the emitter. Since this is the *normal* direction of electron flow through an NP junction, the emitter is said to be "forward biased" with respect to the base.

Reverse Biasing of Collectors

The voltage applied to the collector (right-hand semiconductor) is negative. Its polarity is such that electrons are driven *from* the collector *into* the base. Since this is the reverse of the normal direction of electron flow through a PN junction, the collector is said to be "reverse-biased" with respect to the base.

One fact should be especially noted. The two external voltage sources (batteries in this example) are connected so that they would drive electron current in a counterclockwise direction from the negative terminal of the "reverse bias" voltage and into the collector. From there the electron flow is through the base and emitter, to the positive terminal of the "forward bias" battery, then continues through the rest of the external circuit and returns to the positive terminal of the reverse-bias voltage source. In other words, these two voltage sources are not opposing each other. Rather, they are in series with each other and of like polarity—meaning both are trying to drive electrons in the same direction, or counterclockwise through collector, base, and emitter in that order.

Electrons in Base

In Figs. 1-5, 1-8, 1-9, and 1-10, we are concerned with free electrons from several sources. Since an understanding of each source is essential to an understanding of transistor action, each electron group has been shown .

Fig. 1-5 shows the free electrons of the base material in — . These are bunched along the center line of the base, whereas the positive holes in the adjoining P-type semiconductors, (+) are bunched along the outer edges. The fixed negative acceptor atoms in the P-material, and the fixed positive donor atoms in the N-material, have been omitted in this diagram, and also in Figs. 1-8 through 1-10.

Fig. 1-8 deals with the free electrons which exist within the N-type material; they are — in the solid as in Fig. 1-5. It is the concentration, or density, of these free electrons that determines the *normal* conductivity of the base.

Also shown (in —) are the free electrons driven from right to left through the transistor by the two biasing voltages. When there is no signal voltage applied between base and emitter (as in Fig. 1-8), this current assumes a normal, or "equilibrium," value. It is the main electron stream through the transistor, and corresponds to the plate current in vacuum-tube circuits.

Finally, we have an equilibrium value of electron current (solid) flowing from base to emitter. When electrons from this

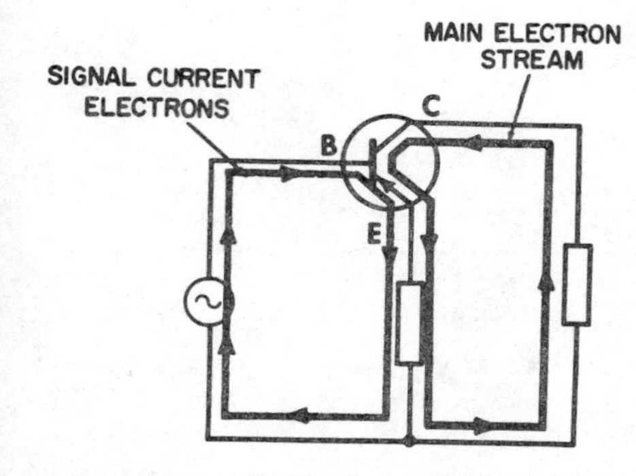

Fig. 1-6. The PNP transistor, with external circuitry.

current are within the base, they increase the electron density in the base, and consequently, the conductivity of the base. The volume of the main electron stream from collector to emitter (solid) is regulated by the conductivity of the base. Therefore, in Fig. 1-9, when the base is "biased" so that additional electrons are driven into it, the volume of current flowing across it from collector to emitter is drastically increased. The attempt has been made in this figure to show the concentration of electrons in the base as relatively unchanged from Fig. 1-8. There are slightly more flow electrons in the base than were shown in Fig. 1-8.

However, because of this slight increase in electron concentration in the base, a considerably larger main collector-to-emitter current flows. These electrons, shown here now flow in some profusion from right to left through the three sections of the transistor.

When the polarity of the signal voltage is reversed (Fig. 1-10), the base-to-emitter current is restricted to a low value. This accounts for the low number of electrons (flow) in the base. The resulting decrease in conductivity of the base substantially reduces the main electron stream through the transistor.

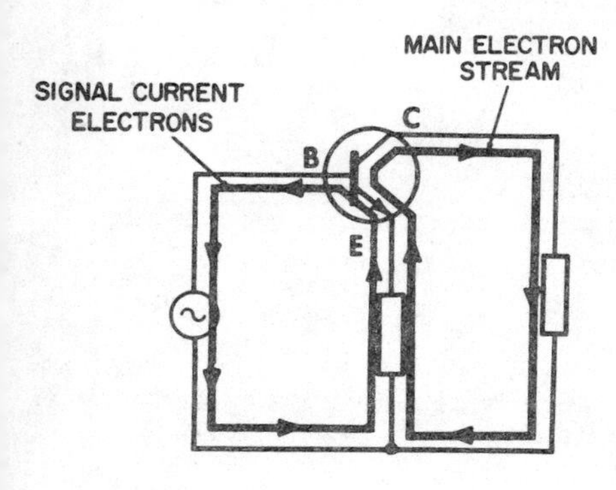

Fig. 1-7. The NPN transistor, with external circuitry.

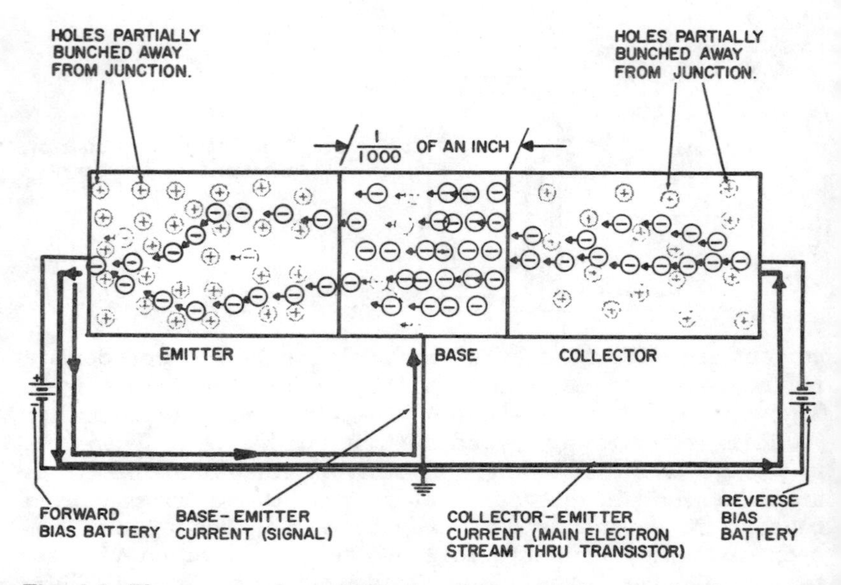

Fig. 1-8. The common-base PNP transistor circuit—no signal current.

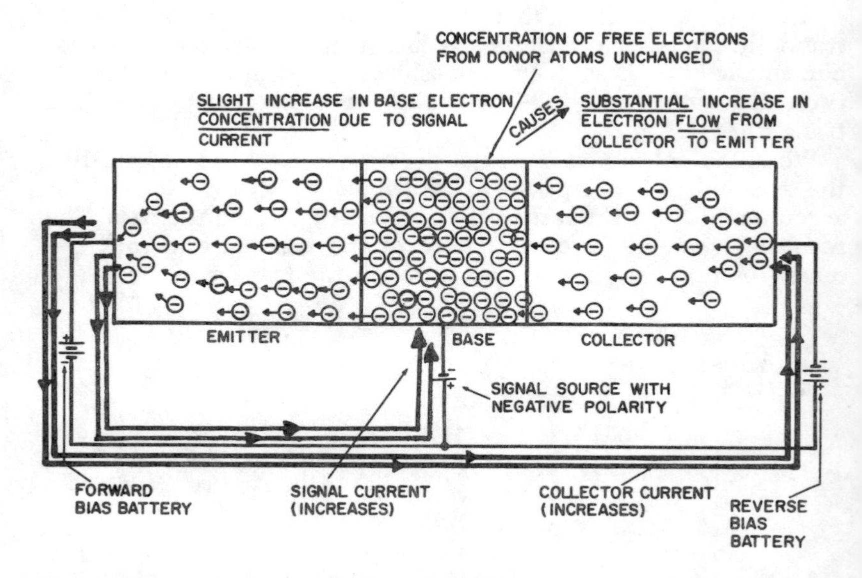

Fig. 1-9. The common-base PNP transistor circuit—negative half-cycle of operation.

It is desirable now to consider the transistor in operation in a sample circuit. This requires us to assign representative values to the applied voltages and to the two resistances in the emitter and collector circuits, and also to provide a means for changing the bias conditions between emitter and base. (In a vacuum-tube circuit, this latter function is known as "driving" the grid.)

TRANSISTORS AS CURRENT AMPLIFIERS

Transistors are referred to as "current-operated" devices, because the driving signal applied between base and emitter is usually referred to as a "current" rather than a voltage. Tubes, on the other hand, are considered to be "voltage-operated" devices, because it is the instantaneous value of voltage at the tube grid that determines the amount of electron current which can pass through the tube.

Transistors, on the other hand, are referred to as "current amplifiers" because a small change in the bias current flowing between emitter and base will cause a much greater change in the amount of current flowing between emitter and collector. The necessary small change in bias current is usually provided by a very weak signal current (on the order of micro-amperes). Obviously, such a signal current will have a companion signal voltage, since one cannot reasonably exist without the other. Convenience and custom, however, have established the signal current as the prime agent in changing the instantaneous bias conditions of the transistor and thus regulating the flow of emitter-to-collector current.

Fig. 1-9 shows the same PNP transistor of Fig. 1-8, but with an additional current source in the line leading to the base. This current source is shown as a small battery in order to establish the desired direction of electron flow from its negative terminal into the base, and then into and out of the emitter. From here the electrons travel into the positive terminal of the "forward-bias" battery (after passing through any resistor included in the emitter circuit), and finally out of the negative terminal and back to the positive terminal of the added battery.

A little reflection will confirm that the emitter-voltage source (battery) and the additional voltage source (shown as a battery) in the base circuit have polarities which drive the electron current in the same direction (counterclockwise), whereas the polarities of the base and collector voltage sources are such that these two voltages *oppose* each other. The negative voltage applied at the base *tends* to drive the electron current *clockwise* through the collector circuit, but is prevented from doing so by the much larger negative voltage applied to the collector; the latter would like to drive the electron current counterclockwise into the base and the emitter.

Signal Current Action

We are now almost down to the water's edge in understanding how a transistor can "amplify" a current, and thus give both voltage and power gain. When this tiny additional electron current is made to flow into the base and emitter, it manages somehow to open the flood gates and let a much larger electron current flow from collector to emitter.

Should the polarity of this signal voltage be reversed, an opposite effect will occur and the electron current flowing from collector to emitter will be reduced considerably. This is depicted in Fig. 1-10, where the new biasing voltage added in the base circuit has a positive rather than negative polarity; consequently, it *tends* to draw electrons out of the base, rather than push new ones in. The consequent reduction in quantity of electrons flowing from collector to emitter is shown pictorially.

Current amplification is considered to have occurred in Figs. 1-8 and 1-9 because the resulting fluctuations in collector-to-emitter current are much larger than the fluctuations in the signal current applied to the base. (The ratio between the two may easily be as high as 50 to 1.)

THE CHARACTERISTIC CURVE

A typical characteristic curve for a PNP transistor is given in Fig. 1-11. The horizontal scale is graduated in volts applied to the collector, and the vertical scale is graduated in milliamperes of collector current. The third variable, the amount of current flowing in the base, is shown by the lines running upward to the right. For any given values of collector voltage and base current, we can locate a single point on the base-current line and, projecting horizontally, find the amount of collector (to emitter) current flowing through the transistor.

The circuit of Fig. 1-12 is known as a "common-emitter" or "grounded-emitter" configuration. It differs somewhat from the examples shown in Figs. 1-8 through 1-10, which were common-base configurations that required two separate biasing voltage sources (the forward-bias and reverse-bias batteries). A common-emitter configuration such as the one in Fig. 1-12 has the special advantage that a single voltage source serves very adequately for both biasing functions.

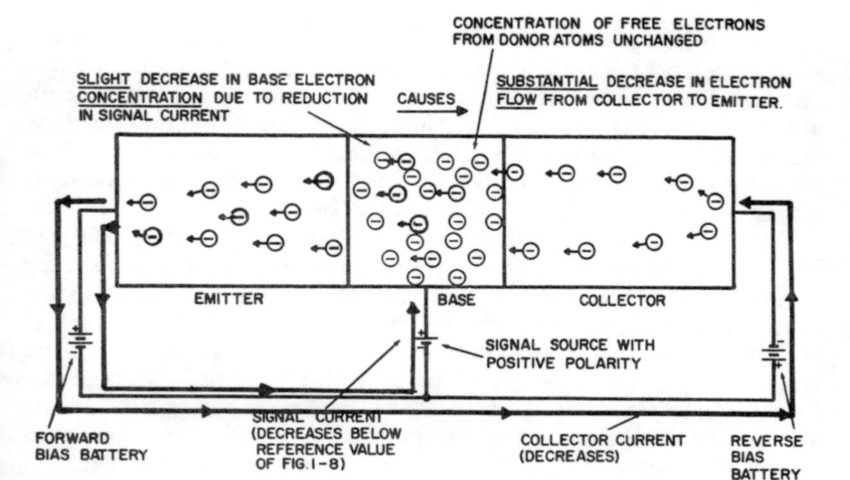

Fig. 1-10. The common-base PNP transistor circuit—positive half-cycle of operation.

The common-emitter circuit corresponds most closely to the conventional vacuum-tube circuit, wherein the cathode of the tube is either grounded directly or is connected to ground through a low-value resistor which provides cathode biasing. Circuit diagrams for the common-emitter configuration are normally drawn with the transistor in the "vertical" position—that is, with the collector at the top, the base in the middle, and the emitter at the bottom. The signal is applied to the base from the left side of the diagram, and is "coupled" away to the next stage on the right side.

Note the convenience of this analogy to conventional vacuum-tube circuitry, where the tube plate is shown at the top of a tube diagram, the grid in the middle, and the cathode at the bottom. This convention of displaying a transistor vertically, with the collector at the top, should ease your mental transition from tube circuits to transistor circuits. That is to say, those who have learned to *visualize* the various electron currents moving around a tube circuit should have a minimum of difficulty in visualizing the comparable electron currents at work in a transistor circuit.

USE OF A "LOAD LINE"

Fig. 1-12 shows the PNP transistor with typical values of applied voltage at emitter and collector, and typical values of resistance in the emitter and collector circuit. Using these values of voltage and resistance, you can work a sample problem on the characteristic curve of Fig. 1-11. First, it is necessary to construct what is known as a "load line" on the characteristic curve. This can be done by determining two important points—corresponding to zero collector current and to zero collector voltage—and drawing a line between them.

The first point is evident from a consideration of Ohm's law, which states that the voltage developed across a resistor is proportional to the current flowing through that resistor. If no current flows through the load resistor, R2, then there is no voltage drop across it. Thus, its two ends must be at the same voltage, which is the value of collector supply voltage E_C. In this example, a value of 4 volts has been chosen, so this is also the voltage at the junction of base and collector when zero collector current is flowing. This value determines the location of the point corresponding to the zero collector current in Fig. 1-11.

Ohm's law can again help us locate the second point. If the collector voltage is truly zero, then the voltage "dropped," or used up by the collector current in flowing through R2, must equal the collector supply voltage. Since, by Ohm's law:

$$\text{Current (in amperes)} = \frac{\text{Voltage (in volts)}}{\text{Resistance (in ohms)}}$$

we can calculate:

$$\begin{aligned} \text{I (collector current)} &= \frac{E_C}{R2} \\ &= \frac{4}{2{,}000} \\ &= 2 \text{ milliamperes} \end{aligned}$$

This determines the point corresponding to zero collector voltage. The line between the two points is the so-called "load line"; at any given instant, it relates the exact voltage at the collector to the exact amount of current flowing through the collector (toward the emitter).

There will always be a small resistance to electron flow through the transistor; consequently, a small "voltage drop" must exist across the transistor. Since this resistance is no more than 10 or 20 ohms at the most, it is negligible in comparison with load resistance (which in this example is assumed to be 2,000 ohms) and may be ignored in a qualitative example like that chosen here.

Operating Point

The next step in using the characteristic curve is to determine the operating point on the load line. This is the point corresponding to no applied signal in the base circuit. Common sense dictates that it should be somewhere near the middle of the load line, so that signal fluctuations in a positive direction will *decrease* the collector-to-emitter current by the same amount that a signal fluctuation in the negative direction will increase this current.

We can arbitrarily choose the point where the 20-microampere base-current line intersects the load line. Also, let us assume that the signal current to be amplified has a peak-to-peak swing of 20 microamperes, and that it is applied in the external circuit between the base and emitter. Any one of several standard coupling methods could be used. The method chosen is capacitive coupling, and the new signal will be driven in and out of the base circuit through capacitor C1.

The special virtue of the load line is that it at all moments will relate the three important variables of collector current, collector voltage, and base current. Thus, when the base current increases to −30 microamperes, the collector current *increases* to −1.6 milliamperes and the collector voltage decreases to −0.9 volt. (This is comparable to vacuum-tube circuitry, where an increase in plate current through a resistive load reduces the plate voltage.)

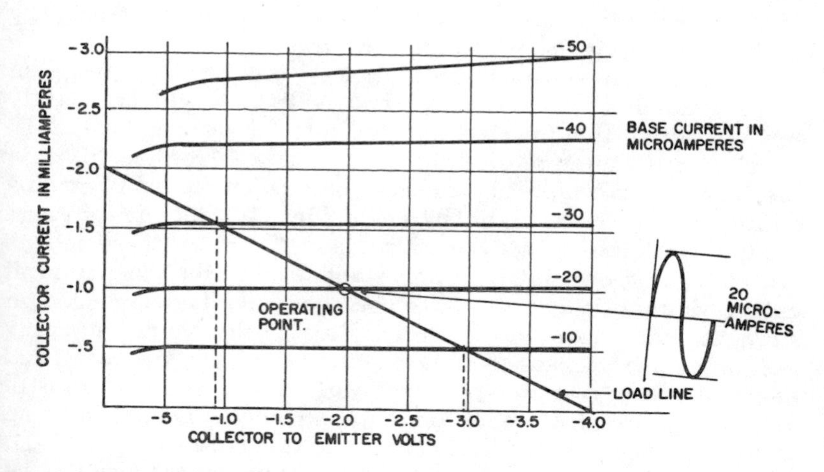

Fig. 1-11. Characteristic curve for a PNP transistor in a common-emitter configuration.

Alternatively, when the signal current reduces the total base current to −10 microamperes, the collector voltage increases to −2.95 volts while the collector current is decreasing to −0.55 milliampere.

THE BETA FACTOR

Current amplification is said to have occurred in the previous example because we started with a signal amplitude of 20 microamperes in the base-emitter circuit and came out with a variation of 1.05 milliamperes in the collector current. The ratio between these two current swings is:

$$\frac{1.05 \times 10^{-3}}{20 \times 10^{-6}} = 52.5$$

Or, slightly over 50-to-1.

Such a high ratio is not uncommon among transistors. It is an important figure of merit known as the "transport factor" and symbolized by the Greek letter β. For this reason it is sometimes called the "beta factor" of a transistor.

To understand *why* a small signal current can subject the collector current to these wide variations, we must further study Fig. 1-5. Obviously, the explanation is not going to be as simple, nor as straightforward, as for vacuum tubes! In Fig. 1-5, note especially how the two junctions will assume certain charge distributions. Within the base section (which is N-type material), the negative free electrons will be repelled slightly from each of the interfaces. Thus, the outer edges of the base are slightly positive and its interior is slightly negative.

Likewise, the so-called free "holes" of positive charge in each of the P-sections (emitter and collector) will be slightly repelled from the interfaces. Both interface areas in the P-material are therefore slightly negative. The free electrons in the base region would like to cross both junctions and combine with the positive holes. Referring to Figs. 1-8 and 1-9 you can see that, with the emitter "forward biased," some of these actions are now going on. But it is important to note that the unusual charge distribution at the interfaces has only been reduced in strength, not wiped out entirely. Free electrons are crossing the junction from base to emitter—*in spite of the fact* that fixed negative charges at the junction edge in the emitter are still exerting an electric repulsion field on them, trying to keep them out. According to the characteristic curve of Fig. 1-11, when the normal base current of −20 microamperes is flowing, the collector current has a value of −1.02 milliamperes.

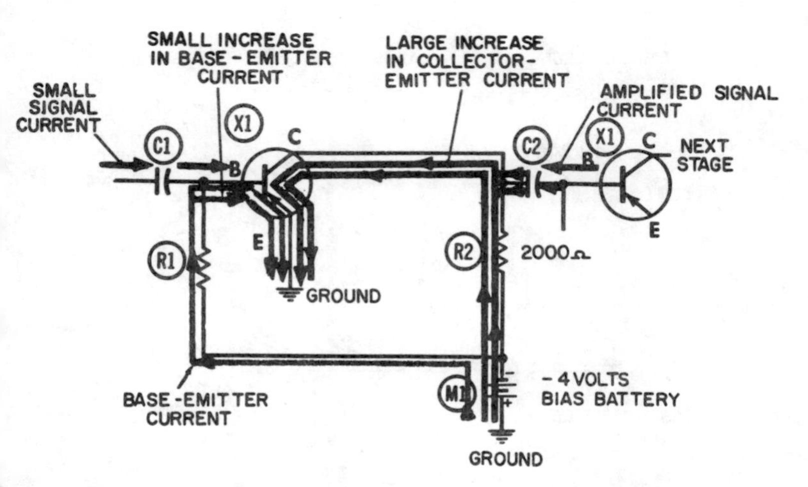

Fig. 1-12. A common-emitter PNP transistor circuit—negative half-cycle.

THE ALPHA FACTOR

Electrons from both the base and collector currents will flow through the emitter. Thus, the total emitter current will be the sum of these two, or 1.04 milliamperes—only slightly larger than the collector current. The ratio between collector and emitter currents, known as the alpha gain, is another important figure of merit for transistors. In a junction transistor it cannot be greater than unity, but is normally very close to it. In this example:

$$\text{Alpha} = \frac{1.02}{1.04}$$

or, about 98%.

In Fig. 1-9, a small external voltage is applied to the base with such a polarity that it drives additional electrons into the base and through it to the emitter. This corresponds roughly to what happens when the base current is increased from −20 to −30 microamperes.

BASE CONDUCTIVITY

What happens within the base itself under these conditions? For one thing, the concentration, or density, of the negative current carriers (electrons) is increased, and the conductivity of any semiconductor varies directly with the density of current carriers in the material. Thus, the base becomes a much better

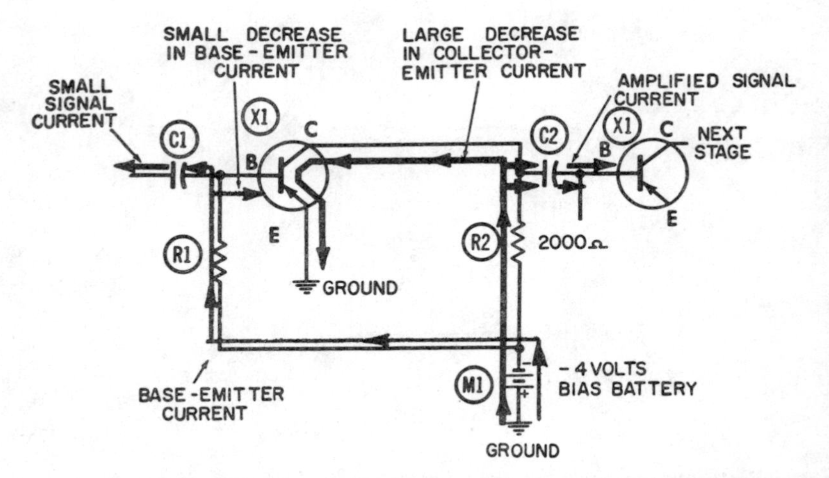

Fig. 1-13. A common-emitter PNP transistor circuit—positive half-cycle.

conductor when more electrons are driven in. Conversely, when fewer base-current electrons flow (as in Fig. 1-10), the conductivity of the base is reduced.

Hence, the base acts as a variable resistor situated between emitter and collector, and having two power sources—the reverse bias and the forward bias.

How is it possible for such a small bias current to actuate this variable resistor? Fig. 1-9 shows the momentary increase in electron density in the base. Thus, the positive carriers (the so-called "holes") in the emitter are repelled less strongly by the "dipole charge layer" at the junction, and will therefore move closer to it. We can assume that this shorter distance between the holes in the emitter and the free electrons in the base greatly expedites the electron drift process.

The "dipole charge layer" between base and collector will also have been diminished in strength, and consequently in width, so that the positive holes in the collector can draw closer to the excess of free electrons in the base. This will also tend to improve the conditions conducive to current flow—the interchange of free electrons from orbit to orbit of positively charged atoms which so closely resembles the electron drift process.

BASE WIDTH

One additional fact of great importance is the width of the base itself. This is normally less than a thousandth of an inch, so that the two sections of P-type material are separated by only this minute distance. Thus, any increase in electron flow between base and emitter will increase the electron density in the base. In turn, more electron current will flow out of the collector, toward the emitter. It is almost as if the increased electron density in the base had created, between the collector and emitter, a series of "electron bridges" across which relatively large quantities of electrons can flow. Of course, this flow is in conformity with the special requirements of the electron drift process.

When the polarity of the applied signal voltage is such that it *reduces* the amount of electron current flowing from base to emitter, (as depicted in Fig. 1-10), the opposite effect can be created. There are fewer intrinsic electrons within the volume of the base; this in turn decreases the conductivity of the base, since conductivity varies directly with the density of current carriers in the material. Even if there were no other contributory effect to consider, the main electron stream through the transistor would still be substantially reduced, by virtue of having to flow through the higher base resistance.

SIZE OF AN ELECTRON

Even though the base is limited to a width of about a thousandth of an inch, it is worthwhile to consider how much "maneuvering room" is provided for the electrons within. The electron has a diameter of less than 10^{-12} centimeter. This means that if electrons could be arrayed end to end, a line one centimeter long would contain *more than one trillion* of them. Thus, a base width of a thousandth of an inch can still accommodate about two and a half billion electrons laid end to end.

This should give some idea of how drastically the number of electrons shown in Figs. 1-8, 1-9, 1-10 and others has been reduced and oversimplified in order to make a pictorial representation that can even be comprehended.

Figs. 1-8, 1-9, and 1-10 suggest a series of possible electron paths through the base of the transistor under three conditions of applied signal current. Fig. 1-8 assumes the condition of normal bias—i.e., zero signal applied to the base, and a forward bias such that an "equilibrium" electron current of −20 microamperes is driven from base to emitter. This permits an "equilibrium" electron current of 1.05 milliamperes through the collector. Note the bunching of free electrons towards the vertical center line of the base, and also that an electron drift process is occurring throughout the three elements of the transistor. From the negative terminal of the reverse-bias battery, the electrons move from right to left into the P-type collector, jumping from hole to hole, then across the first junction and into the base. Here they impart their energy to other free electrons. The latter in turn cross the second junction into the emitter, jumping from hole to hole until they emerge into the external circuit and enter the positive terminal of the forward-bias battery.

Fig. 1-9 shows the same transistor when the signal current from base to emitter is increased from −20 to −30 microamperes. Being composed entirely of free electrons, this larger signal current increases the electron density in the base and also reduces the width of the "dipole charge layer" at the two junction interfaces. Simultaneously, the electron current from collector to emitter is vastly increased. This can be due to the greater ease of the electrons in crossing the existing "electron bridges" in the base, to the opening of new bridges, or a combination of both. If the beta factor of the transistor is 50, then for each additional signal-current electron that passes through the base, a total of about fifty additional electrons will be made to flow across the base from collector to emitter.

Fig. 1-10 shows the same transistor when the signal current from base to emitter is reduced from −20 to −10 microamperes. Now the width of the charged layer at the two junctions will be increased. This restricts rather than encourages the electron drift process across the two junctions, although it still goes on at a reduced rate (−.55 milliampere). The smaller electron current can be due to fewer electron bridges through the base, to the increased difficulty in getting across the existing bridges, or a combination of both.

It is easy, from these analogies, to see how a greater width and consequently greater volume for the base would require more signal current in order to increase the base conductivity by increasing its electron density. It is obvious that a wider base would mean longer electron paths or "bridges" through the base, and that to open up new ones (or improve the conductivity of existing ones) would require many more new electrons driven by the signal current. Consequently, the beta factor—or ratio between a change in collector current for a corresponding change in signal current—would be much lower, and the transistor would not be as good an amplifying device.

MEANING OF NEGATIVE CURRENT VALUES

Those of you who are new to transistors and their circuits should not become confused by the fact that the base and collector currents are both shown as being "negative" in the characteristic curve of Fig. 1-11. Whether a particular current is negative or positive depends on whether it is made up of negative electrons, or of positive charges such as "holes." It also depends on which direction the current is flowing in a circuit.

These matters have already been settled for us—the term "current" in transistor work is normally taken to mean "conventional" current, which is composed of positively charged particles such as "holes." Additionally, the "normal" direction of "conventional" current flow through any PNP transistor is from emitter to base to collector. However, in the NPN transistor, the polarity of the two biasing batteries would be reversed; the direction of *electron* flow would be from emitter to base to collector. Obviously, positively charged particles such as holes will flow in the opposite direction—from collector to base to emitter.

The flow direction of positive hole current through the NPN transistor has been chosen as conventional, or normal, perhaps because the flow directions of electron and hole currents through the NPN transistor correspond to the respective flow directions of electron and conventional or "positive" currents through vacuum tubes. Thus, a characteristic curve for an NPN transistor shows positive values for collector voltage (corresponding to positive plate voltage in a vacuum tube), and also positive values for base current and collector current. Therefore, in making any qualitative analysis of transistor-circuit operation, it is necessary to differentiate between the direction of electron current and that of positive, or hole, current.

THE COMMON-EMITTER CIRCUIT

Fig. 1-12 shows a sample circuit using the same PNP transistor whose characteristic curve appears in Fig. 1-11. This particular transistor, the 2N105, is designed for use at audio frequencies when the currents and powers involved are low. Additional circuit components required are:

R1—Base resistor.
R2—Load resistor.
C1—Input capacitor.
C2—output capacitor.
M1—Bias battery (−4 volts).

There are three electron currents which should be visualized in order to understand the operation of this circuit, which could serve as a low-power amplifier in the audio-frequency range. These currents are:

1. Signal current.
2. Base-emitter biasing current.
3. Collector-emitter current.

The signal current corresponds in many respects to the grid driving current in vacuum-tube operation. This signal current is driven by a signal voltage applied to input capacitor C1. In the so-called negative half-cycle depicted in Fig. 1-12, a negative signal applied at the input point drives a small electron current onto the left plate of capacitor C1. This in turn drives an electron current *of equal size* off the right plate of C1.

Some important facts of transistor operation can be relearned from Figs. 1-12 and 1-13. One is that a *small increase* in the number of electrons flowing from base to emitter encourages a *large increase* in the number of electrons which will flow from collector to emitter. This condition is depicted in Fig. 1-12. During the half-cycle which we have termed "negative," the signal voltage is driving electron current *onto* the left plate of capacitor C1. Normal capacitor action requires that an equal number of electrons be driven *away* from the right plate of this capacitor, and these electrons are shown as being added to the base-emitter current.

The increased base-emitter current causes a substantial increase in the collector-emitter current. The latter current, shown in red, is the main electron stream through the transistor.

This increased electron current flows through load resistor R1 and causes an increase in the voltage "drop" across it, in accordance with Ohm's law. Thus, the voltage at the *top* of the resistor (the output point of the circuit) will become *less* negative during this half-cycle. This has the effect of drawing electrons *onto* the right plate of capacitor C2, as shown in Fig. 1-12.

The flow directions of the original signal current flowing into C1, and the amplified signal current flowing beyond C2 tell us that a phase reversal occurs as the signal passes through this transistor circuit. This is comparable to our experience with vacuum tubes in a grounded-cathode configuration.

Fig. 1-13 is a "positive" half-cycle of the same circuit. The "positive" connotation is arbitrarily chosen to mean that half-cycle during which *less and less* electron current is flowing into the base. The signal voltage applied to capacitor C1 acts as a sort of "pumping" action—alternately drawing off some of the electrons coming up through R1 (on the positive half-cycles), and delivering them to the base (on negative half-cycles). In this way, the total current entering the base can easily be varied between −10 and −30 microamperes.

The main current stream from collector to emitter has been a solid line in these two diagrams. In Fig. 1-13, when the driving current decreases the total current entering the base, the collector-to-emitter current stream is decreased in the same proportion. Using values previously calculated from the characteristic curve, the collector current during this second half-cycle drops to its minimum value of −.55 milliampere, and during the preceding half-cycle it reached its maximum value of −1.6 milliamperes.

The output point of this circuit is connected directly to the collector and, as we have seen, the voltage at this point fluctuates between the extremes of −.75 volt (at the peak of the first half-cycle) to −2.90 volts (at the peak of the second half-cycle). Thus, we have a voltage swing exceeding 2 volts peak-to-peak at the collector.

Both the collector current and the base current are pulsating direct currents. This tells us they are unidirectional (flow in one direction only). Both currents are driven by the same bias battery of 4 volts. The path of the collector current begins at the negative terminal of this battery. Electrons are driven upward through load resistance R2, then to the left and downward through the collector, base, and emitter to ground. From the ground point, the collector current has the necessary free access to the positive terminal of the battery. As is the case with the plate current in a vacuum tube, a closed path must exist through the power supply (battery) and the regulating device (transistor).

The path of the base current also begins at the negative terminal of the battery. Electrons are driven upward through base resistor R1 and through the base and emitter to ground. From here the base current can return freely to the positive (grounded) terminal of the battery. This closed path must also exist; and it is somewhat comparable to the grid return path provided in every vacuum-tube circuit.

One feature that makes the common-emitter configuration particularly attractive is that forward and reverse biasing can both be provided from a single voltage source. The important consideration in any transistor circuit is to assure that the voltage source or sources are wired so that their polarities will drive electrons in the proper direction throughout the circuit. Regardless of whether the transistor is a PNP or an NPN, the collector current and base current must flow in the same direction through the emitter.

The second consideration is that a *negative* voltage must be provided in order to drive these currents in the PNP transistor, whereas those in the NPN require a positive voltage.

a long time, in comparison with the duration of one cycle of the frequency under consideration. The lowest frequency of audio voltage in a normal receiver might be in the neighborhood of 100 cycles per second, so that the time for one cycle is .01 second. (Time for one cycle is always the reciprocal of the frequency, or $T = 1/f$.)

AVC resistors and capacitors are very large components. Typical values for R4 and C5 would be 2 megohms and .05 microfarad, for example.

Calculating the time constant of this combination, we find that:

$$\begin{aligned} T &= R \times C \\ &= (2 \times 10^6) \times 5 \times 10^{-8} \\ &= 0.1 \text{ second} \end{aligned}$$

One-tenth of a second is certainly a long time, in comparison with the one-hundredth of a second of the longest audio cycle considered above.

The AVC capacitor is much larger than C3 (which is much larger than C4). Hence, *many more electrons* are required to charge the AVC capacitor to the average voltage level than are required by C3 and C4. Because the AVC resistor also is large, only a minute amount of electrons will be driven back and forth

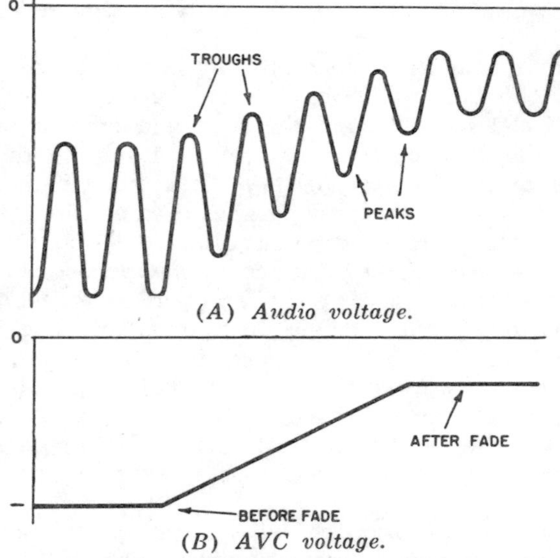

(*A*) *Audio voltage.*

(*B*) *AVC voltage.*

Fig. 5-10. The audio and AVC voltage changes during a signal fade.

Chapter 2

THE THREE BASIC CONFIGURATIONS

A transistor may be connected into a circuit in one of three ways, or configurations—namely, the common base, common emitter, and common collector. Of the three, the common-emitter configuration is by far the most versatile and widely used. However, since the other two configurations have overriding advantages for special applications (such as impedance matching), it is essential that you be able to recognize each of the three circuit configurations when you encounter them.

In the conventional three-terminal transistor, a signal is normally applied *to* one of the three elements (base, emitter, and collector), and the resulting signal extracted *from* one of the two remaining elements. Another way of saying exactly the same thing is that an input signal is applied between two terminals of the transistor and extracted from two terminals, one of which is different and one of which is common to both input and output circuit. The terminal which is common to both input and output circuit will prove, in actual practice, either to be connected directly to ground (grounded, in other words), or to be separated from ground by some fixed voltage source only, such as a battery or power supply. A review of the many circuit diagrams used in this series will confirm that, in the vast majority, the input-signal voltage is developed across some type of impedance between the input terminal and ground. Also, the output-signal voltage is developed across another impedance between the output terminal and ground. ("Ground" in each instance is the point of neutral, or reference, voltage.) Every circuit must have a reference voltage against which all other voltages in the circuit—whether positive or negative, or alternating, pulsating, or direct—can be compared. The most convenient such voltage is that of the earth, or ground, which is normally taken to be zero volts.

It is frequently easier to identify the input and output terminals of a transistor circuit than the terminal common to both circuits. However, the common terminal can be identified by the process of elimination. Once the common terminal has been recognized, it is usually a simple matter to satisfy one's self that it is grounded either directly or through a fixed voltage source.

VOLTAGE, CURRENT, AND POWER GAIN

When the "configuration" of a circuit has been established, it is usually desirable to know what gains in voltage, current, and power the circuit can be expected to deliver. These values of gain are tabulated in Table 2-1, along with the input and output impedances for each type of circuit, since both voltage gain and power gain are directly related to the input and output impedances.

Table 2-1. Typical Gain and Impedance Values.

	Type of Circuit		
	Common Base	Common Emitter	Common Collector
Current gain	Less than unity	Medium—about 50	Medium—about 50
Voltage gain	More than 100	Several hundred	Less than unity
Power gain	Medium	High	Low
Input impedance	Very low	Low	Very high
Output Impedance	Very high	Medium	Very low

THE COMMON-EMITTER CIRCUIT

Figs. 2-1 and 2-2 depict two successive half-cycles in the operation of a typical phase-inverter circuit which uses two PNP transistors, the first in a common-emitter and the second in a common-base configuration. This circuit (and the common collector configuration discussed later) will substantiate the characteristics listed in Table 2-1. Before discussing these characteristics, it is desirable that each circuit operation be understood; and this requires a detailed discussion of the electron currents which flow in each circuit.

Identification of Currents

The following electron currents are flowing continuously in the two-stage phase-inverter circuit of Figs. 2-1 and 2-2.

1. Input driving current for transistor X1.

2. Voltage-divider and biasing current .
3. Base-emitter current for transistor X1 (frequently referred to as base current) .
4. Collector-emitter current for transistor X1 (frequently referred to as collector current) .
5. Base-emitter current (base current) for transistor X2.
6. Collector-emitter current (collector current) for transistor X2 .

Circuit Operation

The input driving current (small dots) flows up and down through resistor R1 at the basic frequency, which is being amplified. During the half-cycle depicted in Fig. 2-1, this current is being drawn *upward* through R1. This tells us the applied input voltage is positive during this half-cycle. The positive component of applied voltage must be subtracted from the negative voltage created at the junction of R2 and R1 by the flow of the voltage-divider and biasing current (solid thin |). This current originates at the negative terminal of battery M1 and flows through R3, R2, and R1 (in that order) in order to reach ground and be able to re-enter the positive battery terminal, which is also grounded. Each point along this path is at a lower negative voltage than any preceding point. In other words, the voltage at the junction of R3 and R2 is less negative than the battery voltage, but more negative than the voltage at the junction of R1 and R2. Likewise, the voltage at the junction of R1 and R2 is less negative than the voltage at the junction of R2 and R3, but it is still negative with respect to ground.

The voltage at the base of X1 is the instantaneous sum of the negative voltage created by the voltage-divider action, and of the input voltage across R1. The latter is positive during the first half-cycle shown in Fig. 2-1, and negative during the second half-cycle of Fig. 2-2. This instantaneous voltage is one of the two important biasing voltages of any transistor, the other being the voltage at the emitter of X1. This is a negative voltage, the result of four separate electron currents flowing downward through R4 and every one varying somewhat during a single cycle of operation. Consequently, an exact computation of emitter voltage at any instant is a difficult process. However, in the PNP transistor the voltage at the emitter of X1 must always be more positive than the base in order for any electron current to flow from base to emitter. This base-emitter current is known as the "biasing current." Another way of saying this is that the base must be more negative than the emitter in order for biasing current to flow. During a positive half-cycle, the base is made less negative, and this restricts the flow of base-emitter biasing current.

Transistor action is such that a slight reduction in base-emitter biasing current generates a much larger reduction in the other electron stream flowing through the transistor—namely, the collector-emitter current, shown a solid line . This phenomenon accounts for the decreases in the two currents indicated in Fig. 2-1.

During a negative half-cycle, the base voltage is made *more negative*. This drives more base-emitter biasing current through the transistor, and thus a much larger collector-emitter current is generated. This phenomenon accounts for the increases in the two currents indicated in Fig. 2-2.

With this somewhat elementary understanding of current and voltage actions around the transistor, we are in a position to explain the meanings of the various characteristics listed in Table 2-1 for the common-emitter configuration—namely, current, voltage, and power gain, and input and output impedance.

$$\text{Current gain} = \frac{\text{Change in collector current}}{\text{Change in base current}}$$

Values of 50 or even 100 are not uncommon for common-emitter configurations. The current gain of a common-emitter circuit is closely related to the beta factor of the transistor discussed in Chapter 1. However, the amount of current gain achieved in any particular circuit will depend not only on the beta factor, but also on the values of resistors external to the transistors.

$$\text{Voltage gain} = \frac{\text{Change in collector voltage}}{\text{Change in base voltage}}$$

During static operation of the circuit of Fig. 2-1, when no input signal voltage is applied (meaning zero driving current is flowing), the difference between emitter and base voltages is normally only a fraction of a volt. Consequently, changing the base voltage by a small fraction of a volt will cause a substantial change in the amount of collector-emitter current. Since this current must flow through a large load resistor, R3, it develops a large voltage drop across this resistor. This voltage drop is always directly proportional to the amount of collector current, in accordance with the Ohm's-law relationship between voltage and current.

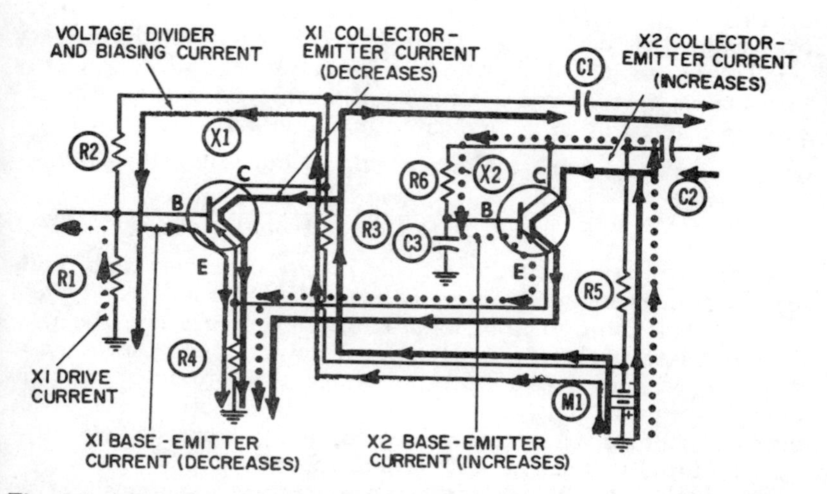

Fig. 2-1. Operation of a two-stage phase inverter using common-emitter and common-base configurations—positive half-cycle.

Because the lower end of resistor R3 is tied to a point of fixed voltage (the negative battery terminal), the voltage at the top of R3 will become more negative when the collector current decreases (Fig. 2-1), and less negative when it increases (Fig. 2-2). Since the upper end of R3 is connected

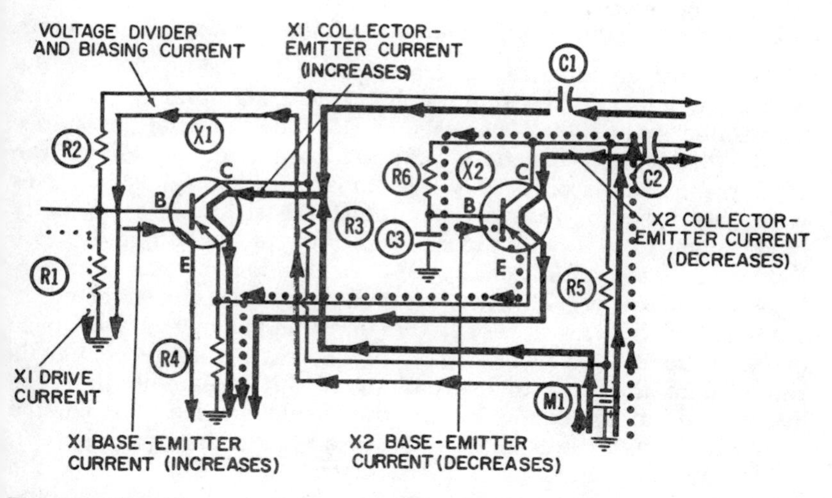

Fig. 2-2. Operation of a two-stage phase inverter using common-emitter and common-base configurations—negative half-cycle.

directly to the collector terminal of the transistor, large changes in voltage are occurring at the collector as a result of very small changes in voltage applied to the base. Thus it is easy to visualize how a large voltage gain is achieved in the common-emitter configuration, as indicated in Table 2-1.

$$\text{Power gain} = \text{Voltage gain} \times \text{current gain}$$

$$\text{Also, power gain} = \frac{\text{Power delivered at load}}{\text{Power delivered to input}}$$

Since we have already satisfied ourselves that high voltage and current gains are both available from the typical junction transistor in a common-emitter configuration, the product of these two gains will obviously give a high power gain. Another way of looking at power gain is provided by the second formula. This leads naturally to a consideration of the two other important characteristics of any transistor circuit—namely, input impedance and output impedance.

Impedance, in the broadest sense, represents *opposition to electron flow*. It can always be expressed as a ratio between any applied voltage and the resulting current flow. This ratio stems directly from Ohm's law, which states that:

$$E = I \times R$$

where,

E is the voltage applied across a component in volts,
I is the resulting flow of electron current through the component, in amperes,
R is the resistance or impedance of the component, in ohms.

The input impedance of transistor X1 in Fig. 2-1 is the resistance between the input point (the base) and the ground connection below emitter resistor R4. The internal resistance in the "forward" direction, between base and emitter of the PNP transistor, is only 20 to 40 ohms. Resistor R4 is in series with this junction resistor, but emitter resistors are normally held down to a few hundred ohms, or at most one or two thousand ohms. Additionally, R4 is shunted by the emitter-base resistance of X2 and the other components connected to the base of X2. As a result, the input impedance at the base of X1 is low. This means that only a small voltage change is required at the base to cause a significant change in the base current. Input impedance might thus be defined as the change in input voltage required to produce a change in input current.

The output impedance of transistor X1 in Fig. 2-1 is the total resistance between the output point (collector) and the ground connection below emitter resistor R4. The bulk of this resistance is made up of the collector-to-base junction resistance within the transistor. This is the so-called "reverse" direction—meaning the electron flow from collector to base is opposite, or against, the normal flow direction for a PN junction. Consequently, the impedance to electron flow from collector to base (this is actually the direction in which collector-emitter current *does* flow within the transistor) is as high as one or two megohms. We can look at the output impedance as a ratio between a change in collector voltage and the resulting change in collector current. Because collectors are biased in the reverse direction, so that batteries such as M1 are trying to drive electrons against the normal flow direction of the collector-to-base junction, a relatively large change in collector voltage will have only an insignificant effect on the amount of collector-emitter current.

Large changes in the collector-emitter current can be generated by changes in the base-emitter biasing current. These changes in collector-emitter current will *cause* substantial voltage changes at the collector, by virtue of the voltage drop which this current develops across R3. However, it does not necessarily follow that these changes in collector-emitter current can be brought about by changing the collector voltage by the same amount. In this respect, the collector terminal is analogous to the plate of a pentode vacuum tube. The plate voltage of a pentode has almost no control over the amount of plate current which flows; likewise the intrinsic value of collector voltage has almost no control over the amount of collector current. Just as the amount of grid-bias voltage regulates electron flow through the pentode, so does the amount of biasing current regulate the collector current through the transistor. Because of these considerations, the output impedance of a common-emitter configuration is very high.

THE COMMON-BASE CIRCUIT

Transistor X2 is connected in a common-base configuration (i.e., the base is common to both the input and output circuits). X2 like X1, is also a PNP transistor, which tells us the direction of electron flow through the transistor is *from* the base and collector, *into* the emitter. The arrows indicate the flow direction of the two transistor currents through X2; the collector-emitter current is a solid line, and the base-emitter current in dotted .

To understand the biasing conditions which regulate the currents flowing through X2, let us refresh our memories on the two important conditions which control the flow of transistor currents. These conditions are the voltages at the emitter and base, and of course the difference between these two voltages.

The base-emitter current of X2 begins at the negative terminal of battery M1, flows through resistors R5 and R6, and then through the very small "forward" resistance of the base-to-emitter junction. From there it continues through resistor R4 to ground, where it has a free return access to the grounded positive terminal of M1. This is in every sense a voltage-divider action, and every point along the current path is at a lesser negative voltage than any preceding point.

We have already seen that during the positive half-cycles of Fig. 2-1, there is a decrease in both the base-emitter and collector-emitter currents through transistor X1. Since both currents flow through resistor R4, toward the ground connection at its lower terminal, any decrease in them will result in a smaller voltage drop across R4. This *smaller negative voltage,* at the upper terminal of R4, is applied directly to the emitter of X2. A smaller negative voltage at this emitter will *increase* the base-emitter biasing current through X2. This increase, and the inevitable increase in collector-emitter current, have been indicated in Fig. 2-1.

During the negative half-cycle shown in Fig. 2-2, both currents flowing through transistor X1 are increased, for reasons previously discussed. Since both must flow downward through R4, a greater voltage drop is generated across this resistor, and the upper terminal becomes more negative during this half-cycle. This more negative voltage is applied to the emittter of X2; hence, as with any PNP transistor, *less* base-emitter biasing current will flow through X2 and automatically reduce the collector-emitter current. These decreases have been indicated in Fig. 2-2.

With this preliminary understanding of voltage and current conditions around transistor X2, let us consider the common-base characteristics listed in Table 2-1. The most important one is the current gain. In the common-base configuration, current gain is similar to the alpha factor of the transistor—it can approach but cannot exceed unity. It can be expressed by this ratio:

$$\text{Current gain} = \frac{\text{Change in collector current}}{\text{Change in emitter current}}$$

Since the base and collector currents of any transistor must also flow through its emitter, the emitter current will always be somewhat greater than the collector current. Also, to achieve any particular change in the quantity of collector current requires a slightly greater change in the emitter current. The emitter is the input point in this type of circuit, and changes in the collector current can only be effected by larger changes in the emitter current. Consequently, the current gain in the common-base configuration, as expressed in the above formula, will always be a fraction less than unity—although in the average transistor, it will usually exceed .95 and will frequently be as high as .98.

$$\text{Voltage gain} = \frac{\text{Change in collector voltage}}{\text{Change in emitter voltage}}$$

Also,

$$\text{Voltage gain} = \text{Current gain} \times \text{resistance gain}$$

In the first formula, the voltage gain in reality is a comparison of output versus input voltages, since the voltage output is taken from the transistor collector, whereas the input is applied to the emitter. You have already seen that extremely small voltage differences between base and emitter voltages will cause significant changes in the amounts of the two currents through the transistor. In the common-base configuration, a small change in emitter voltage is either achieved or accompanied by a change in the base-emitter current flowing through the low-resistance input circuit. Simultaneously, a similar change occurs in the collector current flowing through the collector load resistance in the output circuit. Since this load resistance is usually many times larger than the resistance of the input circuit (through which the base-emitter current must flow), a large change in the voltage at the collector (collector voltage) will occur as a result of a small change in emitter voltage. This is what is meant by voltage gain.

The second formula for voltage gain is based on the Ohm's-law relationship between current, voltage, and resistance. "Resistance gain" has nothing to do with the intrinsic properties of a transistor, but just a convenient way to express the ratio between the resistances of the output circuit and input circuit, respectively. "Current gain" has already been defined as the ratio between output current and input current. Since input current must flow through the small input resistance, and since output current must flow through the large output resistance, the product of current gain and the so-called resistance gain will give the same voltage gain as the first formula.

Because current gain is less than unity in the common-base configuration, the voltage gain achieved by the above formula will be slightly less than the resistance gain.

Table 2-1 tells us that the input impedance of the common-base configured circuit is very low. Input impedance of any circuit can be expressed as a ratio between a given change in applied voltage (in this case, the emitter-voltage) and the resulting change in input current (in this case, the emitter current). We know that the two important biasing voltages of any transistor are those at the base and emitter, and that normally they differ by less than a volt. A small change in either voltage will cause a small change in the base-emitter biasing current, and lead to a much larger change in the collector-emitter current. Thus a small change in the applied voltage at the emitter will eventually produce a *large* change in the collector-emitter current, and the emitter current is composed largely of the collector-emitter current. Therefore, a small change in input voltage leads to a large change in input current (emitter current). This is the definition of a low-impedance circuit.

The output impedance of the common-base circuit is very high. The output impedance is the ratio between any change in applied voltage at the output point (the collector) and the resulting change in output (collector-emitter) current. Because of the "reverse-bias" nature of the collector-base junction, the collector voltage normally has very little control over the collector current. In the common-base circuit of Fig. 2-1, this lack of control is compounded by the fact that any change in collector current (brought about by a change in collector voltage) is largely nullified when the collector current flows downward through resistor R4, thereby developing a change in the important emitter biasing voltage which counter-acts the original change in the collector voltage. Perhaps an example will make this action clear.

Fig. 2-1 shows a half-cycle of operation during which the collector-emitter current of transistor X2 has been increased. Let us forget momentarily about the normal operating conditions responsible for this increase in collector-emitter current, and imagine for a moment that it was brought about by a large increase in the negative voltage applied to the collector (a difficult feat in any transistor!). As this increased collector current flows downward through emitter resistor R4, it develops a larger negative voltage at the upper terminal of R4, and also at the emitter of X2. In a PNP transistor, when the emitter voltage is made more negative (or less positive), less base-emitter biasing current flows and, in turn, less collector-emitter current. This is a form of degeneration. There are many examples of degeneration in tube and transistor circuits, and they all seem to be characteristized by the following two conditions:

1. An increase in electron-current flow through the tube or transistor will change the biasing conditions of the device in such a direction as to reduce the current through the device.

2. A decrease in electron-current flow through the device (usually on an alternate half-cycle) will change the biasing conditions in such a direction as to increase the current flow through the device.

"Phase inversion" of the input signal applied to the base of X1 has been accomplished by the complete circuit, because when the voltage at the collector of X1 becomes *more* negative, (as it did in Fig. 2-1), the voltage at the collector of X2 becomes *less* negative. Relating these important voltage changes to the electron currents which must actually flow in order to couple the voltage changes to the respective output circuits, we see electrons flowing *into* capacitor C1 in Fig. 2-1 and driving other electrons beyond C1, into the external circuit. This action "delivers" a negative voltage to that external circuit. Likewise, we see electrons being drawn out of capacitor C2. This action withdraws other electrons from the external circuit beyond C2 and, by so doing, "delivers" a positive voltage to that external circuit.

In Fig. 2-2 a half-cycle later, the increased collector current through X1 causes a smaller negative voltage to exist at the top of load resistor R3. Some portion of this increased current demand is supplied directly from C1, as electrons are withdrawn from its left plate. This action withdraws other electrons from the external circuit beyond C1, and thereby constitutes a positive voltage to the external circuit. The decreased collector current through X2 (and through R5) causes the voltage at the upper terminal of R5 to become *more* negative. This is symbolized by a flow of electrons onto the left plate of C2, and this action drives other electrons into the external circuit, beyond C2. The external circuit recognized the inflow of electrons as a negative voltage.

THE COMMON-COLLECTOR CIRCUIT

Figs. 2-3 and 2-4 show two successive half-cycles in the operation of a typical common-collector circuit. The common-collector configuration has limited application and for this reason is not widely used. In this example it provides a high-impedance input circuit, in order to "match" this impedance with some equally high-impedance circuit which is providing the input voltage and current (doing the driving, in other words). The common-collector circuit also provides a low-impedance output. The primary purpose for including this type of circuit at this point in the book is to convey some measure of qualitative understanding of the characteristics listed in Table 2-1 for the common-collector circuit. Before these characteristics can be understood, all the currents which flow in the circuit, and the resulting voltages and their changes, should be clear in your mind.

Identification of Components

This circuit includes the following components:

R1—Input resistor.
R2—Voltage-divider and biasing resistor.
R3—Voltage-divider, biasing, and input resistor.
R4—Emitter output resistor.
C1—Input coupling capacitor.
X1—PNP transistor.
M1—Battery or other DC power source.

Circuit Operation

This circuit may quickly be recognized as a common-collector configuration by the process of elimination. The input signal is obviously applied to the base (or more accurately, between the base and ground). The output signal is obviously taken from the emitter (or between the emitter and ground). This leaves the collector, which is automatically "common" to both circuits. In the diagrams of Figs. 2-3 and 2-4, the collector technically is not grounded, but rather is connected to ground through the fixed voltage or power source M1. Power supplies, in practice, are usually bypassed by suitable filter capacitors, so that even at the lowest frequency of operation, signal-frequency currents will be bypassed around the power supply to ground. This filtering process is known as decoupling and is discussed at greater length in other volumes of this series. The presence of such a capacitor effectively places the collector at "ground" voltage to signal frequencies and thus enables us to say that it is a grounded-collector (or the more accurate common-collector) circuit.

Four electron currents are at work in this circuit. They are:

1. Input driving current, usually called the "signal."
2. Voltage-divider and biasing current .
3. Base-emitter biasing current .
4. Collector-emitter current .

Let us momentarily ignore the presence of the signal current in Fig. 2-3, and consider only the currents which would flow during "static" conditions. First is the voltage-divider current, a solid line. It will flow continuously from the negative terminal of battery M1 and downward, through resistors R2 and R3, to ground. This current flow places a certain negative voltage at the junction of R2 and R3. Since the transistor base is connected directly to this point, this voltage constitutes the "base-biasing" voltage, and starts the initial flow of base-emitter current through the transistor. The base-emitter current is shown in small dots; it also originates at the negative terminal of battery M1. Base-emitter current is known as "biasing" current because it regulates the flow of the much larger collector-emitter current (solid) through the transistor. When these two currents are established at their equilibrium values, they develop a voltage across resistor R4 because they must flow downward through R4 on their way to ground. This voltage, which is negative at the upper terminal, becomes the emitter "biasing" voltage. It must be less negative than the base voltage in a PNP transistor in order for any base-emitter biasing current to flow. The difference in the voltages at base and emitter normally is only a fraction of a volt, and only the tiniest change is required in either voltage to change the amount of base-emitter biasing current and consequently to bring about the much larger change in collector-emitter current.

The function of the applied signal voltage is to provide these necessary small changes in the voltage at the base. In the negative half-cycle of Fig. 2-3, the signal current is being driven *downward* through resistors R1 and R3. The small component of negative voltage developed at the top of R3 by the signal current must be added to the fixed negative voltage created at that point by the voltage-divider current. The result is *more* negative voltage at the base and in turn, an *increase* in both the base-emitter biasing current and collector-emitter current. These increases have been indicated in Fig. 2-3.

During the positive half-cycle of Fig. 2-4, the impressed signal voltage becomes positive and the signal current is drawn *upward* through resistors R1 and R3. This creates a small component of *positive* voltage at the upper terminal of R3, and a *less negative* voltage at the base. In a PNP transistor, the lower base voltage *decreases* the flow of both currents through the transistor, as indicated in Fig. 2-4.

The current gain of a transistor in the common-collector configuration approaches the beta value (10 to 100) of the transistor itself. Since only a small change in the base-emitter biasing current is necessary to achieve a much larger change in the collector-emitter current, the current gain is high.

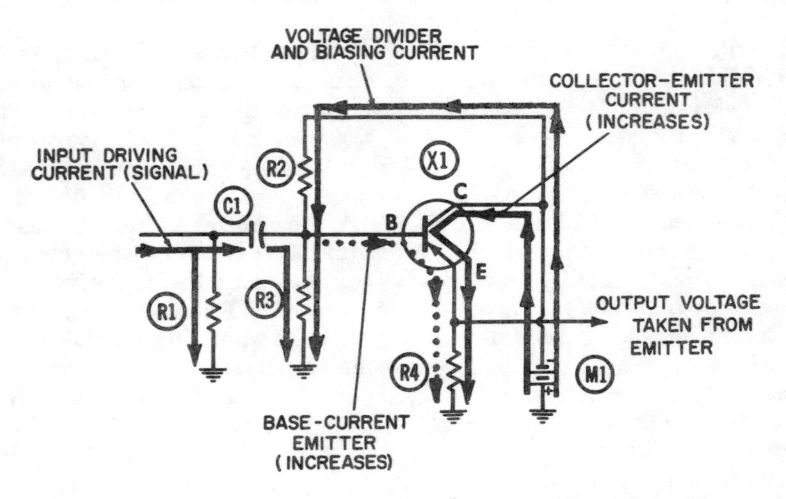

Fig. 2-3. Operation of the common-collector configuration—negative half-cycle.

The voltage gain is a comparison, or ratio, between the voltage change applied to the input and the resulting voltage change developed at the output. This gain cannot exceed unity in the common-collector configuration because the input and output voltages are also the two important biasing voltages for the transistor. It was previously stated that the voltages at the base and emitter are the biasing voltages, and that their

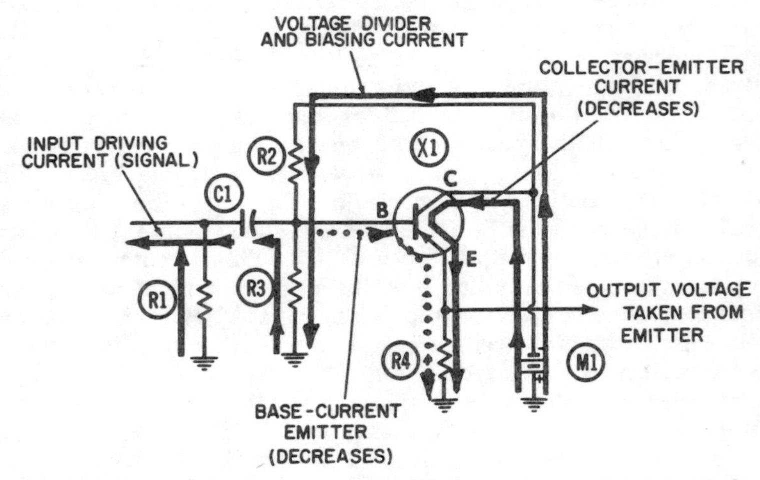

Fig. 2-4. Operation of the common-collector configuration—positive half-cycle.

difference is normally less than a volt. Also, it was stated that a very small change in either voltage will change the amount of the two currents flowing through the transistor.

Let us consider the example in Fig. 2-3, where the applied signal current flowing downward through R3 is of such a quantity that it develops −0.1 volt at the base. This bias is added to the negative voltage already there as a result of the voltage-divider current. A more negative base in a PNP transistor causes additional base-emitter biasing current and, in turn, additional collector-emitter current, to flow. Both of these increased currents must flow downward through emitter resistor R4; and in doing so, they increase the negative voltage already at the emitter.

If this increase in voltage at the emitter were to exceed −0.1 volt, then the *difference* between base and emitter voltage would be a smaller negative voltage than it was before the signal was applied. Consequently, instead of increasing as it should, the base-emitter biasing current would now tend to decrease, and so would the collector-emitter current.

To sum this action up, the voltage change at the emitter is the output voltage, and the voltage change at the base is the input voltage. The output-voltage change cannot exceed the input-voltage change without completely nullifying the latter. Therefore, the voltage gain of a common-collector configuration will always be less than unity.

Even though no voltage gain is available, this is not true for power gain in the common-collector configuration. Power gain is current gain multiplied by voltage gain, and even though voltage gain is slightly less than one, current gain may be high, as previously stated.

Table 2-1 indicates that the input impedance of a common-collector configuration is very high. This is due to the difficulty in effecting any significant change in base-emitter biasing current by making changes in the base voltage. We have already seen that this is true, because the collector current flows through the emitter resistor and acts to nullify any change in the base voltage. Impedance can always be thought of as a ratio between some voltage and an associated current. Input impedance is a ratio between the amount of change needed in the input voltage (base voltage) to change the input current (the base current, which is also the base-emitter biasing current). Because of the "self-compensating" nature of the circuit, it is extremely difficult to effect any change whatsoever in the biasing current. For this reason, the input impedance of the common-collector configuration is very high.

On the other hand, the output impedance of the common-collector is very low. The output impedance is a measure of the relative ease or difficulty with which the output current (collector-emitter current) can be changed by varying the output voltage (emitter voltage). Since the emitter voltage is one of the two important biasing voltages of any transistor, a change of only a volt or two at the emitter will cause a large change in the amount of base-emitter biasing current, and an even larger change in the amount of collector-emitter current.

Chapter 3

OSCILLATOR CIRCUITS

An oscillator is a nonrotating device which generates a signal at a frequency determined by the circuit constants. In this chapter the operation of the Hartley and Colpitts RF oscillators, which generate sinusoidal waveforms, and the multivibrator, which generates a nonsinusoidal waveform, will be discussed.

THE HARTLEY OSCILLATOR

Figs. 3-1 and 3-2 show two alternate half-cycles in the operation of a transistorized version of the Hartley oscillator. This oscillator is widely used in commercial electronic equipment as well as in such household equipment as television and radio receivers.

Identification of Components

The circuit components and their functions are listed below. The manner in which these functions are accomplished—in other words, *how the circuit operates*—will be discussed later in the chapter.

R1—Base-biasing and voltage-divider resistor.
R2—Voltage-divider resistor.
R3—Emitter stabilizing resistor.
R4—Collector load resistor.
C1—Feedback coupling capacitor.
C2—Emitter filter capacitor.
C3—Output coupling and blocking capacitor.
C4—Oscillating tank capacitor.
T1—Oscillating tank inductor and output inductor.
X1—PNP transistor.
M1—Battery power supply.

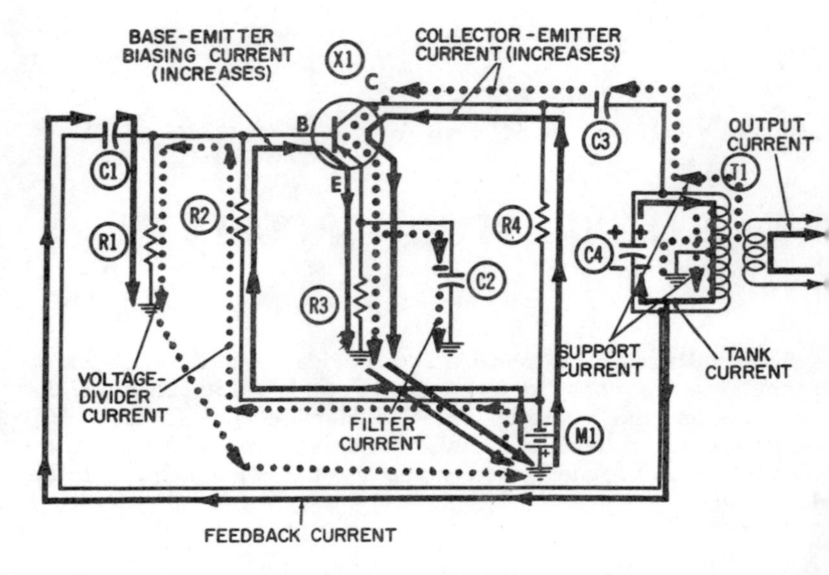

Fig. 3-1. Operation of the Hartley oscillator—negative half-cycle.

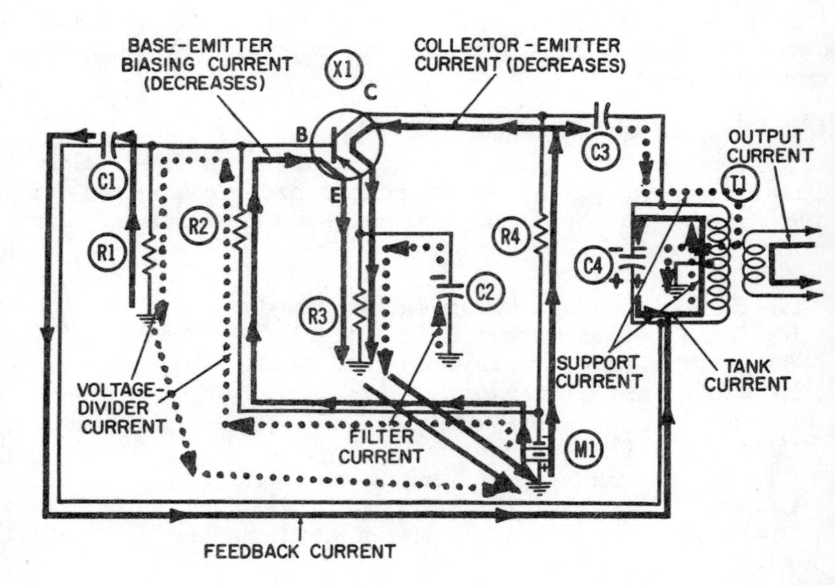

Fig. 3-2. Operation of the Hartley oscillator—positive half-cycle.

Identification of Currents

The following electron currents are at work in this circuit, and it is necessary that their movements be understood before the operation of the circuit as a whole can be comprehended.

1. Voltage-divider current, used for base biasing.
2. Base-emitter biasing current.
3. Collector-emitter current.
4. Oscillating tank current.
5. Feedback current.
6. Replenishment current for support of the oscillation.
7. Output current in transformer secondary.

Circuit Operation — Static Conditions

Fig. 3-1 has been labeled the negative half-cycle of operation, because during this half-cycle the feedback current flows *downward* through R1 and creates a more negative voltage at the upper end of this resistor. This voltage is applied directly to the base of the transistor as part of the base-biasing voltage.

Conversely, Fig. 3-2 is labeled the positive half-cycle, because the feedback current flows *upward* through R1, creating a less negative voltage at the upper end of this resistor.

A single battery, M1, provides electron current for both main current paths through any three-element transistor. These current paths are from base to emitter, and from collector to emitter. The complete path of the base-emitter current, a thin line, includes a trip upward through battery M1 and out through its negative terminal, then to the left, and upward through R2 to the base of the transistor. From the base, the current goes to the emitter *within* the transistor, and then downward through stabilizing resistor R3 to the common ground point. From here it has free access to re-enter the positive terminal of the battery.

This base-emitter current is frequently referred to as the "biasing" current of a transistor, because a small change in its amount will normally cause a much larger change in the amount of collector-emitter current flowing through the transistor.

The base-emitter current normally is not much more than a few microamperes, and it is regulated by the difference in the voltages at the base and emitter. This voltage difference is frequently no more than a tenth of a volt. The origins of these element voltages will be discussed as the other currents are described.

The complete path of the voltage-divider current (in small dots) is from the negative terminal of battery M1, upward through resistor R2 and downward through R1 to common ground, where it has a ready return access to the positive terminal of the battery. Normally, R1 will be considerably smaller than R2 in value, so that most of the battery voltage will be "dropped" across R2. The voltage measured at the top of R1 is the biasing voltage, which is applied to the base of the transistor.

We might assume some typical values as follows:

$$\begin{aligned} \text{Let } R1 &= 1{,}000 \text{ ohms} \\ R2 &= 5{,}000 \text{ ohms} \\ M1 &= \quad -6 \text{ volts} \end{aligned}$$

Then, the voltage-divider current could be calculated from Ohm's law as follows:

$$\begin{aligned} I &= \frac{E}{R} \\ &= \frac{-6V}{5000 + 1000} \\ &= -1 \text{ milliampere} \end{aligned}$$

The voltage across R1 would then be:

$$\begin{aligned} E_1 &= IR_1 \\ &= .001 \times 1000 \\ &= 1 \text{ volt} \end{aligned}$$

and would have a *negative* value at the top of R1. The voltage across R2 would be:

$$\begin{aligned} E_2 &= IR_2 \\ &= .001 \times 5{,}000 \\ &= 5 \text{ volts} \end{aligned}$$

The complete path of the collector-emitter current (a solid line, and usually referred to as the collector current) is from the negative terminal of battery M1, then upward through collector load resistor R4, through the transistor from collector to emitter, and downward through emitter stabilizing or swamping resistor R3 to the common-ground connection. From here it is free to return to the grounded positive terminal of the battery.

The directions in which the two transistor currents flow are of course dictated by the nature of the transistor itself. The one used in this circuit is a PNP, as indicated by the emitter arrow pointing *toward* the base. This is a universal symbol, and the two electron currents always flow into the emitter in the direction *opposite* to that in which the arrow is pointing. The battery must necessarily be connected so that its polarity will support the electron-flow directions dictated by the construction of the transistor itself. With this or any PNP transistor, both the base and collector must be connected to a *negative* biasing voltage.

The base-emitter, collector-emitter, and voltage-divider currents described in the foregoing are "static" currents, in the sense that they flow continuously, whether an oscillation exists in the tank circuit or not. If no oscillation or feedback currents existed, these three currents would all be "pure" DC, with no fluctuations in value. However, if the circuit is properly connected, an oscillating current will spring into existence in the tuned tank circuit as soon as power is applied from the battery. Through the mechanics of the feedback connection, the collector current through the transistor will be increased or decreased in such a fashion that it will in turn support or replenish the oscillation of electrons in the tank current. The circuit is then said to be operating under "dynamic" conditions, during which all the additional currents for accomplishing the oscillation and replenishing it come into existence along with the currents for accomplishing the feedback or filtering action at the emitter and for delivering an output current to the next stage. Additionally, the two currents through the transistor become pulsating rather than pure DC. Let us examine the interrelationships between these currents.

Operation Under Dynamic Conditions

As soon as power is applied to the transistor circuit, a small amount of electron current (solid thin) will begin flowing through the transistor, from base to emitter, and immediately cause a much larger current to flow from collector to emitter. The latter current, both solid and dotted lines, initially is drawn both through load resistor R4 and also from the left plate of coupling capacitor C3.

Capacitor action is always such that when a certain number of electrons are drawn *away* from one plate, an equal number will be drawn onto the opposite plate from the external circuit. That is what happens here; the current being drawn onto the right plate of capacitor C3 is shown in dotted lines. Its complete path is from ground, into the center tap and through the upper half of the transformer primary to the coupling capacitor.

As this current is drawn through part of the transformer, what is known as autotransformer action occurs throughout the entire primary winding. In any inductor, the fundamental electrical action is such that the inductor opposes any change in the amount of current flowing through it. Another way of saying this is that an inductor will try to keep at a constant value the current flowing through it.

We can visualize this fundamental inductor action by looking at the current being drawn upward through the upper half of the winding. During the first half-cycle depicted by Fig. 3-1, this current is flowing upward at an *increasing* rate. In so doing, it *induces* a second current (in large dots) to flow downward through the entire primary winding, also at an increasing rate. Because of its downward flow during the first half-cycle, the induced current delivers electrons to the lower plate of tank capacitor C4 and builds up a negative charge, or voltage, there.

This negative voltage drives electron current through the feedback line to the left plate of coupling capacitor C1, and this action in turn drives an equal amount of electron current *downward,* through biasing-and-driving resistor R1. This makes the voltage at the top of R1 more negative and, in effect, increases the "forward bias" at the base of the transistor, thereby driving *more* electron current from the base through the emitter than would normally flow under static conditions.

Action within the transistor will always be such that a small increase in the base-emitter current is accompanied by a large increase in collector-emitter current. Notice the cumulative nature of all the dynamic conditions described so far during the first half-cycle of Fig. 3-1: The initial surge of electron current (solid and dotted lines) into the collector caused autotransformer action within the primary winding, and the latter action placed a negative voltage on the lower plate of tank capacitor C4. The resulting feedback current through resistor R1 developed a voltage across it of such polarity as to *increase* the base-emitter current and to further increase the current entering the collector terminal.

Two independent circuit actions now occur and prevent these cumulative increases in both currents from continuing indefinitely. As the collector-emitter current increases, the voltage it develops across emitter resistor R3 will also increase. (This voltage will be negative at the top of R3.)

As the negative voltage at the top of R3 increases due to this increased collector current, the voltage-bias conditions between the base and emitter are affected adversely. In any PNP transistor, the voltage at the base must always be slightly *more* negative than the voltage at the emitter, in order for electron current to flow from base to emitter. Thus, the rise in collector current eventually *reduces* the base-emitter current which, in turn, reduces the amount of collector-emitter current flowing through the transistor.

This particular set of cumulative actions eventually reduces the collector-emitter current, bringing us to the conditions shown in Fig. 3-2. This is the logical moment to discuss the second independent action, which operates to keep the collector-emitter current from increasing indefinitely. This is the action of the tuned tank circuit, consisting of capacitor C4 and the primary winding of transformer T1. This tank is tuned to be resonant at the desired frequency of oscillation; and once the lower plate of C4 has been charged to its initial negative voltage, an oscillation of electrons between inductor and capacitor will automatically begin. During the second half-cycle shown in Fig. 3-2, the electrons initially stored on the lower plate of C4 will flow upward through the primary winding of transformer T1, until at the end of this half-cycle practically all of them will have been delivered to the upper plate of C4, placing a negative voltage on it and a positive voltage (a deficiency of electrons) on the lower plate.

This positive voltage on the lower plate of C4 reverses the flow direction of the feedback current, shown a solid line. Electrons are now drawn from the left plate of C1 along the feedback line, toward the tuned tank. This in turn draws electron current *upward* through resistor R1 and creates a small *positive* voltage at its upper terminal. The small positive voltage partially neutralizes the permanent negative voltage which exists at this point because of the flow through R2 and R1 of the voltage-divider current shown in small dots. The algebraic or instantaneous sum of these two voltages across R1—one a fixed negative voltage and the other a voltage which is constantly changing from negative to positive and back again—constitutes the complete biasing voltage present at the base of the transistor.

The voltage at the bottom of the tank circuit reaches its most positive value at the same moment the base-emitter current reaches its minimum value, and the collector-emitter current will in turn be reduced to its minimum value. This reduction is indicated in Fig. 3-2 by the absence of a dotted line passing through the transistor. Electron current flowing upward through the load resistor R4 will be temporarily diverted onto the left plate of capacitor C3 during this second half-cycle, since it is unable to enter the collector terminal. This action accounts for the reversal in flow direction of the support current shown in large dots. It now flows *away* from the right plate of capacitor C3 and downward, through the upper half of the transformer primary winding, to ground.

This replenishment current again supports the oscillation of the tank current during this half-cycle, by inducing in the entire primary winding a current which will again be in phase with the tank current. This induced current (large dots in Fig. 3-2) flows upward through the inductor. Being in phase with the tank current, it is therefore able to replenish the inevitable losses which occur in any oscillation, and thus permit the oscillation to continue indefinitely.

There are three main sources of loss which exist for this particular tank-circuit oscillation. The resistive wire losses within the inductor winding are inherent in any tank circuit. They cause a small percentage of electrons to drop out of oscillation during each half-cycle, so that the number which reach one capacitor plate at the end of any particular half-cycle is never quite as large as the number which left the other capacitor plate at the beginning of the half-cycle.

The electron current, which we call the feedback current and which is driven up and down through the biasing resistor R1 represents another source of loss to the tank-circuit oscillation. Consequently, it is always desirable to keep this feedback current as small as possible. This might be accomplished by making resistor R1 as large as possible. However, the overriding considerations in the choice of resistor values for both R1 and R2 are first, the amount of base-emitter current which the transistor requires for normal operation, and secondly, the amount of normal biasing voltage which should be provided to the base.

The third source of loss to the tank circuit is the support of the output current flowing in the secondary winding of transformer T1. This output current, shown a solid line, flows up and down at the radio frequency generated in the tank circuit.

THE COLPITTS OSCILLATOR

Figs. 3-3 and 3-4 show two successive half-cycles in the operation of a transistorized Colpitts oscillator. The circuit components and their functions are:

R1—Voltage divider or biasing resistor.
R2—Voltage divider or biasing resistor.
R3—Emitter stabilizing resistor.
R4—Collector load resistor.
C1—Filtering or bypass capacitor.
C2—Output coupling and blocking capacitor.
C3 and C4—Voltage-dividing capacitors in tuned tank circuit.
T1—Radio-frequency transformer.
X1—NPN transistor.
M1—Battery power supply.

Identification of Currents

The following electron currents are at work in this typical oscillator circuit, and a thorough understanding of their movements is essential to an understanding of circuit operation, as well as to an ability to troubleshoot and repair the circuit or to modify its design or adapt it to various applications:

1. Voltage-divider current .
2. Base-emitter current .
3. Collector-emitter current .
4. Oscillator tank current .
5. Feedback current to the emitter .
6. Output current in transformer secondary .
7. Support current which supports the tank oscillation .
8. Base-emitter filter current .

The first three are static currents, or DC, and they will flow whenever power is applied to the circuit, whether an oscillation exists in the tuned tank circuit or not. The remaining currents are directly associated with the oscillating tank currents, and are consequently known as "dynamic" currents.

Details of Operation

The transistor base is brought to its desired bias voltage by voltage divider R1-R2 across battery M1. The voltage-divider current (small dots) flows continuously in a clockwise direction around a closed circuit consisting of R1, R2 and M1. This clockwise flow is of course dictated by the polarity of the battery, since electrons must always leave a battery at its negative terminal and re-enter it at the positive terminal. Normally, R2 will be somewhat larger in resistance than R1, so that most of the battery or applied voltage is "dropped" across

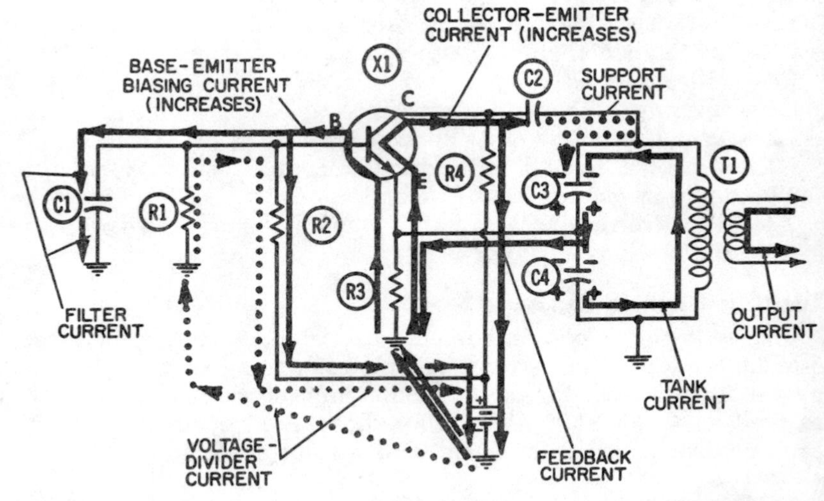

Fig. 3-3. Operation of the Colpitt's oscillator—negative half-cycle.

R2. The voltage at the junction of these two resistors is also the voltage applied to the transistor base.

The base-emitter current (in solid thin in both circuit diagrams) flows around the closed path that begins at the negative terminal of the power supply (which in this circuit

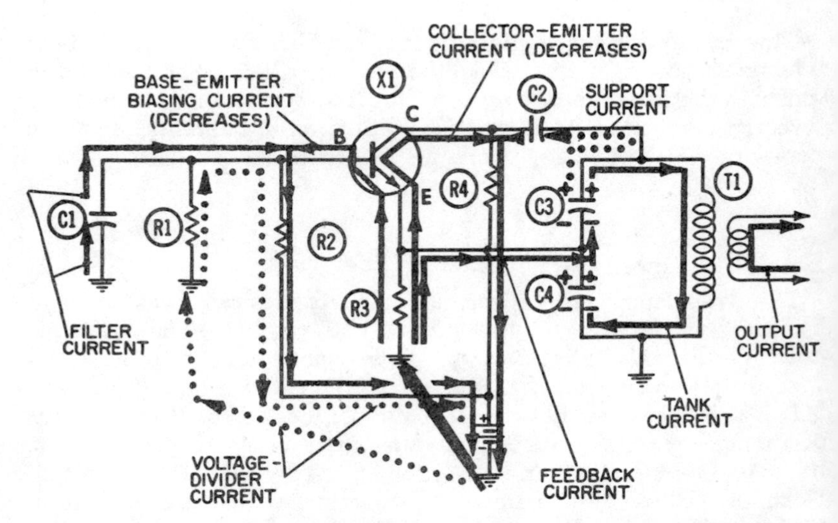

Fig. 3-4. Operation of the Colpitt's oscillator—positive half-cycle.

is connected to ground). This current flows upward through stabilizing resistor R3 and through the transistor, from emitter to base, then downward through resistor R2, and to the right where it re-enters the battery at its positive terminal.

The closed loop around which this base-emitter current flows obviously includes the three circuit elements consisting of R2, R3, and M1. But it also includes the semiconductor junction between the base and the emitter *within* the transistor. The external voltages applied to these two elements really determine how much of this base-emitter current can flow through the transistor and, consequently, around the entire loop. Normally, the base-emitter current will range from perhaps 100 microamperes, down to only a few microamperes.

The third of the three static currents in this circuit is the collector-emitter current, which flows continuously around its own closed loop in a clockwise direction. This current flow (shown a solid line) begins at the negative terminal of the battery. From here, the path is through ground to the lower end of emitter resistor R3 and upward through it to the emitter. The electrons enter the transistor at the emitter and exit at the collector, then flow downward through collector load resistor R4 and enter the positive terminal of the battery.

The fact that an NPN transistor is used here determines the directions of both of the currents which actually flow through the transistor. The emitter arrow points *away* from the base in an NPN transistor, and you will recall that the electron currents through a transistor *always* flow *against* the direction of the emitter arrow.

The amount of collector-emitter electron current which flows around the closed loop discussed previously is determined almost entirely by conditions within the transistor, rather than by the particular voltage value applied to the collector. The overriding condition within the transistor is the amount of base-emitter current flowing. The fluctuations in this small current control the fluctuations in the much larger collector-emitter current. This phenomenon will be discussed in considerably more detail later.

Operation Under Dynamic Conditions

The manner in which an oscillation is set up in the tank circuit can be visualized by referring to Fig. 3-3, which is labeled the negative half-cycle of operation. As soon as power is applied to the circuit, the base-emitter current (in solid thin) will flow through the transistor in the direction shown. As explained in connection with Figs. 1-8 and 1-9 of Chapter 1, the quantity of electrons *within* the base at any instant directly affects and controls the quantity of electrons which can flow through the base between emitter and collector—in other words, the amount of collector-emitter current which can flow.

The initial surge of collector-emitter current delivers electrons to the upper terminal of load resistor R4, and also onto the left plate of coupling capacitor C2. By normal capacitor action, these electrons cannot enter one side of a capacitor unless an equal number are driven away from the opposite side. The electrons which are driven away (large dots) flow down into the tuned tank and set up the oscillation of electrons shown a solid line. Once any voltage or current unbalance is applied to a resonant tank circuit, electrons will be set in oscillation between the tank capacitor(s) and tank inductor. This oscillation, even if unsupported, will continue for many cycles—depending on the strength of the initial disturbance and on the *Q*, or quality, of the tank circuit.

Fig. 3-4 shows a second half-cycle of this tank-circuit oscillation occurring. Electrons now flow downward through the transformer primary, which acts as the necessary inductor for the tank circuit. As a result of this flow, electrons are removed from the upper plate of capacitor C3 and eventually delivered to the lower plate of C4. The current flows downward throughout this entire second, or positive, half-cycle, with maximum voltage (indicated by minus signs on the lower plate of C4 and plus signs on the upper plate of C3) occurring across the tank at the end of the half-cycle.

How Feedback is Accomplished

Feedback is accomplished in this oscillator by connecting the emitter to the point between C3 and C4. These two tank capacitors constitute a capacitive voltage divider. The two voltages across them are in series, and they add to equal the total voltage existing across the tank circuit at any instant. The portion of the total tank voltage across C4 is applied directly to the emitter as the feedback voltage. It is perhaps easier to visualize this feedback voltage if you also visualize the feedback current which must flow in conjunction with it. This current (a solid line), flows to the left and *downward* through emitter resistor R3 during the negative half-cycle of Fig. 3-3, adding a small amount of negative voltage to the positive voltage already existing at the emitter. The latter is produced by the upward flow of the base-emitter and collector-emitter currents through resistor R3. Thus, the effect of the feedback voltage, during the negative half-cycle of Fig. 3-3, is to slightly reduce the positive voltage at the emitter and thereby *increase* the amount of base-emitter current.

This small increase in base-emitter current causes a much larger increase in the collector-emitter current. The extra collector current delivers more electrons onto the left plate of coupling capacitor C2, and in turn adds electrons (large dots) to those being driven down into the tuned tank circuit. Since the electrons in large dots arrive at the tank circuit in the appropriate phase (meaning at the appropriate time) to support or reinforce the oscillation of electrons within the tank, the feedback is said to be "regenerative."

In Fig. 3-4 the total voltage across the tuned tank circuit is positive, as indicated by plus signs on the upper plates of tank capacitors C3 and C4. The fraction of the total tank voltage across C4 now draws electrons upward through emitter resistor R3, and also adds a small increment of positive voltage to the positive voltage already existing at the top of R3. Since the top of R3 is connected directly to the emitter, the positive voltage already at the emitter increases slightly and thereby reduces the base-emitter current.

Transistor action will always be such that a small decrease in base-emitter current will cause a much greater decrease in collector-emitter current. When the collector current decreases, the support current between capacitor C2 and the tank circuit now reverses direction and flows upward, away from the tank and toward capacitor C2. Again, this flow direction supports the oscillation in the tuned tank. Thus, we see that both the increases and decreases in collector current deliver reinforcing impulses to the oscillation in the tuned tank.

Since emitter resistor R3 is not bypassed by a filter capacitor, degeneration will occur as a result of the changes in collector current. Degeneration is synonymous with loss of amplification. To see how it occurs, we need to consider the voltage changes produced at the top of R3 by the variations in collector-emitter current flowing through it.

Three electron currents flow independently through resistor R3—the base-emitter, collector-emitter, and feedback currents. The effects of the feedback current on the collector-emitter current have already been discussed, and need not be reconsidered here. Although the base-emitter current changes in value from half-cycle to half-cycle, the amount—in comparison with the changes in collector-emitter current—is insignificant. Consequently, voltage changes generated by these current changes across R3 are so negligible that they may be disregarded.

This brings us to the collector-emitter current (frequently referred to in many books as the collector current). At the start of the negative half-cycle of Fig. 3-3, the collector current will have been reduced to its minimum; and as a result of its flow through R3, a small component of positive voltage will exist at the top of R3. At the end of this half-cycle, a much larger collector current will be flowing, and hence a much larger positive voltage will exist at the top of R3 and also at the emitter.

This larger positive voltage, at the emitter of an NPN transistor, has an adverse effect on the bias conditions existing between base and emitter. The end result is to *reduce* the base-emitter current. This is contrary to the effect of the feedback current and voltage, which at the end of the same negative half-cycle will *increase* the base-emitter current.

The feedback voltage developed across resistor R3 is in reality the signal, or driving, voltage for the whole transistor circuit. Whenever a feedback voltage and the resulting transistor current are out of phase with each other, then degeneration is said to be occurring.

Degeneration also occurs during the positive half-cycle of Fig. 3-4. At the end of this half-cycle, the feedback voltage across R3 is positive, reducing the two currents through the transistor and R3. The latter flows upward through R3, reducing the positive voltage across this resistor and thus partially counteracting the increase in positive voltage caused by the feedback. This is degeneration; and as stated before, its effect is to lower the amplification from that which the circuit would normally deliver.

Resistor R2 and the battery are bypassed (or filtered) by capacitor C1, so that degeneration does not occur in this part of the circuit. The fluctuations in base-emitter current through R2 would normally change the voltage at the base of the transistor, and these voltage changes would affect the amount of base-emitter current. As an example, the increase in base-emitter current indicated in Fig. 3-3 would normally cause a greater voltage drop across R2 which would *lower* the positive voltage at the top of R2. Since this voltage is applied directly to the base of the transistor, less base-emitter current would flow.

This would be degeneration, which is avoided by having filter capacitor C1 in the circuit. It, along with R2, constitutes a conventional long time-constant circuit. In Fig. 3-3, excess electrons flowing through the transistor from emitter to base (the base-emitter current) will accumulate on the upper plate of capacitor C1 and drive an equal number away from the lower plate. This is the filtering current and is shown a solid line. On the positive half-cycle of Fig. 3-4, when the base-emitter current is reduced, the excess electrons driven onto the upper plate of C1 will now drain off, through resistor R2, to the positive terminal of the power supply or battery.

As long as this filtering is permitted, the base-emitter current will be fairly pure or constant direct current during its passage through R2 and the battery. For the remainder of its journey through resistor R3 and the transistor, it is pulsating DC. Since the base-emitter current through R2 is pure DC, the voltage it develops across R2 will be constant, and the voltage applied to the base will be steady from one half-cycle to the next.

Normal transformer action between the primary and secondary windings of T1 induces an output current in the secondary, as shown a solid line in Figs. 3-3 and 3-4. This output current normally is used to develop the driving voltage for succeeding amplifier stages.

THE FREE-RUNNING MULTIVIBRATOR

Figs. 3-5 and 3-6 show two successive half-cycles in the operating of a typical transistorized multivibrator circuit. The title "free-running" is applied to any multivibrator which oscillates continuously. Such a multivibrator is said to be "bistable" —as opposed to a one-shot, or "monostable," multivibrator where each cycle of oscillation must be initiated by a separate trigger pulse.

Identification of Components

This circuit is composed of the following components:
R1—Voltage-divider and collector load resistor for X1.
R2—Voltage-divider and filter resistor.
R3—Voltage-divider and base biasing resistor for X2.
R4—Voltage-divider and collector load resistor for X2.
R5—Voltage-divider and filter resistor.
R6—Voltage-divider and base-biasing resistor for X1.
R7—Emitter stabilizing resistor for both transistors.
C1—Coupling capacitor between X1 collector and X2 base.
C2—Coupling capacitor between X2 collector and X1 base.
C3—Emitter bypass capacitor for both transistors.
X1—PNP transistor.
X2—PNP transistor.
M1—12-volt battery power supply.

Identification of Currents

At least ten electron currents flow in this circuit. Once their movements and significance are thoroughly understood, their associated voltages likewise become easy to understand. These ten electron currents are:

1. Voltage-divider current, which provides "bias" voltages for the collector of transistor X1 and the base of X2.
2. Collector-emitter current for transistor X1.
3. Base-emitter current for transistor X2.
4. Voltage-divider current, which provides "bias" voltage to the collector of transistor X2 and the base of X1.
5. Collector-emitter current for transistor X2.
6. Base-emitter current for transistor X1.
7. The instantaneous pulsation of current, which flows at the *beginning* of the first half-cycle and cuts off all current flow through transistor X2.
8. The instantaneous pulsation of current which flows at the *beginning* of the second half-cycle and cuts off all current flow through transistor X1.
9. The long time-constant discharge current flowing between capacitor C1 and resistor R2.
10. The long time-constant discharge current flowing between capacitor C2 and resistor R5.

Details of Operation

As soon as power is applied to this circuit, the two voltage-divider currents shown in solid and thin line, respectively, will begin to flow. The flow directions of both currents are as shown in Figs. 3-5 and 3-6. The path of the first current (a solid line) is upward from the negative terminal of battery M1, through resistors R1, R2, and R3 to the common ground connection. The path of the second current (a solid line) is upward from the battery, and through resistors R4, R5, and R6 to ground.

These electron flow directions tell us that the voltage at the left end of resistor R2 (collector voltage of X1) must be more negative than the voltage at the right end of R2, because electrons inevitably flow from more negative to less negative areas.

Fig. 3-7 shows the relationships between the two collector and two base voltages. At the beginning of the first half-cycle, ransistor X1 starts to conduct electrons from collector to emitter. This collector-emitter current (large dots in Fig. 3-5) ollows the expected path for a PNP transistor. This is upward from the negative terminal of M1, through resistor R1, nto the collector and out the emitter of X1, then down through ommon emitter stabilizing resistor R7 to ground. Here it has eady return access to the positive terminal of battery, which s connected to ground.

The fact that two currents are now flowing side by side upward through R1 causes a greater voltage drop across R1 han before. Since the voltage at the battery end of R1 is fixed at –12 volts, the voltage at the top of R1 must become *less* negative. As a rough example, the voltage at the collector of X1 (line 1 of Fig. 3-7) is shown increasing abruptly from –8 to –4 volts.

As soon as this voltage begins to change (at the start of the irst half-cycle), the voltage at the base of transistor X2 also begins to go in the positive direction because of coupling capacior C1. This action quickly cuts off the flow of base-emitter current through transistor X2, because in a PNP transistor the base must be less negative than the emitter in order for this current to flow. And since the base-emitter current is the "biasing" current which causes or permits the other current o flow between collector and emitter, both currents through ransistor X2 are cut off immediately at the first half-cycle.

Now only one current will be flowing upward through resistor R4 instead of two. As a result, the smaller voltage drop across R4 causes the negative voltage at the upper terminal of R4 to become *more* negative.

This increase in negative voltage at the collector of X2 s passed to the base of X1 via coupling capacitor C2. A more negative base voltage on any PNP transistor acts to *increase* the base-emitter biasing current, and also the collector-emitter current. The rise in collector-emitter current through X1 further raises the voltage at the collector of X1, making the collector voltage still less negative and in turn contributing further to the positive voltage at the base of X2.

All the events described in the foregoing are cumulative and occur at the very beginning of the first half-cycle, so that as transistor X1 goes from zero to full conduction, X2 goes from full conduction to zero, or "cutoff."

At the start of the second half-cycle, the opposite sequence will occur—X1, which is conducting, will end up cut off, and X2 will go into full conduction. This sequence is initiated by the discharge action which has been occurring between capacitor C1 and resistor R2 throughout the entire first half-cycle.

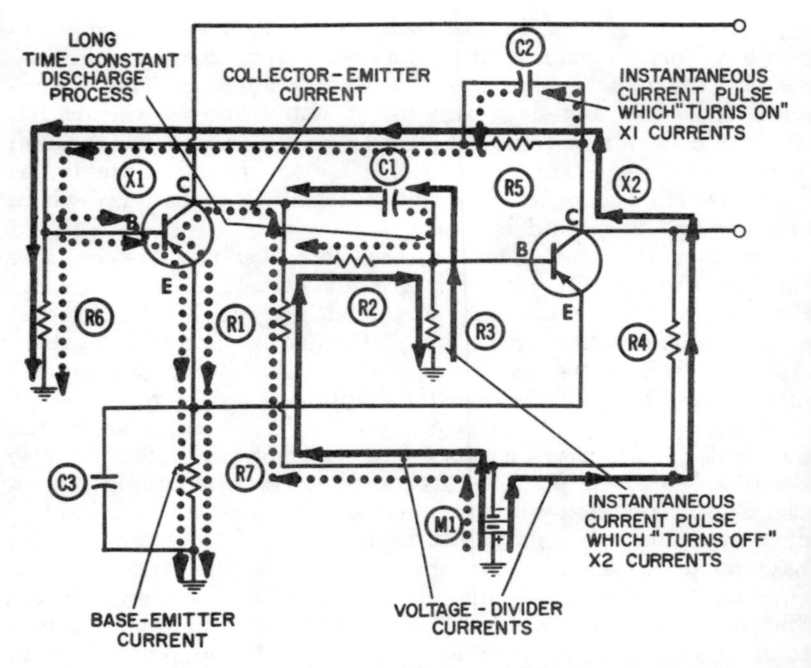

Fig. 3-5. Operataion of the free-running multivibrator—first half-cycle.

Line 2 of Fig. 3-7 shows the voltage at the base of transistor X2 throughout the entire cycle. This waveform, during the first half-cycle, resembles an "exponential" discharge or charging curve and is actually the result of several actions to be described later. There is one significant fact about any free-running multivibrator, whether it uses vacuum tubes or transistors as the switching devices: when one begins to conduct electrons, it changes the bias voltage on the other and thereby cuts it off. The latter will remain cut off until an RC discharge action can take place. After the discharge has been completed, the device which was cut off will begin to conduct again, initiating a new half-cycle. The duration of each half-cycle is therefore regulated by the values of resistors and capacitors being discharged.

It is evident from Figs. 3-5 and 3-6 that transistor X1 conducts its two currents during the first half-cycle only, and that transistor X2 conducts its two currents during the second half-cycle only. The base-emitter current for X1 (small dots in Fig. 3-5) may be considered an offshoot of the voltage-divider current (solid thin) which flows continuously through divider resistors R4, R5, and R6. The base-emitter current,

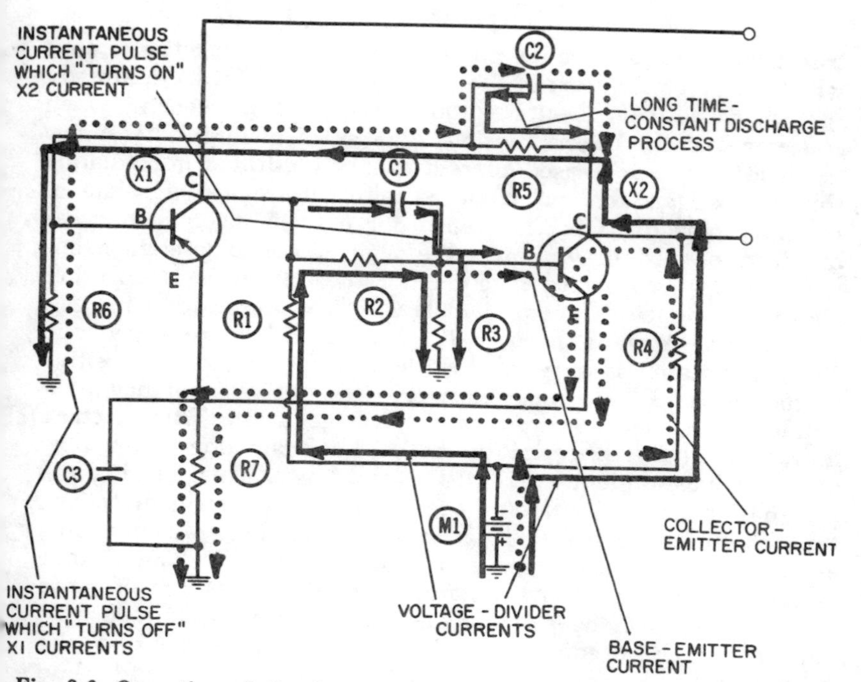

Fig. 3-6. Operation of the free-running multivibrator—second half-cycle.

which is the "biasing" current for X1, enters the base and exits from the emitter (normal flow direction for the PNP transistor). It then flows downward, through stabilizing resistor R7, to ground. From here it can re-enter the grounded positive terminal of battery M1. This current will flow only when the base of X1 is more negative than the emitter.

The base-emitter current of X2 (large dots in Fig. 3-6) may be looked on as an offshoot of the other voltage-divider current which flows continuously through R1, R2, and R3. The former enters the base of X1 and exits from the emitter, then flows down through R7 to ground. Like its companion current in the other transistor, the base-emitter current through X2 can flow only when the base of X2 is more negative than the emitter. This mandatory set of conditions is implied by the descriptive term "forward bias."

The Instantaneous Current Pulses

Figs. 3-5 and 3-6 show two currents which are best described as instantaneous pulses of current. It is necessary to understand the movements of both in order to understand how each transistor is turned on or cut off. Let us consider the actions which occur at the start of the first half-cycle. You have already seen how the X1 collector voltage becomes *less* negative when collector current begins its flow. The extra electrons necessary to make up this increased current cannot be drawn immediately through resistor R1, but must be taken from the left plate of coupling capacitor C1. This can occur only if an equal number of electrons are drawn onto the right plate. These electrons must be drawn upward through resistor R3; there is no other circuit component through which they can possibly come.

In flowing upward through R3, this electron current (a solid line) inevitably is associated with a voltage which is positive at the top of R3, because electrons flow away from negative-voltage areas and toward positive-voltage areas. This electron flow is of course opposite to the continuously downward flow of the voltage-divider current (a solid line). Thus, during the first half-cycle, two currents are flowing in opposite directions through the same resistor, R3, and developing two components of voltage which are opposite in polarity across it. The voltage at the base of X2, at any instant, is the algebraic sum of these two voltages. At the start of the first half-cycle, the positive voltage clearly predominates. This can only mean that the instantaneous current pulse which is shown a solid line and which is equal to the *change* in collector current being drawn into X1, is *much larger* than the voltage-divider current flowing downward through R3.

The discharging action between capacitor C1 and resistor R2 is a difficult one to visualize. It begins at the start of the first half-cycle and continues throughout the half-cycle. The electron current that actually does the discharging has been shown in large dots, to differentiate it from the instantaneous current, a solid line, moving to the left along the capacitive path represented by C2. The path of this discharge current is from the right plate, through resistor R2, to the left plate of C2.

It can be looked upon as an equalizing current, which must flow to correct or redistribute the unbalance of electric charge between the two plates of capacitor C2. It flows at what is called an *exponential* rate, a term derived from higher mathematics and beyond the scope of this book. For our purposes, it describes a discharge process which begins at a high rate and continues at a decreasing rate to zero. Theoretically, no quantity can ever be decreased to zero by this system, because, during each unit of time a certain percentage of the quantity that existed at the beginning of the unit of time will be discharged. Hence, some fraction of the original quantity, however small, will always exist. Practically, five time periods are sufficient for charge redistribution to occur between a capacitor and resistor.

An exponential discharge curve is shown in the first half-cycle of Line 2, Fig. 3-7. This portion of the curve actually represents the intrinsic voltage at the base of X1. However, it also *resembles* the quantity of discharge current flowing between C1 and R2 during the same half-cycle—namely, a large current to start with, and decreasing exponentially throughout the entire half-cycle.

By momentarily considering this RC combination isolated from all other circuit components, it will be fairly simple to visualize the discharge action. When the left plate of C1 is made more positive than the right plate, electrons will flow, or "discharge," through resistor R2 in the direction shown in Fig. 3-5. This discharge will continue until there is no longer any charge unbalance between the plates, meaning there is no voltage across the capacitor. However, this RC combination is *not* isolated from the rest of the circuit. Instead, across resistor R2, there is a permanent, or fixed, voltage difference caused by the flow of voltage-divider current (a solid line). This permanent voltage difference also exists across the capacitor plates, making the left plate more negative than the right.

This permanent voltage across C1 is momentarily upset or modified by the sudden change in collector voltage at the start of the first half-cycle. The collector voltage moves abruptly in the positive direction by an amount assumed to be 4 volts, or from −8 to −4 volts. If the voltage on the right plate of C1, as well as at the base of transistor X2, was assumed to be −1 volt before, it must now increase in the positive direction by the same 4 volts, to a new instantaneous value of +3 volts. This positive peak value of the X2 base voltage (Line 2, Fig. 3-7) will cut off both currents through X2, and they will remain cut off until the base can again be made more negative than the emitter (which does not occur until the end of the first half-cycle).

To sum these actions up, the *total* current through resistor R2 always flows from left to right. The instant before the first half-cycle begins, the total current consists exclusively of the voltage-divider current (a solid line). The amount is determined by the Ohm's-law relationship between the combined series resistances of R2 and R3 and the −8 volts at the collector of X1.

The instant before the first half-cycle ends, and while X1 is still conducting, the total current through R2 again consists exclusively of the voltage-divider current shown a solid line.

Its amount has changed, however, because the voltage at the collector of X1 has dropped from −8 to −4 volts. This would indicate that the voltage-divider current has been reduced by exactly one-half.

This change in amount of current *during* the first half-cycle might be looked on as an exponential *reduction* in the current flowing from left to right. Such a reduction must occur because, the instant before the first half-cycle starts, the voltage difference across R4 is 7 volts (−8 volts at the left end and −1

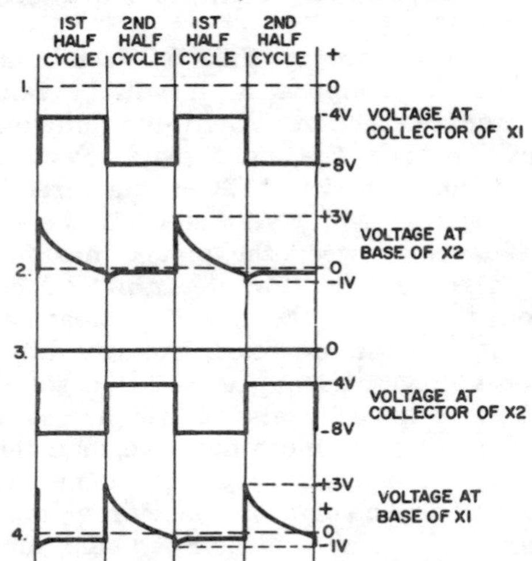

Fig. 3-7. Voltage waveforms in the free-running multivibrator.

volt at the right end), whereas the instant before this half-cycle ends, this voltage difference is only 3 volts (−4 volts at the left end and −1 volt at the right end, after the capacitor discharge).

The Second Half Cycle

How current conduction is initiated through X2 at the start of the second half-cycle has already been discussed. The expected series of cumulative actions occurs almost instantaneously. As the collector voltage of X2 begins to rise from −8 to −4 volts, it raises the base voltage at X1 in the positive direction, cutting off both currents through X1. In turn, the X1 collector voltage is driven from −4 to −8 volts. This tends to drive the base of X2 even more negative, and quickly leads to full electron conduction through X2.

The base voltages cannot be driven as far negative as they can be positive. This is due to the "forward bias" condition of each transistor. As soon as the base of any PNP transistor becomes more negative than the emitter, electrons flow very freely from base to emitter. The reason is that the diode junction, N to P, has very little resistance in the so-called "forward" direction. Another way of saying this is that the forward-biased junction "short-circuits" the base resistors. This does not occur when the base-emitter junctions are reverse-biased. In Fig. 3-6, for example, the instantaneous current (a solid line) flows downward exclusively through R3 as long as the base is less negative than the emitter. The moment the base becomes more negative, this current pulse is largely diverted through the much lower resistance path represented by the diode junction and resistor R7.

In Fig. 3-5, the instantaneous pulse current (in large dots) flows exclusively through R6 until the diode junction between base and emitter becomes forward-biased and offers a much lower resistance path.

Resistor R7 serves to prevent thermal runaway of either transistor. Both transistor currents must flow downward through R7 to ground, and this continuing current flow keeps both emitters at a small negative voltage. Thermal runaway is an undesirable, cumulative condition whereby an overheated transistor begins to conduct larger quantities of both currents, and the increased conduction further aggravates the overheated condition. With an emitter stabilizing resistor such as R7 in the circuit, any runaway current condition is quickly checked by the resulting increase in negative voltage at both emitters because an increase in negative voltage at a PNP emitter will reduce or cut off the currents through the transistor. Thus, a stabilizing resistor automatically prevents thermal runaway.

Relatively little has been said about the instantaneous pulse current, (in large dots) which flows through C2 and R6 and "turns off" the currents through X1; nor about the long time constant discharge current (a solid line) which flows between C2 and R5 and eventually turns the transistor currents back on again. The movements of these currents are associated with the voltages shown in Lines 3 and 4 of Fig. 3-7. Observe that the square wave of voltage generated at the collector of X2 is exactly a half-cycle out of phase with the collector voltage of X1. Also, the exponential voltage waveforms at the two bases are exactly a half-cycle out of phase with each other.

Because of the similarity in physical action between the pulse currents and the exponential discharge currents for the two transistors, there is no need to discuss this set of currents—their action is the same as for the set already discussed.

Chapter 4

AMPLIFIER CIRCUITS

An amplifier receives a signal and "boosts" it to a higher level for application to another amplifier or to an output device. The operation of four types of amplifier circuits is discussed in this chapter. All are audio amplifiers; however the same basic principles apply to RF amplifiers except, of course, the component values are different and transformer coupling is usually employed.

CLASS-A AUDIO AMPLIFIER

Figs. 4-1 and 4-2 show two successive half-cycles in the operation of a typical audio-amplifier circuit which utilizes transformer coupling to the next stage. The circuit is classified as a Class-A amplifier because the collector-emitter current flows continuously during the entire cycle.

Identification of Components

The following components perform necessary functions in this amplifier circuit:

R1—Voltage divider and biasing resistor.
R2—Voltage divider and biasing resistor.
R3—Emitter stabilizing resistor.
C1—Input coupling and blocking capacitor.
C2—Emitter bypass or filter capacitor.
T1—Audio-frequency output transformer.
X1—NPN transistor.
M1—Battery or other DC source.

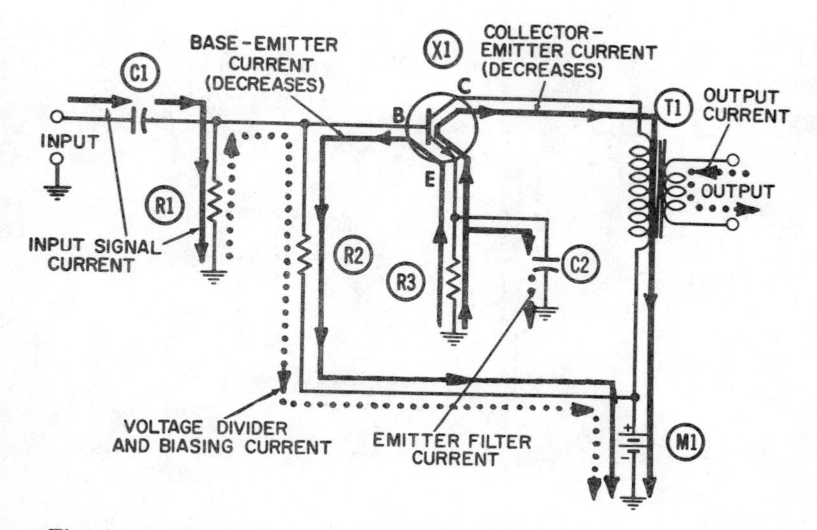

Fig. 4-1. Operation of the Class-A amplifier—negative half-cycle.

Identification of Currents

A total of six different electron currents perform the various essential functions in this circuit. The currents are:

1. Voltage-divider current.
2. Base-emitter current).
3. Collector-emitter current (frequently referred to as the collector current.
4. Input signal current.
5. Output signal current.
6. Emitter filter current.

The first three are static currents, which begin to flow as soon as power is applied to the circuit. In the absence of an applied signal voltage or current, these three static currents will be essentially pure DC. When a signal voltage is applied to the circuit to be amplified, the base-emitter and collector-emitter currents will become pulsating DC.

As soon as a signal voltage is applied, the circuit is then operating under dynamic conditions, and the latter three currents will come into existence.

Details of Operation

The current through resistors R1 and R2 (shown in large dots) flows clockwise into the positive terminal of battery M1

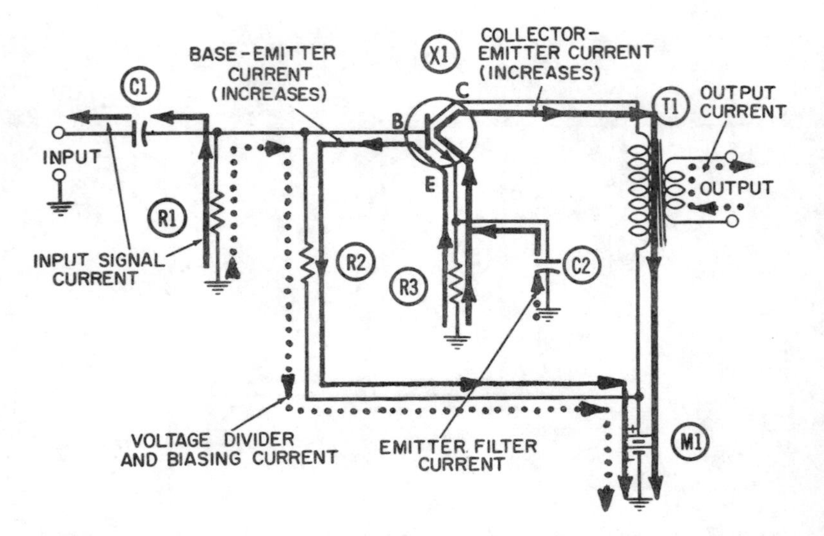

Fig. 4-2. Operation of the Class-A amplifier—positive half-cycle.

and out the negative terminal, returning to the lower terminal of R1. The use of two or more resistors across a fixed voltage source is a very common means of obtaining some fraction of the total voltage available. It is a particularly valuable technique in transistor circuits, which require only 1 or 2 volts for biasing the base with respect to the emitter, and another voltage perhaps three to five times as large—*but always of the same polarity*—for biasing the collector with respect to the emitter.

These requirements can of course be met by two separate voltage sources or batteries, but it is usually desirable to have as few batteries as possible in any circuit. Since the current drain on batteries in transistor circuits is normally very small, the power losses incurred by a voltage-divider current such as this one are well compensated for by the elimination of a separate battery for providing low-voltage bias between base and emitter.

The base-emitter current flowing through a transistor connected like this one will normally be only 10 or 20 microamperes. This current (a thin line) flows upward from ground through resistor R3 and through the NPN transistor, from emitter to base. It then goes to the right, down through resistor R2, and enters the positive terminal of M1. The current then flows through the battery and out its negative terminal, to common ground.

The collector-emitter current flows in a clockwise direction through the closed loop which starts at ground, then upward through resistor R3, through the transistor from emitter to collector, downward through the primary winding of transformer T1, and into the positive terminal of battery M1. Like the two other static currents, this one must also flow *through* the battery, from the positive to the negative terminal (ground).

This circuit does not function as an amplifier until a signal voltage is applied to its input circuit. Its purpose is to amplify, or increase the strength of, this signal voltage. Let us consider now how this important function is accomplished. An alternating signal voltage at an audio frequency will be applied to input capacitor C1. The result is that a signal current (shown a solid line) will be made to flow, at the same frequency, up and down through input driving resistor R1. Fig. 4-1 has been labeled the negative half-cycle, because this input signal current flows *downward* through R1 and develops a small component of negative voltage across this resistor. This negative component must be subtracted from the positive voltage developed across this same resistor by the voltage-divider current. The net result, during the negative half-cycle shown in Fig. 4-1 is a reduction in the positive voltage applied to the base of the transistor. Since this is an NPN, less electron current flows from emitter to base.

Probably the most important single truth in the internal physics of the transistor action is that a *slight* increase (or decrease) in the emitter-to-base (or base-to-emitter) current will produce a much larger increase (or decrease) in the emitter-to-collector (or collector-to-emitter) current. This is the central fact that enables the transistor to be used as an amplifier.

In Fig. 4-1, the voltage to be amplified is negative and therefore drives a signal current (solid line) *downward* through R1, making the top of this resistor negative. As a result of this negative biasing action at the base, the main electron stream through the transistor (the collector-emitter current) is reduced to its minimum value. This *decrease* in current flowing downward through the primary of output transformer T1 causes the current to flow in the same direction in the secondary winding. This is normal transformer action at work, and it accounts for the output current (in large dots) flowing downward in the secondary in Fig. 4-1.

In Fig. 4-2, the phase of the input signal voltage is reversed. Being positive, it now draws the signal-current electrons *upward* through resistor R1, adding to the positive voltage already created there by the upward flow of voltage-divider current (in large dots). The higher positive voltage increases the flow of both transistor currents; and as more collector current flows downward through the primary, more output current flows upward in the secondary.

The transformer action is probably the most difficult of the five basic electronic actions to visualize. When an alternating current is driven through either winding, it becomes the primary winding, and the current through it can be labeled the primary current. The primary current will cause a secondary current to flow in the other winding. The phase relationship between the two currents will always be governed by the following considerations:

1. When the primary current is *increasing* in the downward direction, the secondary current must be *decreasing* in the downward direction, or increasing in the upward direction.
2. When the primary current is *decreasing* in the downward direction, the secondary current must be *increasing* in the downward direction, or decreasing in the upward direction.

Obviously, two similar rules can be written to account for increases or decreases in the primary current when flowing upward.

The quantity relationship between primary and secondary currents is regulated by the turns ratio of the transformer. Expressed as a formula:

$$\frac{N_1}{N_2} = \frac{I_2}{I_1} = \frac{E_1}{E_2}$$

where,

N_1 is the number of turns of wire in the primary,
N_2 is the number of turns of wire in the secondary,
I_1 is the quantity of primary current,
I_2 is the quantity of secondary current,
E_1 is the voltage impressed across the primary winding,
E_2 is the voltage induced across the secondary winding.

In order for the circuit as a whole to function as a voltage amplifier, the alternating instantaneous voltage E_1 across the primary winding must be greater than the instantaneous voltage developed across R1 by the input-signal current.

The classification of this circuit as a Class-A amplifier tells us that the collector-emitter current is never completely cut off during the entire cycle; some current flows continuously through the primary of T1.

Degeneration, or loss of signal strength, is prevented in this circuit by the presence of filter capacitor C2 across emitter resistor R3. C2 keeps the pulsations in collector-emitter current from developing voltage pulsations across R3. During the positive half-cycle of Fig. 4-2, the biasing conditions are such that they encourage or demand more collector-emitter current through the transistor. These extra electrons are drawn from the upper plate of capacitor C2, rather than through resistor R3. An equal number of electrons (the filter current, in large dots) are drawn onto the lower plate of C2 from ground.

The opposite action occurs during the negative half-cycle of Fig. 4-1. Now biasing conditions are such that they discourage or restrict the flow of electrons which make up the collector-emitter current. The electrons flowing upward through R3 are momentarily shunted aside and flow onto the upper plate of capacitor C2, driving an equal number of filter-current electrons downward from the lower plate of C2 to ground.

Thus, C2 is carrying out its normal capacitor action of "passing" an alternating current from one plate to another without permitting any continuous flow in a single direction (DC). The values of R3 and C2 must be so chosen that the resistance of R3 is much greater than the capacitive reactance (opposition to alternating current flow) of C2. The reactance of any capacitor varies *inversely* with the frequency of the current being passed, in accordance with the standard formula:

$$X_C = \frac{1}{2\pi fC}$$

where,

X_C is the reactance in ohms,
f is the frequency in cycles per second,
C is the capacitance in farads.

When this gross inequality between resistance (of R_3) and reactance (of C_2) has been met, then the combination of the two components automatically forms a "long time-constant" circuit. The *product* of the resistance in ohms and the capacitance in farads equals the time constant of the circuit, in accordance with the formula:

$$T = R \times C$$

where,

T is the time constant in seconds,
R is the resistance in ohms,
C is the capacitance in farads.

Any combination where the time constant, T, is at least five times longer than the time required *for a single cycle* of the current being passed to complete itself is defined as a long time-constant circuit. The time for a single cycle of any current is related to the frequency of that current by the simple relationship:

$$T = \frac{1}{f}$$

where,

T is the time duration of a single cycle,
f is the frequency in cycles per second.

The voltage which accumulates on the upper plate of capacitor C2 may be likened to a deep pool of positive ions, out of which electrons are drawn during the positive half-cycle of Fig. 4-2, and to which electrons are added during the negative half-cycle of Fig. 4-1. The true significance of a "long time-constant" RC circuit is that not enough electrons are withdrawn or added to appreciably change the voltage represented by this ion pool. Another way of saying this is that so few electrons are withdrawn during a positive half-cycle, in comparison with the number of positive ions already stored there, that no measurable increase in positive voltage occurs. Likewise, the number of negative electrons added to this ion pool during a negative half-cycle is so small, in comparison with the number of positive ions already stored there, that no measurable decrease in positive voltage occurs.

There is a simple arithmetical relationship between the quantity of electrons (or ions) stored on a capacitor plate, the size of the capacitor, and the resulting voltage across the capacitor. Known as Coulomb's law, it is written as follows:

$$Q = C \times E$$

where,

Q is the quantity of charge in Coulombs (one coulomb is equal to 6.25×10^{18} negative electrons or positive ions),
C is the size of the capacitor in farads,
E is the resulting voltage in volts.

DIRECT-COUPLED AMPLIFIER

Figs. 4-3 and 4-4 show two successive half-cycles in the operation of a direct-coupled (DC) amplifier using transistors. Direct coupling is advantageous in that the inherent losses in capacitive or transformer coupling, at low and high frequencies, respectively, are avoided.

The principle known as complementary symmetry is used in this circuit. Here the collector of an NPN transistor, X1, has been coupled directly to the base of a PNP transistor, X1. The significance of the biasing-voltage polarities and current-flow directions in the two types of transistors will be discussed later in the chapter.

Identification of Components

This circuit is composed of the following components:

R1—Input driving and voltage-dividing resistor.
R2—Voltage-dividing resistor.
R3—Emitter stabilizing resistor.
R4—Collector load resistor.
R5—Emitter stabilizing resistor.
R6—Collector load resistor.
C1—Input capacitor for coupling signal from preceding stage.
C2—Emitter bypass capacitor.
C3—Emitter bypass capacitor.
X1—NPN transistor.
X2—PNP transistor.
M1–Battery or other DC power supply.

Identification of Currents

There are at least eight significant electron currents at work in this circuit. Their movements and interrelationships must be understood by anyone aspiring to know how this circuit operates. These currents are:

1. Input dividing current.
2. Voltage-divider current.
3. Base-emitter current for transistor X1.
4. Collector-emitter current for transistor X1.
5. Base-emitter current for transistor X2.
6. Collector-emitter current for transistor X2.
7. Emitter filter current for transistor X1.
8. Emitter filter current for transistor X2.

Details of Operation

Fig. 4-3 has been labeled a negative half-cycle of operation, because the input driving signal flows *downward* through resistor R1, making the voltage at the base of transistor X1 *more negative* during this half-cycle. Fig. 4-4 has been labeled a positive half-cycle because now the input driving signal flows *upward* through R1, making the base of X1 more positive. The instantaneous voltages which this input-signal current develops at the top of R1 must be added to or subtracted from the more permanent positive voltage developed there by the voltage-divider current (in small dots).

This voltage-divider current flows through the closed path from ground, upward through R1, across and down through R2, and then into the positive terminal of power supply M1. From here, it flows through the battery and out the negative terminal to the common-ground connection. The purpose of this current (and consequently, of the voltage divider itself) is to develop a particular value of positive voltage at the junction of R1 and R2. This becomes the biasing voltage for the base of transistor X1. The term "biasing voltage" requires some explanation. Transistors are normally considered "current-controlled" devices, and the base-emitter current is usually referred to as the "biasing current." However, no biasing current will flow unless certain values of voltage are applied to the base and emitter. It is quite proper to refer to these values as "biasing voltages," because they determine the amount of biasing current which will flow between base and emitter.

The actual biasing current flowing from emitter to base has been shown a thin line. Its electrons flow upward from ground, through emitter stabilizing resistor R3, into the transistor (flowing against the direction of the emitter arrow, of course), then out of the base and downward through resistor R2 to the positive terminal of battery M1. When no signal is applied to input capacitor C1 to be amplified, this biasing current is a pure DC. In the negative half-cycle of Fig. 4-3, the negative voltage which the downward-flowing signal current develops at the top of R1 will *reduce* the biasing current through transistor X1. The lower biasing current will in turn reduce the collector current.

The complete path of this emitter-collector current begins at ground, below resistor R3. The current flows upward, through R3, into the emitter and out the collector of X1, and down through load resistor R4 to the positive terminal of the battery. This electron current then returns to common ground by flowing through the battery and out its negative terminal.

At that moment when the signal current in the negative half-cycle of Fig. 4-3 is flowing downward through R1, at its maximum rate, the biasing and collector currents will have their minimum values. The reduction in collector current causes the voltage at the collector to become *more positive*. The reason is this voltage will always be equal to the power-supply voltage (positive in this case) *minus* the voltage drop occasioned across R4 by the flow of collector-emitter current through it. As this current decreases, so does the resulting voltage drop across R4, and the voltage at the collector will rise toward the full value of power-supply voltage.

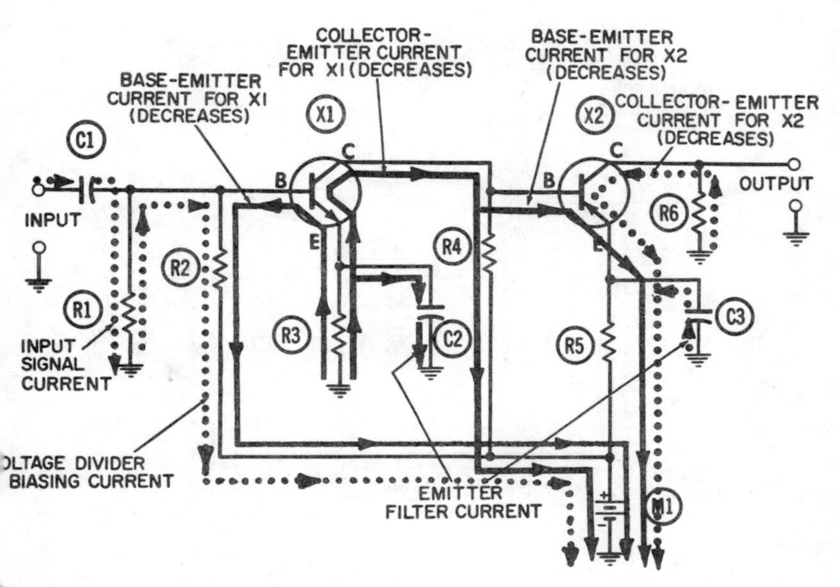

Fig. 4-3. Operation of the direct-coupled amplifier—negative half-cycle.

The exception to the foregoing would be where the collector-emitter current is cut off entirely. Then the voltage difference between the two terminals of R4 (the voltage drop across R4) would be zero, and the collector voltage would necessarily be the same as the power-supply voltage.

Since the collector of X1 is coupled or connected directly to the base of X2, the voltages at these two elements must always be identical. Let us now consider the biasing conditions at the elements of X2, so that we may then predict what will happen to the currents through X2 during this negative half-cycle.

Since X2, is a PNP transistor, electron currents will flow into its collector and out its emitter—instead of from emitter to collector as in NPN transistor X1.

The important biasing voltages for X2 (or any other transistor) are the instantaneous voltages at the base and emitter. You have already seen that the voltage at the base of X2 is determined at all times by the voltage at the collector of X1. The voltage at the emitter of X2 is jointly determined by the power-supply voltage of M1 and by the voltage drop across emitter resistor R5 as the collector-emitter current flows through R5. The complete path of this collector current (in large dots) begins at ground, below resistor R6. It flows upward through R6 into the collector and out the emitter of the transistor, then down through R5 to the

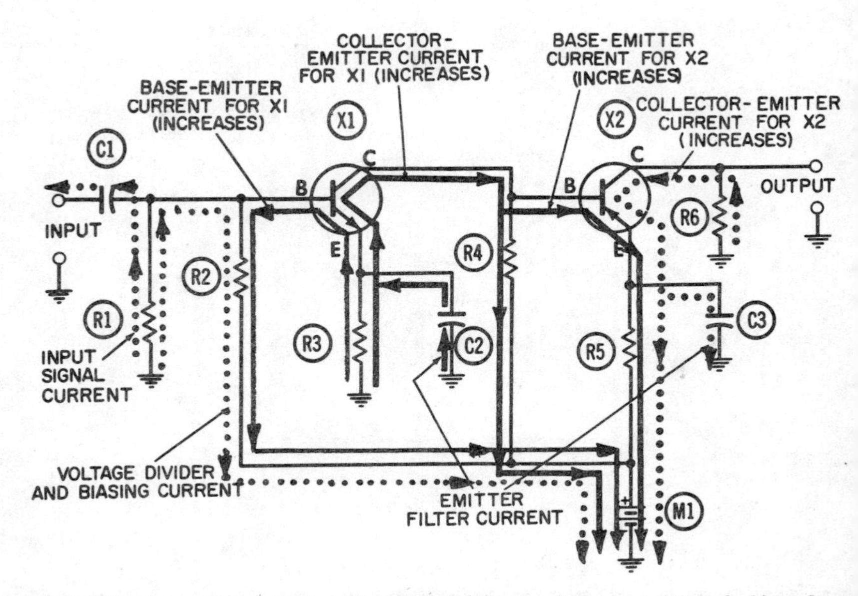

Fig. 4-4. Operation of the direct-coupled amplifier—positive half-cycle.

positive terminal of power supply M1. Its journey is completed to common ground by flowing through the battery and out its negative terminal.

The voltage which this current flow develops across emitter resistor R5 must be subtracted from the battery voltage to determine the exact voltage at the emitter (the emitter biasing voltage). It and the voltage at the base (the base biasing voltage previously discussed) regulate the amount of biasing current flowing from base to emitter. As with all transistors, the amount of biasing current through the transistor exercises direct control over the amount of collector-emitter current. This brings us back to a consideration of how much voltage this collector current will develop across R5 while flowing through it; this voltage affects the emitter biasing voltage, etc.

The complete path of the emitter-base current (a solid line) truly begins at ground, below transistor X1. It flows through resistor R3 and transistor X1 as part of the collector current (a solid line). At the collector of X1 the emitter-base current separates from the main collector current and assumes its individual identity, flowing directly into the base of transistor X1 and out its emitter, then downward through resistor R5 to the positive terminal of battery M1. From here it can return to ground by flowing through the battery.

During the negative half-cycle shown in Fig. 4-3 (collector current through X1 reduced by the biasing conditions at X1), the positive collector voltage is increased and so is the positive voltage at the base of X2. The latter action *reduces* the base-emitter current (biasing current) through X2. It may be difficult to visualize why a more positive base voltage for X2 reduces its base-emitter current. If so, consider again the exception for transistor X1, when we assumed that no collector current at all was flowing through X1. This no-current condition raised the voltage at the collector of X1 until it equaled the power-supply voltage. The base of X2 would necessarily assume the same voltage value. Obviously, no current would then flow between the base and the positive terminal of the battery, since the two points would be at the same voltage. (No electron current can flow between two points unless a difference in voltage exists between them. Electrons will always flow from the more negative point (point of greater electrons) to the more positive point (point of fewer electrons) to make up for the deficiency there.)

During the positive half-cycle depicted in Fig. 4-4, the following changes in voltage polarities and current quantities occur:

1. The signal current (large dots) flows upward through R1, making the voltage at the base of X1 more positive.
2. The base-emitter biasing current (thin line) through X1 increases.
3. The emitter-collector current (solid line) through X1 also increases.
4. The positive voltage at the collector of X1 decreases and lowers the positive voltage at the base of X2.
5. The base-emitter biasing current (solid line) through X2 increases.
6. The collector-emitter current (solid line) through X2 also increases.

Comparison of Output and Input Voltages

The output voltage for this amplifier circuit is taken from the collector of X2. It will vary from a low to a high positive value, depending on the amount of collector current. During the negative half-cycle shown in Fig. 4-3, the collector current is reduced to its minimum value; therefore, the voltage it develops at the top of load resistor R6 will have its lowest positive value, too. This collector voltage will be positive, because the collector current always flows upward through R6, and electron current always flows from a less positive to a more positive point.

During the positive half-cycle of Fig. 4-4, the collector current reaches its maximum value; consequently, the instantaneous output voltage at the top of R6 will have its maximum positive value. The magnitude of the difference in the two collector voltages is the peak-to-peak value of the amplified voltage. With the circuit shown here, this peak-to-peak output might easily reach 12 or 15 volts, even though the input signal voltage developed across resistor R1 by the signal current will normally have a peak to peak value of only a small fraction of a volt. Therefore, we can say that the input-signal voltage has been substantially "amplified."

The output voltage will be "in phase" with the input-signal voltage. This means that in the negative half-cycle of Fig. 4-3 (input voltage negative at the top of resistor R1), the output voltage at the top of resistor R6 will have its least positive value. Conversely, when the input voltage is positive (as in Fig. 4-4), the output voltage will have its highest positive value.

Filter Currents

The emitter resistors, R3, and R5, are bypassed to ground by C2 and C3. These two filter capacitors prevent loss of signal voltage due to degeneration. (Degeneration was described at some length in the previous amplifier discussion.) Without going into extensive detail about the various causative factors, the significant and observable effects that occur in capacitor C2 and resistor R3 can be tabulated:

1. The currents through transistor X1 flow upward through R3 therefore the voltage at the top of R3 has to be positive. The upper plate of capacitor C2 will always assume the same voltage, which can be likened to a pool of positive ions.
2. The combination of R3 and C2 forms a "long time-constant" circuit. Consequently, the pool of positive ions on the upper plate is so everwhelmingly large, in comparison with the number of electrons drawn out of it and into the transistor during the positive half-cycles, that the positive voltage on the plate does not increase during these half-cycles.
3. This pool of positive ions is also overwhelmingly large, in comparison with the number of electrons which flow onto the upper plate of C2 from the top of resistor R3 during the negative half-cycles. Therefore, the positive voltage on this plate does not decrease during these negative half-cycles.

4. When the collector current decreases during the negative half-cycles (Fig. 4-3) electrons flow onto the upper plate of C2, driving the filter current *downward* from the lower plate of C2 to ground.
5. When the collector current increases (Fig. 4-4), electrons are drawn from the upper plate of C2. A like number of electrons are drawn *upward* from ground onto the lower plate of C2.

When all of these actions are permitted to occur, the emitter resistor is said to be bypassed or filtered, and degeneration has been avoided. Otherwise, the following would occur:

1. On negative half-cycles, the positive voltage at the emitter would fall as the collector current through R3 is reduced. The base-emitter and collector-emitter currents would tend to *increase* and thus nullify part of the original decrease in collector current.
2. Degeneration during positive half-cycles would be characterized by a rise in the positive voltage at the emitter as the collector current increases. Now the base-emitter and collector-emitter curents would tend to decrease and thus nullify part of the original increase in collector current.

The filtering action occurring across capacitor C3 is identical in nature but opposite in phase to that just described for C2. A decrease in collector current through X2 during negative half-cycles causes filter current to flow out of ground and onto the bottom plate of C3. An increase in collector current through X2 during positive half-cycles would drive this filter current back into ground.

Complementary Symmetry

The term "complementary symmetry" refers to the fact that currents flow in opposite directions in an NPN and a PNP transistor. Also, it refers to the fact that to increase conduction in an NPN transistor, the signal applied to its base must be positive; whereas in the PNP transistor, the signal applied to its base must be negative.

The principle of complementary symmetry will be utilized later in this chapter in a push-pull circuit.

RC-COUPLED AMPLIFIER WITH NEGATIVE FEEDBACK

Figs. 4-5 and 4-6 show two successive half-cycles in the operation of a two-transistor audio amplifier which employs RC coupling and a negative-feedback network. The important advantage of negative feedback is a reduction in distortion. Other advantages are reduction in the variation in gain provided by different transistors, and an apparent increase in input impedance of the circuit using the feedback. The inevitable disadvantage of negative feedback is some loss in gain—however, this is a small price to pay for the many advantages obtained through its use.

Identification of Components

The following circuit components perform the functions indicated. The manner in which these functions are accomplished will be elaborated on when the individual electron currents are discussed.

R1—Input driving resistor.
R2—Emitter stabilizing resistor for X1.
R3—Collector load resistor for X1.
R4—Voltage-dividing and biasing resistor.
R5—Voltage-dividing and biasing resistor.
R6—Input driving resistor for X2.
R7—Emitter stabilizing resistor for X2.
R8—Collector load resistor for X2.
R9—Feedback resistor.
C1—Input coupling capacitor.
C2—Interstage coupling capacitor.
C3—Output coupling capacitor.
C4—Feedback capacitor.
C5—Emitter filter capacitor.
C6—Voltage-divider filter capacitor.
X1 and X2—PNP transistors.
M1—Battery power supply.

Identification of Currents

This circuit has at least ten separate electron currents at work during normal operation. To understand how such a circuit works, you must be able to visualize each current—what makes it flow and what this flow in turn accomplishes, what its complete flow path is, etc. Once the currents are understood and visualized, the significance of their various functions will be perfectly clear.

1. Input driving current.
2. Voltage-divider current.
3. Base-emitter current through each transistor.

4. Collector-emitter current through X1.
5. Driving current for transistor X2.
6. Collector-emitter current for X2.
7. Feedback current.
8. Emitter filter current for X2.
9. Output current.
10. Filter currents across part of voltage divider.

Details of Operation

As long as no signal is applied to the input capacitor, the driving current shown a solid line will not exist, and the rest of the circuit will be operating under a static condition. Let us consider first the five currents which flow during such a static condition—namely, the voltage-divider current, and the two currents through each transistor.

The voltage-divider current (in small dots) flows continuously in a clockwise path, upward through resistors R5 and R4 and downward through battery M1 from its positive to its negative terminal. The flow of this current through R5 develops a positive voltage at the top of R5, and this voltage is applied directly to the base of each transistor through another set of resistors. Consequently, the positive voltage at the top of R5 can be labeled the "base biasing voltage." Since the full battery voltage (22.5 volts) exists across both of the resistors in series, the amount across R5 can be determined by a simple proportional relationship involving the sizes of R5 and R4 and the battery voltage. The formula is:

$$E_{R5} = \frac{R_5}{R_4 + R_5} \times E_{M1}$$

where,

E_{R5} is the voltage developed across R5 by the voltage-divider currents, in volts,
R_5 is the resistance of R5 in ohms,
R_4 is the resistance of R4 in ohms,
E_{M1} is the voltage of the battery in volts.

The base-emitter current through X1 (solid line) will flow initially in an amount determined by the voltage at the base and emitter. This current flow through emitter resistor R2 develops a small positive voltage at the top of R2, and it may be labeled the "emitter biasing voltage." Flowing downward through resistor R1, this current develops a small component of negative voltage at the top of R1, and this component reduces somewhat the positive voltage applied to the base from the voltage divider. The complete path of this electron current is upward from ground through resistor R2, through the transistor from emitter to base, and downward through R1 to the junction of R4 and R5. Then it goes *upward,* through R4, to the positive terminal of battery M1, where the chemical action of the battery makes it flow out the negative terminal and back to ground. The reason this electron current flows upward rather than downward at the junction of R4 and R5 is the attraction of the positive terminal of the battery. This is the highest positive voltage in the loop made by the current, and consequently the point toward which all electron currents will be drawn.

Since transistors are "current-operated" devices, the flow of an electron current from emitter to base will change the current flow from emitter to collector. (The emitter-to-base current is called the biasing current.) The upward flow of collector current through resistor R2 will develop an additional component of positive voltage at the top of R2. This voltage further modifies the emitter biasing voltage, and consequently affects the total amount of biasing current which will flow through the transistor from emitter to base.

The complete path of the collector current begins at ground, below R2, and flows upward through R2 into the emitter and out the collector. Then it goes downward through the collector load resistor R3, into the positive terminal of battery M1, and returns to ground from the negative terminal.

Within transistor X2, an initial flow of electron current from emitter to base is set in motion by application of the base biasing voltage obtained from the junction of R4 and R5 as a result of the voltage-divider current action previously discussed. Its complete path, shown a thin line is upward from ground through R7, into the emitter and out the base, downward through R6, then upward through R4 to the positive terminal of battery M1. As soon as this initial current begins to flow, it will alter both the emitter and the base voltage. The voltage at the top of R7 will become more positive, and the voltage at the top of R6 will become more negative. Both voltage changes are of such polarity that they *restrict or oppose* the flow of the base-emitter current.

Transistors are very sensitive to temperature changes—a rise in temperature causes *more* current to flow between emitter and base. In turn, more emitter-collector current flows, the transistor becomes still hotter and the current further increases. Such runaway condition will eventually destroy the transistor. The presence of emitter stabilizing resistors such as R2 and R7 prevents this from happening. Since all the current for X2 must

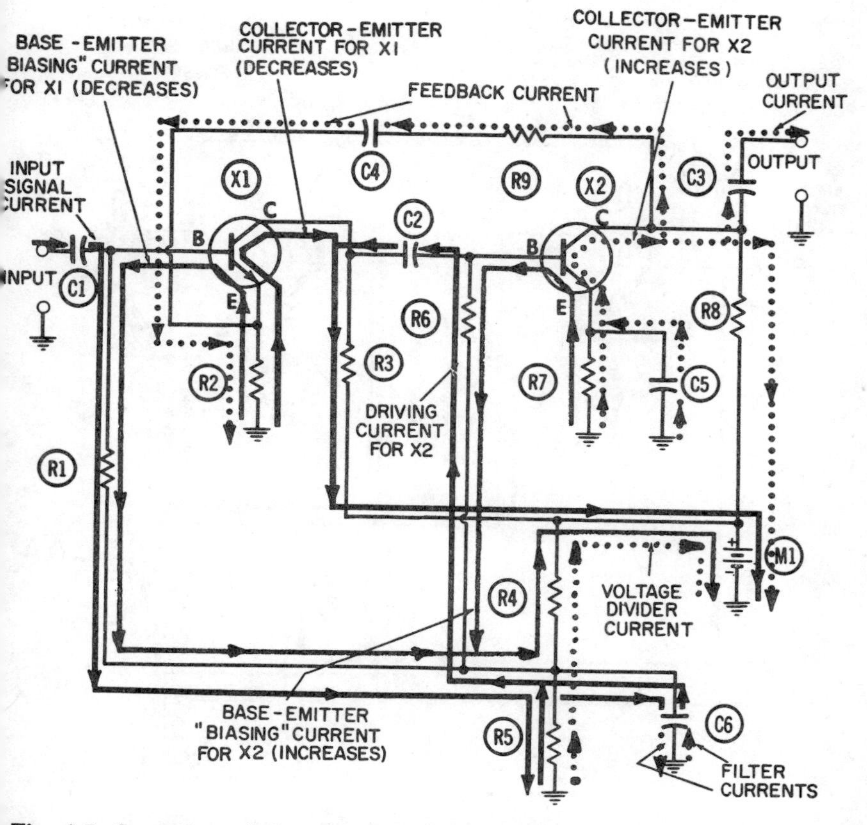

Fig. 4-5. Operation of the resistance-capacitance coupled amplifier with negative feedback—negative half-cycle.

first flow upward through R7, these current increases will rapidly make the voltage at the top of R7 so positive that it will tend to restrict or oppose the flow of these currents. (Remember that the voltage applied to an emitter is one of the two important biasing voltages of a transistor.)

The emitter-collector current for X2 (in large dots) flows upward from ground, through R7, into the emitter and out the collector. Then it heads downward through load resistor R8 and into the positive terminal of the battery.

In summary, it can be verified that all four of the currents which flow through the two transistors are drawn from ground and through their respective paths by the positive voltage of the power supply or battery.

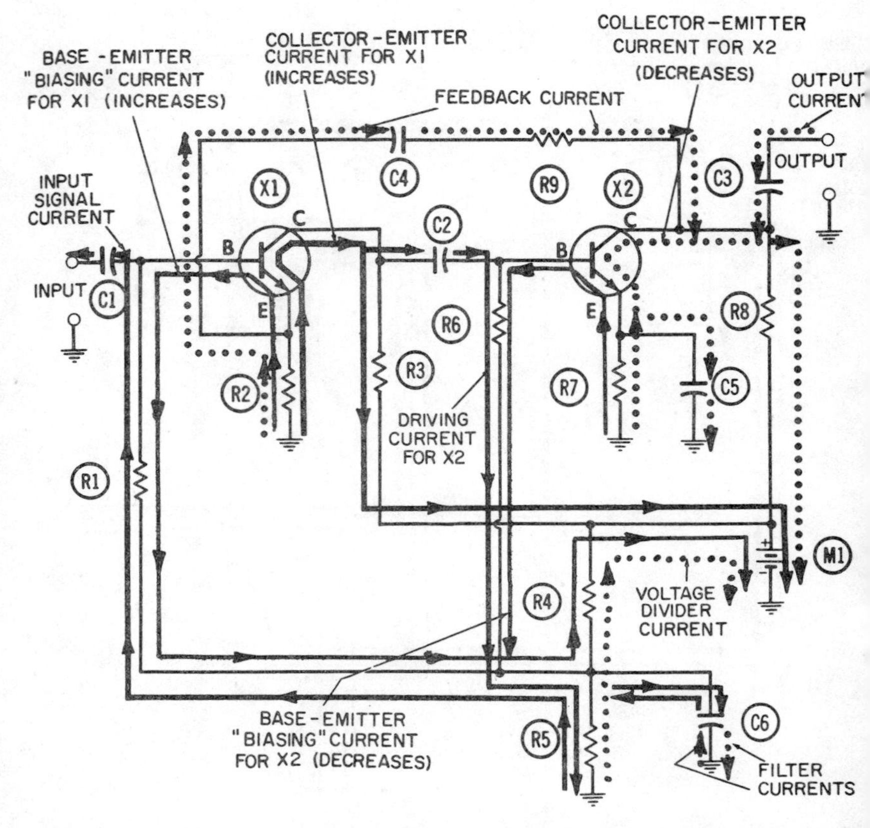

Fig. 4-6. Operation of the resistance-capacitance coupled amplifier with negative feedback—positive half-cycle.

Operation Under Dynamic Conditions

The application of a signal voltage to input capacitor C1 brings six more electron currents into existence, and all of them will be true alternating currents—meaning that they will periodically reverse their directions of flow along their respective paths in accordance with the frequency of the applied signal. Additionally, the presence of an input signal will modify the four currents flowing through the two transistors, so that they will become pulsating rather than pure direct currents.

When an alternating voltage to be amplified is applied to input capacitor C1 the voltage begins to drive an electron current up and down through resistors R1 and R5. Fig. 4-5 has been labeled the negative half-cycle because the input voltage is negative during this period. This negative voltage causes the input driving current (a solid line) to flow *downward* through R1 and R5 to ground. The component of negative voltage now developed at the top of R1 must be subtracted from the positive voltage already applied there by the voltage-divider action at the junction of R4 and R5. Since the upper terminal of R1 is connected directly to the base of transistor X1, this over-all decrease in positive voltage at the base during the negative half-cycle will *oppose* and thereby restrict the flow of emitter-base current.

Any such reduction in the emitter-base biasing current will cause a much larger reduction in the emitter-collector current through the transistor. The latter may be looked upon as the main electron stream through the transistor. A reduction in this current, as it flows through load resistor R3, causes the positive voltage at the top of R3 (and consequently at the collector) to rise. To satisfy yourself that a reduction in collector current causes a rise in collector voltage, suppose the collector current were cut off entirely. Under this condition, there could be no voltage drop whatsoever across R3, so both of its terminals would necessarily assume the full positive voltage of the battery or power supply.

When the collector voltage rises during the negative half-cycles, electrons will be drawn toward this point from any external circuit connected to it. This accounts for the upward movement of the driving current (a solid line) through biasing and driving resistor R6. The component of positive voltage now created at the top of R6 adds to the positive biasing voltage already existing at the base of X2 as a result of the voltage-divider current through R5 and R4.

The resulting increase in positive voltage at the base of X2 acts to encourage, or increase, the flow of emitter-base biasing current through X2 (a thin line). In turn, the emitter-collector current also increases and, in so doing, creates a larger voltage drop across load resistor R8. The amount of this voltage drop must of course be subtracted from the power-supply voltage to determine the instantaneous value of collector voltage. Thus, you can see that as the collector current rises, the positive collector voltage drops in value.

This drop in collector voltage is simultaneously "passed" to the output circuit through capacitor C3 and, via the feedback network, back to the emitter of transistor X1. Physically this is accomplished by electrons being driven *into* each external circuit, as shown in Fig. 4-5. This occurs because excess electrons are momentarily pouring out of the collector but cannot immediately flow downward through load resistor R8; therefore they choose any alternate path available.

Consequently, during this negative half-cycle a feedback current (in large dots) flows to the left, through resistor R9, and into feedback capacitor C4. An equal number of electrons are driven away from the left plate of C4 and downward through emitter resistor R2. The downward movement through R2 places a small component of negative voltage at the emitter of X1, which must be subtracted from the positive voltage normally existing there as a result of the two transistor currents flowing upward through R2. Any reduction in the positive voltage at this point (previously identified as the "emitter biasing voltage") will increase the electron current flowing from emitter to base through the transistor (the "biasing" current of the transistor).

Since the negative portion of the signal had originally *reduced* the biasing current through transistor X1, this feedback action must be classified as negative because it opposes the signal action and thereby lowers the total gain available from the transistor.

Recall earlier that when the negative portion of the signal was applied to the base, less collector current flowed. As a result, an entirely independent negative-feedback action occurs simultaneously across emitter resistor R2. The decrease in both the emitter and collector currents flowing upward through R2 lowers the voltage drop across R2 and also reduces the positive voltage at the emitter (the emitter biasing voltage). This change in the bias conditions of the transistor will tend to *increase* the biasing current flowing from emitter to base. Thus the biasing current is acting in opposition to the negative signal at the base because, during this same negative half-cycle, the signal is trying to *reduce* the flow of this current through the transistor.

The common name for this particular negative feedback is degeneration. One drawback is that it reduces the gain available from the transistor.

Actions During Positive Half-Cycle

During the positive half-cycle shown in Fig. 4-6, the following changes in current directions and voltage polarities occur:

1. The signal current (in solid blue) flows upward through R1, developing a component of positive voltage at the base. As a result, more emitter-base biasing current (a thin line) flows through the transistor.
2. The emitter-collector current (a solid line) is correspond-

ingly increased, causing a driving current (a solid line) to flow *downward* through resistor R6. The latter develops a small component of negative voltage which alters the bias of transistor X2 in such a manner that *less* base-emitter biasing current now flows through the transistor.

3. In turn, the collector-emitter current through X2 also decreases, and as it falls, the positive voltage at the collector rises.
4. The higher positive voltage draws electrons toward the collector from any external circuit connected to it. This accounts for the reversal in flow of the output and feedback currents.
5. Since feedback current is drawn toward the collector, it must flow upward through emitter resistor R2. As it does, it creates a small component of positive voltage at the top of R2. Now the emitter biasing conditions of X1 are altered in such a manner that *less* base-emitter biasing current flows through X1. Since the positive portion of the applied signal is simultaneously trying to *increase* this biasing current, the feedback can again be identified as negative.
6. Negative feedback known as degeneration also occurs across resistor R2 during this half-cycle. The signal increases the two currents through transistor X1, but in flowing upward through R2, the now higher currents will increase the positive voltage at the emitter. In an NPN transistor, a more positive emitter will reduce the currents through the transistor.

Filter Currents

Emitter resistor R7 for transistor X2 is bypassed with filter capacitor C5, so that degeneration does not occur across R7. The fluctuations from half-cycle to half-cycle in the current going into the emitter of X2 are in a sense "absorbed" by the filter capacitor. As a result, a constant current flows upward through R7 throughout the entire cycle. Likewise, the voltage it develops across R7 is constant rather than fluctuating. The capacitor is able to smooth out the current by delivering extra electrons to the emitter during the negative half-cycles and receiving extra electrons (from the current flowing up through R7) during the positive half-cycles. The filter current which flows between the lower plate of C5 and ground follows the movements of electron flow on and off the upper plate. For instance, extra electrons are drawn into the transistor from the upper plate of C5 during the negative half-cycles, so the filter current flows up from ground into the lower plate. During the positive half-cycles, the upper plate of C5 "recharges" by receiving additional electrons from R7, so the filter current flows down from the lower plate of C5 to ground.

Resistor R5 in the voltage-divider circuit is also bypassed with a filter capacitor, C6. This is necessary because the driving currents for the two transistors (both solid lines) flow through R5 in opposite directions. Without this capacitor, the voltage each current would develop across R5 would oppose the other voltage. Two separate filter currents flow side by side between the lower plate of capacitor C6 and ground. While the driving current through resistor R1 draws a filter current downward from C6 during negative half-cycles, the driving current through R6 is drawing its own filter current upward from ground into C6.

Summary

There are two electron currents which flow through the average transistor and the amount of each is closely regulated by the voltage at the base and emitter terminals. These are the biasing or controlling current, which flows from base to emitter (or emitter to base in NPN transistors); and the main electron stream (called collector-emitter current, or more commonly, the collector current), which flows from emitter to collector in NPN transistors.

The amount of biasing current which flows from emitter to base is regulated by the instantaneous voltages at the emitter and base. The voltage at the collector, like the voltage at the plate of a pentode amplifier, has an almost insignificant effect on the amount of collector current which flows.

The voltage at the base of transistor X1 is the algebraic sum of three voltages (one fixed and two variable). The fixed voltage is that developed across R5 by the voltage-divider current. The variable voltages are the ones developed across R1—one by the downward flow of base-emitter current, and the other by the signal current.

The instantaneous voltage at the emitter of transistor X1 is the algebraic sum of three voltages, all variable. Two of them are pulsating direct currents and are the two transistor currents —namely, the emitter-base and emitter-collector. The one which is alternating is the feedback current (shown in large dots) that flows up and down through resistor R2.

The instantaneous voltage at the base of transistor X2 is the algebraic sum of three independent voltages. The one fixed voltage is developed across resistor R5 by the voltage-divider current. One of the two variable voltages is developed across driving resistor R6 by the amplified signal current, which actually flows in two directions through the resistor. The second variable voltage is developed across this same resistor by the emitter-base biasing current (a thin line)—a pulsating direct current which flows downward through R6.

The voltage at the emitter of X2 is the sum of two fixed voltages; therefore, it can probably be described as a "pure" direct voltage. The two fixed voltages which contribute to it are those developed across emitter resistor R7 by the base-emitter and collector-emitter current through the transistor. Both currents pulsate through the transistor; but thanks to the filtering action of capacitor C5, they are "pure" DC when flowing through R7.

COMPLEMENTARY SYMMETRY PUSH-PULL AMPLIFIER

Figs. 4-7, 4-8, and 4-9 show three different conditions in the operation of a push-pull amplifier using the complementary symmetry of an NPN and a PNP transistor. This feature permits using the push-pull connection to drive a speaker without the necessity of an output transformer. Both are classed as power transistors, because of the relatively large collector current each must deliver for driving the speaker.

Fig. 4-7 shows the currents which flow in this circuit under static conditions only (while no signal voltage is being amplified).

Fig. 4-8 has been labeled a negative half-cycle, because the signal current (a solid line) flows in such directions through driving resistors R1 and R4 that it makes the bases of the two transistors negative. (The bases are their input points.)

Fig. 4-9 has been labeled the positive half-cycle, because the signal current causes positive voltages to exist at the two bases.

Identification of Components

The following components perform the indicated functions in this circuit:

R1—Voltage-divider and biasing resistor.
R2—Voltage-divider and biasing resistor.
R3—Emitter stabilizing resistor.
R4—Voltage-divider and biasing resistor.
R5—Voltage-divider and biasing resistor.
R6—Emitter stabilizing resistor.
C1—Input capacitor.
C2—Input capacitor.

C3—Emitter bypass capacitor.
C4—Emitter bypass capacitor.
L1—Voice coil of the speaker.
X1—NPN power transistor.
X2—PNP power transistor.
M1—Battery or power supply.
M2—Battery or power supply.

Identification of Currents

During the static period depicted by Fig. 4-7, no signal currents are flowing in this circuit; but the following electron currents, all pure DC, will flow:

1. Voltage-divider and biasing currents for both transistors.
2. Base-emitter current through both transistors.
3. Collector-emitter current through both transistors.

During the period of dynamic operation depicted by Figs. 4-8 and 4-9, the following additional electron currents will come into existence:

4. Input signal current.
5. Fluctuations in the two base-emitter currents.
6. Fluctuations in the currents through the two voltage-divider circuits.
7. Fluctuations in the two collector-emitter currents.
8. Current through the speaker voice coil.
9. Two emitter filter currents.

Details of Operation

Being fairly straightforward the three electron currents shown in Fig. 4-7 for the static period of operation can be easily explained. The voltage-divider current (small dots) flows continuously counterclockwise, upward through the two battery power supplies and to the left through resistor R2. Next it heads downward through resistors R1 and R4, then to the right through resistor R5, where it re-enters the positive terminal of the battery M2.

Electron current will flow around the base-emitter closed-loop circuitry of any transistor, in accordance with the voltage-biasing conditions at the base and emitter. Transistor X1, an NPN, has an initial negative voltage applied to its emitter from the negative terminal of battery M1. Because of the voltage-divider current action at the junction of resistors R1 and R2, however, a less negative voltage is applied to its base.

The voltage polarities around the circuit loop consisting of M1, R2, and R1 may be estimated qualitatively by recognizing that the voltage at the right terminal of resistor R2 will have the same voltage as the battery M1 (−22.5 volts). The voltage at the left terminal of R2 (this is the voltage applied directly to the base of X1) must be somewhat lower than this value (meaning somewhat less negative) because the voltage-divider current flows to the left through this resistor. Recall that the terminal from which electron current leaves a resistor is *always* more positive (or less negative) than the terminal at which it enters the resistor.

Since the junction of resistors R1 and R4 is connected directly to ground, the voltage at this point must always be zero. The

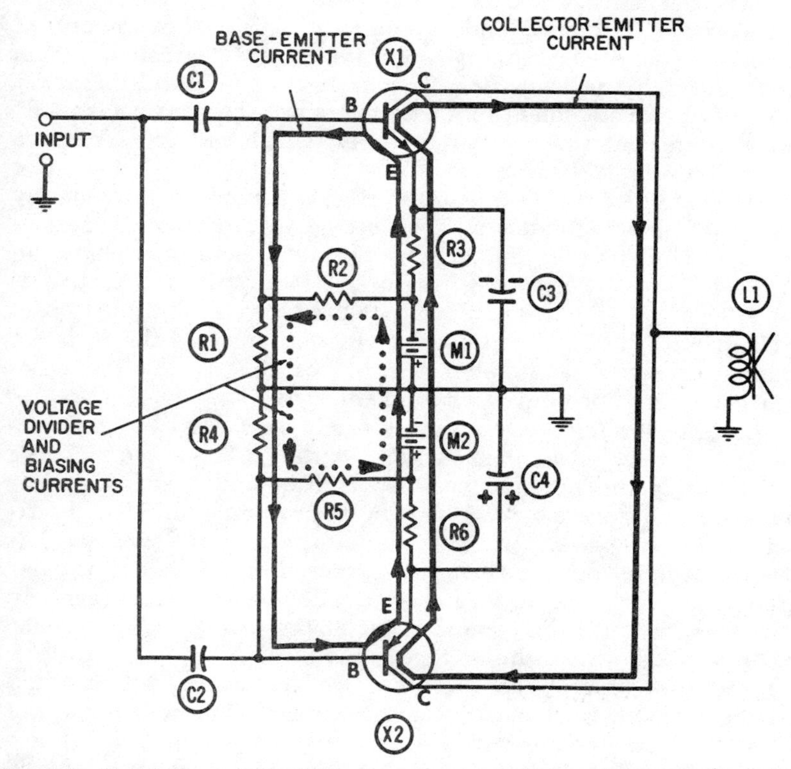

Fig. 4-7. Current conditions in the complementary symmetry push-pull amplifier—static operation with no signal currents.

junction of the two batteries is also grounded. As the voltage-divider current continues around this closed loop, it moves into a region of progressively higher positive voltages. The junction of R4 and R5 is positive with respect to ground, and the right terminal of R5 will always be at the full +22.5 volts of the battery.

Under no-signal conditions, the base-emitter current for both transistors (thin line) also flows continuously in a counter-clockwise loop. Its path might be considered to begin at the ground point between the two batteries. From here it flows upward through M1 and resistor R3, into the emitter and out the base of X1. Heading downward through R1 and R4, it goes into the base and out the emitter of transistor X2, then through R6 and into the positive terminal of M2. Here, it flows through M2 back to the reference point or reference voltage which we call ground.

This base-emitter current of a transistor is usually called the "biasing" current, and it controls closely the amount of collector current (meaning collector-emitter current) which flows through the transistor. This phenomenon is the basis for the descriptive statement that transistors are "current-controlled" devices—in contrast to vacuum tubes, which are considered to be "voltage-controlled" devices.

The amount of this base-emitter biasing current is itself closely controlled and regulated by the voltage values at the base and emitter. Under no-signal conditions such as are depicted in Fig. 4-7, the voltage at the base of X1 is negative. In fact, it is the sum of (1) the negative voltage produced at the junction of R1 and R2 by the voltage divider current, and (2) the negative voltage caused at the upper end of resistor R1, by the base-emitter current flowing downward through R1.

The latter voltage may be more easily visualized by recognizing the current path which includes battery M1, resistor R3, the resistance of the junction between emitter and base within the transistor, and resistor R1 as another voltage divider. Starting at the negative terminal of battery M1, the voltage has its maximum negative value. Proceeding counterclockwise around this loop, the voltage at each point becomes progressively less negative until the ground point at the lower end of resistor R1 is reached, where the voltage is zero.

Under the assumption that the two transistors will conduct identical amounts of biasing current during this no-signal condition, the same amount of base-emitter current (a solid line) continues to flow through resistor R4, the resistance of the junction between base and emitter within transistor X1, resistor

R6, and battery M2. This loop constitutes another voltage divider, the voltage again becoming progressively more positive (i.e., less negative) as this electron current proceeds around the loop. Electrons move in the direction indicated because they flow inescapably *away* from more negative and *toward* more positive voltages.

Thus the "biasing" voltage at the base of transistor X2 is positive and is the sum of two separate components of positive voltage, both produced across resistor R4 by the two currents which flow downward through it.

The voltage at the emitter of transistor X1 is negative and is the sum of two separate components of negative voltage. These are produced at the upper terminal of resistor R3 by the two electron currents (the base-emitter current, a solid line; and the collector-emitter current, in solid red) flowing upward through R3.

The voltage at the emitter of transistor X2 is positive, and is the sum of two separate positive voltages. Both exist at the lower terminal of resistor R6 as a result of the same two electron currents—the base-emitter and collector-emitter currents which flow upward through it.

The path for the combined collector currents (under these no-signal conditions) might be considered to start at ground (the reference point) between the two batteries, and to flow upward through M1 and R3 into the emitter and out the collector of X1. Next it heads downward to the collector of X1, through X1 and out the emitter, then upward through R6 to the positive terminal of M2. This electron current is finally returned to ground by flowing through M2 to its negative terminal.

Operation Under Dynamic Conditions

Once these three static currents have been visualized, you will be in a much better position to understand how this circuit operates when a signal is being amplified. Three additional currents come into existance. Also, the three currents previously discussed will now be changed from pure to pulsating DC. These pulsations represent the variations in the audio-frequency signal.

During the negative half-cycle depicted by Fig. 4-8, a signal voltage of negative polarity is applied to the left plates of capacitors C1 and C2. This negative voltage drives electron current (in solid blue) *onto* the left plates of these capacitors. In turn equal amounts of electron currents are driven away from the right plates. The current driven away from C1 flows downward through resistor R1 to ground and, in so doing, develops an instantaneous component of negative voltage at the upper terminal of R1. This component of negative voltage adds to the negative voltage already existing at that point as a result of the two currents which flow through R1. The *increase* in negative voltage at the base of X1 (or any other NPN transistor) will decrease the base-emitter biasing current flowing through the transistor. This, in turn, decreases the collector-emitter current through X1.

Unlike the collector current through transistor X1 the collector current through transistor X2 increases during negative half-cycles. The reasons why it does may be summarized briefly, as follows:

1. The input signal current (in solid blue) flows *upward* through resistor R4 to ground. The component of negative voltage it creates at the lower terminal of R4 must be subtracted from the positive voltage already existing at that point as a result of the two voltage-divider actions previously described. The net result is a reduction in the positive voltage at the base of transistor X1.
2. A reduction in positive voltage at the base of any PNP transistor will always increase the flow of biasing current. Therefore, more base-emitter current flows through transistor X1.
3. The increase in biasing current causes a companion increase in collector-emitter current, which is also the load current.

It should be obvious that this additional collector current for X2 cannot all be drawn through X1, since the collector current through X1 is decreasing during this negative half-cycle, as previously explained. Consequently, the extra collector current for X2 is drawn upward from ground and through the voice coil of the speaker. This current (in large dots) will set up magnetic lines of force which will draw the speaker diaphram in one direction only, such as to the right in Fig. 4-8. The complete path of this current is through the voice coil, then into the collector and out the emitter of X2 upward through resistor R6 to the positive terminal of battery M2. Flowing through the battery to its negative terminal, the current re-enters the common ground, which provides a ready return access to the voice coil.

During the positive half-cycles depicted by Fig. 4-9, a similarly long series of events (which will be discussed later) increases the collector current through X1 and decreases the collector current through X2. This increase in collector current through X1 cannot be accepted by X2 during this same time period. Therefore, the current is driven through the speaker

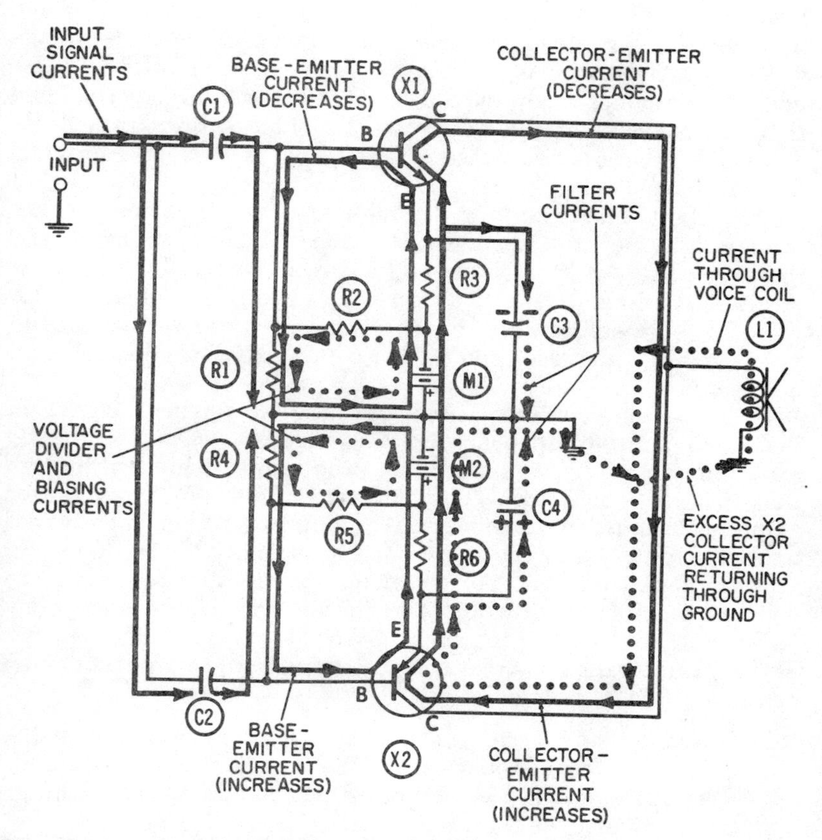

Fig. 4-8. Operation of the complementary symmetry push-pull amplifier—negative half-cycle.

voice coil and into the ground connection. Thus, current flows through the voice coil in one direction during negative half-cycles, and in the opposite direction during positive half-cycles. This causes the resultant magnetic lines of force in the voice coil to change direction every half-cycle. As a result, the permanent magnet (and speaker diaphram connected to it) will be alternately attracted to and repelled by the voice coil during each cycle. The back-and-forth movements of this diaphragm set up the air vibrations we know as sound waves.

To consider all the current and voltage changes which occur during a positive half-cycle of operation, we should start with the input signal. As long as its polarity is positive, electrons will be drawn onto the right plates of C1 and C2. As it flows upward through resistor R1 and downward through resistor R4, this electron current places small components of positive voltage at the upper terminal of R1 and the lower terminal of R4. These voltage components are added to the voltage already existing there as a result of the two voltage-divider actions previously discussed (the ones associated with the currents in solid and dotted lines).

This signal current flow through the two resistors makes the normally negative voltage at the top of R1 less negative, and the normally positive voltage at the lower terminal of R4 more positive. As a result, the two currents flowing through transistor X1 (the base-emitter biasing current and the collector-emitter load current) increase, and the two flowing through transistor X2 decrease.

Thus, by application of a small audio-frequency signal at the input point of this circuit, it is possible to cause a fairly heavy flow of alternating current at the same frequency through the speaker voice coil. This circuit is classed as a power amplifier because much more audio-frequency power is delivered to the output (speaker) than is consumed in the input circuit (resistors R1 and R4). As an example, the speaker might be delivering one or more watts of audio power, whereas only a few milliwatts are being consumed in input resistors R1 and R4. Several important facts about transistors make this phenomenon possible:

1. In a transistor, a change of only a fraction of a volt in the voltage difference between base and emitter will cause a small change in the amount of biasing current and a much larger change in the amount of collector load current.
2. This fraction of a volt required can be developed at the transistor base by causing a current as small as a fraction of a milliampere to flow through an input resistor such as R1 or R4.

Moreover, the collector-emitter load current flowing through these power transistors may be hundreds of times greater than the signal current flowing up and down through input resistors R1 and R4.

Because of the voltage changes across these two resistors while the signal current is flowing, the amount of voltage-divider current (in small dots) will be disturbed slightly. During the negative half-cycles (Fig. 4-8), the voltage at the junction of R1 and R2 is made more negative; consequently, the smaller voltage difference between the two terminals of resistor R2 will draw less current through R2 from the negative terminal of the battery.

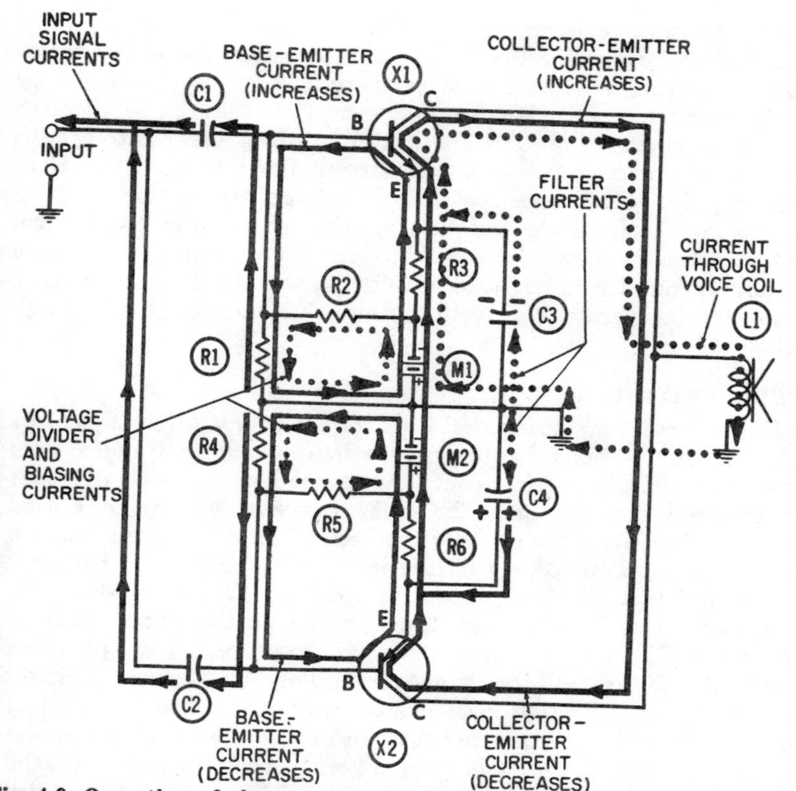

Fig. 4-9. Operation of the complementary symmetry push-pull amplifier—positive half-cycle.

During the positive half-cycles, this junction is made less negative and now the opposite is true—a larger voltage-divider current will be driven through R2 by the greater voltage difference across its terminals.

By similar reasoning, it can be shown that the voltage-divider current flowing through the lower network (consisting of R4, R5, and M2) also fluctuates with the applied signal voltage. Here, however, it *increases* during the negative and decreases during the positive half-cycles.

For this reason, the voltage-divider current (in small dots) has been shown as two separate currents—flowing in two separate networks—while a signal is applied. Any difference in amount between these two currents will automatically be compensated for by a flow of electron current through the common ground connection between the junction of power sources M1 and M2, and the junction of resistors R1 and R4.

Likewise, any difference in amounts between the base-emitter currents through the two transistors will be compensated for by an appropriate flow of electron current (a thin line) through this same common ground connection. During the positive half-cycles (when more base-emitter current flows through X1 than through X2), this current will flow through ground from left to right—that is, from the junction of resistors R1 and R4, toward the junction between the two batteries. Conversely, it will flow through ground from right to left during the negative half-cycles, when the base-emitter current through X2 exceeds that through X1.

Filter Currents

Emitter resistors R3 and R6 are bypassed by filter capacitors C3 and C4, respectively, to prevent loss of signal strength from that type of negative feedback known as degeneration. In order to forestall degeneration, the filter currents shown in dotted red must be able to flow freely between C3 and C4 to the nearest ground point. The voltage on the upper plate of C3 must always be identical to the one at the emitter of X1. Since this is a negative voltage, the accumulated charge on the upper plate of capacitor C3 can be conveniently represented as a "pool" of electrons. There will be a continual flow of electron current driven upward through resistor R3 by the negative terminal of battery M1. (This is the point of most negative voltage in the entire circuit.) If no collector current could escape into the transistor, the upper plate of capacitor C3 would soon acquire enough electrons that its voltage would equal the full negative battery voltage. At that time, the upward flow of electron current through R3 would cease. However, since transistor X1 is being operated under what are called Class-A conditions, some collector current flows throughout the entire cycle. Therefore the negative voltage on the upper plate of C3 will never attain the full negative voltage power source M1.

During the positive half-cycle such as in Fig. 4-9, the biasing conditions of transistor X1 cause an increase in the collector current. In turn, the electrons which make up this current are drawn quite easily from the electron pool on the upper plate of capacitor C3, and an equal number flows upward from ground to the lower plate. This is the essence of capacitor action . . . it is how capacitors will appear to "pass" an alternating current.

When the capacitor has sufficient size, or capacity, the quantity of extra electrons demanded by the transistor collector current during the positive half-cycles is such an infinitesimal percentage of the total number of electrons stored in the capacitor that the voltage does not change appreciably from half-cycle to half-cycle.

This is how a filter capacitor avoids the loss in signal strength known as degeneration. If emitter resistor R3 were not bypassed with a filter capacitor, then the increased collector current during the positive half-cycles would have to be drawn directly upward through R3. This would make the voltage difference, or "drop," across R3 larger, and would *lower* the negative voltage at the upper terminal of R3. Since the emitter of X1 is connected directly to this point, this lower negative voltage would constitute a change in the biasing conditions of the transistor, and would ultimately reduce the collector current flowing through the transistor. This would nullify at least part of the original increase in collector current, and would constitute a loss in the amount of amplification the circuit delivers. This is degeneration.

The voltage on the lower plate of capacitor C4 is always identical to the positive voltage at the emitter of transistor X1, since they are connected together. It is convenient to visualize a positive capacitor voltage as a pool of positive ions, out of which a continual flow of electrons will move upward and through resistor R6, toward the higher positive voltage of power source M2. If no collector-emitter current flowed through X2, the positive voltage on the lower plate of C4 would eventually equal the +22.5 volts of the battery at the lower terminal of M2. Instead, the continual inflow of collector-emitter current from X2 keeps it at a lower positive value.

Assuming capacitor C4 has sufficient size, or capacity, the quantity of excess electrons which flow onto its lower plate from the collector-emitter current during the negative half-cycles (Fig. 4-8) is such an infinitesimal percentage of the total number of positive ions already stored there that the voltage across capacitor C4 does not change appreciably from half-cycle to half-cycle.

If C4 were not in the circuit, or if for any reason the filter current between it and ground were prevented from flowing, then degeneration (loss of signal strength) would occur across R6. The increased collector-emitter current through X2 during negative half-cycles would flow directly through R6 and thereby increase the voltage drop across this resistor. The lower positive voltage now produced at its lower terminal and at the emitter would be a fundamental change in the biasing conditions of the transistor. The resultant decrease in collector-emitter current would nullify part of the original increase in this current and thereby lead to some loss in amplification.

Chapter 5

DETECTOR CIRCUITS

Like their triode vacuum-tube counterparts, transistors can be employed to detect, or demodulate, a modulated RF carrier signal and thereby develop an audio voltage. The first circuit discussed in this chapter is for a PNP transistorized detector. Crystal diodes are also employed as detectors in many transistor receivers. The second circuit discussed in this chapter is for an IF amplifier and diode detector, with AGC applied to the IF amplifier.

PNP DETECTOR CIRCUIT

Figs. 5-1 and 5-2 show two successive audio half-cycles in the operation of a transistor detector. The necessary components of this circuit include:

R1—Voltage-divider resistor.
R2—Voltage-divider and base-bias resistor.
R3—Emitter stabilizing resistor.
R4—Collector load resistor.
C1—IF tank capacitor.
C2—IF filter capacitor.
C3—Emitter bypass capacitor.
C4—IF filter capacitor across load resistor R4.
C5—Output coupling and blocking capacitor.
T1—IF tank transformer.
X1—PNP transistor.
M1—Battery or other power supply.

Identification of Currents

The following separate and distinct electron currents are at work in this circuit. The actions occurring in this or any circuit cannot be understood until the movements of these currents are clearly understood.

1. IF tank current.
2. Base-emitter biasing current, a direct current which pulsates at the intermediate frequency.
3. Voltage-divider current.
4. Two intermediate-frequency filter currents.
5. Collector-emitter current.
6. Audio filter current.

Details of Operation

The intermediate-frequency input current (a solid line) oscillates in the tank circuit consisting of C1 and the primary of transformer T1. This IF tank current induces a similar current in the secondary winding. With this secondary current there will always be associated the so-called secondary voltage, also known as "back electromotive-force" (back emf). Once during every cycle of this intermediate-frequency current, the upper terminal of the secondary winding will be driven to a peak of negative voltage. When this happens, the transistor will conduct the maximum base-emitter biasing current because, in the PNP transistor, electrons invariably flow *from* the base *to* the emitter against the direction of the emitter arrow). Hence, a more negative base voltage will naturally increase this flow.

Between each negative peak of induced voltage there will be an equal-sized peak of positive voltage. On these positive peaks, the transistor will conduct the minimum base-emitter current. Since this current always flows in the same direction through the transistor, the voltage alternations which were induced across the secondary winding of T1 are converted into current pulsations through the transistor.

The complete path of this base-emitter biasing current, which is shown a solid line and which pulsates at the intermediate frequency, begins at the negative terminal of battery M1. It flows to the left and upward through resistor R2 and the secondary winding of T1. Then it heads to the right into the base of the transistor and out the emitter and downward through resistor R3, to the common ground connection, where it has free access to re-enter the positive terminal of the battery.

In addition to the pulsations which occur at the basic intermediate frequency, this base-emitter current also varies in amount according to the modulation carried by the carrier signal. Fig. 5-3A shows a fairly conventional graphical representation of the so-called modulated carrier signal, or waveform. Many hundreds of even thousands of the carrier cycles will occur during one audio cycle, and their *strength* will periodically rise and fall in accordance with the modulation imposed on the carrier at the transmitter. These variations in carrier strength make up what is commonly known as the "modulation envelope." As is customary in waveform diagrams of this type, the modulation envelope has been indicated on Fig. 5-3A. This so-called "modulation envelope" in such diagrams is nothing more than a convenient graphical device which relates the relative strength of individual cycles of the carrier signal and, of course, points up the fact that the strength of these carrier cycles varies in accordance with the audio modulation imposed on the carrier at the transmitter.

Each pulsation of base-emitter current causes one cycle of IF filter current to flow through filter capacitor C2 to ground. During the negative peak portions of the secondary induced voltage across T1, maximum base-emitter current flows through the transistor X1. Inevitably some electron current is drawn from the upper plate of capacitor C2, and in turn the same quantity of electrons is drawn onto the lower plate. On the positive peak portion of secondary induced voltage, the base-emitter current is reduced to minimum. During these half-cycles of IF, the filter current flowing on either side of capacitor C2 heads downward again to ground.

The amount of base-emitter biasing current which flows through any transistor is determined by the two important biasing voltages, (at the base and emitter) and, of course, the difference between them. The voltage at the base of transistor X1 is determined primarily by the flow of voltage-divider current (in small dots) upward through battery M1, to the left and upward through resistor R2, and downward through R1 to the common ground. From here, it has easy access back to the positive terminal of battery M1. The resulting voltage at the junction of resistors R1 and R2, being less negative than the power-supply voltage, will cause a flow of electron current across the junction between base and emitter. The amount of this so-called biasing current controls, or regulates, the flow of electron current from collector to base, within the transistor. (Once this current crosses the difficult "reverse" junction from collector to base, it flows fairly easily from base to emitter, and exits from the transistor at the emitter terminal.)

This second current through the transistor is usually called merely the collector current, and will be from 25 to 100 times larger than the biasing current (frequently referred to merely as the "base" current). The collector current, shown a solid

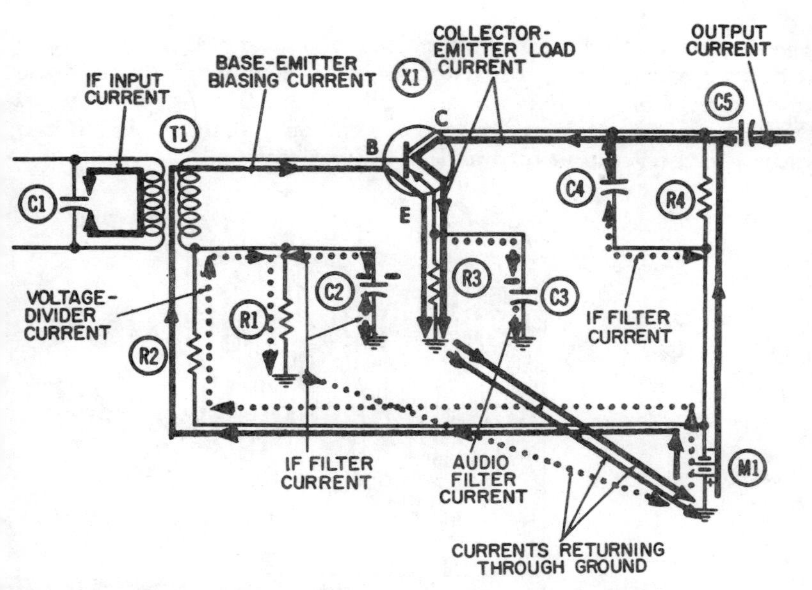

Fig. 5-1. The transistorized detector circuit—conditions leading up to an audio modulation trough.

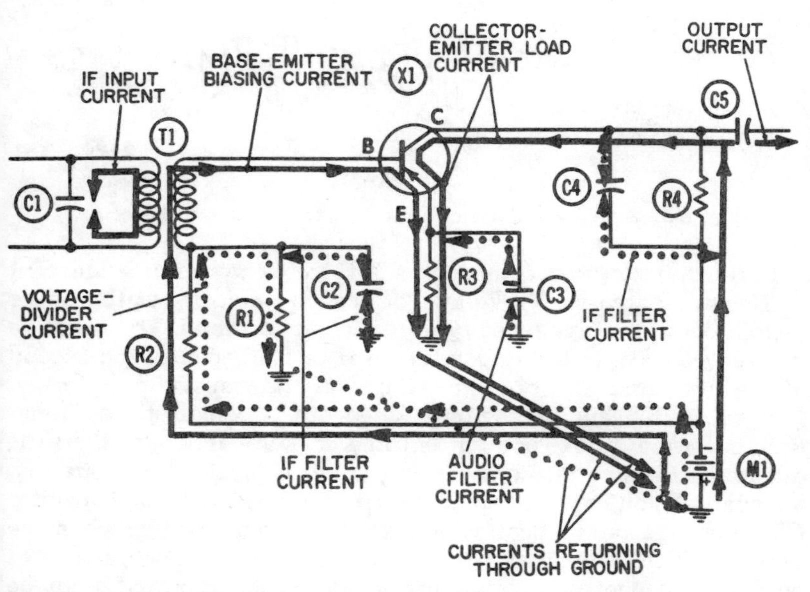

Fig. 5-2. The transistorized detector circuit—conditions leading up to an audio modulation peak.

line, begins at the negative terminal of battery M1. From here it flows upward through resistor R4, then downward through the transistor from collector to emitter, continuing downward through emitter resistor R3 to the common ground, where it can return to the positive terminal of the battery.

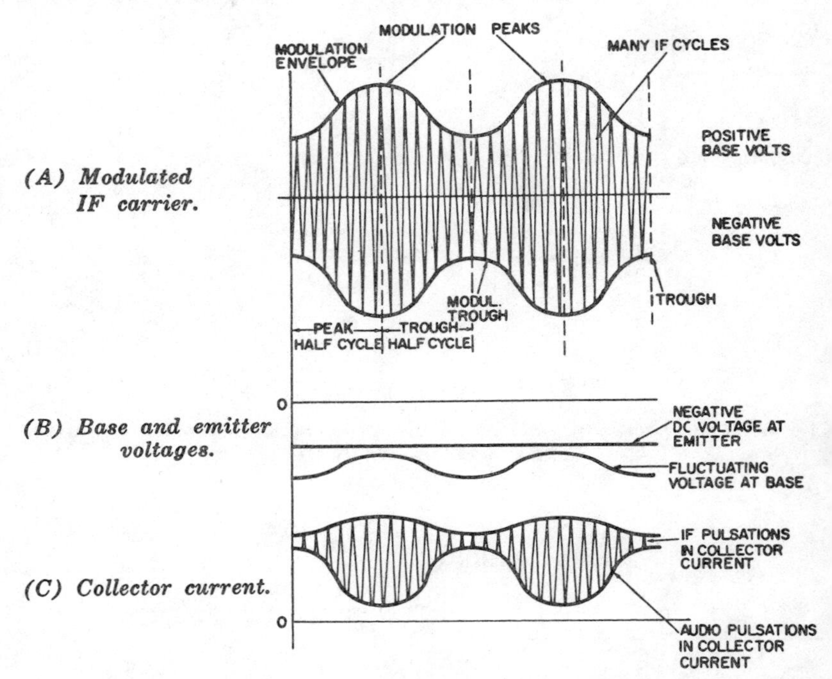

Fig. 5-3. Waveform diagrams for the transistorized detector.

In flowing downward through R3, the collector-emitter current will develop across R3 a voltage of negative polarity at the upper terminal and positive polarity at the lower terminal. The negative voltage polarity at the upper terminal of R3 becomes the second of the two important biasing voltages of the transistor.

Resistor R1 and capacitor C2 together constitute a "long-time constant" filter—one whose time constant is longer than the duration of a single intermediate-frequency cycle. Consequently, a negative voltage will appear on the upper plate of capacitor C2. This negative voltage, which remains unchanged between the individual IF cycles, has been indicated by the deep minus signs on C2. (A negative voltage stored on a capacitor plate can be most conveniently represented as a group or pool of electrons, the symbol for which is one or more minus signs.)

In Fig. 5-2, which represents a "Peak" half-cycle of audio voltage, the single minus sign on the upper plate of C2 indicates the negative base voltage which would correspond to a modulation peak. Fig. 5-3A tells us that during a modulation peak, the individual IF cycles have their maximum value, or strength. Each such cycle drives the transistor base to a negative voltage of such value that maximum base-emitter biasing current is permitted to flow through the transistor. This extra quantity of base-emitter biasing current must flow through resistor R2. In so doing, it lowers the negative voltage at the upper terminal of R2, because of the increased voltage drop across the resistor. The reduced negative voltage at the junction of R1 and R2 is also the voltage applied to the base. Thus, during the modulation peak of Fig. 5-2 a smaller negative biasing voltage is applied to the base of the transistor. As a result, the base-emitter biasing current which flows continuously through the transistor is reduced.

During the modulation trough of Fig. 5-1, the individual IF cycles reach smaller peak values and thereby reduce the base-emitter current during each cycle. Since this current must flow through resistor R2, the decreased voltage drop across R2 makes the voltage at the junction of R1 and R2 *more* negative. This is one of the two transistor biasing voltages, and it causes a general increase in the continuous flow of base-emitter current throughout the entire cycle.

The voltage which appears on the upper plate of capacitor C2 marks the first appearance of an audio voltage when a carrier signal is being modulated. It varies between two negative values, as indicated in Fig. 5-3B, and consequently regulates the flow of the two currents through the transistor so that they pulsate at an audio rate. The transistor currents also have pulsations, which occur at the intermediate frequency, as shown in Fig. 5-3C, and are filtered out by capacitors C2 and C4. These IF filtering currents have been shown as dotted lines.

Since the collector current varies, or pulsates, at the audio frequency being demodulated, an audio voltage is developed across collector load resistor R4. This audio voltage is coupled to the next amplifier stage by capacitor C5, which also serves to "block" the fixed negative voltage of battery M1 and thereby keep it from reaching the next stage. Fig. 5-1 depicts circuit conditions leading up to a modulation trough. Here the collector current is increasing, and this extra current is drawn from the left plate of C5. This action draws an equal amount of electron current onto the right plate from the external circuit beyond C5 which recognizes this surge of electrons as a positive voltage.

Under the conditions leading up to the modulation peak of Fig. 5-2, the collector current is decreasing. Thus, the excess electrons being driven through resistor R4 from battery M1 will flow onto the left plate of C5, driving an equal number off the right plate and through the external circuit. The external circuit beyond capacitor C5 recognizes this surge of electrons through it as a negative voltage.

The combination of emitter resistor R3 and filter capacitor C3 is deliberately chosen to have a longer time constant than the duration of one cycle of the lowest audio frequency being demodulated. Consequently, even though the collector-emitter current is pulsating as it comes through the transistor, it is prevented from flowing through R3 in pulsations and thereby developing an audio voltage across R3. Instead, the additional collector current which flows during the modulation trough is shunted momentarily onto the top plate of capacitor C3, and an equal quantity of electrons flows harmlessly from the lower plate into ground. This is one half-cycle of audio filter current. If there were no capacitor, this extra collector current would have to flow immediately through resistor R3, and, in so doing, would cause an additional component of negative voltage at the upper terminal of R3. Since this is one of the two important biasing voltages of the transistor, a more negative voltage at the emitter would restrict or reduce the flow of both currents through the transistor, just when they were trying to increase. This would be degeneration.

The decrease in collector current during the modulation peak of Fig. 5-2 would likewise cause a smaller voltage drop across R3 and, in turn, a less negative voltage at the emitter. As a result, both currents through the transistor would increase, just when they were supposed to decrease. This would be another half-cycle of degeneration. When a large enough filter capacitor is connected across R3, the excess electrons stored on its upper plate during the previous half-cycle will not be drawn off the top plate, and will flow down through R3 to ground, along with the regular flow of collector current. While this action is occurring, the filter current below C3 will be drawn upward from ground, toward the capacitor. This constitutes the second half-cycle of filter action.

AMPLIFIER WITH AGC DIODE

Figs. 5-4 and 5-5 show two successive half-cycles of a single radio-frequency cycle of operation for a fairly typical IF amplifier and detector circuit. The diode also provides automatic

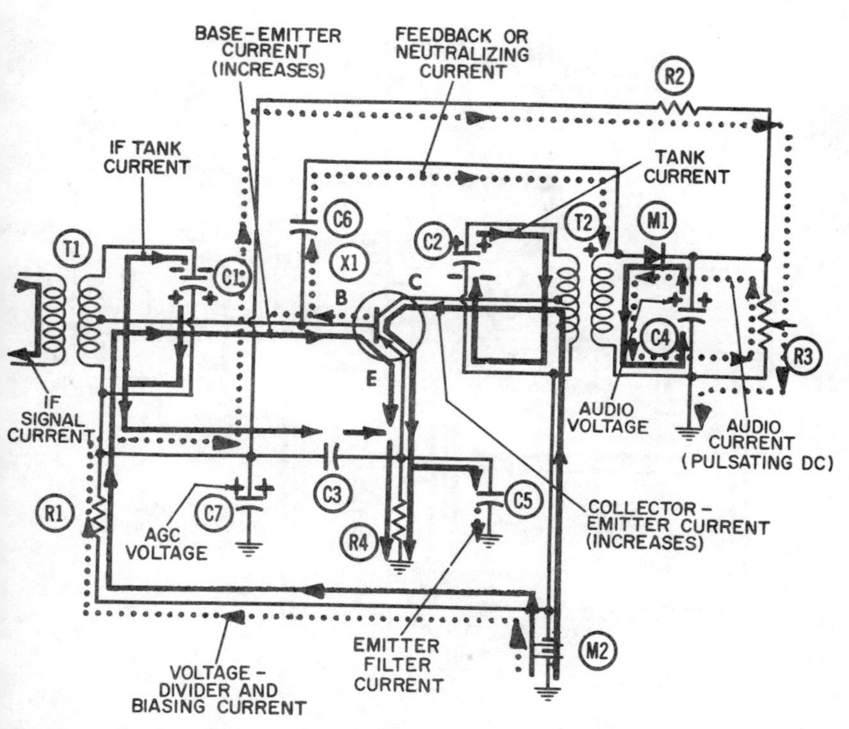

Fig. 5-4. An IF amplifier with AGC diode—negative half-cycle during a period of normal signal strength.

gain control (AGC), frequently called automatic volume control (AVC).

A fairly common intermediate frequency in broadcast receivers is 455 kilocycles per second. (In electronics terminology, "per second" is understood and so is omitted, and "kilocycles" is abbreviated to kc.) An intermediate frequency is normally much lower than the original carrier or radio frequency, and as such is somewhat easier to handle—meaning that the circuits which amplify it are less susceptible to losses from parasitic oscillations, radio-frequency interference (RFI), and other such disturbances. At the same time, a frequency as high as 455 kilocycles per second is high enough to be classed as a "radio frequency," so that tuned circuits of reasonable size can be put together which will resonate at the basic frequency, with many attendant advantages of signal strength gain, frequency selectivity, etc.

The components which make up this completed circuit are as follows:

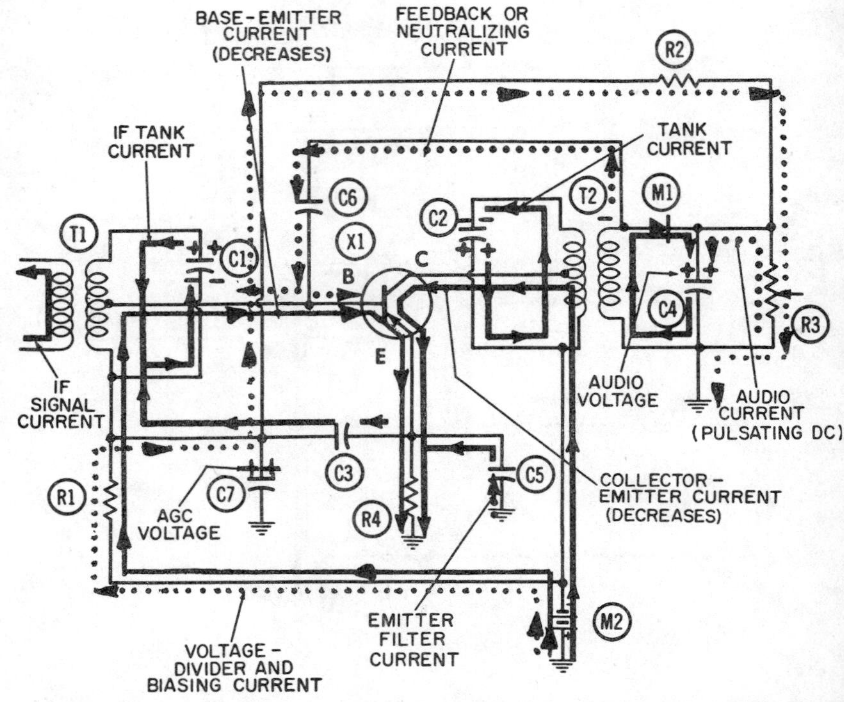

Fig. 5-5. An IF amplifier with AGC diode—positive half-cycle during a period of normal signal strength.

R1—Voltage-divider and biasing resistor.
R2—Voltage-divider and AGC resistor.
R3—Voltage-divider and audio-output resistor.
R4—Emitter stabilizing resistor.
C1—Input tank-circuit capacitor.
C2—Output tank-circuit capacitor.
C3—RF bypass capacitor.
C4—Audio-output capacitor.
C5—Emitter bypass capacitor.
C6—IF neutralizing capacitor.
C7—AGC capacitor.
T1—Input IF transformer.
T2—Output IF transformer.
X1—PNP transistor.
M1—Crystal diode.
M2—Battery or other DC power supply.

Identification of Currents

There are a great many separate electron currents flowing in this circuit. Each current should be clearly identified in your mind before you can hope to understand the movements, and more important, the functions performed by these currents. First, there are the usual three currents which will flow in the average transistor circuit under "static" conditions, irrespective of whether a signal voltage or current is actually being amplified. These static currents are:

1. Voltage-divider current.
2. Base-emitter current.
3. Collector-emitter current.

In addition to these static currents, several additional currents come into existence when a signal voltage or current is being amplified. These currents include:

4. Input signal current.
5. IF tank current in both the input and output tank circuits.
6. Output secondary current.
7. IF neutralizing current.
8. Unidirectional diode current.
9. Emitter filter current.
10. AGC current.

Details of Operation

The movements of most of these currents can be understood from Figs. 5-4 and 5-5. In order that the actual generation of an audio frequency current and voltage can be visualized however, it is helpful to resort to additional circuit diagrams, of a type which will depict audio rather than intermediate-frequency half-cycles. Also, the generation of an automatic gain-control current and voltage can best be visualized by using extra diagrams to show how circuit conditions change as the signal strength does (signal fade or buildup). Figs. 5-6 through 5-9 depict these current and voltage changes. These extra diagrams are necessary because of the three widely separated frequencies that are always involved in the demodulation of an RF or IF carrier signal, and then in the development of a DC voltage which will be proportional to the strength of the original carrier signal. The latter may consequently be used to provide automatic control of the transistor gain and thus of the volume of the audio signal being delivered by the speaker.

These three frequencies are:

1. The carrier or intermediate frequency—in this case, 455,000 cycles per second.
2. The modulation frequency which is carried by the IF carrier and which, after demodulation, becomes the audio frequency from the speaker. A good average audio frequency in the "listening" range of frequencies is represented by the key of middle C, whose pitch is 256 cycles per second.
3. The frequency at which signal fades or buildups will occur because of anomalous propagation conditions. Fades and buildups occur independently of each other, so it is inaccurate to imply that such a thing as a single whole cycle of signal fade and signal buildup exists. It is more accurate to consider individual half-cycles, such as *either* a fade or a buildup, and to understand within what length of time such an event must occur. A signal fade or buildup may require several seconds or minutes to complete itself.

The time of one IF cycle of operation is always equal to the reciprocal of the frequency, in this case 1/455,000 second.

Thus, one half of one cycle will require slightly longer than one millionth of a second to complete itself. This is about the length of time required for the actions occurring in the tank circuits of Fig. 5-4. In the tank circuit consisting of the secondary winding of T1 and capacitor C1, the electrons which make up the tank current or circulating current have moved *upward* through T1 and are amassed on the upper plate of capacitor C1. Thus the voltage across the entire tank circuit has its maximum negative value at the end of this first half-cycle. The amount, or value, of this tank voltage exists to a lesser and lesser degree across portions of the secondary winding. Halfway down this coil, the instantaneous negative voltage at that point will be half of what it is at the top, and so on.

The base of the emitter is connected directly to a point that is quite far down on the secondary winding. This is done for impedance-matching purposes. The concept of impedance matching is a difficult one to visualize qualitatively. Impedance, like resistance, is a ratio between an existing voltage *and* the curent which this voltage will set in motion. The concept of impedance also represents the ratio of a *change* in an existing voltage across a circuit, to the *change* in current through this same circuit as a result of this voltage change.

A transistor connected in a common-emitter configuration such as this one, is said to have a very low input impedance.

This means that the ratio between a *change* in input voltage and the resulting change in current flow is low. In other words, only a small change in voltage will cause a substantial change in current. As always, in discussing and using the term "impedance," it is not merely helpful but actually mandatory that we understand exactly which voltage and which current we are talking about.

The input impedance of a transistor refers to the amount of change in voltage necessary between emitter and base in order to produce the desired change in base-emitter current through the transistor. The voltage difference between base and emitter is one of the fundamental biasing conditions of a transistor, and you have seen that this voltage difference is normally only a small fraction of a volt. For the desired fluctuations in base-emitter current to occur, it is only necessary to vary this existing base-emitter voltage difference by the tiniest fraction of a volt. Thus, a circuit in which a very small voltage change causes a substantial current change is a "low-impedance" circuit.

It will be helpful to remember the Ohm's-law relationship between resistance, voltage, and current whenever the impedance of a circuit is under discussion. Impedance, like resistance, is nothing more than a measure of opposition to the flow of electron current. For this reason, it can always be expressed mathematically as a ratio between voltage and current, just like resistance.

Transistors which are used in common-emitter configurations like this one are considered to have a high "output impedance." The output impedance of a transistor is again a means of expressing a ratio between a particular voltage and current. In an output circuit, we are interested in knowing what effect a change in collector voltage will have on the amount of collector-emitter current. Normally, it has relatively little effect. In this respect, the transistor is quite comparable to the pentode vacuum tube, where the plate voltage has only a small effect on the amount of plate current. As a result, the plate circuit of the pentode is considered to have a high output impedance.

For the transistor we can conclude, if we keep the Ohm's-law relationship in mind, that the output impedance is high by recognizing that an average change in the voltage at the collector will produce only a very small change in the amount of collector current. The important biasing conditions of a transistor are the voltages at the base and emitter. The difference between these two voltages exercises an overriding influence on the amount of the two currents which flow through a transistor.

Returning to the input circuit, it is necessary to use only a small portion of the voltage developed across the tank circuit consisting of capacitor C1 and the secondary winding of T1. This is why the inductor is tapped across only a small portion of its length. One might well ask why a tuned circuit is used, when the tank voltage it develops is much higher than is necessary or than can be used for amplification. The answer is that a tuned circuit provides the important feature of *selectivity*, or discrimination between signals of different frequencies. Depending on its values of inductance and capacitance, a highly tuned circuit will oscillate strongly at one particular frequency and will "reject" all others including those close to its own frequency of oscillation.

The selectivity of a tuned circuit varies in accordance with the Q of the coil which is part of that circuit. Q in this usage refers to "quality"—unlike in Coulomb's law, where Q refers to the "quantity" of electric charge.

The Q of a coil is the ratio between its reactance and resistance and is written arithmetically as:

$$Q = \frac{2\pi f L}{R}$$

where,

Q is the coil "quality,"
f is the frequency of operation in cycles per second,
L is the coil inductance in henrys,
R is the coil resistance in ohms.

This relationship tells us that by increasing the inductance, L, in a coil without changing its resistance, R, we can greatly increase its Q and thus improve the selectivity of the tuned circuit of which the coil is a part.

The "output impedance" of the transistor collector is also "matched" to an appropriate point on the primary winding of output transformer T2. The pulsations of collector current flowing upward through the lower portion of this primary winding will sustain an oscillation of tank-current electrons throughout the entire tank circuit. The exact phase relationships between these two currents have been discussed in greater detail in a preceding chapter on the Hartley oscillator. The cases are similar, because in a Hartley oscillator, current is drawn through a small portion of an inductor and, by autotransformer action, in turn supports an oscillation in the entire tank circuit.

The appropriate phase relations between the collector current, tank current, and tank voltage have been depicted in Figs. 5-4 and 5-5. In Fig. 5-4, while the collector-emitter current (a solid line) is increasing in the upward direction through the lower part of the primary winding of T2, it induces another current to flow at an ever-increasing rate in the downward direction through the entire primary winding. This induced current supports the tank current (a solid line) and is essentially in phase with it. The tank current flows downward through the primary winding of T2 during the negative half-cycle of Fig. 5-4, and this accounts for the negative voltage shown on the lower plate of capacitor C2 at the end of this negative half-cycle.

During the positive half-cycle of Fig. 5-5, the collector current is still flowing upward through the lower portion of the primary winding of T2, but is now decreasing. Autotransformer action will now cause an induced current to flow in the same upward direction through the entire winding, but at an *increasing* rate. This induced current supports the main tank-circuit oscillation, which can be seen flowing upward through the winding in Fig. 5-5 and delivering electrons to the upper plate of capacitor C2. So, at the end of this positive half-cycle the upper plate of capacitor C2 exhibits a negative charge or voltage.

Probably the most important feature of a tuned tank circuit is that a relatively small amount of replenishment or support can set a sizable amount of electron current in oscillation and maintain it in oscillation. Thus, the current flow induced by the pulsations of collector current is intrinsically quite small in comparison with the amount of tank current which it maintains in oscillation.

This large tank current induces a current (solid line) at the same frequency in the secondary winding of T2. Whenever alternating current flows through an inductor, an alternating voltage (known by such names as "induced emf," "counter emf," or "back emf") must exist across the inductor terminals. Its instantaneous polarity is always related directly to the direction in which its associated current flows, and also depends on whether this current is increasing or decreasing. (For a fuller discussion of the phase relationships existing between applied and induced voltages and current, refer to the introductory chapter of the book on oscillators in this "Basic Electronics" series.)

The Principle of Neutralization

Neutralization is a technique used for coupling some of the energy from an output circuit of a tube or transistor back to its input circuit for the express purpose of preventing the circuit from breaking into self-sustained oscillations. It is a form of negative feedback between output and input, and is provided to counteract the effects of positive feedback which may be inherent in the tube or transistor. Let us examine the nature of this positive feedback which is inherent within transistors, and then see how capacitor C6 provides the desired negative feedback to neutralize the positive feedback.

The normal flow path for collector current in the PNP transistor is into the collector and out the emitter. In Fig. 5-4, you can see the increase in collector current following this path. However, there is inevitably some capacitance between the interior of a transistor and the external wires which lead up to it. Because of these inherent capacitances, the increase in collector current flowing through the base and toward the emitter in Fig. 5-4 will drive some electron current *away* from the base and into the external circuit leading up to the base. This current (in large dots) flows in a direction that makes the transistor base *more* negative. This follows from the universal fact that when electron current is moving through a conductor, the terminal or point *from* which the electrons move is more negative than the point towards which they move.

In Fig. 5-5 the decrease in the collector current passing through the base again exerts a capacitive effect on the electrons within the wire leading up to the base, this time drawing them *toward* the base. Since these electrons are being drawn through the entire external circuit between base and ground, they develop a component of positive voltage at the base.

Both of these current actions in the circuit external to the base are classified as positive, or regenerative, feedback because they tend to reinforce the very conditions which caused them. During the negative half-cycle of Fig. 5-4, this component of negative voltage at the base further *increases* the two currents through the transistor. The resulting additional increase in collector current will make the component of negative voltage at the base still more negative, and so on. During the positive half-cycle of Fig. 5-5, the small component of positive voltage at the base *decreases* the two currents through the transistors. Because of the capacitive effect, the resulting additional decrease in collector current will make the component of positive voltage at the base still more positive.

These positive-feedback actions can be nullified or neutralized by causing another current to flow side by side in the external base circuit with this positive feedback current, but in the opposite direction. This is the neutralizing current, shown in large dots and it can be obtained from the appropriate side of the secondary winding of T2. (A connection to the wrong end of the coil would give more positive feedback and probably lead to self-sustained oscillations.) In Fig. 5-4, the top of the secondary winding of T2 is assumed to be at a positive voltage as indicated by the deep + sign. This voltage is "coupled," via neutralizing capacitor C6, to the base of the transistor. Thus, a component of feedback current is drawn *toward* the transistor base at the same time the undesired feedback current is being driven *away* from the base.

In Fig. 5-5, when the top of the secondary winding of T2 is negative, electrons are driven to the left, through capacitor C6, and thus flow *away* from the base at the same time the unwanted feedback current is flowing *toward* it. In this fashion, neutralization automatically and continually compensates, during each cycle, for the positive feedback caused by inherent capacitances within the transistor.

The Demodulation Process

Diode M1 in this circuit is a unidirectional device (one that permits current to flow in only one direction through it). Electrons can flow with relative ease *against* the direction of the arrow (to the left, in other words), but only with very great difficulty can they be made to flow in the opposite direction. This inherent property makes the diode a useful rectifying device for the demodulation of a modulated signal. When the top of the secondary winding of T2 has a positive voltage induced on it (such as during the half-cycle of Fig. 5-4), electrons (in large dots) will flow up from ground, through resistor R3 and the diode. This current flow creates a positive voltage at the top of R3, as indicated by the plus signs on capacitor C4.

During alternate half-cycles as in Fig. 5-5, the top of the secondary winding of T2 has a negative voltage induced on it, so

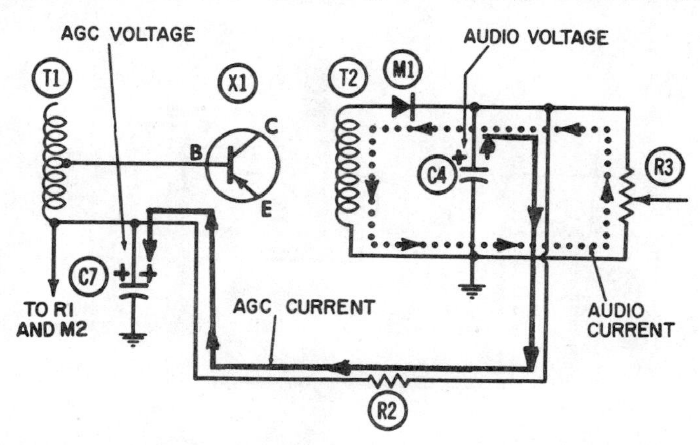

Fig. 5-6. AGC portion of the circuit in Figs. 5-4 and 5-5—modulation trough during a period of weak signal strength.

no current flows through diode M1 in either direction. The positive voltage on the upper plate of capacitor C4 persists throughout this positive half-cycle, because of the "integrating" action of the long time-constant RC combination consisting of R3 and C4. Any RC combination is a long time-constant combination to a particular frequency if the resistance in ohms multiplied by the capacitance in farads is more than five or six times greater than the time required for a single whole cycle of current movement to occur at the particular frequency. This is in accordance with the time-constant formula:

$$T = R \times C$$

where,

T is the time constant of the combination in seconds,
R is the resistance in ohms,
C is the capacitance in farads.

If an unmodulated carrier signal were being received, every RF cycle would have the same strength, and the voltage on the upper plate of C4 would be essentially DC. When a modulated signal is being received the strength of the individual RF cycles is not constant. Rather, it varies in accordance with modulation, and so does the voltage on C4. When a strong RF cycle is being received (this happens during a modulation peak), the induced voltage across the secondary winding of T2 will be higher. Thus, more electron current will be drawn upward through R3 and the diode. This means that a higher positive voltage will exist across R3 and on the upper plate of C4.

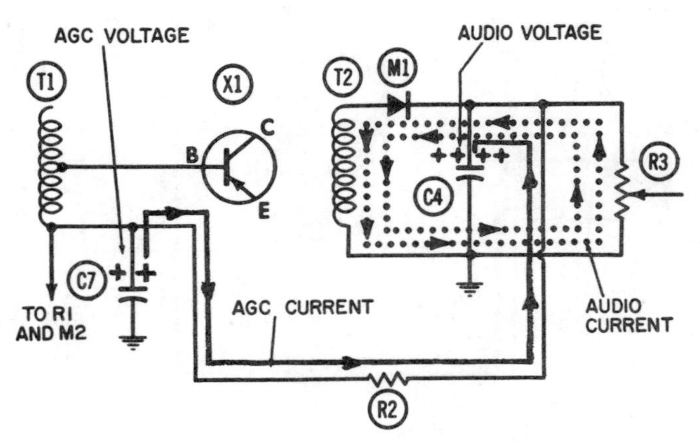

Fig. 5-7. AGC portion of the circuit in Figs. 5-4 and 5-5—modulation peak during a period of weak signal strength.

When a weak RF cycle is being received (this happens during a modulation trough), the induced voltage across the secondary winding of T2 will be lower. Now, less electron current will be drawn upward through R3, and a lower positive voltage will exist across R3 and on the upper plate of C4. Since the strength of the individual cycles of RF depends on the amount of modulation, the voltage produced at the top of resistor R3 is an audio-frequency voltage.

Figs. 5-6 and 5-7 show two successive half-cycles of the audio voltage as it appears for the first time, immediately after demodulation, in a typical receiver. Circuit components in these figures have been numbered to correspond to their counterparts in Figs. 5-4 and 5-5. The radio-frequency currents have not been shown in Figs. 5-6 and 5-7. These illustrations depict one entire audio cycle, consisting of a modulation trough followed by a modulation peak, while a weakened carrier signal is being received. The audio currents are shown in large dots the same as in Figs. 5-4 and 5-5. The AGC current (so labeled, and shown a thin line) flows in either direction along part of the same path used by the voltage-divider current (in small dots in Figs. 5-4 and 5-5). During a modulation trough, when a low positive voltage exists on the upper plate of C4, the AGC current is drawn to the left, through resistor R2, by the higher positive voltage stored on the upper plate of C7.

During a modulation peak as in Fig. 5-7, this AGC current is drawn to the right, through resistor R2, by the higher positive voltage now stored on the upper plate of capacitor C4.

The intrinsic amount of positive voltage stored on the upper plate of AGC capacitor C7 does not change during a single audio cycle, even though electrons flow onto it during a modulation trough (Fig. 5-6) and out of it during a modulation peak (Fig. 5-7). The reason is that the combination of resistor R2 and capacitor C7 forms a long time-constant to the lowest audio frequency likely to be encountered. This means these components are large enough that their product is several times the period of a single low-frequency audio cycle. When these components are made large enough, the *amount* of electrons which flow through R2 on successive half-cycles will be insignificant, compared with the number of positive ions already stored on the upper plate of C7. Thus, by appropriate choice of the sizes of these two components, a DC voltage can be developed on the upper plate of C7 that does not vary with the modulation represented by the audio signal.

The voltage stored on an AGC capacitor such as C7 will always adjust itself to the average value of the trough and peak voltages on the upper plate of the audio-output capacitor—in this case, C4. This average value will remain the same unless the over-all strength of the RF carrier changes because of propagation anomalies, which cause what are known as signal fades or buildups. Figs. 5-8 and 5-9 depict two successive audio half-cycles during a period of excessive carrier-signal strength, or what is called a signal buildup. All the radio-frequency tank currents shown a solid line in Figs. 5-4 and 5-5 will become proportionately stronger during such a period. As a result, more audio current will be drawn upward through R3 and diode M1 during both a modulation peak and a modulation trough. This is depicted by the additional dotted lines in Figs. 5-8 and 5-9. This additional current flow increases the trough and peak voltages stored on C4 and so, increases their average value.

The voltage stored on capacitor C7 is "replenished" by the voltage stored on C4. During a signal buildup, when both the audio modulation trough voltage (Fig. 5-8) and the audio modulation peak voltage (Fig. 5-9) are increased, their average value is also increased. Consequently, the positive voltage on the upper plate of C7 must increase to this average value. This important action is accomplished by the AGC current shown in Figs. 5-6 through 5-9. During a period of normal or unchanging signal strength such as are shown in the two half-cycles of Figs. 5-4 and 5-5, an AGC current will flow, *at the audio frequency,* back and forth between capacitors C7 and C4, through resistor R2. The amount of current flowing to the left during each modulation trough *will be exactly equal* to the amount flowing to the right during the next modulation peak. Because of this, the positive voltage stored on C7 is maintained at the average value of the high and low positive voltages on C4.

When signal strength is increased by a signal buildup, both the high and low positive voltages will be increased in value. When this happens, the two half-cycles of AGC current flowing through resistor R2 will become "unbalanced." In other words, *more* electrons will flow away from C7 during the modulation peak of Fig. 5-9, and *fewer* electrons will flow into C7 during the modulation trough of Fig. 5-8. This condition of current unbalance will continue until the AGC voltage stored on C7 becomes sufficiently positive to just equal the average value of the trough and peak voltages on C4.

The circuit actions involved in achieving automatic gain control in a transistor circuit are fundamentally similar to those in vacuum-tube circuitry. These principles have been discussed with greater detail in the chapter on automatic volume control in the book about detector and rectifier circuits in this "Basic Electronics" series. Also, waveform diagrams appearing in that book may help to clarify the meaning and significance of terms such as modulation trough, modulation peak, signal fade, and signal buildup. The most important difference between AVC circuitry using tubes and transistors is that in tube circuitry a *negative* AVC voltage normally is developed whereas in the example of this chapter, the AVC voltage developed on capacitor C7 is positive instead.

How AGC Voltage Controls Transistor Gain

In this type of circuit, the AGC voltage controls, or regulates, the transistor gain by exercising some control over the base-emitter current which flows through the transistor. Since this is a PNP transistor, any positive component of voltage applied to the base of the transistor will *decrease* the base-emitter current (called "biasing" current). Its quantity depends on the biasing conditions (meaning the biasing voltages) at the base and emitter. As was pointed out earlier in this chapter, a certain negative voltage is applied to the emitter of X1 (as a result of current flow through resistor R4), and a slightly higher negative voltage is applied to the base (as a result of voltage-divider action through R1, R2, and R3). The difference between these two applied voltages is one of the fundamental biasing conditions of the transistor (the most important one, in fact), and it controls the amount of electron current flowing from base to emitter (the base-emitter biasing current).

The existence of a permanent positive voltage on the upper plate of capacitor C7 will reduce the negative voltage created

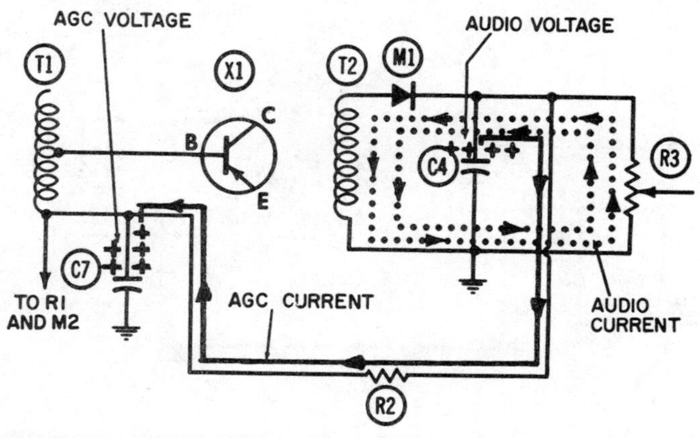

Fig. 5-8. AGC portion of the circuit in Figs. 5-4 and 5-5—modulating trough during a period of strong signal.

at the base of the transistor by the flow of voltage-divider current through R1, R2, and R3. The positive AGC voltage shown on C7 is not a composite value of all the voltages at the base of the transistor, but only the result of the AGC filter action occurring between R2 and C7. You have already seen that four other voltages exist at the base, each contributing in some small degree to the amount of base-emitter biasing current flowing. The AGC voltage makes a fifth one. These five voltages are:

1. The voltage resulting from voltage-divider current through resistors R1, R2, and R3. This is a fixed, or DC, voltage which is negative at the junction of R1 and R2, and consequently negative at the base.
2. The signal voltage, which is alternately positive and negative at the intermediate frequency, and which therefore lowers and raises the negative voltage at the base.
3. The voltage resulting from positive feedback caused within the transistor by capacitance between the internal elements (B, C, and E) and the external wiring of the transistor (IF).
4. The voltage due to negative feedback from output circuit to input circuit, through capacitor C6. (IF)
5. The AGC voltage on C7. Although essentially a DC voltage, it does vary if and when the signal strength of RF carrier varies.

The combination of resistor R2 and capacitor C7 is chosen so that their product will be a "long time-constant" when compared with the duration of a single low-frequency audio cycle. If the lowest audio frequency being amplified were 50 cycles per

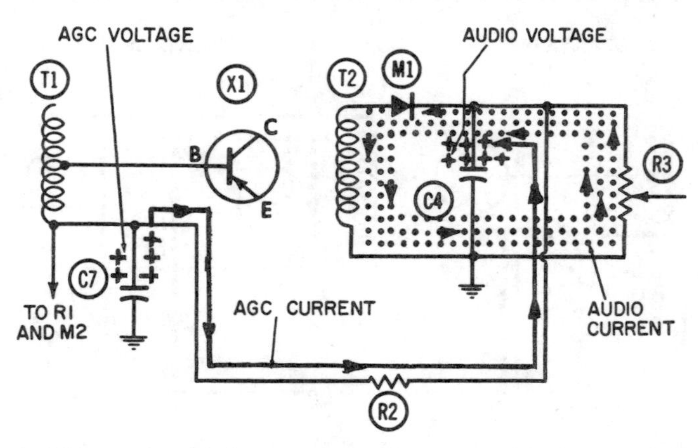

Fig. 5-9. AGC portion of the circuit in Figs. 5-4 and 5-5—modulation peak during a period of strong signal.

second, with a period of .02 second for each cycle, the R2-C7 combination would be a "long time-constant" combination to this frequency if their product were more than five times longer than this period—i.e., a tenth of a second or longer.

The Audio-Output Voltage

An audio-output voltage is developed across resistor R3, by the pulsating DC which continuously flows upward through it. This current (in large dots) also flows through M1, and as a result of the demodulation process previously described, its pulsations occur at the audio frequency. The arrow pointing into R3 indicates that this is a potentiometer, or variable resistor, with which any desired fraction of the total voltage may be tapped off for amplification by the next succeeding audio-amplifier stage. Potentiometer R3 thus constitutes a manual volume control for the receiver in which it is installed.

SECTION 5

RADIO CIRCUITS

Chapter 1

HETERODYNING ACTIONS

The mixer circuit combines two different radio-frequency currents in order to obtain a third current, usually of lower frequency, whose modulations will be a true representation of the modulations carried by one of the two original currents. In receivers one of these two original currents is called the *carrier,* or *signal,* current. This current is the one which is being received from a transmitting station. The other of the two original currents is generated within the receiver by a separate oscillator circuit and is usually referred to as the *local oscillator* current, or oscillator current.

In general, it is less difficult to construct circuits that will amplify low-frequency currents than it is to construct them for high-frequency currents. Thus, in receiver work it is almost universal practice to "step down" the carrier frequency by some process such as mixing two currents and getting a third current, before amplifying the signal.

TRIODE MIXER

The triode mixer circuit, while not too common in present-day receivers, is the basic mixer circuit. Figs. 1-1, 1-2, and 1-3 demonstrate the operation of this circuit.

Identification of Components

The components which make up this circuit are identified as follows:

C1—Grid tuned-tank capacitor.
C2—RF filter capacitor.
C3—Plate tuned-tank capacitor.
T1—First RF transformer.
T2—Second RF transformer.
T3—IF transformer.
V1—Triode vacuum tube used as mixer.
M1—Antenna.
M2—Grid biasing-voltage source.

Identification of Currents

Four different electron currents, each flowing at a different frequency are present in Figs. 1-1, 1-2, and 1-3. These electron currents are as follows:

1. Current at carrier or signal frequency.
2. Current at oscillator frequency.
3. Current at difference frequency.
4. Current at sum frequency.

Circuit Operation

The first current to be discussed is the radio-frequency carrier current being received at the antenna. This current carries the modulation, or intelligence, which is intended to be amplified and demodulated. Although the primary of T1 has been shown as connected directly to the antenna, the signal current may be passed through an RF amplifier stage before being subjected to the mixing or frequency-changing process. (The mixer stage was formerly referred to as the "first detector," to differentiate it from the demodulator stage; however, this terminology is seldom used in modern practice.)

Fig. 1-1 shows the currents which will flow in this circuit *when the local oscillator is not energized.* (The local oscillator is connected to the primary winding of T2.) Only a single half-cycle of operation is shown in Fig. 1-1; it is the half-cycle when the signal current makes the control grid negative, thereby cutting off any current flow through the tube. However, the previous half-cycle would have permitted a pulsation of current to flow through the tube, and consequently, the remnants of this tube current are seen flowing off filter capacitor C2 and exiting downward through the primary winding of T3 to the power supply. At the same time a component of filter current at this frequency is shown re-entering the lower plate of C2.

The tuned tank, composed of C3 and the primary of T3 in the plate circuit, is not resonant at this carrier frequency; therefore no oscillation will be set up in the tank due to a pulsation of current which repeats itself at this frequency.

Fig. 1-2 shows a half-cycle of circuit operation *when there is a local oscillator current but no signal current*. The local oscillator current is shown (in solid) flowing through the lower, or primary, winding of T2, thereby supporting a current at the same frequency in the grid tank circuit.

The half-cycle chosen for Fig. 1-2 is one which makes the grid positive so that plate current is flowing through the tube. This current is shown *entering* the top plate of filter capacitor C2 before flowing to the power supply through the primary of T3.

Fig 1-3 shows the conditions when the carrier and oscillator currents both exist simultaneously in the grid circuit. As an example of the frequencies involved, if the carrier signal being received were from a broadcast station operating at 1,000 kc per second, the local oscillator would be operating at a higher frequency, such as 1,500 kc per second. Thus, a million complete cycles of the current shown will flow back and forth in the grid tank each second. Also, it is known that one-million-five-hundred-thousand cycles of the other current shown will flow in the grid-tank circuit each second. Each one will attempt to act independently in turning the electron stream on and off in the tube. When the two currents are momentarily in phase they will be aiding each other. Conversely, when they are momentarily out of phase with each other, or in "opposite phase," they will oppose and neutralize each other.

Pulsations of current are released through the tube at both frequencies, although the sizes of the pulsations will vary from cycle to cycle, depending on the extent to which each of the grid driving current-voltage combinations is augmented or cancelled by partial combinations with the other. The size of filter capacitor C2 is chosen so that it will have low reactance at both frequencies. Thus pulsations are effectively filtered back to ground through this capacitor, while the electrons themselves may continue their journey to the power supply after passing through the primary winding of T3.

The formula for determining the reactance of a capacitor at any frequency is:

$$X_c = \frac{1}{2\pi fC}$$

where,

X_c is the capacitive reactance in ohms,
f is the frequency of the current in question in cycles per second,
C is the capacitance in farads.

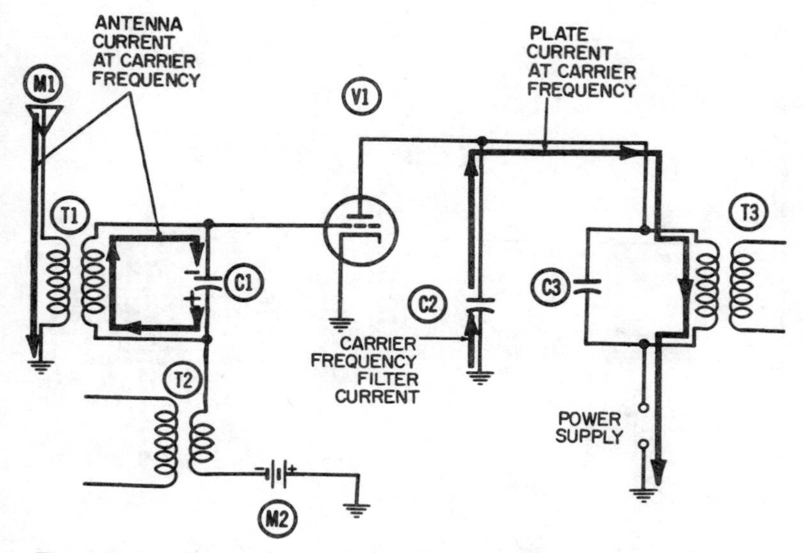

Fig. 1-1. Operation of the triode mixer circuit—negative half-cycle, no oscillator current.

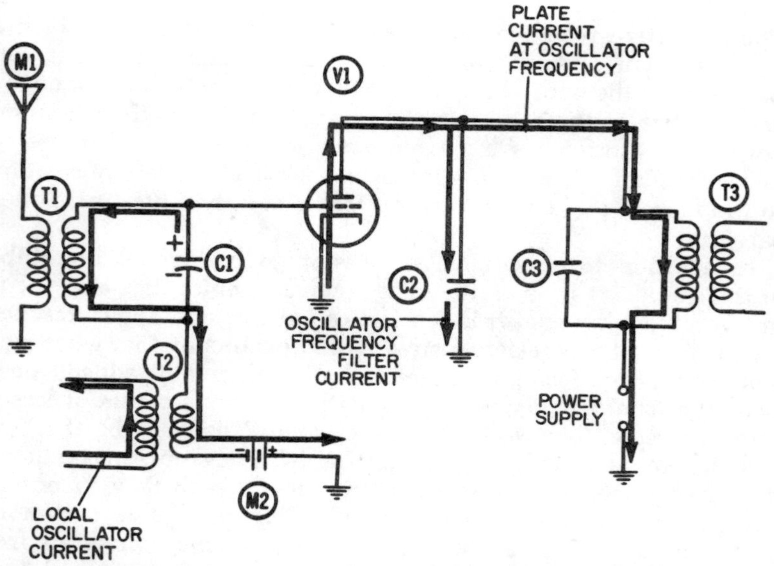

Fig. 1-2. Operation of the triode mixer circuit—positive half-cycle, no carrier current.

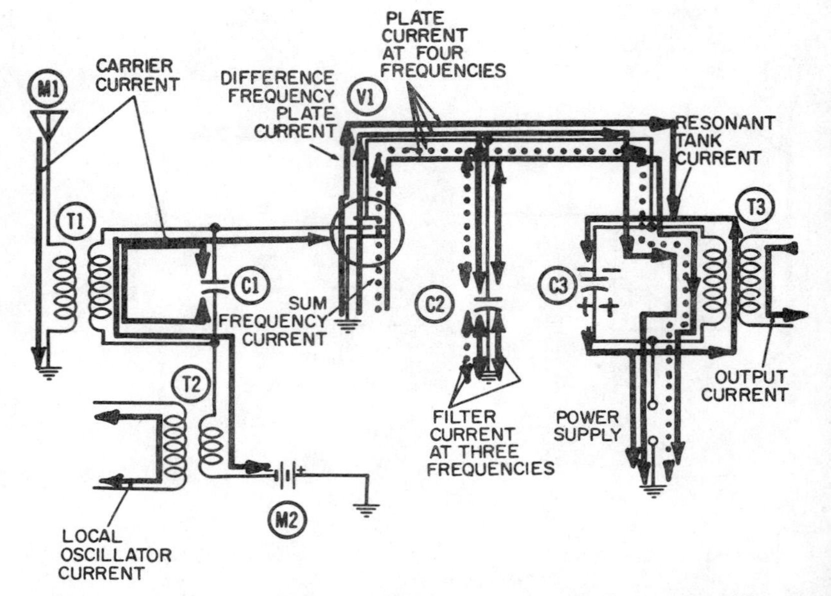

Fig. 1-3. Operation of the triode mixer circuit—carrier and oscillator currents both present.

The capacitive reactance is a measure of the opposition which a capacitor will offer to electron flow. From this formula it can be seen that the opposition varies inversely to both the frequency and the size of the capacitor. Thus, any capacitor offers less opposition to the flow of high-frequency currents than to those of lower frequency, and also a larger capacitor offers less opposition to the flow of any current at any frequency than does a smaller capacitor.

In addition to pulsations of current going through the tube at the carrier and the oscillator frequencies, pulsations also occur at two other principal frequencies, called the sum and difference frequencies. It can be demonstrated mathematically that when two sine waves of differing frequencies are combined, or added, these two additional frequencies, along with numerous others of lesser importance in this case, will be created. The sum of the two original frequencies is 2,500 kc. In Fig. 1-3 a new current is shown completing the filtering process through C2, and flowing out to the power supply essentially as DC. This current is shown in small dots, and is labeled as the *sum* frequency. Since its frequency is considerably higher than either of the original frequencies, we know from the foregoing reactance formula that the pulsations at this frequency will be filtered by C2 with even greater ease than are the original frequencies.

A fourth current has been shown a thin line in Fig. 1-3. This current is intended to represent the *difference* frequency. Using the 1,000-kc carrier frequency and the 1,500-kc oscillator frequency, as before, the difference frequency is 500 kc. Capacitor C3 and the primary of T3 are tuned to be resonant at this frequency; hence, the recurring pulsations will build up a sizable tank current at this frequency. Transformer action across T3 will build up a current-voltage combination at this same difference frequency in the grid circuit of the next amplifier stage.

All four currents are shown crossing the tube in Fig. 1-3. This signifies that the plate current carries pulsations at all four of the frequencies under consideration. The pulsations at the three highest frequencies are shown entering the upper plate of filter capacitor C2; filtering currents at these frequencies are shown moving between the lower plate and ground. However, the fourth current (a thin line), representing the important difference frequency, is shown flowing directly to the tuned plate circuit, where it arrives in the proper phase to reinforce the oscillation which exists there at that frequency. The plate tank is constructed and tuned to be resonant only at this difference frequency.

It is a characteristic of mixing or conversion circuits of this type that the modulation of the carrier signal, which represents the desired intelligence being received is conveyed or transplanted intact to the new lower-frequency current, called the difference-frequency. Thus, in later stages of amplification, even though a lower frequency is being amplified, the intelligence of modulation it carries is a faithful reproduction of the original modulation which will eventually be demodulated from the difference frequency. The output current shown in Fig. 1-3 is caused to flow by the transformer action of T3. It flows at the same frequency as the difference frequency and has the same modulation characteristic, or "envelope," as existed on the original signal current received from the station.

The output current-voltage combination from a mixer circuit is usually referred to as the intermediate frequency, or IF, because it is between the signal current and the audio range in the frequency spectrum.

Unlike most of the circuit diagrams in this text, Fig. 1-3 depicts *many* cycles of the three higher frequencies, while showing a single half-cycle of the lowest, or difference frequency, in the plate circuit resonant tank. It is for this reason that arrows are shown in both directions in the grid and filter circuits.

PENTAGRID MIXER CIRCUIT

Another effective means of mixing voltages of two different radio frequencies is shown in Figs. 1-4, 1-5, and 1-6. This arrangement involves the use of two vacuum tubes.

Identification of Components

The pentagrid mixer circuit contains the following components:
R1—Grid-drive and grid-return resistor.
R2—Cathode-biasing resistor.
R3—Screen-grid dropping resistor.
R5—Power-supply decoupling resistor.
C1—RF tank capacitor, working in conjunction with L2.
C2—Coupling capacitor between local oscillator and mixer.
C3—Screen-grid filter capacitor.
C4—Plate-tank capacitor.
C7—Power-supply decoupling capacitor.
L1—Primary winding of input RF transformer.
L2—Secondary winding of same transformer.
L3—Plate-tank inductor.
V1—Pentagrid mixer tube.

The local oscillator circuit contains the following components:
R4—Grid-leak biasing resistor.
C5—Oscillator tank capacitor.
C6—Grid-leak capacitor.
L4—Oscillator tickler coil.
L5—Oscillator tank coil.
V2—Triode oscillator tube.

Identification of Currents

There are a total of 11 significant currents during normal operation of the circuit in Figs. 1-4, 1-5, and 1-6. These currents are:

1. Input or "signal" radio-frequency current.
2. V1 plate current.
3. Plate-tank current.
4. V1 screen-grid current.
5. V1 screen-grid filter current.
6. Oscillator-circuit grid-tank current.
7. Grid-driving current for both tubes.
8. Grid-leakage current from oscillator tube.
9. V2 plate current.

10. Feedback current.
11. High-frequency decoupling currents.

Details of Operation

As is the case with all frequency-mixing, or *heterodyning*, circuits, one of the two voltages to be mixed is the signal voltage; and the other one is a voltage generated locally in the oscillator portion of the circuit. The purpose of the mixing process is to obtain a new and lower frequency, because lower frequencies are easier to handle and amplify.

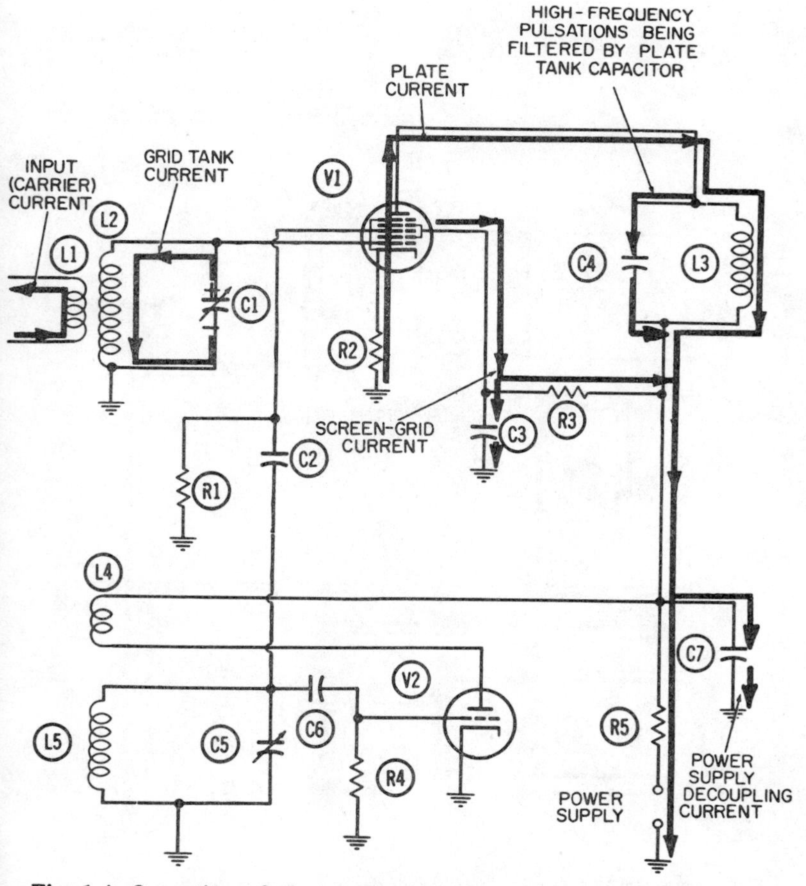

Fig. 1-4. Operation of the pentagrid mixer circuit—local oscillator not operating.

Local Oscillator Not Operating—Fig. 1-4 shows the relatively few currents which flow when the local oscillator is not operating and a signal current is being received. The signal current is received from the antenna and flows in inductor L1. In Fig. 1-4 it is shown flowing upward through L1 and inducing a companion current to flow downward in L2. Since the tank circuit composed of L2 and C1 is tuned to the particular frequency being received, the current induced in L2 will quickly set up an oscillation of electrons in the tuned circuit.

Fig. 1-4 depicts the half-cycle of oscillation when the upper plate of C1 is made positive. Since the first control grid of V1 is connected directly to the top of this tank, the voltage on this

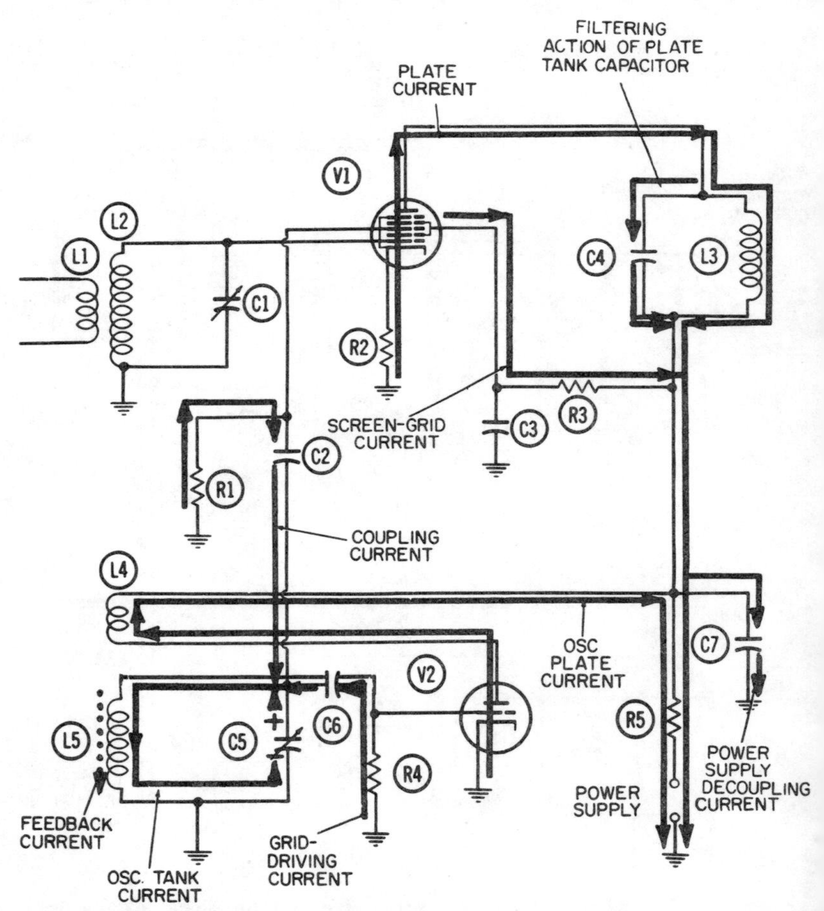

Fig. 1-5. Operation of the pentagrid mixer circuit—no input carrier signal.

grid will be identical with the voltage at the top of the tank at all times. During positive half-cycles, such as in Fig. 1-4, plate current through mixer tube V1 will be increased. During the next succeeding half-cycle, when the top of the tuned tank exhibits a negative voltage, this same plate current will be reduced.

Thus the voltage variations of the tuned tank, which are occurring at the frequency of the carrier signal being received, will impose pulsations on the plate current stream at this same frequency. The plate current flows through plate-tank inductor L3 and on through the power supply and back to ground. Inductor L3 and capacitor C4 form a resonant tank circuit at the difference frequency, or IF. Consequently, the pulsations in plate current which occur at the signal frequency are unable to excite the tank circuit into oscillation.

Since the screen grid which surrounds the second control grid within the tube is connected through a resistor to the positive power supply, it will attract and "capture" a large number of electrons from the plate-current stream going through the tube. These captured electrons become the screen-grid current (also shown a solid line in Fig. 1-4) which flows through R3 and rejoins the plate current as it enters the power supply.

No Signal Conditions—Fig. 1-5 shows the currents which flow when the local oscillator is operating, but no carrier signal is being received. The oscillator is a standard type, known as a tickler-coil oscillator, in which energy is fed back from the plate to the grid circuit in the appropriate phase to support the oscillation.

The oscillator works in the following manner. When positive voltage is first applied to the plate of V2, it draws an initial surge of plate-current electrons across the tube. This current must flow through L4 on its way to the power supply. As it surges, or accelerates, upward through L4, it induces, by the electromagnetic induction process which occurs between inductors, a companion current to surge *downward* through L5. This is the tickler action, and the induced, or companion, current becomes the feedback current which is shown in small dots . This feedback current moves in the downward direction and tends to remove electrons from the upper plate of tank capacitor C5 and deliver them to the lower plate. The preceeding makes the upper plate positive; the control grid of V2 is made positive at the same time by the action of the grid driving current which is attracted upward through grid resistor R4 by the positive tank voltage.

The initial surge of feedback current in L5 sets up the condition of resonance in the tank circuit, which is tuned to a new frequency that is known as the *local oscillator frequency*. Thus, on the next succeeding half-cycle, the top of the tank will be at a negative voltage, and the control grid will stop the flow of plate current through V2. The tube thus conducts intermittently rather than continuously, and the plate current is a special case of pulsating DC.

The feedback current generates and supports the oscillating tank current (shown a thin line). In addition to driving the control grid of V2, this tank current and its companion tank voltage also drive the second control grid of mixer tube V1. When the voltage on the top of C5 is positive (Fig. 1-5), it attracts electrons toward it from both directions, namely, upward through resistors R1 and R4. Thus, the control grids to which these resistors are attached will be made positive. On a negative half-cycle, when the voltage at the top of C5 is negative, both grid driving currents will be driven away from the tank, and consequently, will flow downward through the two resistors, making both control grids negative.

The action of the grid driving current in flowing up and down through R1 affords a means of imposing pulsations on the plate-current stream through tube V1; these pulsations will occur at the local oscillator frequency. Since the plate tank is tuned to a much lower frequency, these pulsations of plate current do not excite the plate tank into oscillation. All of the plate current flows through the inductor and on to the power supply, being joined at the entrance to the power supply by the screen-grid current previously described.

The pulsations in the plate current are filtered by plate-tank capacitor C4. This applies to both of the special cases just described. The local oscillator and the carrier signal frequencies are considerably higher than the resonant frequency of the plate-tank circuit. Resonance is defined as a condition wherein the capacitive reactance is equal to the inductive reactance. As the frequency is increased, the reactance of the capacitor decreases and the reactance of the inductor increases. This is stated by the two reactance formulas, which tell us that capacitive reactance is inversely proportional to frequency and that inductive reactance is directly proportional to frequency.

Figs. 1-4 and 1-5 show an individual pulsation of plate current flowing momentarily onto the upper plate of tank capacitor C4 and driving an equal number of electrons out of the lower plate toward the power supply. During the time period between two successive positive half-cycles, the actual electrons which made up the pulsation in the first place will be drawn off the upper plate of C4 and into the power supply by flowing through inductor L3. Also, during this period, the filter action completes itself, and electrons flow back into the lower plate of C4.

The filter current in Fig. 1-4 is shown solid, since it flows at the same frequency as the carrier signal being received. In Fig. 1-5 it is shown thin, since it flows at the frequency of the local oscillator.

Some arrangement must be provided for bypassing these pulsations around the power supply. The most common arrangement for performing this function is the simple resistor-capacitor combination known as a decoupling network. R5 and C7 form the decoupling network in this example. When the unfiltered pulsations reach the junction of these two components, the path through the resistor appears as a relatively high impedance, whereas C7

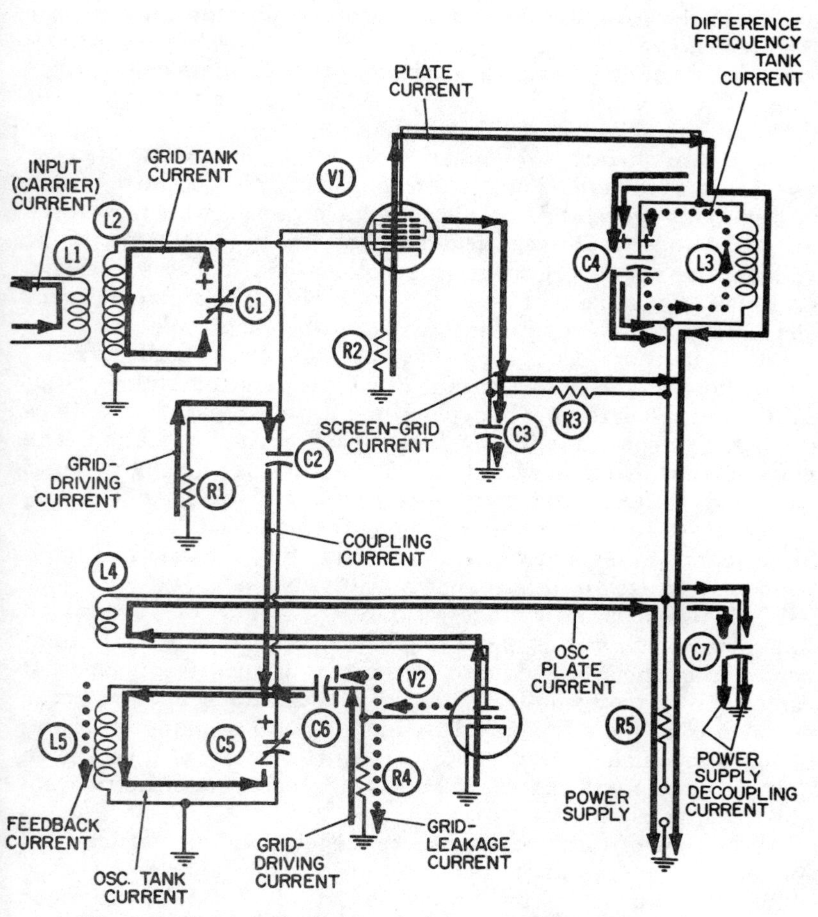

Fig. 1-6. Normal operation of the pentagrid mixer circuit.

has been chosen to have almost negligible reactance or impedance at all radio frequencies. Consequently, when each pulsation is filtered through the plate tank by C4, it is again filtered past the power supply by C7 (Fig. 1-4 and 1-5).

The triode oscillator tube operates under Class-C conditions, which means that the tube conducts less than 50% of each cycle. This is accomplished by using grid-leak biasing. The grid-leakage current has been shown in large dots in Fig. 1-6. During each positive half-cycle in the oscillator tank, the grid of V2 will be made positive, attracting some electrons from the plate-current stream going through the tube. These electrons, once they strike the control grid, cannot be re-emitted within the tube; they must exit from the tube and flow back to ground through grid resistor R4. Because of the high value of R4, the electrons cannot flow immediately to ground; instead, they will accumulate on the right hand plate of C6, building up a permanent biasing voltage until they can leak downward through R4.

The instantaneous grid voltage varies around this value of negative biasing voltage. Some electrons flow into C6 during each positive half-cycle, and the discharging process through R4 goes on continuously. During negative half-cycles of the oscillator voltage, the negative grid-driving voltage added to the negative grid-biasing voltage will be sufficient to cut off the plate current and hold it cut off for more than a half of each cycle.

Full Operation—Fig. 1-6 shows the sum total of all currents which flow in the mixer circuit during normal operation. Basically, all the currents which flow in the two separate examples discussed previously will also flow during normal operation. The plate current through V1 will be affected by the varying voltages on its two control grids, and the plate current will pulsate simultaneously at each of these two input frequencies. Thus, the two filter actions in C4 and C7 will occur side by side with and independently of each other, as shown in Fig. 1-6.

When two separate voltages of different frequencies are mixed in a circuit such as this one, the plate current pulsates at a number of frequencies in addition to the two applied frequencies. It can be shown mathematically that these pulsations will also occur at the *sum* of the two applied frequencies and at many multiples of this sum, such as twice, three times, etc. It can also be shown that pulsations will occur at the *difference* between the two applied frequencies.

All but one of these additional frequencies will be considerably higher than the resonant frequency of the plate-tank circuit. The diffrence frequency is the lowest one at which pulsations occur in the plate-current stream. Hence, if the plate tank is tuned to resonate at the difference frequency, all the other frequencies will be filtered to ground through C4 and C7, in the same manner that the two primary frequencies are filtered.

An oscillation of electrons (shown in large dots in Fig. 1-6) will be set up in the plate-tank circuit at the difference frequency. Each cycle of the oscillation will be reinforced by a single pulsation of plate current. The amplifier stages which follow the mixer tube are tuned to this same frequency so that it is the only one which will be amplified.

In AM broadcast reception a difference frequency or intermediate frequency of 455 kc per second is fairly standard. In higher-frequency reception, such as FM, television, and radar, intermediate frequencies ranging from 1 or 2 mc up to 50 or more are not uncommon.

This feature is important when you consider that most receivers are designed to receive any one of several transmitters which are radiating over a band of frequencies, rather than on a single frequency. Thus, the input tank circuit, consisting of L2 and C1, has to be tunable; at least one of the components must be variable. The normal practice is to use a variable capacitor, as indicated in Fig. 1-4, 1-5, and 1-6.

In the oscillator tuned circuit, we find another variable capacitor, C5. C5 and C1 are usually ganged and varied simultaneously by a single tuning control. All circuit elements are designed so that no matter what frequency of carrier is being received, the oscillator tank-circuit frequency will always differ from it by the same amount, such as 455 kc. When this condition is met, the plate-tank circuit and all subsequent amplifier-tank circuits can be manufactured to resonate at this single frequency without the necessity of any tuning adjustments on the operator's part. This feature allows the following circuits to be designed for maximum gain, sensitivity, etc.

PENTAGRID CONVERTER CIRCUIT

Figs. 1-7 and 1-8 show two alternate moments in the operation of the pentagrid converter circuit of a radio. This circuit performs several complex and important functions, as follows:

1. It receives the incoming radio-frequency signal from the antenna.
2. It generates an entirely separate oscillation frequency.
3. It mixes or combines these two frequencies into a third frequency, known as the intermediate frequency (IF). This combining process is referred to as *frequency conversion.*

4. It amplifies or increases the strength of this new frequency, to a level much higher than either of the original frequencies.

Identification of Components

The various components which make up this circuit, along with their functional titles, are as follows:

R1—Grid-leak biasing and driving resistor.
C1—Automatic volume-control (AVC) storage capacitor.
C2—Variable capacitor, controlled by the tuning dial.
C3—Trimmer capacitor.
C4—Variable capacitor in oscillator circuit, controlled by the tuning dial.
C5—Fixed oscillator tank capacitor.
C6—Coupling and isolating capacitor.
L1—Radio-frequency transformer, also called the antenna transformer.
L2—Oscillator inductor, which also serves as an auto-transformer.
T1—IF transformer.
V1—Five-grid vacuum tube used as a frequency converter.

Identification of Electron Currents

In order to understand everything that is happening inside this circuit you must be able to visualize each electron current at work in it. These currents are as follows:

1. Antenna current.
2. RF tank current.
3. Driving current for second control grid.
4. Oscillator tank current.
5. Oscillator feedback current.
6. Driving current for first control grid.
7. Tube plate current.
8. Grid-leakage current.
9. Screen-grid current.
10. IF plate-tank current.
11. Grid-tank current for next stage.
12. AVC current.

Details of Operation

Transformer Action—The signal or antenna current, a solid line in Fig. 1-7, is caused to flow up and down through the primary winding of antenna transformer L1 by the so-called radio waves transmitted by a radio station. The frequency with which it changes direction is, of course, equal to the frequency of the signal being received. For example, if your radio is tuned to a station that is broadcasting on a frequency of 1,000 kc, the antenna current makes a million complete journeys up and down through the primary winding of L1 *every second.*

Because of the transformer action between the primary and secondary windings of L1, the RF tank current (also shown a solid line) is driven by this antenna current. Inductance is a sort of electrical inertia and can be compared with the inertia of mechanical devices. For example, it requires extra effort to get any large stationary object, such as an automobile, moving. However, once it is moving, extra effort is required to bring it to a stop.

Both of the foregoing effects result from mechanical inertia. In electrical inertia, which is called inductance, it is not the mass of the electron that concerns us, but the electrical charge carried by each electron. The principal characteristic of any inductance is that it tries to keep the amount of current flowing through it at a constant value. This property leads us to the transformer action which occurs between the primary and secondary windings of L1. As the amount of antenna current shown in Fig. 1-7 increases or builds up in the upward direction, the amount of RF tank current increases in the downward direction in the secondary winding.

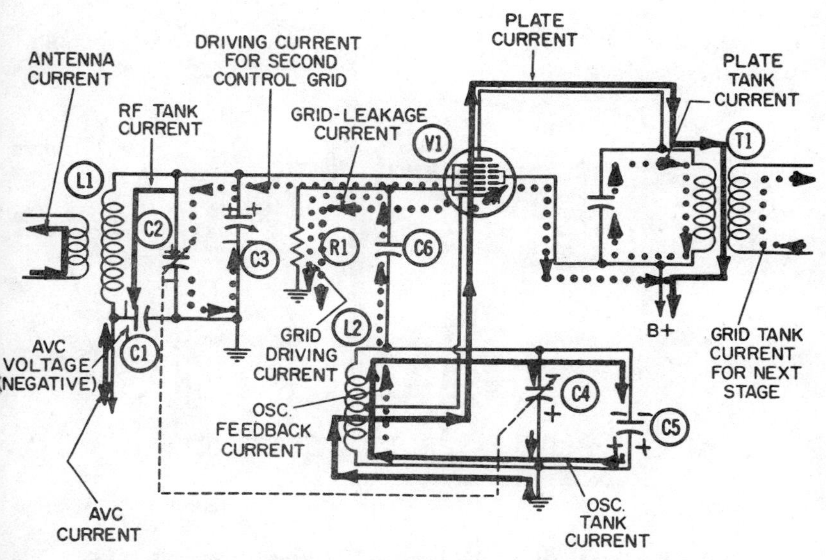

Fig. 1-7. Operation of the pentagrid converter circuit—first control grid negative, second control grid positive.

Similarly, in Fig. 1-8, when the antenna current increases in the downward direction in the primary winding, the RF tank current in the secondary winding increases in the upward direction. An inductance or a transformer responds to *changes* in the amount of driving current, rather than to the intrinsic amount itself. Since the antenna current is constantly changing direction and amount, making a million complete round trips up and down through the primary winding each second, it drives the tank current down and up through the secondary winding at the same frequency.

The RF tank circuit consists of capacitors C2 and C3 in parallel with the secondary winding of transformer L1. In Fig. 1-7 we see a condition where a positive voltage, indicated by plus signs, exists on the upper plates of these two capacitors. This positive voltage exists there because during the previous half cycle, the oscillating electrons have all migrated downward through the second winding, as indicated by the arrows.

In Fig. 1-8 a negative voltage exists on the upper plates of these capacitors. This negative voltage is indicated by minus signs; it results from the fact that during the preceding half-cycle, the tank current electrons migrated upward through the secondary winding, thus delivering a surplus of electrons at the top of the tank. The tank current oscillates continuously between these

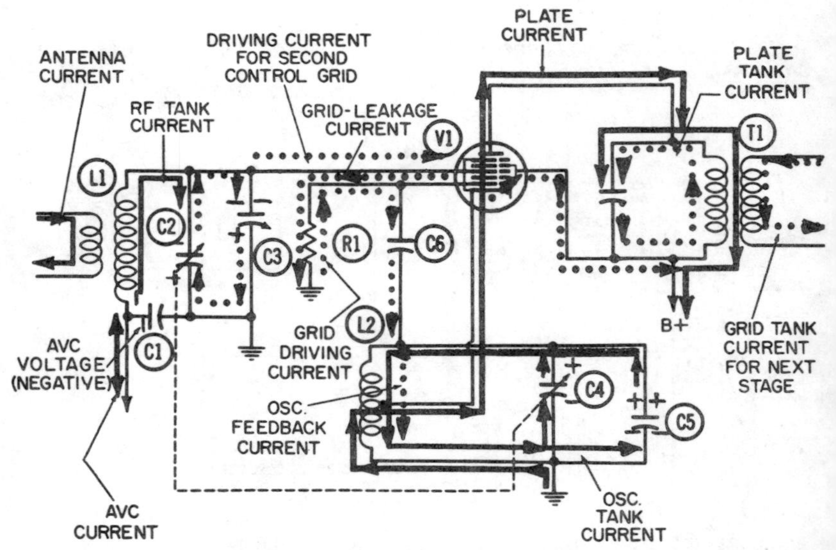

Fig. 1-8. Operation of the pentagrid connector circuit—first control grid positive, second control grid negative.

capacitors and the secondary winding. It is driven or supported in its movements by the antenna current in the primary winding. This tank current oscillation results in the top of the tank exhibiting a voltage that fluctuates from positive to negative at the same frequency as the antenna or signal current. Since the top of the tank is connected directly to the second control grid of the tube V1, this alternating voltage also exists at that control grid and can be used to control, or regulate, the flow of plate-current electrons through V1. The control grid acts like a throttle valve on the electron stream. When the control grid is positive, as in Fig. 1-7, more electrons are allowed to pass through the tube.

When the control grid is negative, as depicted in Fig. 1-8, the quantity of plate-current electrons flowing through the tube will be reduced accordingly.

Plate Current—The complete path of plate current, which has been shown solid in both Figs. 1-7 and 1-8, starts at the ground connection below the three components, L2, C4, and C5. The plate current flows through the lower half of inductor L2, then out through the center tap and to the cathode of the tube. The electrons which make up the plate current are then *emitted* into the vacuum of the tube from the heated cathode.

The plate current passes between the wires of all five grids of the tube and strikes the plate of the tube, which absorbs them. They are drawn downward through the primary winding of IF transformer T1, and on to the power supply. The high positive voltage of the power supply is the attractive force which draws the electrons of the plate current (shown a solid line) along the entire path.

Grid-Leakage Current—Some electrons which are emitted by the cathode do not reach the plate of the tube, but strike various grid wires and leave the tube as grid current. The grid leakage current (shown in large dots) which flows downward through resistor R1 is one such grid current. The complete path of this grid-leakage current takes it downward through R1 to ground and then upward through the lower half of L2 to the cathode of the tube, where it is again emitted into the tube.

In continuously flowing downward through R1, this electron current developes a voltage across R1 which is known as a *grid-leak bias voltage*. This voltage is more negative at the top of R1 than at the bottom. This is confirmed by the fact that electrons *always* tend to flow away from an area of more negative voltage toward an area of less negative voltage. This grid-leak voltage provides a stable and fixed negative voltage at the first control grid.

Oscillating Tank Voltage—The grid leak bias voltage is not the only one on the first grid. There is an oscillation of electrons occurring in the tank circuit which consists of inductor L2 in parallel with capacitors C4 and C5. The oscillating current has been shown a thin line . Since the top of this tank circuit is connected to the first control grid by means of C6, the oscillating tank voltage is said to be coupled to this grid. In Fig. 1-7 when the voltage at the top of this tank circuit is negative, electrons are driven upward into the lower plate of C6; this action, in turn, drives other electrons away from the upper plate of C6 and downward through R1. The downward flow of electron current through R1 tells us that the top of the resistor is more negative than the bottom during the particular instant represented by the diagram.

An opposite set of conditions is depicted in Fig. 1-8. The tank current (shown a thin line) has reversed itself and flows downward through L2 to the bottom plates of C4 and C5, making them negative with respect to the upper plates. An electron deficiency has been created on the upper plates so that they now exhibit a positive voltage (indicated by plus signs). This positive voltage draws electrons downward from the lower plate of C6, and in turn, draws other electrons upward through biasing and driving resistor R1 and downward to the upper plate of C6. The upward flow of electron current through R1 tells us that the top of this resistor is more positive than the bottom during this particular half-cycle of oscillation.

The current (shown in small dots) which flows up and down through R1 is labeled as the grid driving current because it "drives" the first control grid by developing an alternating voltage at that grid. This current flows simultaneously with the grid-leakage current, but is completely independent of it. Each current develops its own voltage at the grid; the total voltage at this grid at any instant of time is the algebraic sum of these two separate voltages. The resulting voltage at the first control grid will fluctuate at the same frequency as the oscillator tank current. The result will be that the electron stream flowing through V1 will be increased and decreased at this same frequency. During the half-cycle represented by Fig. 1-7 when the control grid is made most negative, the electron stream, which is in reality the plate current through the tube, will be "throttled back" to a minimum amount.

During the half-cycles represented by Fig. 1-8 when the first control grid is made least negative, this plate current stream will be "turned up" to a maximum value. Hence, the plate current through V1 fluctuates at the oscillator frequency.

Sustaining Oscillations—The tank circuit, consisting of L2, C4, and C5, along with the cathode and first control grid of the tube, make up the most essential parts of a conventional Hartley oscillator. No oscillation of electrons can continue to exist unless there is some form of feedback between the output and the input circuits to replenish the inevitable losses. In Figs. 1-7 and 1-8 feedback is obtained by means of the autotransformer action which occurs between the lower portion of inductor L2 and the entire inductor.

Oscillator feedback current has been shown in small dots in Figs. 1-7 and 1-8. In Fig. 1-8, it is shown in phase with the oscillator tank current (thin line) flowing downward through entire inductor L2. Because the two currents are in phase, the feedback current reinforces or strengthens the oscillator tank current.

Autotransformer Action—It is important to understand what causes feedback current to flow in the first place. In Fig. 1-8 when the top of the tank is positive, the control grid reaches its least negative (or most positive) voltage. This permits a surge of plate-current electrons to flow from the cathode into the tube. These electrons must first be drawn upward from ground and through the lower portion of L2. Any inductor will always oppose any increase or decrease in the amount of current which is flowing through it. Consequently, when the plate current flowing upward through the lower portion of L2 is increased, a separate electron current will be generated in the entire inductor, increasing in the downward direction. It is by autotransformer action such as this that an inductor tries to keep the total current through it from changing. This new current is the feedback current. In Fig. 1-7 when the top of the oscillator tank is negative, the voltage at the first control grid has its most negative value. This restricts the flow of plate current through the tube, and of course, it reduces the upward flow of current through the lower portion of L2 by the same amount. The inductor responds in the conventional and expected manner and this time generates a current which flows upward through the entire inductor at an increasing rate. In this way the inductor succeeds (at least momentarily) in keeping the total current from decreasing. This newly generated current again acts as the oscillator feedback current; since it is flowing in the same direction as the tank current, it reinforces it. Thus, the tank current is strengthened or replenished during each half-cycle of operation by autotransformer action.

The tank circuit is considered to be the input circuit, because it drives the control grid. The lower portion of L2 is considered part of the output circuit, because plate current flows through it. Thus, the lower portion of L2 is a part of both the input and output circuits. The Hartley oscillator, of which this circuit is an example, satisfies the general requirement that there be some feedback of energy from the output circuit to the input circuit for an oscillation to sustain itself.

Obtaining the Intermediate Frequency—We have seen that two different control grids act as individual throttle valves on the plate-current electron stream which passes through the tube. The oscillations of electrons in the two tank circuits driving these grids are occurring at different frequencies. Both circuits are tuned by turning the familiar tuning dial on the front of the radio. The circuit components are chosen so that the oscillator tank which drives grid number 1 (the lower grid) will always oscillate at a frequency which is 455 kc higher (faster) than that of the antenna current being received. This antenna current, of course, supports the RF tank current which drives the second control grid (grid number 3).

The plate current through the tube will obviously fluctuate at each one of the two basic frequencies; however, it will also fluctuate at many other frequencies which *depend* on these frequencies. Most important are the sum of the two basic frequencies and the difference between them (455 kc). In addition, it will fluctuate at a frequency which is twice their sum, twice their difference, three times their sum or difference, etc.

We are interested in only one of these many new frequencies—the difference frequency of 455 kc—since we have already chosen this as the intermediate frequency, or IF, of our radio. The advantage in using a single or fixed IF is that each of the tuned circuits (normally 4) in the IF stage can be tuned to this one frequency and thereafter will require little or no attention.

Since the plate-tank circuit is tuned to the IF frequency, the pulsations in plate current which occurs 455,000 times each second will very quickly excite an oscillation in the tank. This oscillating plate-tank current has been shown in large dots. Fig. 1-8 depicts an instantaneous set of conditions when the electrons of the plate-tank current are moving upward through the primary winding of T1, thereby making the upper plate of the tank capacitor negative with respect to the lower plate. At the same instant, a pulsation of plate-current electrons is arriving from the tube and making this upper capacitor plate still more negative with respect to the lower plate. In this manner the pulsations of plate current *reinforce* the oscillation in the tank current. This sequence of events repeats itself 455,000 times every second when the radio is tuned to a station.

The plate-tank current supports another oscillation of electrons in the grid-tank circuit of the next stage. This is accomplished by transformer action between the primary and secondary windings of T1 and will be discussed more fully in a later chapter.

Screen-Grid Current—The wires of grids number 2 and 4 constitute a screen grid, which "screens" or isolates the second control grid from the first control grid and from the plate circuit of the tube. The screen grid is connected to the high positive voltage of the power supply; therefore it also attracts electrons from the area around the cathode. Most of these electrons pass through the screen-grid wires and eventually strike the plate. However, some of the electrons passing through the tube will actually strike the screen-grid wires and will exit from the tube as screen-grid current, or screen current. This current has been shown in large dots up to the point where it rejoins the plate current. Then both currents are shown as solid as they proceed along the B+ line of the radio to the power supply.

AVC Current and Voltage—The final current which flows in the circuit of Figs. 1-7 and 1-8 is the one associated with the function known as "Automatic Volume Control" (AVC). This current and the resulting AVC voltage which exists on the left plate of C1 have been shown a thin line. The AVC voltage acts as a permanent biasing voltage on the second control grid of V1. The means by which it is obtained will be discussed more fully in Chapter 2.

The AVC voltage varies from a low to a high value, but it is always negative. When the strength of the radio signal being received is low, or weak, the antenna current will be weak, and the AVC voltage stored on the lower plate of capacitor C1 will be a low negative voltage.

When the received signal strength is high, or strong, the antenna current will be strong, and the AVC voltage will be a high negative voltage.

A weak AVC voltage will *increase* the gain of the converter stage. A strong AVC voltage will *reduce* the gain of this stage.

BEAT-FREQUENCY OSCILLATOR

The beat-frequency oscillator, or BFO, is a special example of the heterodyning process by which two voltages of different frequencies are heterodyned to produce a third voltage of a much lower frequency. In the examples previously studied the input carrier signal was amplitude-modulated and that the resultant difference frequency also carried the same intelligence modulation on it.

A beat-frequency oscillator is required when the input carrier signal has been keyed or coded with dots and dashes. This is a special type of modulation known as interrupted continuous-wave (ICW). The signal can be detected in the ordinary sense by a simple diode detector, such as V3 in Fig. 1-9. However, after the detection process is completed, there would be no audio voltages by which the listener could tell that demodulation or detection had occurred. In such a case, the detection process will have been wasted.

When an RF or IF signal which has been keyed or interrupted to form the dots and dashes of the well-known Morse code or similar intelligence reaches a detector circuit, the diode detector will conduct electrons at a constant rate during the periods when dots and dashes occur. During the periods when these pulses are not occurring, it will not conduct. Electrons flowing at a constant rate downward through load resistor R4 will generate a constant or DC voltage for as long as they flow (during each dot and dash, for example) but this flow will not result in an audio voltage in the headphones. Instead, a single surge of current will flow into the headphones (downward) at the start of each pulse, and another surge of current will flow upward through the headphones at the end of each pulse. The beginning and end of each dot and dash will cause a single cycle of "noise" to be heard in the headphones.

Identification of Components

The following circuit components perform the indicated functions in this simplified BFO circuit:

Mixer Components

R1—Cathode biasing resistor for V1.
R2—Screen-grid dropping resistor.
C1—Input-tank capacitor (variable).
C2—Screen-grid filter capacitor.
C3—Plate-tank capacitor.
L1—Input-tank inductor.
L2—Plate-tank and coupling inductor.

Oscillator Components

R3—Grid driving and biasing resistor.
C4—Oscillator tank capacitor (variable).
C5—Grid coupling and biasing capacitor.
C6—Oscillator coupling capacitor.
C7—Plate-filter capacitor.
L3—Oscillator tank inductor.
L4—Radio-frequency choke.
V2—Triode oscillator tube.

Detector Components

R4—Variable resistor (volume control).

R5—Cathode resistor.
C8—RF filter capacitor.
C9—Additional RF filter capacitor.
C10—Audio coupling capacitor.
L5—Output inductor.
V3—Diode detector.
M1—Output meter for indicating zero beat.
M2—Headphones for listening to code.

Identification of Currents

There are at least 15 different electron currents working in this BFO circuit; these include:

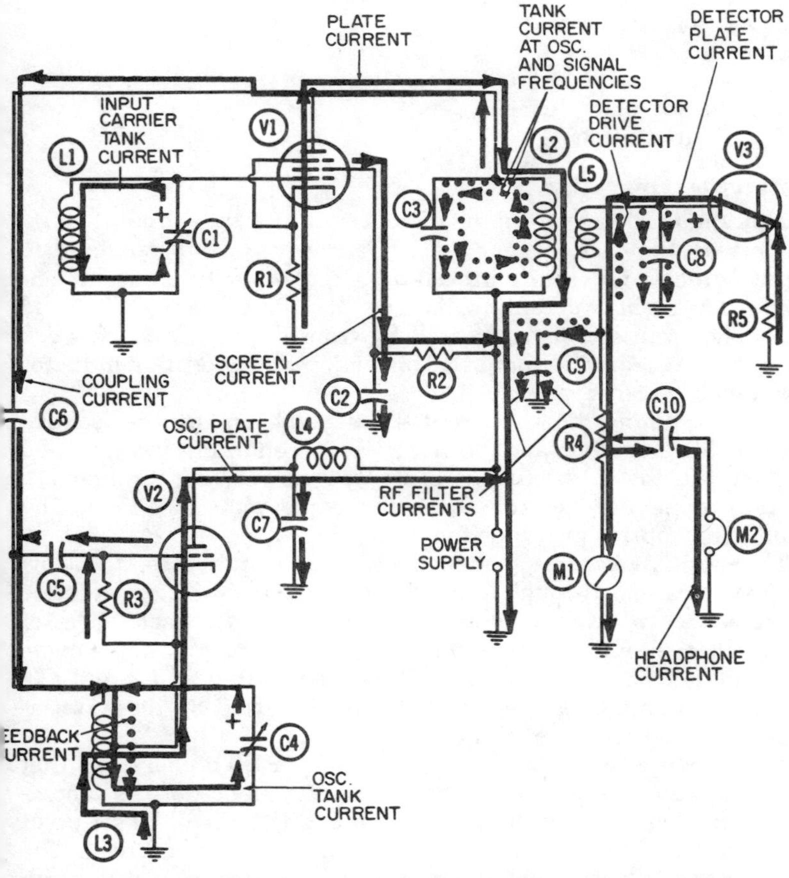

Fig. 1-9. Normal operation of the beat-frequency oscillator circuit.

Mixer Currents

1. Input signal tank current.
2. Pentode plate current.
3. Pentode screen current.
4. Screen-grid filter current.
5. Plate-tank current at signal frequency.
6. Plate-tank current at oscillator frequency.

Oscillator Currents

7. Oscillator tank current.
8. Oscillator grid driving current.
9. Coupling current to plate tank.
10. Triode plate current.
11. Feedback current in oscillator tank.
12. Plate-filter current.

Detector Currents

13. Detector drive current.
14. Detector plate current.
15. RF filter currents.

Details of Operation

Any qualitative analysis of the operation of this circuit should begin with the carrier signal which is received at or delivered to input inductor L1 from a preceding amplifier or the receiver antenna. The signal current (solid line) oscillates between L1 and C1. (The oscillation is supported by transformer action between L1 and a preceding inductor.) This action and the prior inductor are not shown in Fig. 1-9.

The oscillating tank current alternately makes the control grid of V1 negative and positive. With conditions as depicted in Fig. 1-9, the tank current is flowing downward through L1, thereby removing electrons from the upper plate of C1, making it and the control grid, positive.

The positive control grid encourages, or increases, the flow of plate current through the tube so that a *pulsation* of plate current occurs, one such pulsation occurring for each cycle of the oscillation. Each of these pulsations will support a single cycle of oscillation in the plate tank circuit, consisting of L2 and C3. The oscillation which is generated and supported by these pulsations has been shown in large dots.

The values of plate-tank components have been chosen so that they will resonate at approximately the frequency of the incoming carrier signal. Therefore, a natural oscillation will be set up at this frequency whenever a carrier signal is being received. The plate current, which has been a solid line, continues through L2 and enters the positive terminal of the power supply. It is joined at the junction of L2 and R2 by the screen-grid current, also a solid line, which exits from the tube at the screen grid and flows through screen dropping resistor R2. Any pulsations, or surges, which may characterize this screen current as it comes from the tube, will be filtered or bypassed harmlessly to ground through C2. Fig. 1-9 shows a half-cycle of this filter current flowing downward from the lower plate of C2, because a pulsation of electrons has just come from the screen grid. On the next succeeding half-cycle, when no pulsation is occurring, this filter current will flow upward onto the lower plate of C2.

The oscillator circuit which is constructed around triode V2 is a conventional Hartley circuit. The tank circuit, composed of L3 and C4, is tuned by variable capacitor C4. Component values are chosen so that they will resonate at a frequency very close to the frequency of the incoming carrier signal. In this respect, it differs from the circuits previously described, whose purpose was to produce a new frequency (IF) which still falls in what is known as the radio-frequency range. In the beat-frequency oscillator the two input frequencies—the carrier and the local oscillator—produce a difference frequency which is in the low audio range, from a few cycles to a few thousand cycles per second. Furthermore, C4 may be varied so that the difference frequency may be varied over a small range.

The oscillator tank current (thin line) moves continuously between L3 and C4. Fig. 1-9 depicts a moment when the electrons have all moved to the lower plate of the capacitor, thereby making the upper end of the tank positive. This action draws electron current (also shown as thin) upward through the grid resistor R3. This makes the top of the resistor and the control grid of V2 positive, and a pulsation of plate current (solid line) will be released to flow through V2. It originates at ground below L3, flows through the lower portion of L3 then through the tube and the radio-frequency choke L5 to the power supply. This current flows intermittently rather than continuously. During that portion of a cycle when it is *increasing* in amount, it induces a feedback current in L3 which is also increasing, but in the opposite direction. This feedback current (dotted line) increases in the downward direction through L3 when the plate current through the lower part is increasing in the upward direction.

Since the tank current is moving downward through L3 at this same time, the feedback current and the tank current are in the appropriate phase with each other so that the tank circuit oscillation will be sustained by the feedback action.

In addition to driving the control grid of V2, this oscillating tank voltage is coupled via C6 to the plate-tank circuit of tube

V1. A single half-cycle of this coupling action is depicted in Fig. 1-9. Since the top of the oscillator tank is positive at this instant, electrons are drawn toward this voltage from the lower plate of C6. This action draws an equal number of electrons onto the upper plate and away from the plate tank of V1. Since the plate tank and the oscillator tank are tuned to almost the same frequency, this coupling action will excite a second oscillation in the plate tank circuit at the oscillator frequency. This oscillating current has been indicated in dotted green.

The coupling action between the oscillator tank and the plate tank could be performed by a straight wire instead of with coupling capacitor C6. The real function of the capacitor is to isolate or block the positive voltage of the power supply from having direct access to ground through L3 and L2.

As we stated in the previous example of a mixer circuit, when two frequencies are mixed, a number of new frequencies are created—namely, the sum and difference, twice the sum and twice the difference, etc. All but one of these are unwanted frequencies, and arrangements must be provided for filtering them all to ground. All of these new frequencies will be inductively coupled between L2 and L5.

The only one of these many frequencies which is wanted is the difference frequency. All of the other frequencies, including the two original frequencies, have an important distinguishing characteristic in common, namely, that they are radio-frequencies and may therefore be easily separated from the difference frequency by simple capacitive filtering. C8 and C9, situated on either side of output inductor L5, provide this filtering.

When the plate of V3 is made positive with respect to its cathode, the diode will conduct electrons. This diode current is shown a solid line. It flows upward from ground through R5, through the diode, downward through L5 and R4, and finally through indicating meter M1.

It is desired that this diode plate current flow *only* in response to the low-frequency "beat note"—which is the difference between the two input frequencies. Neither of these input frequencies, nor any of the other resulting frequencies, will cause the diode to conduct. C8 and C9 both provide a very low impedance path for currents at these frequencies. Fig. 1-9 shows the two original frequencies being filtered through the capacitors to ground. R5 has a low value, but nonetheless any current which flows through the diode must also flow through R5. Consequently, R5 constitutes a load to anyone of these higher frequencies which may attempt to draw current through the diode. Since C8 has lower impedance than R5 to these higher frequencies, each of the higher-frequencies will draw current upward into C8 during its own positive half-cycle, and discharge them downward from C8 during its negative half-cycle.

Capacitor C4 can be varied to adjust the oscillator frequency within small limits. This action varies the difference frequency which the listener hears in the headphones to suit his tastes, or even to permit the best readability under various conditions of static, noise, etc. The difference frequency can be reduced to zero by exactly matching the oscillator and carrier signals to equal each other. When this is done, the diode conducts no plate current, and milliammeter M1 indicates no current flow through it. This condition is known as "zero beat," and is one of the earliest and still one of the best methods for precisely calibrating unknown frequencies against a known or standard frequency. When the two frequencies drift apart by even a few cycles per second, the resulting low beat note can be heard by the operator; this enables him to readjust the oscillator frequency as necessary to attain the condition of zero beat.

The diode plate current flows only when dots or dashes are being received. At those times it is a pulsating DC rather than a pure DC. The pulsations occur at the so-called difference frequency. Each individual pulsation drives a surge of electron current onto the left-hand plate of capacitor C10, and this action drives an equal amount of electrons downward through the headphone circuit. Between each two successive pulsations, the electron current flows back out of the left hand plate of C10 and downward through R4 and the output meter M1. This action draws electron current upwards through the headphone circuit.

In communications receivers it is the usual practice to beat the oscillator frequency against a fixed carrier frequency, rather than against a variable frequency. The commonest example of a fixed carrier frequency is the intermediate frequency or IF which is generated by the mixer circuit. As an example, if an IF of 450 kc were in use, the oscillator might prove to be adjustable over a 10-kc range—5 kc on either side of the basic, IF, in other words from 445 to 455 kc. When the oscillator oscillates exactly at 450 kc conditions for producing zero beat are said to exist.

Chapter 2

SIGNAL DEMODULATION AND AUTOMATIC CONTROL OF VOLUME

After the IF signal obtained at the mixer output is amplified by the IF amplifier, it must be changed to an audio signal which corresponds to the audio that produced the original modulation at the transmitter. This process is called demodulation or detection. (Actually, detection is a misnomer which stems from the early days of radio when circuits were devised to "detect" or "discover" the presence of a signal from a distant station. Today the term detector is widely used and can be considered as synonymous with demodulator. The term "second detector" is also used when referring to the demodulator stage.)

It is desirable to provide an automatic means of changing the amount of amplification of the signal to compensate for varying signal strengths because of differing propagation characteristics. The AVC voltage and current were discussed in conjunction with the mixer circuit of Figs. 1-7 and 1-8.

The weak audio signal obtained at the demodulator stage requires additional amplification before it is usable. In modern radio receivers all of the foregoing functions—demodulation, automatic control of volume (AVC), and audio amplification—are performed by a single tube.

DETECTOR, AVC, AND AUDIO AMPLIFIER

The 12AV6 tube (V3) in Figs. 2-1 and 2-2 is capable of performing like two separate and distinct tubes. It is a dual diode-triode tube; independent electron currents can and do flow from the cathode to the diode plates and to the triode plate. In this particular circuit the two diode plates are shown connected together so that they function as if they were a single plate.

Identification of Components

The components which make up Figs. 2-1 and 2-2, and the portion of the circuit which they function in are as follows:

Detector Circuit
R4—Detector load resistor.
R5—Detector load resistor and volume control.
C8—IF filter capacitor.
C9—IF filter capacitor.
T2—(Secondary) Final IF transformer and tank capacitor.
V3—(Diode portion) Detector tube.

AVC Circuit
R3—AVC resistor.
C1—AVC storage capacitor.

Audio Circuit
R6—Grid driving resistor.
R7—Triode plate-load resistor.
C10—Audio coupling capacitor.
C11—IF filter capacitor.
C12—Coupling capacitor to next stage.
V3—(Triode portion) Audio amplifier tube.

Identification of Currents

The following separate and distinct electron currents are at work in these three circuits:

Detector Circuit
1. Final IF tank current.
2. Rectified current which flows through the diode.
3. IF filter current.

AVC Circuit
4. AVC current.

Amplifier Circuit
5. Audio grid driving current.
6. Triode plate current.
7. IF filter current.

Detector Operation

The final IF tank current (dotted line) oscillates continuously between the second winding of T2 and the unnumbered capacitor in parallel with it. The plate-tank current of the preceding IF stage flows up and down through the primary winding of T2 and provides the necessary energy to sustain the current flowing through the secondary winding.

The final tank current alternately makes the top of the tank vary between negative and positive voltage values. Fig. 2-1 depicts an instant when the tank current is flowing upward through the secondary winding and delivering electrons to the upper plate of the tank capacitor, thus making it negative. Fig. 2-2 depicts conditions a half-cycle later when the tank current is flowing downward through the secondary winding, taking electrons away from the upper plate of the capacitor and creating a deficiency of electrons, or, more simply, a positive voltage at the upper plate of the tank capacitor.

These voltage changes (negative to positive and back to negative again) occur 455,000 times each second—the intermediate frequency of your radio. On the half-cycle depicted in Fig. 2-2 the voltage at the top of the tank makes the two diode plates of V3 positive; therefore they attract electrons from the cathode within the tube. These electrons become the so-called diode current, or rectified current, of the circuit. This current is shown a solid line; it flows through the diode portion of the tube *only* when the diode plates are more positive than the cathode. The complete path of these electrons takes them downward through the secondary winding of T2, and on through resistors R4 and R5 to the common ground connection, from where they have a ready return access to the cathode of the tube. This diode current is a pulsating DC.

The electrons which make up the diode current are prevented from flowing immediately downward through R4 and R5; instead they accumulate on the upper plate of C9, building up a small negative voltage there. This negative voltage is indicated by minus signs in both Figs. 2-1 and 2-2. Even though diode current is not flowing in Fig. 2-1, this negative voltage persists on capacitor C9 and continues to discharge its electrons downward through the resistors. Consequently, although electron current flows intermittently through the diode with 455,000 pulsations occurring each second, it flows continuously downward through the resistors.

IF Filter Current—The pulsations in the diode current which are occurring at the intermediate frequency are filtered out be-

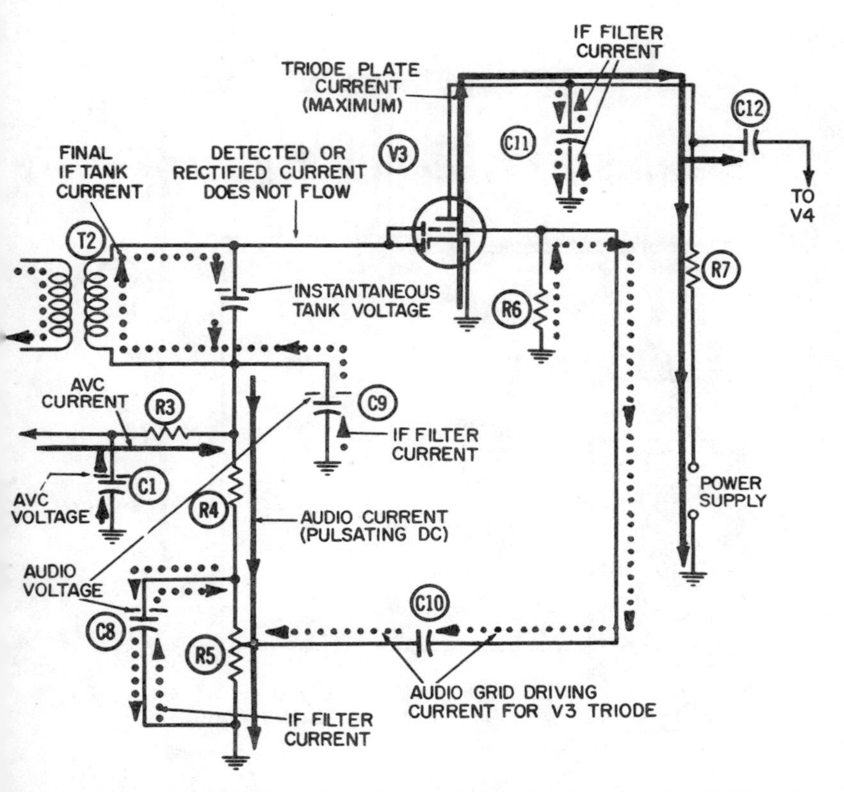

Fig. 2-1. Detector, AVC, and audio-amplifier circuit—negative half-cycle of IF tank voltage, positive half-cycle of audio voltage.

tween C9 and the ground. Fig. 2-2 shows a single half-cycle of this current being driven into ground, as a pulsation of diode current flows onto the upper plate of C9. Fig. 2-1 shows the next succeeding half-cycle of filter current flowing back onto the lower plate of C9.

The Audio Voltage—The negative voltage which accumulates on the upper plate of C9 is important, since it marks the first appearance of the audio voltage—in other words, the intelligence or the message, you are trying to receive. This voltage can be compared to a pool of negative electrons, with the depth of the pool representing the amount of negative voltage. The depth of the pool does not change significantly during a single cycle of the IF current even though electrons are discharging or "draining" continuously downward through R4 and R5 to ground. This is because the quantity of electrons coming into or going out of the

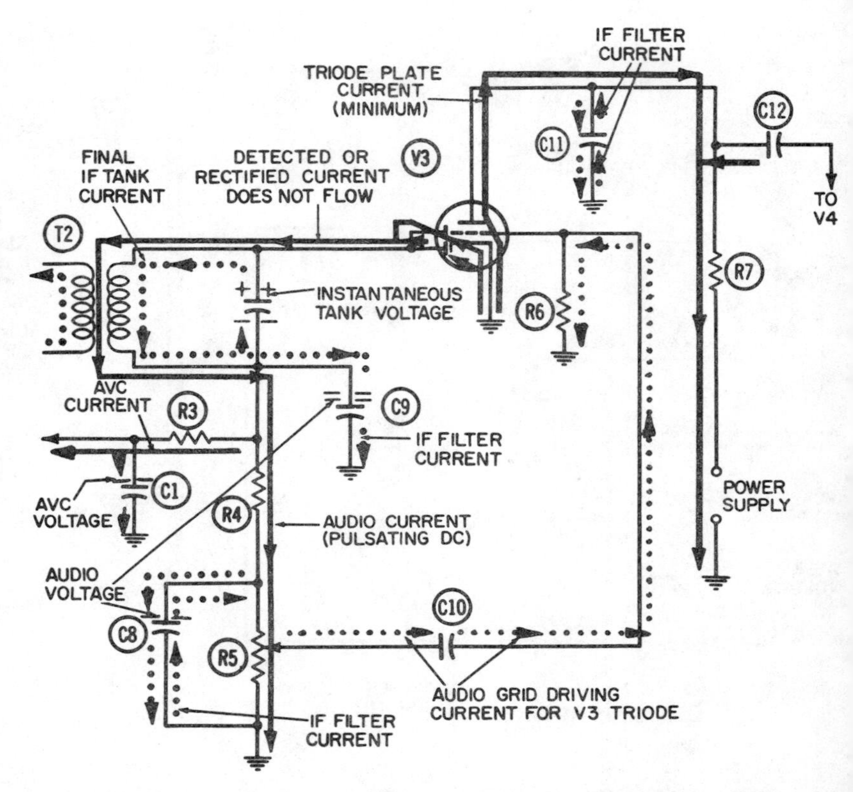

Fig. 2-2. Detector, AVC, and audio-amplifier circuit—positive half-cycle of IF tank voltage, negative half-cycle of audio voltage.

capacitor during a single half-cycle is an insignificant percentage of the quantity which is already stored or accumulated there

Modulation—When the incoming radio wave carries intelligence, such as speech or music, it is said to be *modulated*. The term modulation is often used to describe or refer to this intelligence, and the purpose of a detector (also called demodulator) is to extract this information or modulation from the radio signal current. As mentioned previously, the point at which this extraction process occurs is at the junction of R4 and C9, and the electron pool on the upper plate of C9 marks the first appearance of the *modulation voltage*. This voltage varies at frequencies which are within the range of the human ear—in other words, audio frequencies.

Fig. 2-3 shows a typical modulated waveform. The modulation process which occurs at the transmitter consists of a periodically varying or changing of the strength of the individual cycles of the signal current so that the radio-frequency signal current can be made to "carry" the audio information from the transmitter to the receiver. This carrying process has given rise to the almost universal practice of calling the modulated radio-frequency signal the *carrier*.

The final IF tank current which causes the diode portions of V3 to conduct electrons varies in strength from one cycle to the next in accordance with the modulation pattern. When the tank current is strong, as it is during the modulation peaks shown in Fig. 2-3, larger pulsations of diode current will flow. These larger pulsations deliver more electrons into the pool of electrons accumulated on C9. Since there will be several hundred or thousand individual pulsations of this rectified current flowing through the tube during the period occupied by a single audio modulating cycle, the depth of the electron pool on C9 will increase as a modulation peak approaches.

When the individual cycles of the final IF tank current are reduced in strength by the approach of a modulation trough, the corresponding pulsations of rectified current flowing through the diode portion of V3 will also be reduced. This results in *fewer* electrons being delivered into the storage pool on C9. Thus, more electrons discharge downward through R4 and R5 and flow in during this period, and the negative voltage on C9 *decreases* as the modulation trough approaches.

This process continues as long as a modulated signal is being received, with the result that the negative voltage on C9 (represented by the electrons in storage on the upper plate) increases with each modulation peak and decreases with each modulation trough. In other words, this negative voltage rises and falls at an audio rate, and is therefore, by definition, an audio voltage. This audio voltage first appears on the upper plate of C9.

The current which drains continuously downward through R4 and R5 from this point is driven downward by the amount of this voltage. Consequently, it pulsates at the same audio rate or frequency. This is another of the many possible forms of pulsating DC which exist in a radio during normal operation. It is interesting to note that the electrons which accumulate on C9 arrive there as pulsating DC from the diode plates of tube V3. These pulsations are occurring at the intermediate frequency, or in other words, 455,000 pulsations each second. These same electrons must eventually leave there, also as pulsating DC, but now the pulsations are occurring at whatever audio frequency is being received at the moment—a few hundred or at most a few thousand pulsations each second.

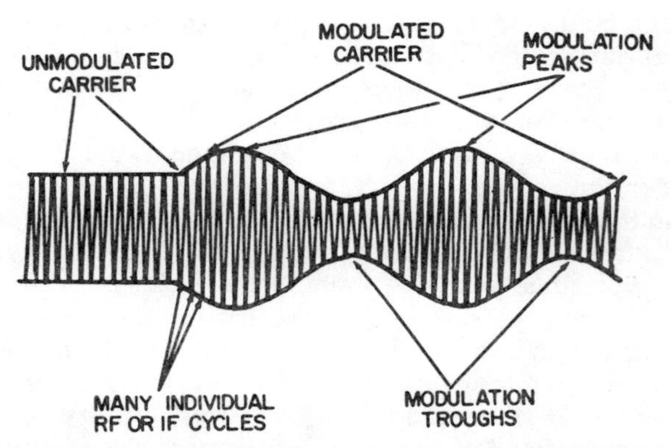

Fig. 2-3. Sine-wave representation of carrier-signal current or voltage.

Ohm's Law—Since an audio current is flowing downward through R4 and R5, an audio voltage must be developed across these resistors. This is in accordance with Ohm's law, which states that the *current through* any resistor must always be proportional to the voltage across it. The formula for this is:

$$E = I \times R$$

where,

E is the voltage across a resistor in volts,
I is the current through the resistor in amperes,
R is the resistance of the resistor in ohms.

Coulombs Law—There is another convenient and simple formula known as Coulomb's law which tells us that the quantity of electrons stored in a capacitor must always be proportional to the voltage across the capacitor. The formula for this is:

$$Q = C \times E$$

where,

Q is the quantity of electrons in storage in coulombs (one coulomb equals about 6×10^{18} electrons),
C is the capacitor size in farads,
E is the voltage between the capacitor plates in volts.

The voltage across R4 and R5 which you would compute using Ohm's law would be identical to the amount of voltage stored on the capacitor. This voltage could also be computed using Coulomb's law, if you knew the number of electrons accumulated there. At all lower points on R4 and R5 a proportionately smaller voltage will exist, depending on the distance of the point from the top of R4. Resistor R5 is variable; therefore, almost any proportional value of this audio voltage can be tapped off and coupled to the control grid of the triode portion of V3. R5 serves as the manual volume control and is controlled by the knob on the front panel of the radio.

Amplifier Operation

Coupling Action—During a modulation peak, when the audio current is flowing downward through R5 at its maximum rate, some portion of this current will be diverted onto the left-hand plate of C10; this action will drive other electrons away from the right-hand plate upward to the control grid of the tube and down through grid driving resistor R6 to the common ground. This condition is depicted in small dots in Fig. 2-2. This would constitute a negative half-cycle of audio voltage, because the voltage developed across R6 by electrons flowing downward through it will be negative at the top and positive at the bottom (since electrons always flow from negative to positive).

During a modulation trough, when the audio current is flowing downward through R5 at its minimum rate, electrons which were previously diverted into the left-hand plate of C10 will now flow back out, and downward through R5 to ground (Fig. 2-1). This action will draw the grid driving current *upward* through R6, and into the right-hand plate of C10. The upward flow of electrons through R6 makes the voltage at the top of this resistor positive; consequently, a modulation trough would lead to a positive half-cycle of operation of the audio amplifier.

Triode Plate Current—When the grid voltage is made negative during a modulation peak, the flow of plate current through the triode will be reduced to its minimum value. When the grid voltage is made positive during a modulation trough, this plate current will be increased to its maximum value. This latter condition is depicted in Fig. 2-1.

The triode plate current is another one of the several pulsating direct currents which flow in a radio. After it exits at the plate of the tube it flows to the left-hand plate of C12, where its pulsations are coupled to the control grid of the following stage. The plate current then flows downward through R7 and to the power supply. It eventually will pass through the power supply and be returned to the common ground, from where it has ready return access to the cathode of V3.

It is worth remembering that all tube currents must inevitably be provided with a return path to the cathodes of their respective tubes. Normally, the radio chassis provides these return paths.

IF Filter Currents—There are two additional points in Figs. 2-1 and 2-2 where filtering action occurs. Capacitors C8 and

C11 act as filter capacitors for any IF pulsations which were not filtered to ground by C9. We have learned that the electron current which flows downward through R4 and R5 is flowing in pulsations which occur at the audio frequency being demodulated. However, there will inevitably be some small pulsations occurring at the intermediate frequency of 455 kc. These pulsations will divide between C8 and R5 in *inverse proportion* to the impedances offered by these two components. Impedance may be defined as the *opposition to electron* flow which a component offers.

Since C8 is deliberately selected to be large enough to offer *very low impedance* to a current pulsating at the intermediate frequency, most of these pulsations (shown in large dots) will be diverted onto the top plate of C8, and each such pulsation will drive an equal number of electrons from the lower plate of C8 to ground. During the periods between each two successive pulsations, electrons will flow off the upper plate of C8 and downward (through R5) to ground, and an equal number will flow up from ground onto the lower plate of C8.

For purposes of analogy, you might assume that 90% of the strength of each IF pulsation which reaches the junction of C9, and R4 will be filtered to ground through C9, and the remaining 10% will enter R4. Also, assume that 90% of the *remaining strength* of each IF pulsation which reaches the junction of capacitor C8 and resistor R5 will be "filtered" to ground through C8, and the remaining 10% will enter R5.

From the foregoing approximations it can be concluded that 99% of the strength of the original IF pulsations is filtered out by C9 and C8, and only 1% flows through R5. This 1%, however, is coupled to the control grid of the triode section of V3 through coupling capacitor C10. Consequently, the plate current stream flowing through this triode, in addition to pulsating at the audio frequency, will also exhibit extremely small pulsations at the basic intermediate frequency. Thus, 90% of the strength of these individual pulsations will be filtered to ground through capacitor C11.

AVC Operation

In the discussion of the detector circuit operation it was pointed out that the negative voltage pool on the upper plate of C9 would rise and fall at an audio rate and that it represented the first appearance of the audio voltage in a receiver. Since this voltage is never positive, but always negative, it continuously drives electrons downward through R4 and R5. Now examine what happens along the AVC line, which connects to the junction of C9 and R4.

The most important thing you should understand about the AVC line is that it *does not* lead to ground, except through R4 and R5. Therefore none of the diode rectifier current, which is shown a solid line, can flow to ground through the AVC line.

AVC Current Flow—Since the AVC line does not lead to ground, any current which flows through R3 has to be *a two-way current*—a true alternating current. During modulation peaks, when the electron pool on C9 is large, electrons will be driven to the left from C9 into the AVC line, and through large-value AVC resistor R3. This condition has been depicted in Fig. 2-2. It is this current which delivers electrons to the upper plate of C1 and builds up the stored charge which becomes known as the AVC voltage.

During modulation troughs, when the electron pool on C9 is small, the AVC current will flow to the right through the AVC resistor. This action tends to discharge the electrons stored on C1.

From the foregoing, we see that the AVC current is an alternating current which flows through R3 at the audio frequency being demodulated or detected.

AVC Voltage—Before a radio is turned on, of course, there is no AVC voltage stored on the upper plate of C1. However, once the detection process begins, electrons will begin accumulating on the upper plate of C1. This accumulation process can result from only one sequence of events, namely, that the AVC current which flows to the left through the AVC resistor R3 during modulation peaks (Fig. 2-2), is larger than the current which flows through the resistor to the right during the modulation troughs depicted in Fig. 2-1.

After several hundred audio cycles have occurred, the AVC voltage on C1 will assume the *average* value of the peak and trough voltages which occur on the upper plate of C9. When this happens, the imbalance between the two half-cycles of AVC current will disappear. In other words, the incoming AVC current of Fig. 2-2 will bring just as many electrons into the storage pool on AVC capacitor C1 as the outgoing current of Fig. 2-1 takes away.

Resistor R3 is a very large resistor—usually 2.2 or 3.3 megohms. C1 is also large—approximately .05 mfd. Therefore the product of these two values (R × C) will give a very long time constant. Using 3.3 megohms, and .05 mfd, the time constant is .165 second; thus, it will require perhaps five or ten times this long to discharge the AVC voltage, once it has accumulated on the upper plate of C1. Obviously, since a single audio cycle completes itself in a small fraction of a second, the AVC voltage cannot discharge itself during a single modulation trough, or even several hundred of them.

Another way of saying this is that the *quantity* of electrons stored on C1 is so great in comparison to the number which flow in or out during a single half of an audio cycle, that the amount of the AVC voltage cannot be changed by a single modulation peak or trough. Still another way of saying this is that the AVC voltage does not "respond" to the audio modulation.

Signal Fading—The AVC circuit is provided to protect the listener against undesirable changes in volume as a result of changes in signal strength due to propagation anomalies.

During a signal fade the incoming carrier signal received at the antenna will be significantly reduced in strength. This causes a proportional reduction in all subsequent derivatives of this same current. Therefore, the final IF tank current shown in dotted line in Figs. 2-1 and 2-2 will be reduced. When this happens, fewer electrons will be drawn across the diode portion of V3 each positive half-cycle, and the audio voltage which appears on the upper plate of C9 will fluctuate between lower values of negative voltage. In other words, both the peak and the trough voltages will be reduced in size when a signal fade—which may persist for several minutes—occurs.

During a fade, a new imbalance is created between the AVC current which comes into C1 during the modulation peaks and which flows away from C1 during the modulation troughs. More electron current will flow away from C1 during a modulation trough than flows into it during a modulation peak. After many cycles have elapsed, this imbalance partially discharges the negative AVC voltage stored on C1, until it eventually again assumes the average value of the peak and trough voltages which are occurring on the upper plate of C9. When it reaches this new average value, the imbalance between incoming and outgoing AVC current again disappears.

A signal fade would ordinarily reduce the audio output of the receiver and require adjustment of the volume control in order to maintain a comfortable sound level. The negative AVC voltage on the upper plate of C1 does this job automatically. Since it is connected to the control grids of the preceding tubes, it acts as a biasing voltage for these tubes. A less negative AVC voltage, which results from a signal fade, will *increase* the gain (amplification) of these tubes and will thereby largely nullify the effect of the signal fade.

Signal Build-Up—A signal build-up, which is also caused by propagation anomalies, will increase the strength of the antenna current; this will cause proportionate increases in the strength of each of the subsequent derivatives of that current, including the final IF tank current. Each individual pulsation of detector current through the diode portion of V3 will therefore increase, and the quantities of electrons stored on C9 during each modulation peak and trough will increase. When this occurs, a new imbalance is created between the two alternate half-cycles of AVC current. More current will be driven to the left along the AVC line during modulation peaks than will flow to the right during modulation troughs. This has the effect, after hundreds of audio cycles, of delivering more electrons into storage on C1, than are taken away so that it recharges C1 to a more negative voltage. This imbalance will persist until the AVC voltage again equals the *average* value of the peak and trough voltages appearing on C9.

A signal build would normally increase the output of the receiver, again requiring adjustment of the volume control. With an AVC circuit, the more negative AVC voltage biases the preceding tubes more negatively, reducing their gain and partially nullifying the original effect of the signal build.

As long as a radio is turned on and tuned to a station, there will be some electrons in storage on the AVC capacitor; consequently, the control grids of the converter and IF amplifier tubes will always have some negative biasing voltage applied to them. The amount of this voltage will vary directly with the strength of the incoming signal; therefore the gain of the stages preceding the detector will vary inversely with the strength of the incoming signal. The result is that once the audio output level has been chosen by the volume control, the output level will be maintained or adhered to very closely despite fairly wide variations in the incoming signal strength. Automatic volume control is sometimes called *automatic gain control*.

When a radio is tuned so that it is not receiving any station, or when it is turned off completely, the electrons stored on C9 will discharge to ground through R4 and R5 in a fraction of a second. The electrons stored on C1 will also discharge to ground, but they must first flow through the large resistor R3, and then through resistors R4 and R5. This discharging process will therefore require several seconds for completion.

GENERATION OF POSITIVE AVC VOLTAGES

One of the many significant differences between vacuum-tube receivers and transistor receivers is in the generation and application of voltages for the automatic control of receiver volume (AVC), or gain (AGC).

With vacuum-tube amplifier stages, gain is most readily varied and controlled by controlling the grid-cathode voltage relationship. A decrease in the negative grid-to-cathode voltage will increase the gain of a vacuum-tube amplifier. A conventional diode rectifier may generate either a negative or a positive voltage, depending on whether the output is taken from the plate or the cathode. In either case, this voltage will vary proportionately with the variations in carrier signal strength.

A positive control voltage, which becomes *more* positive as carrier signal strength increases, cannot be used for automatic gain control in a vacuum-tube stage, because a more positive voltage applied to a control grid will increase rather than decrease the gain of the stage. However, a negative control voltage which becomes more negative as signal strength increases can be used for AVC, since it decreases the gain of a tube amplifier stage. This is the essence of the AVC action in vacuum-tube receivers.

The foregoing restriction does not apply to PNP transistor amplifier stages. Because of the directions of electron flow through a PNP transistor, a positive control voltage, which becomes more positive as carrier strength increases, can be used for AVC purposes. An increase in the positive biasing voltage applied to the base of a PNP transistor will *decrease* the current through the transistor, thereby decreasing the gain of the stage.

Identification of Components

In the circuit of Figs. 2-4, 2-5, and 2-6, the following components perform the indicated functions:

R1—Emitter biasing resistor.
R2—Base biasing resistor.
R3—First IF filter resistor.
R4—Variable resistor used as volume control.
R5—AVC resistor.
C1—Collector tank capacitor.
C2—First IF filter capacitor.
C3—Second IF filter capacitor.
C4—AVC capacitor.
L1—Primary winding of first IF transformer.
L2—Secondary winding of the same transformer.
L3—Collector tank inductor (primary of IF output transformer).
L4—Secondary of IF output transformer.
X1—PNP transistor.
M1—Detector diode.
M2—Battery power supply.

Identification of Currents

The electron currents listed in the following flow during normal operation of this circuit. The reader is again reminded that his best chance for understanding how a particular circuit operates is to be able to visualize the various electron currents in motion, and to relate the movements of each current to its neighbor. The student can properly claim that he understands how a circuit operates only when each current has (1) been properly identified, (2) its complete path through the circuit traced out, (3) the action which makes it flow understood, and (4) the job it does visualized.

The following 10 electron currents are at work in the circuit of Figs. 2-4, 2-5, and 2-6:

1. Four IF currents.
2. Base-emitter current.
3. Collector-emitter current.
4. Diode current.
5. AVC current.
6. Voltage-divider current.

Details of Operation

In a transistor, electron flow between the collector and emitter is always *against* the direction the arrow of the symbol points. Likewise, electron current from base to emitter must also flow against the direction of the emitter arrow. Figs. 2-4, 2-5, and 2-6 all show these two transistor currents flowing against the emitter arrow.

Two important conditions must be fulfilled for a PNP transistor to conduct these electron currents. These conditions are: (1) the base must be negative with respect to the emitter, and (2) the collector must be negative with respect to both the base and the emitter.

These conditions are fulfilled in transistor X1 by connecting both the collector and the base to the negative terminal of biasing battery M2. Since the emitter is connected to ground (through resistor R1), the negative bias voltage will tend to drive electron current from the base and from the collector into the emitter and through R1 to ground. These currents are appropriately indicated in Figs. 2-4, 2-5, and 2-6.

The intrinsic voltage difference which exists between the base and the emitter at any instant is the most important factor in determining the amounts of these two currents which will flow. To understand what causes this voltage difference to exist or to vary, consider (initially) the movements of three separate electron currents. These currents are:

1. The voltage-divider current.
2. The base-emitter current.
3. The collector-emitter current.

When this circuit is at rest, meaning when no currents are flowing and when the biasing power supply is disconnected, the three terminals of the transistor will be at ground or zero voltage, since they are all connected to ground through various combinations of resistors and inductors. However, when power is applied to the transistor, the three previously mentioned currents begin to flow; each contributes a substantial change in the voltage or voltages which exist at one or more of the electrodes. It is absolutely necessary to recognize and understand the voltage changes which occur at each terminal. In order to do this, each of the currents which are associated with these voltages must be visualized.

The voltage-divider current, shown in large dots, is the simplest one of the three to understand, since it does not flow through the transistor. This electron current leaves the negative terminal of the battery, flows to the left through R2, then down, and to the right, and upward through R5 and downward through R4 to ground. Each point along this path is progressively less negative than any point preceding it.

The base-emitter current, shown a solid line, is also driven by the battery. It flows to the left through R2, upward through L2, and into the base of the transistor and out through the emitter, then downward through emitter resistor R1 to ground. Like the preceding current, each point along this path is progressively less negative than any point preceding it. Therefore the emitter is *more* negative than ground because of the current flow through R1, but it is *less* negative than the base, because there is a small amount of resistance between the base and the emitter of the transistor.

The collector-emitter current is also driven by the negative power-supply voltage (although it is regulated or controlled by the amount of current flowing from base to emitter). This current, which is commonly called the "collector" current, flows from the negative terminal of the battery upward through L3, then through the transistor from collector to emitter, and downward through R1 to ground. In flowing through R1, this current develops an additional component of negative voltage across R1, which alters or modifies the biasing conditions existing between the base and the emitter.

In the absence of a received carrier signal, these three currents will very quickly stabilize at "equilibrium" values. Each one will exert its own particular effect on the biasing voltages of the transistor, and each current will become a pure DC.

When a carrier signal is being received, all biasing conditions and all currents are changed. The carrier signal is shown a solid line, and in Fig. 2-4 it is flowing upward through inductor L1. The applied emf which causes this upward flow has been indicated by a plus sign at the top of L1. The resulting induced current in L2 is also shown a solid line, and flows downward. The "back emf" or "counter-emf" associated with this current is indicated by a deep minus (negative) sign at the top of L2. It is this induced emf which acts as an additional biasing voltage at the base of the transistor.

In any PNP transistor, such as this one, electrons will flow *from* the base and *into* the emitter. In order for this to happen the base must be more negative than the emitter. The base-emitter current, shown a solid line, contributes a small amount

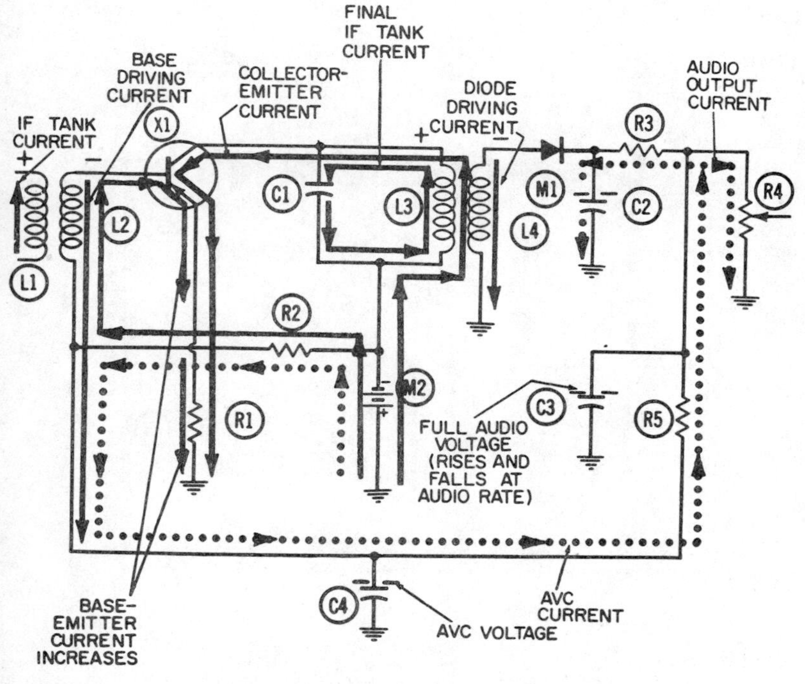

Fig. 2-4. Transistorized IF amplifier, detector, and AVC circuit—negative half-cycle of IF.

of this voltage difference. A much larger component of this voltage difference is contributed by the negative voltage induced at the top of L2 during any negative half-cycle, such as that depicted in Fig. 2-4. Thus the amount of electron flow between base and emitter is greatly increased during a negative half-cycle. The amount of collector current which flows is greatly influenced and regulated by the amount of base-emitter current (usually called emitter current) which flows. Thus, the two transistor currents (base-emitter and collector-emitter) are both substantially increased during a negative half-cycle.

During a positive half-cycle of RF or IF, both of the transistor currents are substantially reduced. This comes about initially when the input carrier current (shown a solid line) induces a positive component of voltage at the top of L2. This positive component is added to the negative voltage existing at the bottom of L2 by virtue of the voltage divider current shown in large dots. The net result is a reduced value in the negative voltage at the base of the transistor. Since the PNP transistor requires a negative voltage at the base for electrons to flow from the base to the emitter, it should be evident that any reduction

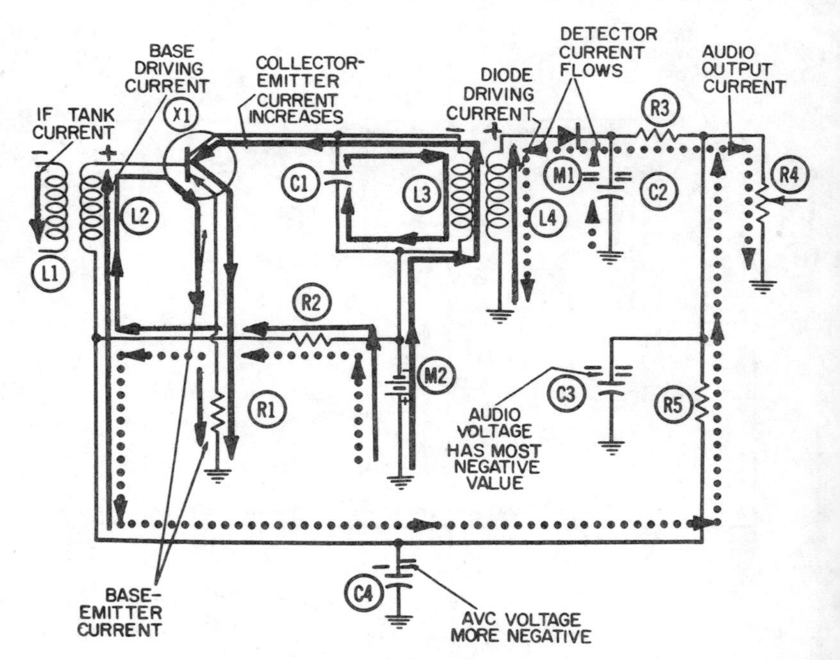

Fig. 2-5. Transistorized IF amplifier, detector, and AVC circuit—signal strength decreased.

in the negative base voltage will reduce this flow of base-emitter current.

When the base-emitter current is reduced, collector-emitter current is reduced proportionately. Normally, the amount of collector-emitter current which flows is between 10 and 50 times as great as the amount of base-emitter current. It is this property of the transistor which enables it to be used as an amplifier, because a relatively small amount of change in the base-emitter current will cause a substantial amount of change in the collector-emitter current.

Fig. 2-5 shows a positive half-cycle when a relatively weak carrier signal is being received. Fig. 2-6 shows a positive half-cycle when the strong carrier signal is being received. Inspection of Figs. 2-5 and 2-6 will reveal very few differences in current directions or voltage polarities throughout the entire circuit. However, the *amounts* of most currents and voltages will change significantly as we go from a weak signal to a strong signal, or vice versa. It is only by visualizing how these changes occur, and by recognizing their significance, that the AVC action can be understood.

Referring to Fig. 2-5, the pulsations in the collector current as it flows upward through L3 will first shock-excite the tank circuit, C1 and L3, into oscillation, and then each individual pulsation will replenish or sustain a single cycle of this oscillation of electrons in the tank. Of course, the pulsations also occur during the negative half-cycles. In Fig. 2-4 a pulsation of electrons is shown (a solid line) flowing upward in the tank circuit. In Figs. 2-5 and 2-6 the tank-circuit electrons are returning downward through L3.

A companion current is induced in secondary winding L4 by each primary winding. During the negative half-cycles (Fig. 2-4), this secondary current is shown as flowing downward through L4; during the positive half-cycles (Figs. 2-5 and 2-6), it is flowing upward. The back emf associated with the secondary current has polarities which are indicated by the appropriate signs at the top of L4. The voltage polarities at the top of L4 determine whether or not diode M1 will conduct electrons. When the top of L4 is positive, as it is during positive half-cycles of RF (or IF), M1 will conduct from right to left. This is the normal flow direction for a solid-state rectifier from cathode to anode. This diode current is shown in large dots. These electrons are drawn initially out of the upper plate of filter capacitor C2, and flow through the diode as indicated, then downward through L4 to ground.

When no carrier signal is present, the upper plate of C2 will charge to a value of negative voltage that is determined by the voltage divider current shown in large dots. This current flows continuously from the negative battery terminal through R1, R5, and R4, in that order, to ground. The amount of the initial voltage on C2 can be calculated by simple arithmetic using Ohm's law.

During the weak-signal positive half-cycles of Fig. 2-5, when the diode anode is only slightly positive, a small quantity of electrons are drawn out of this "electron pool" on C2 and flow through the diode. This action reduces the negative voltage on C2 so that it becomes less negative than the voltage at the right-hand end of R3. Consequently, electrons will flow from right to left through R3 to equalize this voltage imbalance. Initially, these electrons come from the upper plate of C3, also making the voltage stored there less negative by a small amount.

During the strong-signal positive half-cycles of Fig. 2-6, when the diode anode is made very positive by the induced voltage at the top of L4, a larger quantity of electrons are drawn out of the electron pool on C2 and flow through the diode. This action substantially reduces the negative voltage on C2 so that a greater amount of electrons must flow from right to left through R3 to equalize the voltage imbalance. These electrons come from the upper plate of filter capacitor C3, reducing the negative voltage stored there by a significant amount.

There are three separate and distinct time periods which must be considered when we try to analyze the manner in which an AVC circuit operates. These time periods are as follows:

1. The time required for a single cycle of RF or IF to complete itself—one or two millionths of a second (microseconds, in other words).
2. The time required for a single cycle of audio-frequency to complete itself—a few thousandths of a second (milliseconds, in other words). This time period is several orders of magnitude longer than one that is measured in microseconds.
3. The time required for a signal fade or a signal build to occur, due to atmospheric or propagation anomalies. This will require several seconds or even several minutes to occur. This time period is obviously many orders of magnitude longer than one that is measured in milliseconds.

There are three important resistor-capacitor filter combinations in this circuit, each one of which is designed to respond to a current/voltage action occurring in a different one of these three time periods. These RC filter combinations are as follows:

1. R3 and C2, which respond only to RF or IF.
2. R4 and C3, which respond only to audio frequencies.

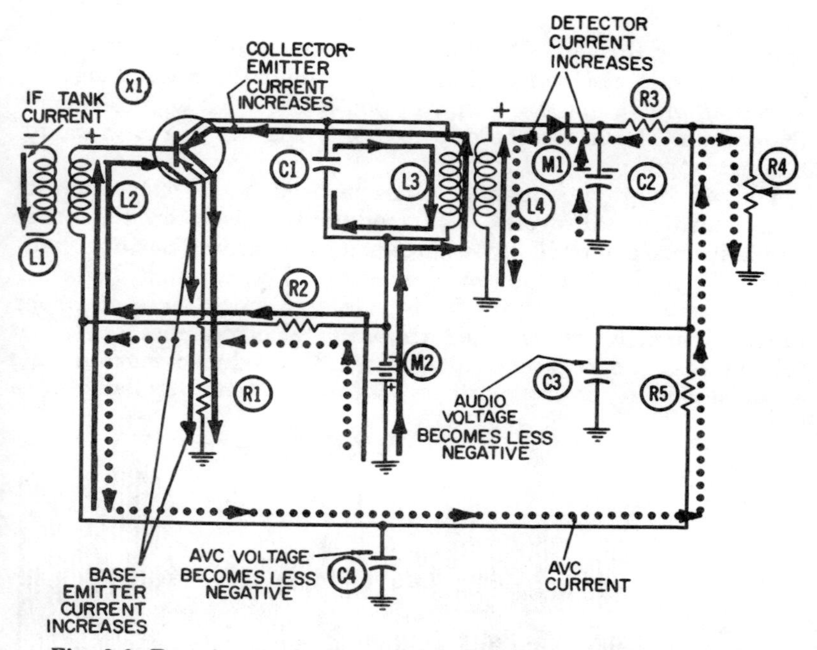

Fig. 2-6. Transistorized IF amplifier, detector, and AVC circuit—signal strength increased.

3. R5 and C4, which respond only to sustained changes in received signal strength.

It is evident in Figs. 2-4, 2-5, and 2-6 that negative voltages (meaning electrons) are stored on the upper plates of C2, C3, and C4. The voltage level on C2 rises and falls at the intermediate frequency to which the final amplifier tank is tuned—455,000 cps is a typical example. On positive half-cycles, electrons are drawn out of this capacitor, and an equal number flow up from ground and onto the lower plate of C2. On negative half-cycles, electrons flow downward onto C2 from R3, recharging C2 and driving an equal number of electrons from the lower plate of C2 back into ground. The electron current which flows between the lower plate of C2 and ground is the principal component of IF filter current. (A lesser component which has not been shown will flow between the lower plate of C3 and ground.)

The voltage level on C3 rises and falls at an audio rate. This voltage marks the first appearance of the audio voltage in the receiver system. Fig. 2-7 shows typical waveforms which relate an IF carrier signal to the audio intelligence which it carries. During modulation troughs, the carrier pulsations are relatively weak, and each pulsation will cause only a small number of electrons to flow through the diode on the positive half-cycles. Consequently, the electron pool (negative voltage) on C2 is not depleted as much, and the electron pool on C3, which replenishes the one on C2 will also be depleted only slightly during a modulation trough.

The modulation trough is characterized by a succession of weak positive half-cycles. The modulation peak, on the other hand, is characterized by a succession of strong positive half-cycles. Since large numbers of electrons flow through diode M1 during each strong positive half-cycle, and since these electrons **must eventually be supplied from the electron pool on C3, it follows that the negative voltage on C3 is reduced *more* during a succession of strong positive half-cycles than it is during a**

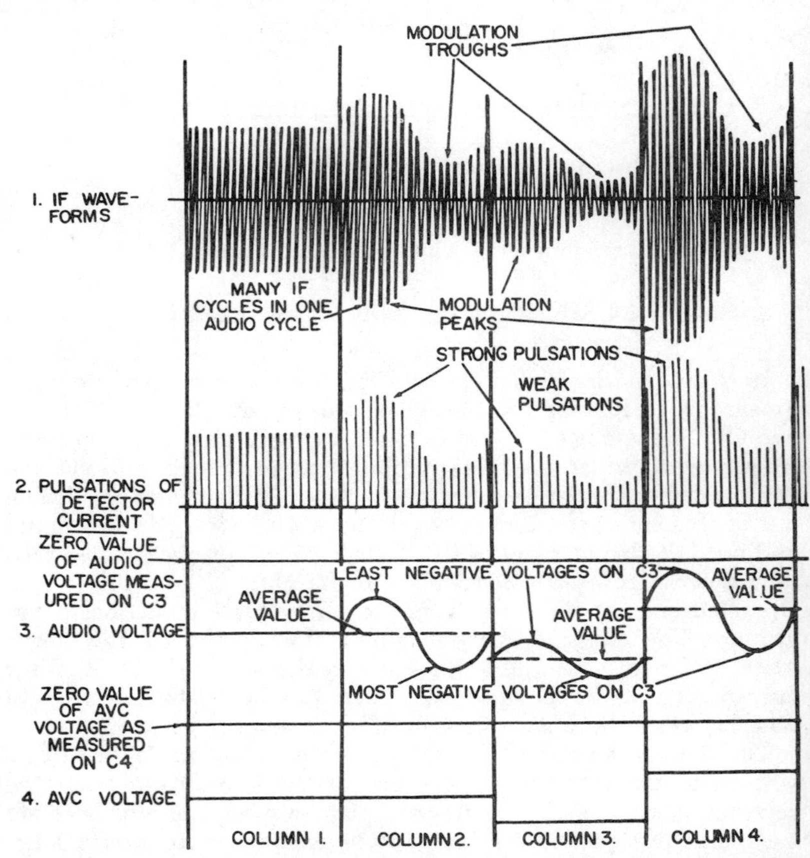

Fig. 2-7. Voltage waveforms at four significant points under four different operating conditions of the circuit in Figs. 2-4, 2-5, and 2-6.

succession of weak positive half-cycles. Therefore, the voltage on the upper plate of C3 is *less* negative during modulation peaks, and *more* negative during modulation troughs.

The negative voltage stored on C4 does not change with each modulation peak and trough. This voltage will remain constant as long as the average signal strength being received is constant. If this signal strength fades out or builds up due to propagation anomalies, the voltage on C4 will change proportionately, and affect the transistor biasing conditions. Thus, the amplification provided by the transistor is varied. Consider how this voltage change on C4 can be brought about.

First, consider a signal fade. The waveforms of column 3 in Fig. 2-7 indicate that a signal fade is characterized by a long succession of weakened IF cycles so that the modulation peaks as well as the troughs are reduced in strength. Consequently, during the entire period of a signal fade, fewer electrons must be drawn away from C3 to replenish the current which flows through M1. Therefore the *average* voltage on C3 remains at a fairly high negative value.

The negative voltage on C4, which is the AVC voltage, always assumes the average value of the peak and trough voltages which are occurring on C3 (after allowing for the steady component of voltage developed across R5 by the continuous flow of voltage divider current).

Now consider a signal "build." The waveforms of column 4 in Fig. 2-7 indicate that a signal build is characterized by a long succession of strengthened IF pulses. The modulation peaks and troughs are both stronger, placing greater demand for electrons on the negative voltages stored on C2 and C3. The end result of a signal build is that the *average* voltage on C3 will be considerably *less* negative than it is during a signal fade.

During a signal build, the AVC voltage stored on the upper plate of capacitor C5, also becomes less negative. Since the base of the transistor is connected directly to this point, the amount of base-emitter current through the transistor will be *reduced*, and the total amplification which the transistor can provide will also be reduced. This decrease in amplification nullifies or compensates for the adverse effects of the signal build.

During the signal fade previously discussed, the AVC voltage on C4 is more negative than it is during a signal build. This serves as a biasing voltage at the base of the transistor and *increases* the amount of base-emitter current which will flow during any positive half-cycles. This increases the amplification which the transistor provides, and thereby nullifies or compensates for the loss in signal strength due to the signal fade.

Chapter 3

NOISE-LIMITING PRINCIPLES

In the transmission of a signal between the station and the receiver, noise pulses are often superimposed on the signal. These pulses, which may be caused by atmospheric or man-made conditions, will cause "static" in the output if allowed to pass through the receiver. Most amateur and communications receivers employ circuits for removing these pulses so that they will not appear in the output; however, noise-limiting circuits are seldom employed in "entertainment-type" home receivers. In this chapter, three types of noise limiters—shunt diode, series diode, and dual diode—as well as pentode squelch circuit will be discussed.

SHUNT-DIODE NOISE LIMITER

Figs. 3-1 and 3-2 show two separate moments during the operation of a simple noise-limiting circuit, which places a diode tube, V1, across the grid input circuit of a conventional pentode audio amplifier, V2. In the absence of an undesirable noise pulse, the diode tube does not conduct; this condition might be labeled as the *normal* mode of operation (Figs. 3-1 and 3-2). When a noise pulse is present, the diode tube conducts, as shown in Fig. 3-3. This conduction biases the control grid of the pentode tube beyond cutoff, cutting off the tube for the duration of the noise pulse.

Identification of Components

The following circuit components in Fig. 3-1, 3-2, and 3-3 perform the functions indicated:

R1—Variable resistor (volume control).
R2—Variable resistor used as voltage divider to set the noise level.

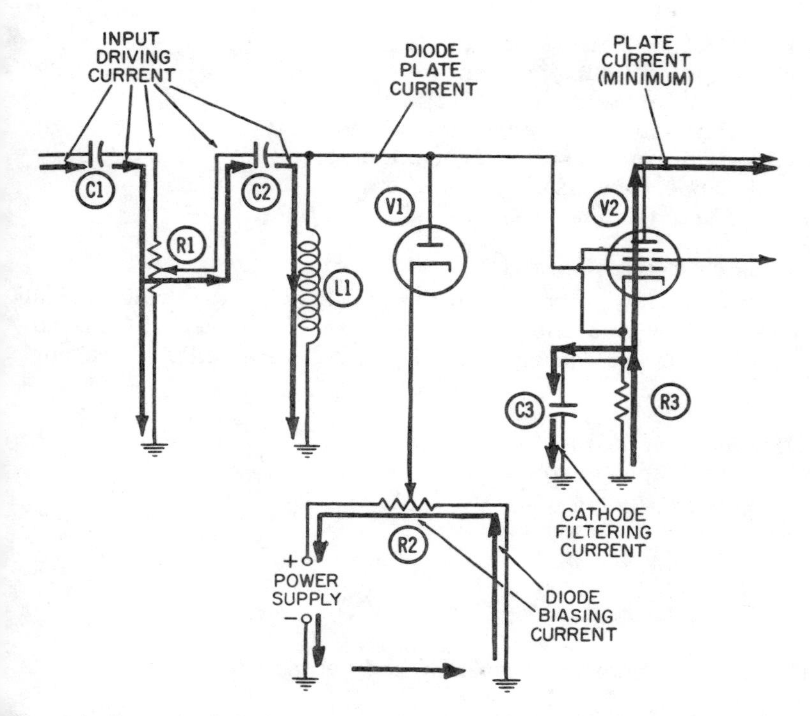

Fig. 3-1. Operation of the shunt-diode noise limiter—negative half-cycle.

such as is shown in Fig. 3-1, the plate current will be reduced to its minimum value.

Fig. 3-2 shows a positive half-cycle during normal operation of this circuit. The input audio current is now being drawn out of the left-hand plate of C1; this action draws electron current upward through R1 and also out of the left-hand plate of C2 and upward through L1. This action places a positive voltage on the control grid of V2 and causes the maximum amount of plate current to flow through the tube.

During both the negative and positive half-cycles of operation shown in these two illustrations a voltage-divider current (shown a solid line) will be flowing continuously through R2 from right to left. This current flows continuously in a counterclockwise direction through R2 and the power supply in ground, then out of ground and back into R2. As the potentiometer arm is moved from right to left, a succession of higher and higher positive voltages will be encountered, and applied to the cathode of V1. This potentiometer arm is used to set the *noise level* at which the circuit operates.

R3—Cathode-biasing resistor for V2.
C1—Input coupling capacitor.
C2—Coupling and biasing capacitor.
C3—Cathode-bypass capacitor for V2.
L1—Grid-driving and biasing inductor.
V1—Diode tube used as noise limiter.
V2—Pentode tube used as audio amplifier.

Identification of Currents

Three electron currents will flow in this circuit during normal operation. Two additional currents are introduced during abnormal operation (when a noise pulse is received). These currents are:

Normal Operation.
1. Input audio current.
2. Voltage divider current.
3. Pentode plate current.

Abnormal Operation

4. Noise current.
5. Diode current.

Also during abnormal operation, the pentode plate current does *not* flow.

Details of Operation

Fig. 3-1 shows a negative half-cycle of operation of the noise-limiter circuit when no undesired noise pulse is present. The input audio signal from the demodulator circuit reaches input capacitor C1 and drives an electron current *downward* through R1, producing a negative voltage at the upper end of R1. Each point below the top of R1 will exhibit a lesser negative voltage than that at the top during this negative half-cycle, depending on the distance. Thus, a movement of the potentiometer arm taps off any desired amount of the audio voltage for coupling to V2.

The physical means by which this coupling action is accomplished is indicated in Fig. 3-1. When current is driven downward through R1, it is also driven onto the left-hand plate of C2. This action drives an equal number of electrons out of the right-hand plate of C2 and downward through L1. During this downward motion of electrons through L1, the top of the inductor will be negative in voltage; this is the voltage applied to the grid of V2.

This pentode will be conducting throughout an entire audio cycle, i.e., continuously, during normal operation. This is usually referred to as Class-A operation. During a negative half-cycle,

The most important single condition for a diode tube to conduct electron current is that the instantaneous voltage at the diode plate must be more positive than the instantaneous voltage at the cathode. The cathode of V1 is held at a certain value of positive voltage, depending on the position of the potentiometer arm, so that the tube is normally nonconducting. Even the positive half-cycle of audio voltage depicted in Fig. 3-2 is assumed to be insufficiently strong to make the diode plate more positive than this cathode voltage. Consequently, the diode acts as an open circuit during the entire audio cycle. The term "open circuit" is frequently used in this kind of situation and should be considered as synonomous with infinite resistance. Thus, it has no effect on the normal operation of the pentode amplifier input circuit.

Fig. 3-4A shows the RF (or IF) waveform during normal operation. This illustration also shows how the same waveform would be modified and distorted by a typical noise pulse. Noise pulses are characteristically of extremely short time duration and usually of very high amplitude. These pulses may be the result of either natural or artificial interference—lightning bursts, automobile-ignition systems, X-ray equipment, or any one of dozens of other types of industrial electronic equipment.

When the positive half-cycle of even a single cycle of such a noise pulse reaches the plate of V1, it makes this plate more positive than the cathode, and the diode conducts electrons strongly from cathode to plate. The "noise current" which causes this condition is shown being drawn upward through L1 (in large dots) in Fig. 3-3 (the resulting diode current is shown in small dots). This diode current is drawn out of ground below the right-hand end of R2 and flows in a short burst (or a series of short bursts, depending on the number of cycles of noise voltage present) through the diode from cathode to plate. From the diode plate, the current flows onto the right-hand plate of C2, where it accumulates and very quickly forms a "pool" of negative voltage which biases both the diode plate and the pentode control grid negatively. This negative bias at the control grid cuts off the flow of electron current through the pentode until the noise pulse has ended. Thus, no audio occurring during the period of a noise pulse will be reproduced at all. Fig. 3-4B shows the resultant output waveform of V2 when the noise pulses of Fig. 3-4A occur. Here, it can be observed that there is no output at all during the noise pulse.

Since the noise pulse is of such short duration, lasting only for a portion of an audio cycle, or at the most, for just a few audio cycles, the absence is normally not noticeable to the listener. When the noise pulse passes, the electron pool which accumulated on the right-hand plate of C2 will very quickly discharge to ground through L1, and the diode will again be able to conduct electrons on the succeeding positive half-cycles of audio voltage should another noise pulse occur.

The statement appears frequently in the literature on noise-limiting and noise-cancellation circuits that the noise-limiting action punches a hole in the signal. This refers to the fact which is portrayed graphically in Fig. 3-4B, namely, that no audio output signal is delivered while a strong noise pulse is being received.

L1 and C2 act in much the same manner as a long time-constant RC filter. Noise pulses will occur at high frequencies, well above the audio range, so that the electrons in storage on C2 cannot discharge downward through L1 after one noise cycle before the next such cycle occurs.

SERIES-DIODE NOISE LIMITER

The circuit shown in Figs. 3-5 and 3-6 is another popular noise-cancellation, or noise-limiting circuit. It derives its name from the fact that noise-limiting diode V1 is in series with the audio signal path. During normal operation, the diode conducts continuously, and an output signal is developed across R5 for coupling to the pentode amplifier stage. During abnormal or noise-limiting operation, the diode does not conduct so that no audio signal can be developed across R5 for the duration of the noise pulse.

Identification of Components

As far as possible, the components in Figs. 3-5, 3-6, and 3-7 have been labeled to coincide with their counterparts in the shunt noise limiter of Fig. 3-1, 3-2, and 3-3. The various components with their functional titles are as follows:

R1—Variable resistor (volume control).
R2—Voltage-dividing resistor for setting the noise level.
R3—Cathode biasing resistor.
R4—Noise-pulse filtering resistor.
R5—Diode load or output resistor.
R6—Grid-driving and grid-return resistor.
C1—Audio storage capacitor.
C2—Blocking and coupling capacitor.
C3—Cathode bypass capacitor.
C4—Coupling and blocking capacitor.
V1—Noise-limiting diode.
V2—Pentode audio-amplifier tube.
V3—Diode detector or demodulator.

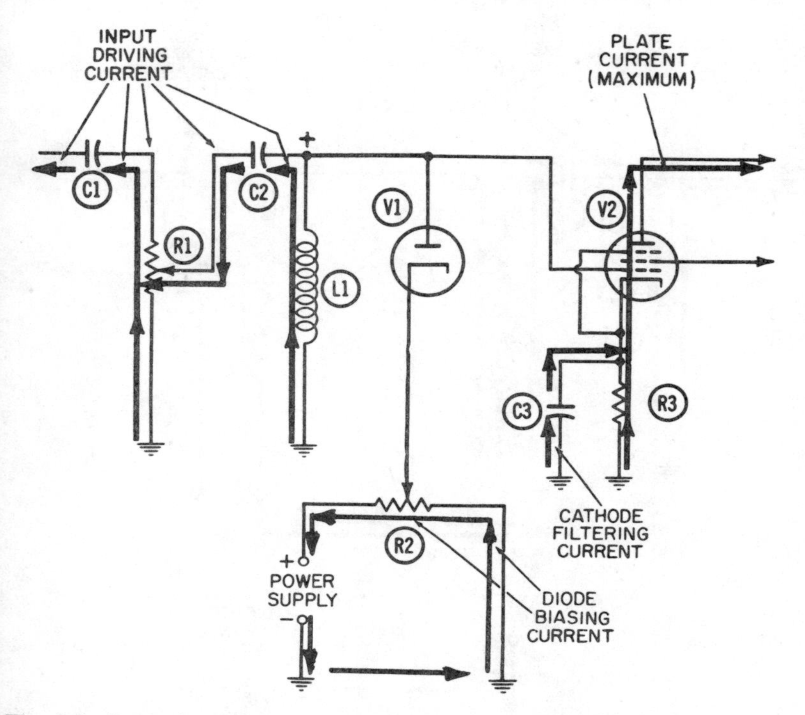

Fig. 3-2. Operation of the shunt-diode noise limiter—positive half-cycle.

Identification of Currents

The following electron currents will flow in this circuit during normal operation (meaning during conditions when no noise pulse is present):

1. Diode-detector load current, which is also the input audio signal.
2. Voltage-divider current.
3. Noise-pulse current.
4. Diode current.
5. Pentode grid-driving current.
6. Pentode-plate current.
7. Cathode-filter current.

During abnormal operation, meaning when an unwanted noise pulse is being received, the last four of the currents listed—the diode current, the grid-driving current, the pentode-plate current, and the cathode-filter current—*do not* flow. Since the plate current is invariably being used to deliver an audio signal to the next amplifier stage, or to some output device such as a speaker or

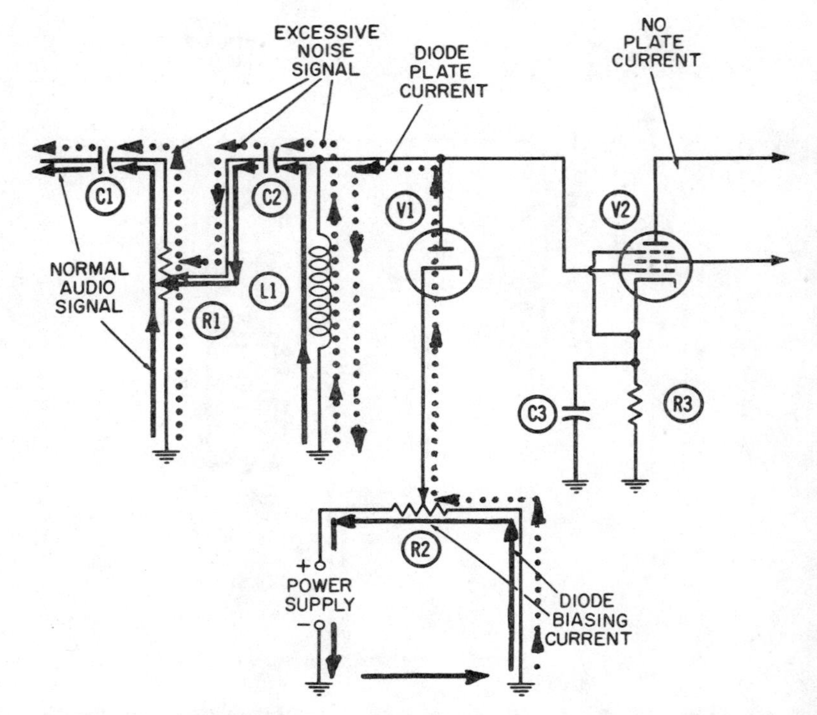

Fig. 3-3. Operation of the shunt-diode noise limiter—excessive noise pulse being received.

headphones, the cutting off of plate current during a noise pulse effectively cancels out any other adverse effects of that noise pulse.

Details of Operation

Fig. 3-5 depicts the current actions which occur during a negative half-cycle of audio operation. The input current (shown a solid line) is flowing at an audio frequency and represents the output of the V3 diode-detector circuit. This current is being driven downward through R1, causing a flow into the left plate of C2, and downward through R4. As a result of this downward movement of electrons, the voltages at the tops of R1 and R4 will be negative.

The variable tap on R1 enables any desired portion of this negative voltage to be coupled to the diode noise limiter. This variable feature regulates the amount of electron current which is being driven into C2 and downward through R4; therefore it regulates the amount of negative voltage developed at the top of R4 during this negative half-cycle.

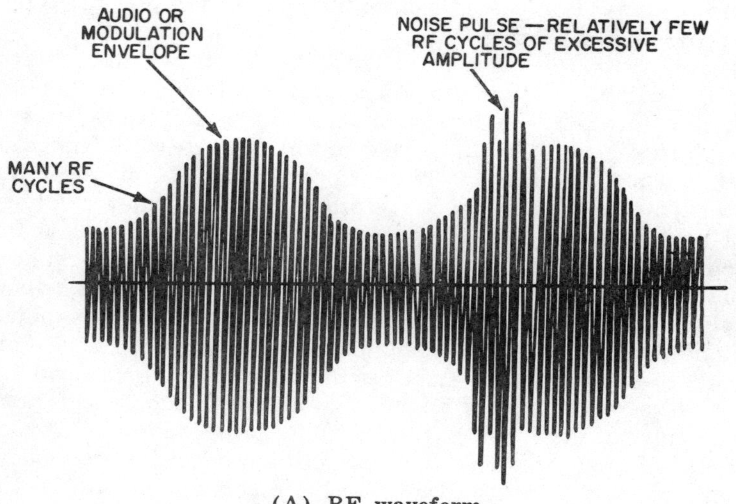

(A) RF waveform.

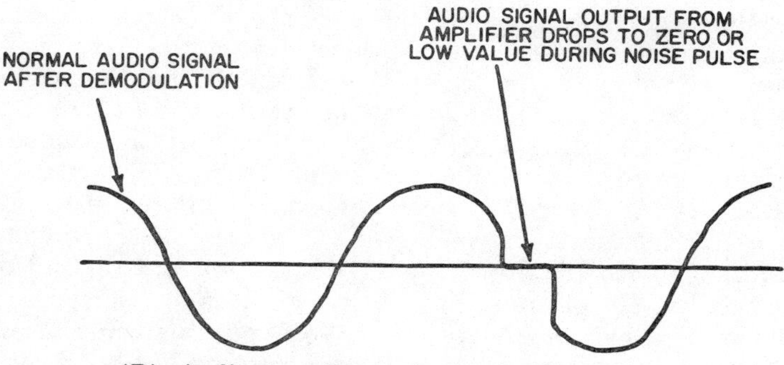

(B) Audio waveform after noise limiting.

Fig. 3-4. Effect of noise pulses on the RF and audio waveform.

V1 is biased by the voltage existing at voltage divider R2 so that it conducts continuously during normal operation. This is accomplished when the voltage divider current through R2 (shown a solid line) flows continuously in the counterclockwise direction, being drawn upward from ground and through R2 to the positive terminal of the power supply. The diode plate is connected to this positive terminal, whereas its cathode is connected to some point of lower positive voltage on R2. Because the plate is more positive than the cathode, the diode current shown in small dots will also tend to flow continuously. Its complete path begins at the ground connection at the right-hand end of R2. It then flows through part of R2, upward through R5, through the diode, downward through R4, into the positive terminal of the power supply, and through it to ground.

Because of this diode current flow, the voltage at the plate of the diode will be less than the power supply voltage by the amount of current "dropped" or developed across R4 by this same current. Also, during a negative half-cycle of audio, such as is shown in Fig. 3-5, a negative voltage is developed across R4 by the input current (shown a solid line). The amount of this negative voltage must be subtracted from the positive voltage which exists at the plate because of the biasing actions just described. Therefore, during a negative half-cycle of audio, the positive voltage at the diode plate will be reduced. This will cause a reduction in the amount of diode current (shown in small dots); this reduction causes a smaller voltage drop to exist across the diode cathode load R5.

The voltage at the top of R5 and the cathode of the diode is positive in polarity at all times. During noise-pulse reception (which is described later) when no diode current flows, the lowest positive voltage which the cathode can attain is reached; this will be the same amount of positive voltage as that which exists at the point on voltage divider R2 where the variable tap is placed. When a small amount of diode current flows during the negative half-cycles, the voltage at the top of R5 will be only slightly more positive than this value. When a large amount of plate current flows during positive half-cycles, the voltage at the top of R5 will be considerably more positive than this value.

Thus, it can be seen that the voltage at the diode cathode and the top of R5 fluctuates between two values of positive voltage, in accordance with the flow of audio-driving current up and down through R4. On negative half-cycles, such as are shown in Fig. 3-5, the diode cathode voltage has its least positive value. On these half-cycles, electron current will flow into the left-hand plate of C4. This action will drive an equal number of electrons downward through grid-driving resistor R6. This is the grid-driving current for V2 (shown in small dots). The downward flow of current through R6 places a negative voltage on the grid of V2, thereby reducing the plate current through this tube to a minimum value.

During a positive half-cycle of audio, such as that shown in Fig. 3-6, the detector current flowing through V3 and R1 cannot reverse its direction, since a diode is a unidirectional device. However, this current is reduced to a low value on positive half-cycles, causing a reversal of current direction in capacitors C1 and C2 and in resistor R4. During the negative half-cycles of Fig. 3-5, the upper plate of C1 "fills up" with accumulated electrons. During the positive half-cycles of Fig. 3-6, this reservoir becomes depleted.

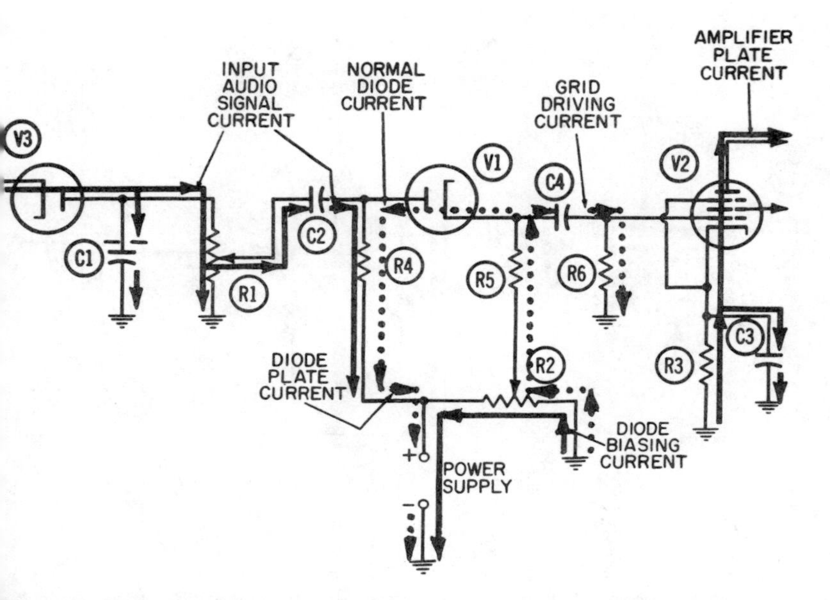

Fig. 3-5. Operation of the series-diode noise limiter—negative half-cycle.

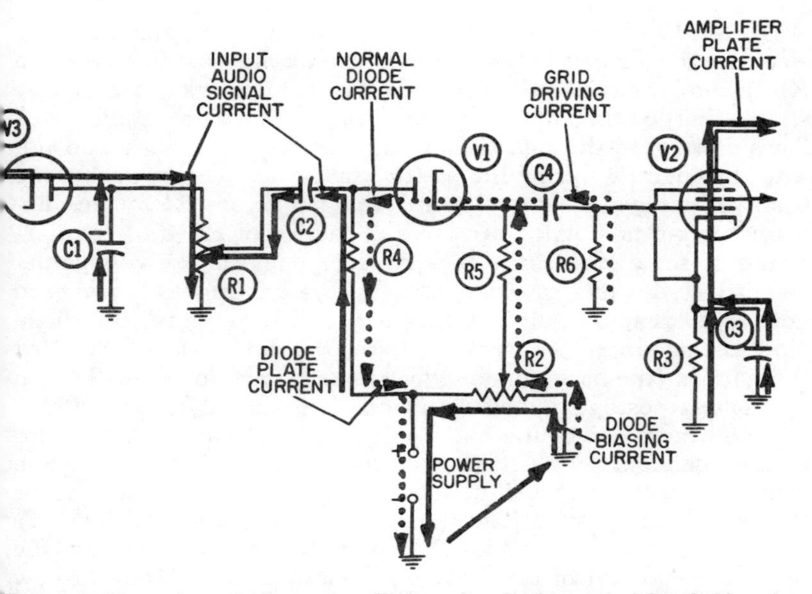

Fig. 3-6. Operation of the series-diode noise limiter—positive half-cycle.

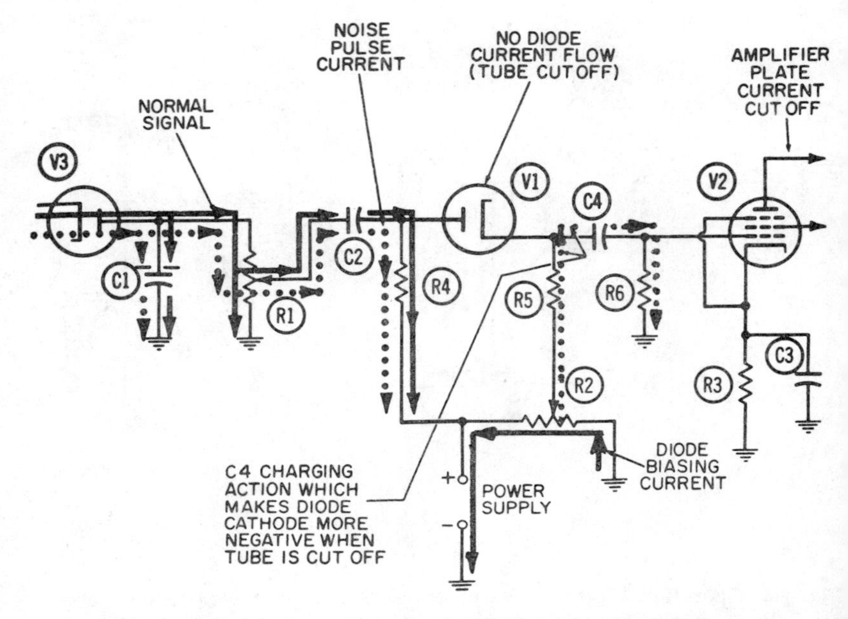

Fig. 3-7. Operation of the series-diode noise limiter—strong noise pulse being received.

The left-hand plate of C2 acts in a similar manner, accumulating electrons during negative half-cycles and discharging them back to R1 during positive half-cycles. In consonance with the charge and discharge action of C2, the input current (shown a solid line) flows downward through R4 during negative half-cycles, and upward through R4 during the positive half-cycles. Thus, during the positive half-cycles, the voltage developed across R4 by this upward current flow will be positive at the top of R4, thus counteracting, to some extent, the negative component of the voltage developed across this same resistor by the continuous downward flow of current coming through diode V1. Therefore, the diode plate is made more positive, *increasing* the diode current flow. At these times (the positive half-cycles) the cathode of V1 will reach its highest positive voltage; and on these half-cycles, electrons will be drawn out of the left-hand plate of C4 to supply the increased demand for electrons flowing into the diode. This action draws an equal number of electrons upward through resistor R6, and their upward flow (shown in small dots in Fig. 3-6) makes the grid of pentode V2 positive, thereby causing the maximum amount of plate current (shown a solid line) to flow through V2.

Operation During Noise-Pulse Reception

Fig. 3-4A indicates a typical noise pulse and the manner in which it will distort a normal carrier wave. Resistor R2 acts as a noise-level control in this circuit and will normally be set to a position so that the diode will conduct electrons during the entire range of any audio cycle which might be received. However, when a noise pulse is received, it is desirable that the diode not conduct. Fig. 3-7 shows the additional noise current that flows during receipt of a noise pulse; from this illustration we can see how the noise current cuts off the diode and further results in cutting off audio-amplifier tube V2.

The noise current (shown in large dots in Fig. 3-7) is flowing downward through R4. Since the noise signal is by nature much stronger in amplitude than the normal audio signal, this noise current develops a much stronger component of negative voltage at the top of R4. When this negative voltage is large enough, it will exceed the positive voltage which is simultaneously developed between the diode plate and cathode by the voltage-divider current flowing from right to left through R2. When this happens, the diode stops conducting.

When the diode stops conducting, the cathode voltage drops to its lowest positive value; this voltage drop is "coupled" across C4 to grid-driving resistor R6. Translated into terms of current flow, we find electrons flowing *onto* the left-hand plate of C4, driving other electrons *downward* through R6, and making the voltage at the top of R6 negative enough to cut off the electron flow through the pentode amplifier entirely. This portion of the action is identical (except in degree) to that which is depicted in Fig. 3-5 for C4 and R6, when a negative half-cycle of audio occurs.

DUAL-DIODE NOISE LIMITER

Figs. 3-8 and 3-9 show two separate moments in the operation of a dual-diode noise limiter. Fig. 3-10 shows a typical audio waveform, as distorted by a strong noise pulse. This particular circuit may be used in the audio section of a receiver to protect against a strong negative or positive noise pulse.

Identification of Components

The circuit of Figs. 3-8 and 3-9 is composed of the following components:

R1—Grid-driving resistor.
R2—Plate-load resistor.

R3—Grid-driving resistor.
R4—Cathode-biasing resistor.
C1—Coupling and blocking capacitor.
C2—Cathode-filter capacitor.
V1—Triode audio amplifier.
V2—Positive limiter diode.
V3—Negative limiter diode.
V4—Triode audio amplifier.
M1—Bias battery (or other voltage source).
M2—Bias battery (or other voltage source).

Identification of Currents

The following currents are at work in this circuit during normal (no noise pulses) operation:

1. Two grid driving currents.
2. Two triode plate currents.
3. Cathode filter for V4.

During abnormal operation when a noise pulse is present, one of the following additional currents will flow.

4. Positive limiting diode current through V2.
5. Negative limiting diode current through V3.

(Both of these currents will flow if the noise pulse has both positive and negative components, but the currents cannot flow simultaneously—they must flow in sequence, or consecutively.)

Details of Operation

With the two diodes and bias batteries removed from the circuit, it becomes a conventional RC coupled audio amplifier and, in normal operation, acts like one. In Fig. 3-8, if you disregard the current which flows through V2 (small dots), you see what appears to be a positive half-cycle in the operation of V1. (The term *positive* as applied to a half-cycle of operation is an arbitrary one and can be taken to refer either to the instantaneous grid voltage or the instantaneous plate voltage. It refers to the plate voltage of V1 in this example.)

The grid voltage is negative during this half-cycle, as shown by the downward movement of grid-driving electrons through R1 (solid line). The negative grid voltage reduces the flow of plate current through V1 and the downward flow of plate current through load resistor R2. Consequently, the voltage at the plate of V1 must become more positive. This rise in plate voltage draws an electron current upward through grid resistor R3 so that the

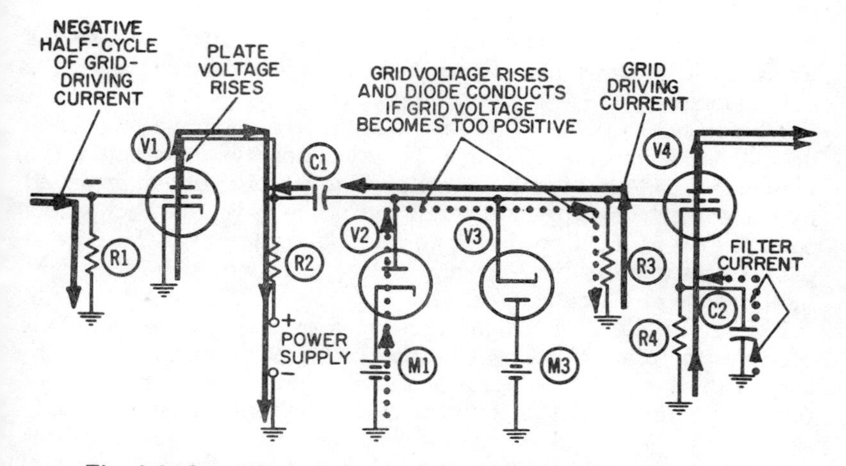

Fig. 3-8. Operation of the dual-diode noise limiter—positive noise pulse being received.

top of the resistor becomes positive in voltage. (The grid-driving current in R3 has also been shown as a solid line.)

In Fig. 3-10, that portion of the audio waveform which is above the center line has been arbitrarily labeled as "positive." That portion of the waveform which appears sinusoidal in nature is considered to be within the normal operating limits of the circuit. The bias battery, or voltage source shown below the cathode of V2 must be chosen so that it is *equal to or greater than* the voltage represented by the normal operating limit voltage. When this is

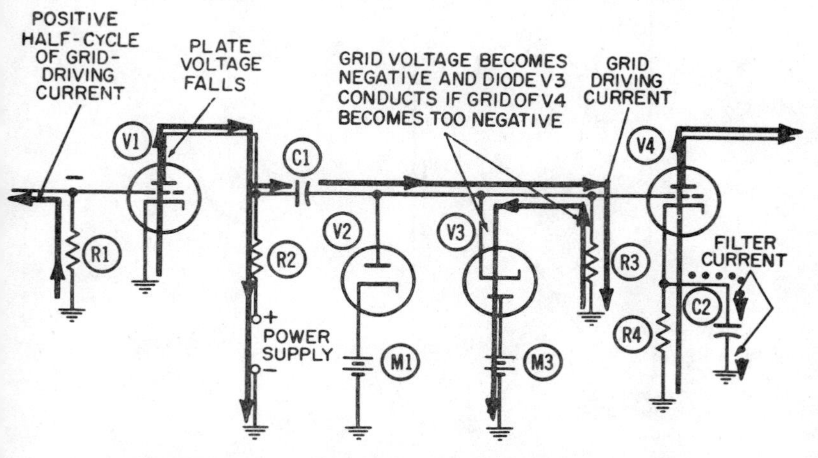

Fig. 3-9. Operation of the dual-diode noise limiter—negative noise pulse being received.

done, under normal conditions, V2 cannot conduct because the diode plate voltage can never become more positive than the cathode voltage.

When a strong noise pulse having a positive polarity is received, normal conditions are exceeded. The most positive voltage that the plate of V1 can attain is the value of plate-supply voltage provided by the power supply. It will reach this value only if and when the control grid of V1 is made negative enough to cut off the flow of plate current entirely. A strong noise pulse would be the most likely cause of V1 cutting off.

Before the plate of V1 (and the grid of V2) can become this positive, the plate of V2 will become more positive than its cathode, with the result that diode current will flow. This current (shown in small dots) flows along the path shown from cathode to plate within V2 and downward through R3 to ground. Since it is flowing through R3 in a direction *opposite* to the flow of the grid-driving current, it partially neutralizes the high positive voltage which would otherwise be developed across R3.

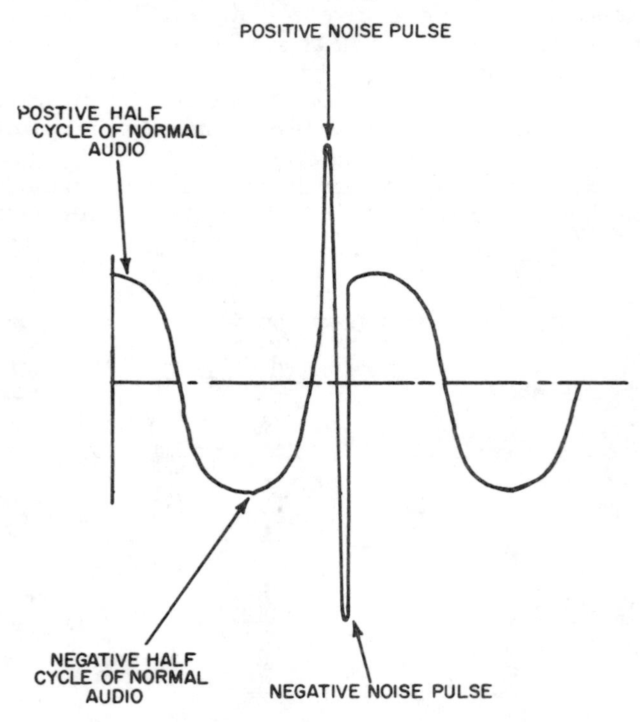

Fig. 3-10. Amplitude relationships between normal audio cycles and unwanted noise pulses.

As the noise pulse increases in strength, it tends to draw more electron current upward through R3; this, in turn, makes the diode plate even more positive and thereby increases the amount of diode current. Thus, a larger amount of diode current is available to flow downward through R3, tending to neutralize any increased positive voltage at the grid of V4 due to a stronger noise pulse.

Fig. 3-9 shows what might be called a negative half-cycle in the operation of the circuit (the term "negative" being used to describe the direction in which the plate voltage of V1 changes). The lowest point on the audio sine wave of Fig. 3-10 is well within the operating limits of the circuit; no current will flow through diode V3. The grid of V1 is made positive during this half-cycle as a result of current being drawn *upward* through R1. This releases a large pulsation of plate current into tube V1, and by virtue of their downward flow through load resistor R2, the plate voltage is lowered. This drop in voltage at the plate of V1 drives electron current downward through grid resistor R3, making the grid of V4 negative.

If a strong negative noise pulse is being received, the voltage at the grid of V1 will be made excessively positive, probably causing "saturation" current to flow in V1. This would lower the plate voltage of V1 to its lowest possible value, and thus, in turn, drive the grid of V4 so negative that it would probably be cut off. This is an undesirable condition, and one which exceeds the normal operating limits of the circuit. The bias voltage below the plate of diode V3 has a value which is so chosen as to be considerably less than this value of cutoff grid voltage. Consequently, before the grid voltage can reach such a negative voltage, the cathode of V3 will become more negative than its plate, and V3 will conduct an overload, or limiting current (a thin line). It flows upward through R3 and downward from the cathode to the plate of diode V3.

Since it flows upward through R3 while the grid-driving current is flowing downward through the same resistor, it partially offsets, or neutralizes, the large negative voltage which would otherwise exist at the grid as a result of the negative noise pulse. If the noise pulse becomes stronger, the diode conducts more electron current and thus tends to counteract the effects of the noise pulse flowing through R3.

Since both the positive and negative noise pulses cannot be occurring simultaneously (or they would cancel each other out) the two noise limiting diodes cannot conduct at the same moment. A noise pulse may be either positive or negative at any given instant, but not both.

SQUELCH CIRCUIT

In the reception of certain types of communications, it is necessary for someone to be listening to the receiver at all times, even when no signal is being received. This is done so that when a signal does come in, it will not be missed. A Federal airways ground station guarding several different channels or frequencies on several different receivers, all simultaneously, is a good example of this. There is a certain disagreeable background noise or hissing which comes from a receiver under conditions of no-signal reception. With two or more channels being guarded at the same time, this combination of background noises becomes most unpleasant, often leading even to inattentiveness on the operator's part and ultimate loss in communications.

The squelch circuit is a simple combination of parts which is designed to eliminate this undesirable condition. The squelch circuit is a carrier-operated device, or switch, which turns the audio amplifier off when no carrier signal is being received and turns it on when a carrier signal is present. The switching action uses the negative voltage which is associated with the AVC action to accomplish this function.

Two identical circuit diagrams have been selected to illustrate how this function has been accomplished. Fig. 3-11 shows a typical half-cycle of audio operation when a signal is being received. Fig. 3-12 depicts operating conditions when no signal is being received and the squelch circuit cuts off the amplifier tube.

Identification of Components

The squelch circuit and the audio amplifier contain the following individual components, with functions as indicated:

R1—Isolating resistor.
R2—Grid-return resistor for squelch tube.
R3—Variable resistor used as Squelch Control.
R4—Load resistor for V1; also serves as grid-driving and grid-return resistor for V2.
R5—Additional grid-return resistor.
R6—Voltage-divider resistor.
R7—Cathode-biasing resistor.
R8—Amplifier-load resistor.
C1—AVC storage capacitor.
C2—Screen-filtering capacitor.
C3—Audio-input coupling capacitor.
C4—Cathode-bypass capacitor.
C5—Output-coupling capacitor.

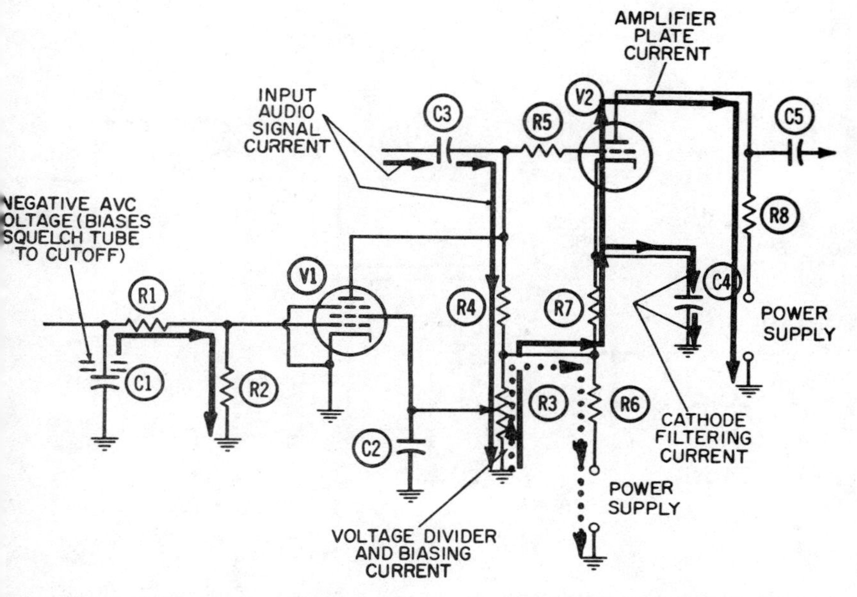

Fig. 3-11. Operation of the squelch circuit—negative half-cycle of audio being received.

V1—Pentode used for squelching purposes.
V2—Triode audio amplifier.

Identification of Currents

During the normal operation, when a carrier signal is being received, the following electron currents will flow:

1. Current discharge from AVC storage capacitor.
2. Input audio current.
3. Voltage-divider current.
4. Amplifier-plate current.
5. Cathode-filter current.

When no carrier signal is present, the voltage divider current (shown in small dots) is the only one of the foregoing currents that will flow. However, plate and screen current (both shown as solid line) will then flow in squelch tube V1.

Details of Operation

When a carrier signal is being received in a typical receiver, it will normally be used, among other things, to generate an AVC voltage. This voltage is stored on a large-value AVC storage capacitor. C1 in Figs. 3-11 and 3-12 provides this function.

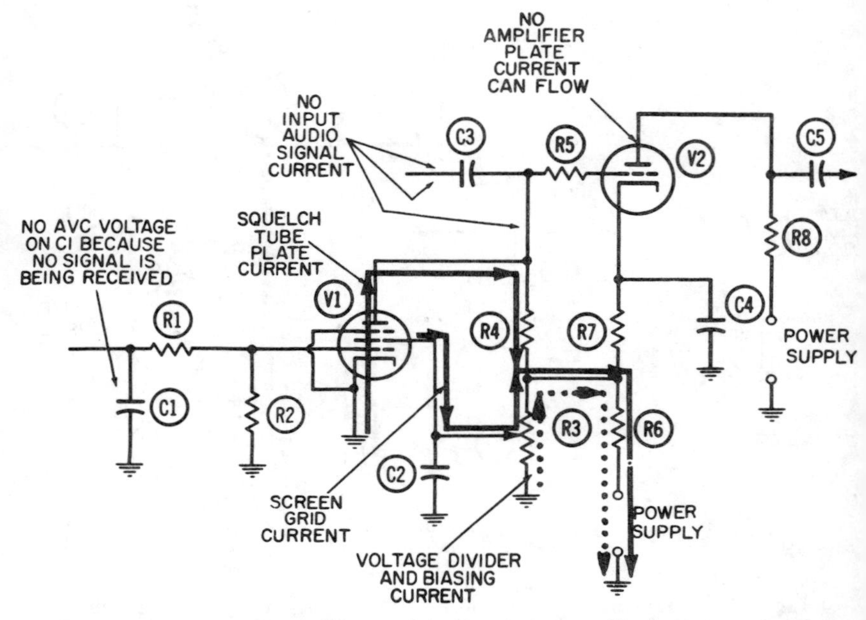

Fig. 3-12. Operation of the squelch circuit—no audio being received.

The AVC voltage is a negative voltage, consisting of an accumulation of electrons in storage on one plate of the capacitor. When the received carrier signal increases in strength, additional electrons are placed in storage on the capacitor, thereby increasing the amount of the negative AVC voltage. When the received carrier signal is reduced in strength, some electrons are given up from storage, thereby reducing the amount of the negative AVC voltage. When no carrier signal is present, all electrons in storage on the AVC capacitor will discharge to ground through R1 and R2 (Fig. 3-12).

None of the circuitry necessary for the generation of the AVC voltage has been shown in Figs. 3-11 and 3-12, nor discussed in this chapter. Refer to Chapter 2 of this book or to an earlier text, *Detector and Rectifier Circuits,* for a full discussion of this circuitry.

The negative voltage stored on the upper plate of C1 causes a continual discharge of electrons through R1 and R2, as shown in Fig. 3-11. This places the grid of squelch tube V1 at a sufficiently negative voltage to cut off the normal flow of plate current through this tube. Thus, whenever a carrier signal is being received, the squelch tube does not conduct. Under these conditions V2 is biased by the voltage divider current (shown in small dots) as well as by its own cathode-to-plate current (shown a solid line). It is driven by the audio-input signal (shown a solid line).

The voltage-divider current flows continuously along the path shown, upward through R3 and downward through R6 and into the power supply. Thus, a fixed positive voltage is provided at the junction of these two resistors. It will be noted that both the cathode and the control grid of V2 are returned to this same point, therefore the positive voltage resulting from the flow of voltage divider current biases the cathode and grid of V2 equally. This intrinsic voltage could be almost any value up to full B+, depending on the choice of sizes for R3 and R6.

When plate current flows through V2, it must follow the complete path indicated as a line in Fig. 3-11. This current originates at ground below R3, and flows upward through R3 before entering cathode resistor R7 and flowing upward through it to the cathode. It then flows through the tube from cathode to plate, downward through R8, and into the power supply, then through it to the common ground where it has ready return access to R3. In flowing upward through R7, an even more positive voltage is created at the cathode. Thus, the cathode is biased positively with respect to control grid, or stated differently the control grid is biased negatively with respect to the cathode—the normal operating condition for an amplifier.

The audio-input current (shown as a line) is coupled to the amplifier circuit via C3. Its complete path is, into the left-hand plate of C3, driving an equal number of electrons out of the right-hand plate and downward through R4 and R3 to ground. These two resistors, R4 and R3, consequently serve as grid-driving resistors for V2 in addition to their other functions as a plate-load resistor (R4) for V1 and a biasing resistor (R3) for the screen grid of V1 and the control grid and cathode of V2.

During the negative half-cycle of operation depicted in Fig. 3-11, the control grid of V2 will be made more negative than usual, and the plate current flowing through the tube will be reduced to a low value. Also, at this time, the cathode filter current for V2 will be flowing downward to filter capacitor C4.

During the positive half-cycles of audio operation, the audio current and the cathode-filter current will reverse directions. That is, audio current will be drawn out of C3; this action will, in turn, draw an equal number of electrons upward through R3 and R4 and into the right-hand plate of C3. This action adds a component of positive voltage at the control grid of V2, or, stated differently, it reduces the amount of negative bias existing between the grid and cathode of V2. Thus, more plate current will flow through the tube during the positive half cycles.

Operation When No Carrier Signal Is Present

Fig. 3-12 depicts the two currents which flow in the squelch circuit when no carrier signal is being received. When there is no carrier signal, there is, of course, no audio signal to be demodulated, and no AVC voltage can be developed from it.

When the negative AVC voltage normally stored on C1 is removed, V1 will begin to conduct. This plate current begins at the ground connection below the cathode, passes through the tube, and downward through R4 and R6, through the power supply and back to ground.

This downward flow of electron current through R4 makes the upper end of this resistor negative with respect to the lower end. Since the grid of V2 is connected to the upper end of R4, while the cathode of V2 is connected to the lower end of R4, conditions are suitable for restricting or cutting off the flow of plate current through V2. Normally, the amount of plate current through V1 will be of sufficient magnitude that it develops a large enough component of negative voltage across R4 to cut off V2. This, of course, is the basic purpose of a squelch circuit—to prevent the audio amplifier from conducting when no carrier is being received.

The variable tap on squelch control R3 can be used to vary the amount of positive voltage on the screen grid of V1. If this tap is placed near the lower end of R3, a fairly low positive voltage will be applied to the screen grid. Thus, it would be possible for a small AVC voltage to exist on the upper plate of C1, due to the reception of a weak or faraway signal, without the squelch tube conducting. If the variable tap should be moved upward on R3, the screen grid would eventually become sufficiently positive for V1 to start conducting and cut off V2.

The squelch control provides a simple means of selecting or rejecting signals of any desired strength or intensity. This feature is particularly attractive for adapting to variable atmospheric phenomena, as well as varying operating conditions or criteria.

Chapter 4

HALF-WAVE POWER SUPPLY

The primary function of a rectifier power-supply circuit in a typical piece of electronic equipment, such as a radio, is to convert the alternating current (AC) which is supplied to homes to direct current (DC). This is a necessary function, because vacuum-tube circuits require the application of fairly high and stable voltages to the tube plates and screens. The purpose of this chapter is to clarify the difference between an alternating current, or voltage, and a direct current, or voltage, and then to show how this direct voltage is achieved with this circuit.

HALF-WAVE RECTIFIER CIRCUIT

Figs. 4-1 and 4-2 show the two half-cycles of operation in a half-wave rectifier circuit. While a transformer has been employed in this circuit, it is often omitted and the power line is connected (through the switch) directly to the rectifier tube plate. In either case, operation of the circuit is the same.

Identification of Components

The components of the power supply circuit shown in Figs. 4-1 and 4-2 and their function are as follows:

R10—Filter resistor.
C14—Filter capacitor.
C15—Filter capacitor.
T4—Power transformer.
V5—Half-wave rectifier tube.

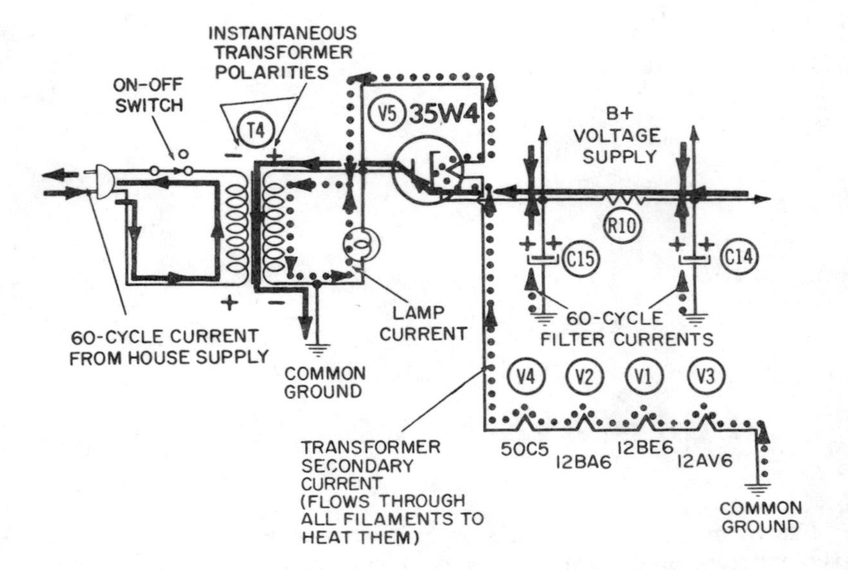

Fig. 4-1. Operation of a half-wave power supply—positive half cycle.

In addition, the filament circuits for all tubes in the radio (V1 through V5) are included in Figs. 4-1 and 4-2. Each of these filaments heats the cathode of its respective tube so that the cathodes may emit electrons within the tubes.

Identification of Currents

There are four separate and distinct electron currents at work in the circuit of Figs. 4-1 and 4-2 during normal operation. These electron currents are:

1. Transformer primary current.
2. Transformer secondary current.
3. Rectifier plate current.
4. 60-cycle filter currents.

Details of Operation

The On-Off switch of the radio is shown in the upper left-hand corner of Figs. 4-1 and 4-2. When this switch is opened, or "Off," the radio is isolated from the house electrical supply, and no electron currents will flow in any of the circuit components. When the switch is closed, or "On," (Figs. 4-1 and 4-2) the house current (solid line) will flow back and forth through the primary winding of T4.

A half-cycle when the house current is flowing upward through the primary winding of T4 is depicted in Fig. 4-1. The instantaneous voltage polarities across the primary winding are as shown, namely, the upper end of the winding is negative and the lower positive. These are the polarities of the applied voltage, meaning that these voltages are applied from the house supply and, in turn, cause the primary current to flow.

The secondary current is shown as flowing downward through the secondary winding. Associated with this current are the instantaneous polarities of the induced voltage, which are positive on the upper end and negative at the lower end of the secondary. This current also flows through the filaments of the five vacuum tubes, which are connected in series. During the half-cycle represented by Fig. 4-1, this filament current flows to the left through all of the filaments. Each of the filaments is heated by this process very much as the coil of an electric toaster is heated.

During the half-cycle represented by Fig. 4-2, the direction of flow of both the primary and secondary currents is changed. The primary current flows downward, and the secondary current flows upward and to the right through the tube filaments. The voltage polarities across the primary and secondary windings are also reversed during this half-cycle. Since we have assumed the supply current to be conventional 60-cycle, the positive and negative half-cycles shown in these two illustrations are repeated 60 times each second.

Diode Tube Operation

Referring again to Fig. 4-1, the diode plate current (solid line) is shown flowing from the cathode to the plate of V5. The complete path of this current is from the filter system made up of resistor R10 and capacitors C14 and C15, through V5, and through the secondary winding of the T4 to ground.

In order for a vacuum tube, such as V5, to allow electron current to flow across the open space within the tube, two essential conditions must be met:

1. The cathode must be heated by some external means so that it will emit electrons into the tube. This heating is accomplished by placing the cathode very close to the filament, which is heated by the flow of transformer secondary current described previously. The emission process is analogous to the action that occurs at the surface of boiling water, when very small droplets of water appear to jump free from the surface.
2. The plate of the tube must be at a more positive potential

than the cathode in order to attract the negative electrons across the tube. Since every electron possesses one unit of negative charge, it will be repelled by any negative voltage and attracted by any positive voltage.

In V5 the first of these conditions is fulfilled whenever the On-Off switch is closed. The second condition is fulfilled only during the positive half-cycles depicted in Fig. 4-1. Thus, diode current (solid line) flows only during the positive half-cycles.

It is important to understand the complete diode-current path. All of the plate and screen-grid currents from the four other tubes eventually join together at the junction of R10 and C15 to make up the diode current. This current flows through diode V5 on positive half-cycles only; therefore, it can be described as a pulsating DC. The purpose of the filter system composed of R10, C14, and C15, is to smooth out this pulsating flow of electron current so that the currents flowing from the other tube plates and screen grids will flow smoothly up to the filter circuit.

The Filter Circuit

R10 and C15 form a simple RC filter circuit, which operates to create a high positive voltage on the upper plate of C15. This high positive voltage is obtained as a result of the departure of many electrons which are drawn into V5. Electrons leave the upper plate of C15 in pulsations (whenever the plate of V5 is more positive than the cathode), creating an electron deficiency on the upper plate of C15. It is this electron deficiency, which is the same thing as a positive voltage, that attracts electrons from the plates and screen grids of other tubes.

As electrons flow in from the other tubes, they would tend to equalize or neutralize the positive voltage on C15. However, other electrons are being drawn into V5 as fast as they are arriving from the other tubes, with the result that a positive voltage remains on the upper plate of C15 as long as the radio is turned on. Thus, we have a situation where electrons continuously flow onto the upper plate of C15, and intermittently flow from this same capacitor and into the diode tube V5.

The amounts of electrons involved in these two current patterns will eventually stabilize and be equal to each other, resulting in a voltage on the upper plate of C15 which is almost as positive as the peak voltage on the upper terminal of the transformer secondary winding.

It is the positive voltage at the upper end of the transformer secondary winding that attracts electrons across V5 during the positive half-cycles. Likewise, it is the resulting positive voltage on the upper plate of C15 that draws electrons from the plates and screen grids of the other tubes. The positive voltage on the upper plate of C15 is a DC voltage, as a result of the filter action which occurs between R10 and C15. These two components form what is known as a long time-constant circuit. This is a mathematical term, but its significance can be explained and understood with the use of some simple arithmetic. When any resistor and capacitor are connected together, they have a time constant, which is determined by the product of the values of the two components. This relationship is expressed by the formula:

$$T = R \times C$$

where,

T is the time constant of the combination in seconds,
R is the resistor value in ohms,
C is the capacitor values in farads.

The time constant of any circuit is considered to be long when it is at least several times longer than the time or period of one cycle of the current which is passing through the combination. Assuming the value of R10 is 680 ohms and C15 is 80 mfd, the time constant of the two components is:

$$\begin{aligned} T &= R \times C \\ &= 680 \times 80 \times 10^{-6} \\ &= 54{,}400 \times 10^{-6} \\ &= .0544 \text{ second} \\ &= \text{approximately } 1/18\text{th of a second.} \end{aligned}$$

Since the frequency of the current this filter is trying to handle is 60 cps, the time or *period* of one of these cycles is one sixtieth of a second. Thus, the time constant of the combination of R10 and C15 is more than three times as long as one of these periods.

The positive voltage on the upper plate of C15 can be likened to a pool of positive ions. As more negative electrons are drawn away from this pool and flow through V5, the number of positive ions on C15 will increase by the same amount. Also, as electrons flow into this pool of ions from the plates and screen grids of the other tubes. (Fig. 4-2), the number of positive ions will be reduced accordingly.

Positive ions in concentration represent a positive voltage the amount of which is directly proportional to the number of ions present. The voltage at the cathode of V5 and at the upper plate of C15 in a typical radio will be +125 volts. This voltage does not change significantly from half-cycle to half-cycle, because the quantity of electrons which leave the capacitor on the positive half-cycles is an insignificant percentage of the quantity of positive

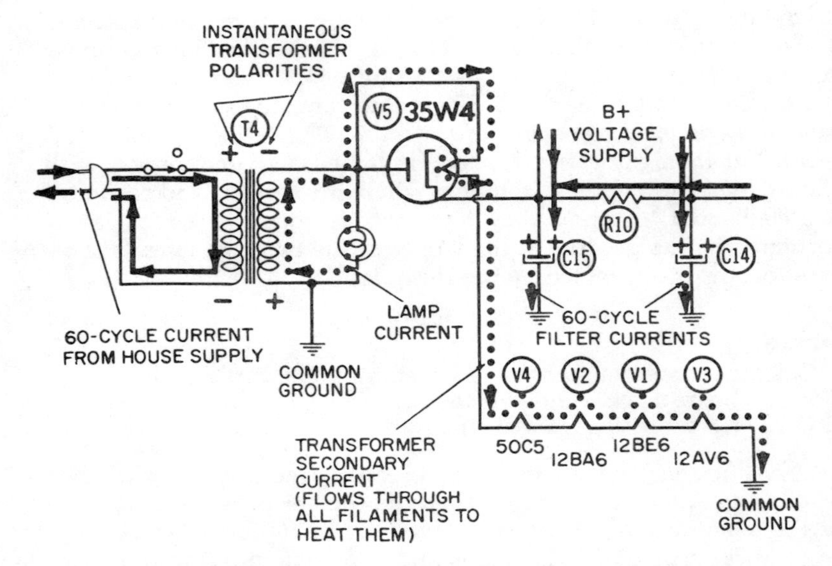

Fig. 4-2. Operation of a half-wave power supply—negative half cycle.

ions already stored there, and the electrons which do leave are replenished during the negative half-cycles.

Ripple Voltage

All power-supply filter circuits exhibit a *ripple voltage*. The ripple voltage is the minute fluctuations in output voltage that exist as a result of electrons being drawn away from the voltage pool on C15 on the positive half-cycles. In a very long time-constant circuit, this ripple voltage will be an exceedingly small fraction of a volt. In a less sophisticated system, such as the typical radio, it will be much larger. However, if the ripple factor becomes too large, the operation of a radio will be adversely affected, and you will hear a 60-cycle hum along with the regular program.

A filter capacitor, such as C15, operates like a mechanical shock absorber. On each positive half-cycle, such as is depicted in Fig. 4-1, electrons are drawn away from the upper plate of C15 and through V5; an equal number will also be drawn upward from ground onto the lower plate of the capacitor. Likewise, on the negative half-cycles, such as are depicted in Fig. 4-2, V5 is not conducting, but the positive voltage on the upper plate of C15 will continue to draw electrons onto it from the plates and screen grids of all other tubes; this action will drive an equal number of filter current electrons downward from the lower plate of C15 to ground. This filter current has been shown in large dots in Figs. 4-1 and 4-2.

Other Tube Currents

The electron currents which flow in and out of C14 are regulated and controlled by the same events which drive the currents in and out of C15, but to a lesser degree. All of the tube currents (plate and screen grid currents) from V1, V2, and V3, plus the screen-grid current from V4, must flow through R10 on their way to the power supply. (The plate of V4 is connected to the junction of C15 and R10.) These currents flow toward V5 because the diode plate is made positive on the positive half-cycles drawing electron current from the upper plates of both C15 and C14. Fig. 4-1 depicts these positive half-cycles, and shows electron current being drawn *out* of the upper plate of capacitor C14, and to the left through R10 to the cathode of the power supply diode V5.

Fig. 4-2 depicts the negative half-cycle of operation, when the diode plate is negative, and no electron current crosses V5. During these half-cycles, the positive voltage on the upper plates of C15 and C14 will continue to draw electron current from the plates and screen grid currents of the other four tubes. During this half-cycle, these electron currents will flow *onto* the upper plates of these two capacitors, and will drive filter currents from the lower plates to ground. These flow directions have been indicated by arrows in Fig. 4-2.

Summary

The principal function of any rectifier power-supply circuit is to utilize the alternating voltage from a typical wall-plug circuit, and from it to obtain the high positive voltage required for vacuum-tube operation. This high positive voltage exists on the upper plates of C14 and C15 and is depicted by the plus signs on the upper plates of these capacitors.

Although currents regularly flow onto the upper plates of these capacitors during the negative half-cycles and out of these same capacitors during the positive half-cycles, the resulting *changes* in the amount of positive ions stored on the capacitors are insignificantly small. Therefore, the positive voltage which these stored ions represent does not change. All of the plate and screen grid currents from the other tubes in the radio must eventually flow through diode V5.

Chapter 5

REGENERATION

Regenerative circuits were widely used in the early days of radio. For example, a regenerative detector was often employed to detect, or demodulate, very weak radio-frequency signals because of its ability to amplify the signal as it was demodulated. Such signals might have been continuous wave (CW) signals used for the transmission of code, or amplitude-modulated (AM) signals carrying voice or entertainment-type information.

REGENERATIVE DETECTOR

A typical regenerative detector circuit is given in Figs. 5-1, 5-2, and 5-3. Gains in signal strength of 10,000 or 12,000 are common with this type of circuits. In modern entertainment-type equipment, the regenerative detector is seldom encountered; however, it is often employed in small communications-type receivers.

The development of new RF amplification techniques has led to the abandonment of the regenerative detector in entertainment-type equipment. For example, an RF amplifier circuit with a voltage gain, or amplification, of 30 times is not uncommon. Three such stages in series, or cascade, would have an over-all gain of 27,000. While such a circuit would require three tubes instead of one and appear to invite more complexity and cost, it eliminates the inherent disadvantages of the regenerative detector, namely, the need for critical adjustment. A slight misadjustment in a "regen" detector can cause it to go into self-sustained oscillation, even in the absence of a radio-frequency signal.

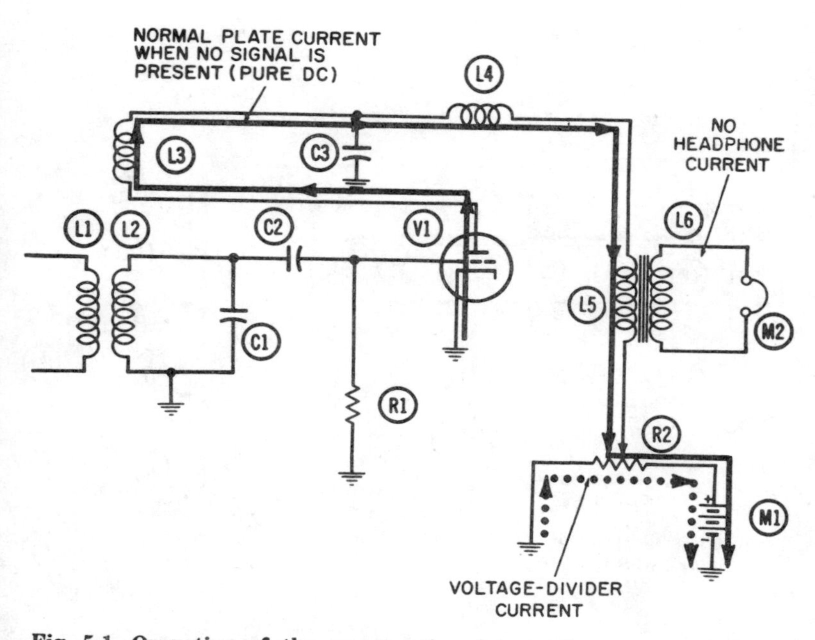

Fig. 5-1. Operation of the regenerative detector—no received signal.

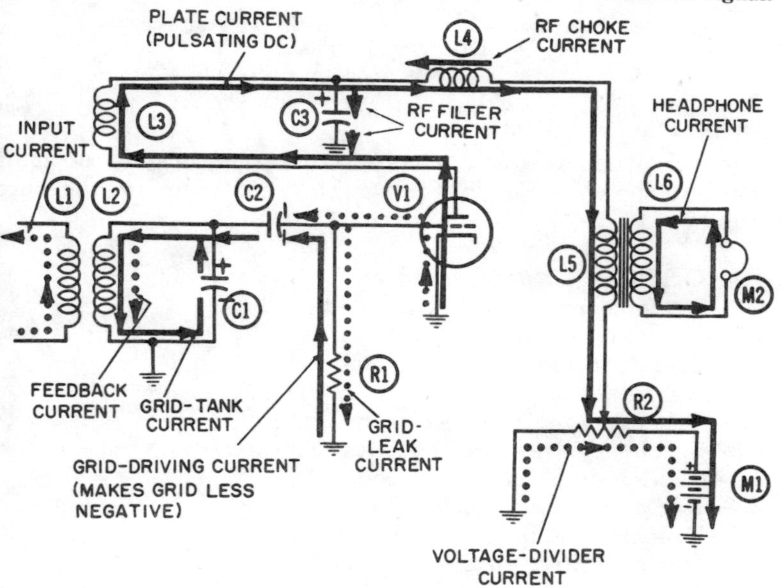

Fig. 5-2. Operation of the regenerative detector—positive half-cycle of signal.

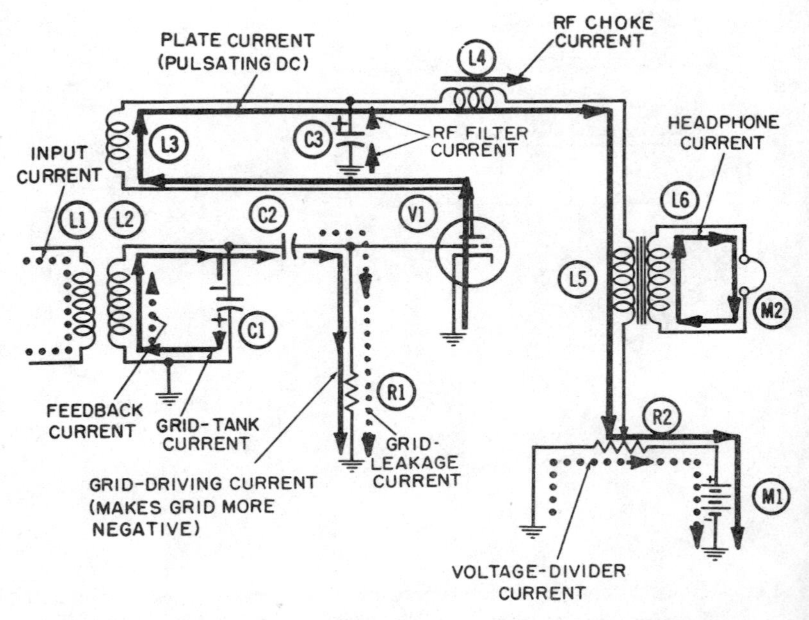

Fig. 5-3. Operation of the regenerative detector—negative half-cycle of signal.

Identification of Components

The components which make up the regenerative detector circuit are shown in Fig. 5-1, 5-2, and 5-3, and their functions are as follows:

R1—Grid-leak biasing resistor.
R2—Variable resistor used as voltage divider.
C1—Tuned-tank capacitor.
C2—Coupling and biasing capacitor.
C3—RF filter capacitor.
L1—Primary winding of RF transformer.
L2—Secondary winding of RF transformer.
L3—Tickler or regeneration coil.
L4—Radio-frequency choke coil.
L5—Primary winding of AF transformer.
L6—Secondary winding of AF transformer.
V1—Detector-amplifier tube.
M1—Power supply.
M2—Headphones.

Identification of Currents

In the absence of an RF signal, only two significant electron currents will flow in this circuit. These currents are:

1. Plate current through V1.
2. Voltage-divider current through R2.

When an RF signal is being received, seven *additional* currents flow in the circuit. These currents are as follows:

3. RF input or signal current.
4. RF tank current.
5. RF grid-driving current.
6. Grid-leakage current.
7. RF filter current through C3.
8. Audio current through headphones.
9. Feedback, or regenerative, current.

Details of Operation

In Fig. 5-1, when no RF signal is being received, only two electron currents—the voltage-divider and the tube current—are flowing. The voltage divider is placed across the power supply to provide the operator a means of controlling the amount of regeneration and, consequently, the amount of amplification available from the tube. The voltage-divider current (shown in large dots) flows continuously from ground, from left to right through R2 and enters the positive terminal of the power supply, returning through the power supply to ground. Because of this continuous current movement through R2, a progressively higher positive voltage exists at each successive point from left to right. Thus, the position of the movable arm on R2 determines the amount of positive voltage applied to the plate of V1; this directly affects the amount of tube current that will flow.

The tube current (shown as a line) consists of electrons which are drawn out of ground below the cathode. The heated cathode causes them to be emitted into the tube where they are attracted across the tube by the positive voltage on the plate. From the plate, they flow successively through L3, L4, L5, and R2. From R2 they are drawn into the positive terminal of the power supply, through which they must be delivered back to ground in order to have ready return access to the cathode of the tube.

Once stable conditions exist and when no signal is being received, this plate current is a pure DC. Consequently, no feedback or regeneration can exist between L3 and L2, or L5 and L6.

In Fig. 5-2 the additional currents which come into existence when a signal is being received from some transmitting station are shown. L1 may be considered as connected directly to an antenna; therefore the signal current induced in the antenna flows directly up and down through L1. This current is shown in large dots; in Fig. 5-2, it is flowing upward through the coil. The continual up and down flow of this current through L1 induces a companion current to flow down and up through secondary winding L2. This current is shown as a line; in Fig. 5-2 it is flowing downward. This action delivers electrons to the lower plate of C1, making it negative, and withdraws electrons from the upper plate of C1, making it positive. Whenever the voltage on the upper plate of C1 is positive, it draws electrons toward it from any external circuit to which it may be connected. In this circuit it draws electrons upward through grid resistor R1 and onto the left-hand plate of C2. This current (shown as a line) becomes the electron current which drives the grid; in Fig. 5-2 it drives the grid to a positive peak of voltage.

(A simple rationale for correctly relating current directions to the resulting voltage polarities exists. Since electrons are themselves negative in nature, they will always flow away from a more negative area and toward a less negative area. Thus, an upward flow of these grid-driving electrons through R1 tells us that the voltage at the upper end of the resistor is more positive than that at the lower end.)

When the grid of V1 is made positive by this flow of grid-driving current, two important things happen within the tube: first, the amount of plate current flowing through the tube is increased; and second, grid leakage electrons flow out of the tube at the control grid.

The additional surge of plate current must flow upward through L3 on its way to the power supply. As this plate current increases, it induces a separate current in L2, to which coil L3 is inductively coupled. This new induced current, which is shown in large dots, is the feedback current which provides the basic regenerative action which gives the circuit its name. Since it is flowing downward simultaneously with the downward flow of the tank current, the feedback current reinforces the tank current.

Fig. 5-3 shows the current conditions a half-cycle later. It is called a negative half-cycle because the grid-driving current is flowing downward through R1, making the voltage at the top of this resistor negative. This grid-driving current is itself being driven by the negative voltage on the upper plate of tank capacitor C1. This negative voltage results from the fact that the tank-current electron flow has reversed during this half-cycle, and electrons are flowing upward through L2 to the upper plate of C1, making it negative.

The negative voltage at the control grid during this half-cycle restricts or reduces the plate current flowing through the tube. A *decrease* in the amount of current flowing upward through L3 can induce a decrease in the current flowing downward through L2, or it can induce an increase in a current flowing upward through L2. For convenience, the latter case has been depicted in Fig. 5-3. Since the feedback and tank currents are both flowing upward simultaneously, one reinforces the other, and regeneration of the received signal occurs on both half-cycles.

Because of regeneration, an extremely weak input signal flowing through L1 may be amplified many thousands of times. With the appropriate adjustment of R2 (which controls the amount of plate current through the tube), the circuit can be operated just below the point of oscillation. A slight increase in the coupling between L2 and L3, or in the plate voltage applied to the tube, would cause the circuit to go into self-sustained oscillations, even in the absence of an input signal. This adjustment is one of the major disadvantages of the regenerative detector.

On the positive half-cycles depicted by Fig. 5-2, grid-leakage current flows. The complete path of this current (shown in small dots) begins as usual at the ground connection below the cathode of the tube. Whenever a control grid has positive voltage on it, it will attract some of the electrons from the electron stream passing through the tube. Once these electrons strike the control-grid wires, they cannot be re-emitted into the tube; therefore, they must exit via the control grid. Eventually, the grid-leakage electrons will flow downward through grid resistor R1 and back to ground. If R1 is made large enough (as it usually is in practical circuit design), the electrons cannot flow immediately to ground but will first accumulate on the right-hand plate of C2, thereby building up a negative voltage, known as a *grid-leak bias voltage*.

During the negative half-cycles represented by Fig. 5-3, the grid is negative; consequently, it does not attract any electrons from the plate-current stream crossing the tube. However, those electrons which had previously accumulated on the right-hand plate of C2 constitute the negative grid-bias voltage and will continue to discharge downward through R1. This action is made possible by the relationship between the large sizes of C2 and R1 and the time duration of an individual cycle of the signal current. The combination of C2 and R2 in this type of circuit will invariably form a long time-constant network.

C3 acts in conjunction with radio-frequency choke coil L4 as a filter whose purpose is to filter or remove the RF pulsations which characterize the plate-current stream. As each individual pulsation reaches the entrance to L4, it has a choice of flowing into the high impedance represented by the choke coil or the low impedance represented by the capacitor. Most of the strength of each pulsation will flow downward into C3 during the positive half-cycles of Fig. 5-2, driving an equal number of electrons downward from the lower plate of C3 into the ground connection. During the negative half-cycles depicted in Fig. 5-3, the plate current has diminished; therefore, the filter current through C3 reverses and flows upward.

The action of the radio-frequency choke, when confronted by RF pulsations in plate current, is interesting. The purpose of any such choke is, of course, to prevent or inhibit the passage of high-frequency current. As each pulsation of plate current enters this choke coil at its left hand terminal, the increase in current which it represents generates a small current flowing in the opposite direction. This current is shown as a solid line, and, in

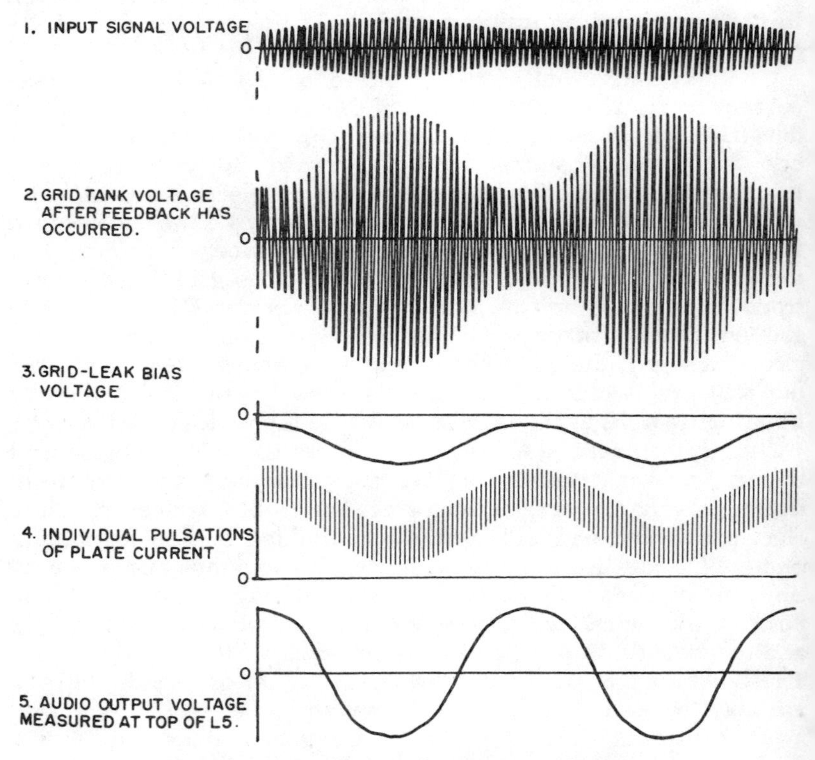

Fig. 5-4. Regenerative detector waveforms.

Fig. 5-2, it flows from right to left, thereby reducing the total change in current which would otherwise occur, and also satisfying the basic electrical property possessed by any inductance, namely, that it always reacts to any change in applied current flow in such a manner as to oppose that change.

In Fig. 5-3, when the negative grid-driving voltage causes a reduction in the amount of plate current through the tube and eventually through coil L4, the choke coil again reacts in such a manner as to oppose this reduction in current. Thus, it brings into existence the small choke-coil current (shown a solid line), this time flowing in the same direction as the plate current, namely, from left to right.

The actions of C3 and L4 are essentially independent actions. The capacitor action is intended to filter as much of the high-frequency component (the pulsations) of the plate current as possible to ground before it reaches the load, which is the primary winding of audio transformer L5. The choke-coil action is provided to prevent these pulsations from reaching the audio transformer. Thus, when the two components are put together they provide a highly effective filter combination, with a low-impedance path in the desired flow direction (to ground in this example) and a high-impedance path in the undesired flow direction (the output load and the power supply).

Fig. 5-4 shows a graphical representation of certain current and voltage waveforms at various points in the regenerative detector circuit. Line 1 of Fig. 5-4 shows the very weak nature of the input signal as it is received in L1. Line 2 shows essentially the same waveform, after it has been vastly amplified or regenerated by amplifier tube action plus feedback action. Line 2 represents the strength or amount of tank current which flows in the tuned circuit (L2 and C1).

Line 3 (Fig. 5-4) indicates the relative strength of the grid-leak bias voltage which is detected by the grid-leak circuit combination of C2 and R1. Since any grid-leak bias voltage always consists of "trapped" or stored electrons, this voltage is always negative. We can see from a comparison of Line 3 with Line 2 that when a modulation peak occurs in the input signal, the individual RF cycles exhibit their maximum strength, and the leakage bias voltage reaches its maximum negative value. This is because the stronger RF cycles drive the grid to higher positive voltage values, and these higher positive voltages draw more grid leakage electrons out of the tube each cycle, and into storage on the right hand plate of C2.

Line 4 of Fig. 5-4 indicates the complex nature of the plate current flowing through V1. Since tube current is always a one-way current, or unidirectional in nature, the term pulsating DC is normally used to describe it. Inspection of Line 4 reveals that the plate current is pulsating at both a radio frequency and an audio frequency. Each cycle of the RF signal applied to the grid causes a small pulsation to occur in the plate-current stream. This action is accomplished primarily by the grid-driving current flowing up and down in R1. Each cycle of audio-modulating voltage which is demodulated by the grid-leak bias arrangement causes a larger (and slower) pulsation to occur in the plate current.

Since L3 and L4 have extremely small amounts of inductance, the slow audio pulsations will pass through them without being affected. (The reactance of any inductor is directly proportional to the frequency of the current flowing through it.)

Lines 3 and 4 point out that when the maximum negative grid-leak bias voltage exists on capacitor C2, the audio pulsation in the plate current stream is at a minimum. We can arbitrarily decide to label this period as a negative half-cycle of audio. The plate current flows downward at all times through the primary winding of the audio transformer. When plate current is approaching its minimum value, the steady decrease in plate current will induce a current flow downward in winding L6. This is the current which is shown as a thin line in Figs. 5-2 and 5-3.

When the plate current is approaching its maximum value, the steady increase in plate current will induce a current to flow upward in the secondary winding. Thus, a current which flows up and down at the basic modulation frequency through the headphones is brought into existence.

SUPERREGENERATIVE RECEIVER

It is only a short and logical step from the regenerative detector circuit of Figs. 5-1, 5-2, and 5-3 to the superregenerative receiver shown in Figs. 5-5 and 5-6. There are two important difference between the two circuits: (1) the regenerative detector portion of the superregenerative receiver is permitted to oscillate at the frequency of the signal being received, and (2) a separate circuit, called a *quenching oscillator*, is provided to stop the oscillations.

Identification of Components

The upper portion of Fig. 5-5 is a regenerative detector, similar in most respects (but not all) to the circuit explained previously. The components in Fig. 5-5 have been numbered wherever possible to coincide with their counterparts in Fig. 5-1. The additional components which make up the quench oscillator are as follows:

R3—Grid-leak biasing resistor.
C5—Tuned-tank capacitor.
C6—Grid-leak storage capacitor.
C7—Coupling capacitor.
L7—Coupling inductor.
L8—Tuned-tank inductor (autotransformer).
V2—Quench-oscillator tube.

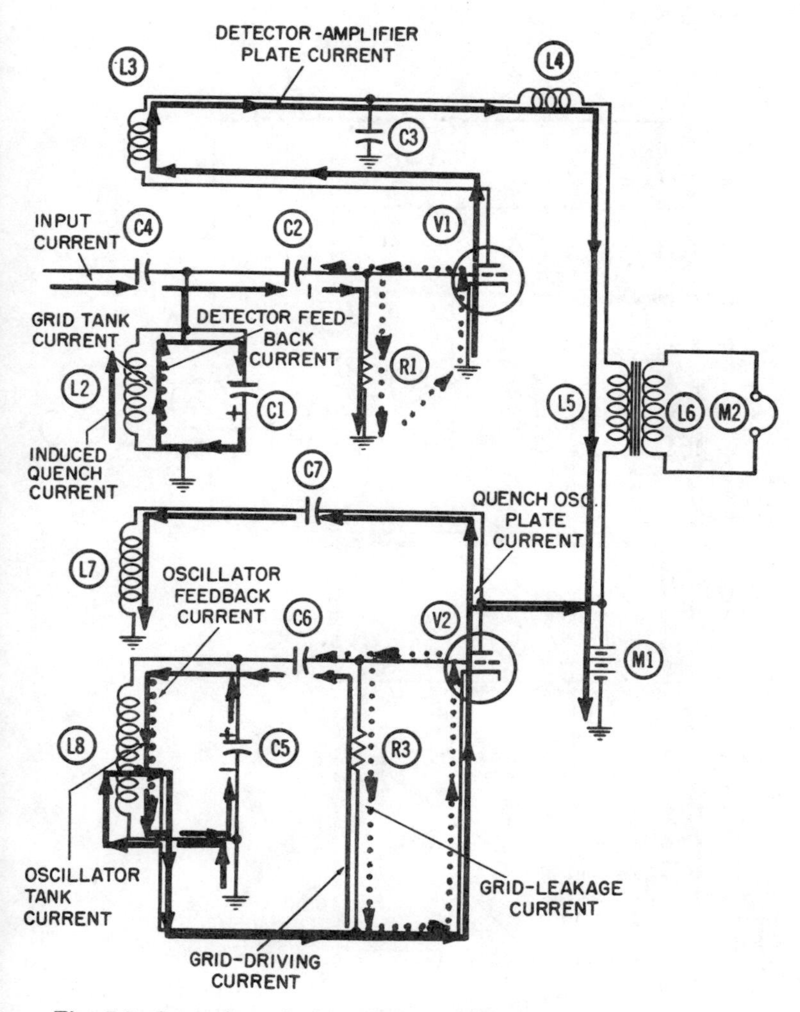

Fig. 5-5. Operation of the superregenerative receiver—negative half-cycle of quench oscillator.

Identification of Currents

The quenched-oscillator portion of this circuit operates at a much lower frequency than the signal frequency being received. A quench frequency of about 20 kc is normal. The quenching oscillator is a conventional Hartley oscillator. The various currents are as follows:

1. Tank current.
2. Grid-driving current.

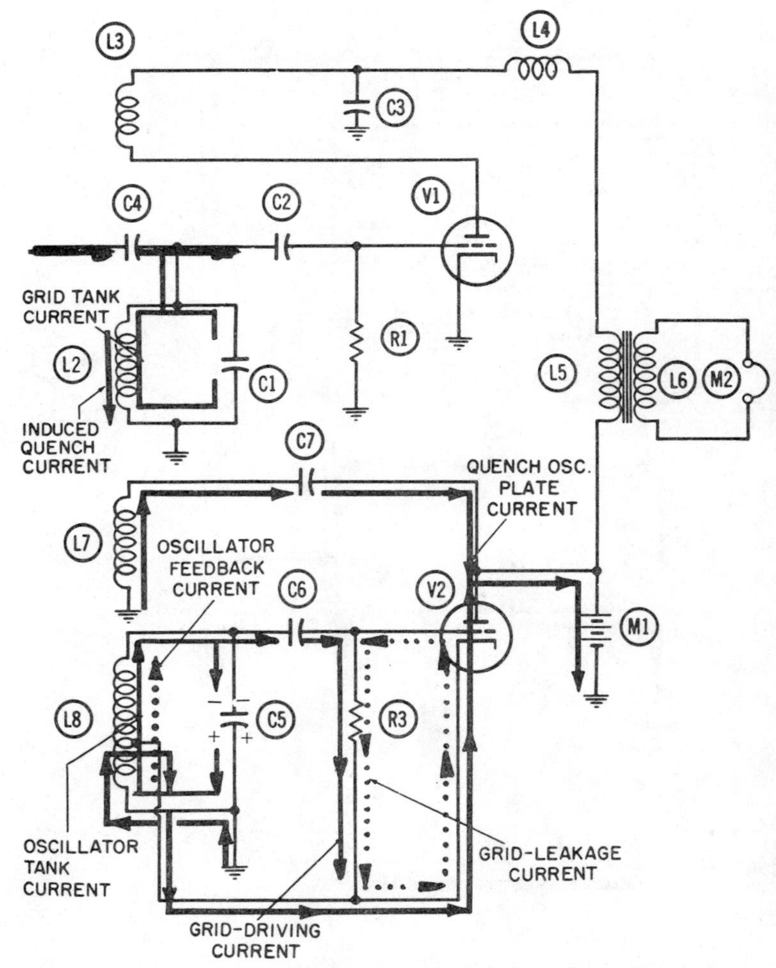

Fig. 5-6. Operation of the superregenerative receiver—positive half-cycle of quench oscillator.

3. Grid-leak biasing current.
4. Plate current.
5. Oscillator feedback current.
6. Autotransformer feedback current.
7. Quenching current.

Details of Operation

Figs. 5-5 and 5-6 are chosen to represent two alternate half-cycles in the operation of the quenching oscillator. It must be recognized that because of the much higher frequency of the received signal (to which the regenerative detector is tuned), many cycles of operation of the detector portion of the circuit will occur during a single cycle of the quench oscillator. The regenerative detector is normally operated just below the point of oscillation so that a slight increase in the amount of inductive coupling between L2 and L3 will provide sufficient regeneration and feedback to cause the detector circuit to oscillate.

This condition represents the maximum possible amount of gain, or amplification of the received signal. By itself, it is not an acceptable circuit technique, because once the oscillation is started it will continue even if the received signal goes off the air. Thus, the continued oscillation would indicate a signal which is actually not present. To protect against this eventuality, the quench oscillator turns the oscillating detector circuit off at its own basic frequency, namely, 20,000 times each second. Once the oscillating detector has been stopped from oscillating, it cannot be started up again unless the desired input signal is again on the air.

Fig. 5-5 depicts a positive half-cycle of the quench oscillator, when V2 is delivering a maximum surge of plate current. The oscillating tank current (a solid thin line) has moved downward through L3, thus delivering the electrons to the lower plate of C5, charging it to a negative voltage and making the upper plate positive. Whenever the upper plate is positive, it will draw the electrons of the grid-driving current upward through R3, making the upper end of R3 and the control grid of V2 positive. This releases a large pulsation of plate current into the tube.

This plate current (a solid line) must first be drawn upward through the lower portion of L8. An increase of this plate current in the upward direction through the lower portion of L8 causes a separate feedback current to flow at an increasing rate in the downward direction through the entire winding. This feedback current (shown in large dots) flows in phase with the tank current, reinforcing it and supporting the quench oscillation.

The pulsation of plate current through V2 initially flows into the right-hand plate of C7, driving an equal number of electrons out of the left-hand plate and downward through L7 to ground. Because L7 is inductively coupled to L2, an upward flow of electrons will be induced in L2 during this half-cycle—this current is shown in large dots. The polarity of the resulting "back electromotive force" or back emf associated with this new current flow will be positive at the top of L2 and negative at the bottom. This positive polarity is also applied to the grid of V1, enabling it to conduct electrons from cathode to plate. Consequently, if a signal is being received through C4, oscillations will be set up in the tank circuit made up of L2 and C1.

Fig. 5-6 depicts a negative half-cycle in the operation of the quench oscillator circuit. The electrons which make up the tank current in the quench oscillator have oscillated upward through L8, thereby delivering a large negative voltage on the upper plate of capacitor C5, as indicated by the thin minus signs. Whenever this voltage is negative, it will drive electrons away from it along any available path. In this case, the only such path is downward through grid-driving resistor R3. (Electrons do not actually flow *through* C6 nor through any other capacitor for that matter, but normal capacitor action is such that when the negative voltage on C5 drives electrons onto the left-hand plate of C6, an equal number must flow out of the right-hand plate and downward through R3.) This downward movement of electrons through R3 causes a negative voltage at the upper terminal of R3 and the grid of V2, causing an inevitable reduction in the plate current flowing through the tube.

Two important actions stem from this reduction in plate current. First, the feedback current flowing in L8, (shown in large dots), reverses its direction of flow so that it again flows approximately in phase with the tank current, thereby replenishing or sustaining its oscillation. Second, the electron current which was previously driven *downward* through L7 will now be drawn upward through this coil, and onto the left-hand plate of C7. This upward flow of current through L7 induces a counter current in L2, which is shown (a solid line in Fig. 5-6) flowing downward through the inductor.

The back emf associated with this induced current in L2 is assumed to be negative at the top of L2 and positive at the bottom. A negative voltage at the top of L2 will cut off the electron flow through V1, and stop all of the current movements previously existing in the regenerative detector portion of the circuit.

The input carrier signal may still be in existence and flowing in and out through C4; however, when it lacks the support of V1 and the feedback arrangement between coils L3 and L2, it cannot sustain an oscillation in the tank circuit made up of L2 and C1.

Chapter 6

TYPICAL SUPERHETERODYNE RECEIVER

In this chapter the final two stages in a typical superheterodyne receiver—the IF amplifier and the audio power amplifier—will be discussed. Then the methods for checking voltages and making signal-substitution tests for the entire receiver will be presented. Voltage checks and signal substitution tests are the two most common methods employed to isolate a trouble when servicing a receiver.

IF AMPLIFIER

The function of the IF amplifier stage is to increase the strength of (amplify) the signal voltage which is supplied by the mixer to the control grid of the tube to the level required by the detector.

Identification of Components

A typical IF amplifier circuit is depicted in Figs. 6-1 and 6-2. This circuit is composed of the following individual components, with functional titles as indicated:

R12—Cathode-biasing resistor.
C7—RF and IF filter or decoupling capacitor.
T1—IF input transformer (the secondary winding and the capacitor in parallel with it are part of this circuit).
T2—IF output transformer (the primary winding and the capacitor in parallel with it, are part of this circuit).
V2—Pentode tube used as IF amplifier.

Identification of Currents

The following electron currents are at work in this amplifier circuit:

1. Grid-tank current.
2. Plate current.
3. Screen-grid current.
4. Plate-tank current.
5. IF filter current.

Details of Operation

The oscillation of electrons (shown in large dots) which occurs in the grid-tank circuit drives the control grid of V2 to alternate positive and negative values. It is supported, or replenished, each half-cycle by the movements of the tank current which flows up and down through the primary winding of T1. Fig. 6-1 shows the current in the primary winding flowing downward and the current in the secondary winding flowing upward. This action delivers electrons to the top of the tank, where they become concentrated on the upper plate of the tank capacitor, constituting a negative voltage at this point. Consequently, this half-cycle has been labeled as a negative half-cycle of operation.

Fig. 6-2 shows the current in the primary winding moving upward and the tank current moving downward through secondary winding of T1. This action delivers electrons to the lower plate of the tank capacitor, creating a deficiency of electrons on the upper plate. This deficiency constitutes a positive voltage; therefore Fig. 6-2 has been labeled as a positive half-cycle of operation.

Plate Current—During negative half-cycles (Fig. 6-1), the control grid of V2 will have its most negative voltage, and the plate current electron stream will be reduced to its minimum value. During the positive half-cycles (Fig. 6-2), the plate current stream is increased to its maximum value. Thus, the plate current is a form of pulsating DC, which flows continuously from cathode to plate within the tube, then downward through the primary winding of T2 to the power supply.

The pulsations in this plate current coming out of the tube support a new oscillation of electrons in the plate-tank circuit. This electron current is shown a solid line. In Fig. 6-2 when another pulsation of electrons arrives from the tube, it reaches the upper plate of the tank capacitor simultaneously with the tank current which has moved upward through the primary winding

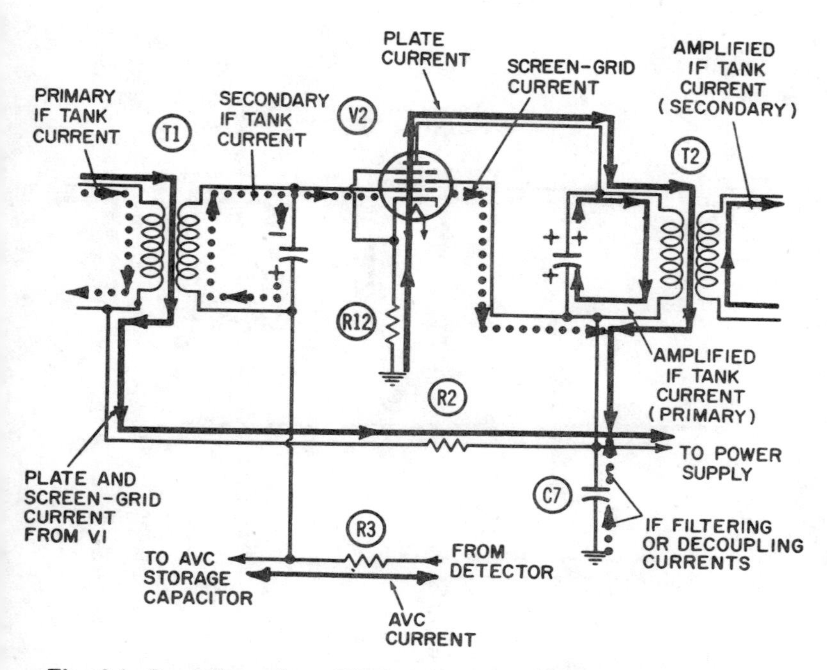

Fig. 6-1. Operation of an IF-amplifier circuit—negative half-cycle.

of T2, thereby reinforcing this tank current. The plate-tank current is an amplified version of the signal current which flows in the grid-tank circuit. This means that it is a stronger oscillation, or in other words that a larger quantity of electrons is oscillating in the plate tank than in the grid tank.

Another current (shown as a solid line) is shown flowing in the secondary winding of T2 in Fig. 6-2. It is sustained by the primary tank current which oscillates up and down through the primary winding. As the primary current moves upward (Fig. 6-2), the secondary current moves downward; as the primary current moves downward (Fig. 6-1), the secondary current moves upward.

Screen-Grid Currents—The screen-grid current, and its associated filter current (both shown in large dots) are the final set of electron currents in this circuit. The screen-grid current consists of electrons which are captured from the plate current stream within the vacuum tube. The screen grid has a fairly high positive voltage on it so that some of the negative electrons of the plate-current stream adhere to the wires of the screen grid. These electrons exit from the tube and rejoin the plate current below T2 and continue on to the power supply.

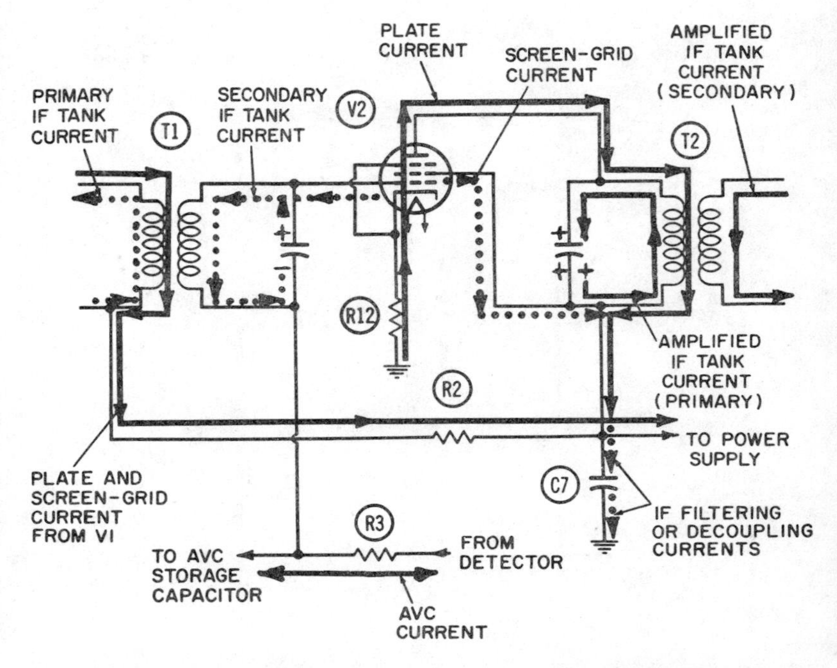

Fig. 6-2. Operation of an IF-amplifier circuit—positive half-cycle.

Like the plate current, the screen-grid current is also a pulsating DC. The pulsations occur during the positive half-cycles shown in Fig. 6-2. In Fig. 6-1 when the control grid is negative, the electron stream within the tube is reduced to its minimum value, and the plate and screen-grid currents are also reduced.

C7 filters out the fluctuations, or pulsations, in the plate and screen-grid currents. Fig. 6-2 shows such a filter action occurring. When a pulsation occurs in the screen-grid current, electrons flow onto the upper plate of C7, and other electrons are driven away from the lower plate to ground.

In Fig. 6-1, when no such pulsation occurs, electrons flow off the upper plate of C7 and into the power supply. This permits other electrons to be withdrawn from ground, and flow onto the lower plate of C7. The action of filtering out the pulsations in tube currents before they reach the power supply is frequently referred to as *decoupling* of the power supply.

AVC Voltage—The bottom of the grid-tank circuit is connected directly to the AVC storage capacitor (explained in Chapter 2). Since this capacitor has a permanent negative voltage stored on it, this voltage is also applied to the grid of V2 and acts as a permanent biasing voltage on that tube. Thus, as the AVC voltage varies, the gain of V2 is changed. The variations in grid voltage caused by the oscillation of electrons in the grid tank of this tube will alternately add to or subtract from this permanent biasing voltage.

AUDIO POWER AMPLIFIER

The basic function of an audio power amplifier is to increase the strength of the audio signal delivered to it by the preceding amplifier, and to generate a heavy electron current at these same audio frequencies in order to operate the speaker.

Identification of Components

A typical audio power-amplifier circuit is illustrated in Figs. 6-3 and 6-4. This circuit is composed of the following individual components, with the functions indicated:

R8—Grid-driving resistor.
R9—Cathode-biasing resistor.
R10—B+ dropping resistor (actually part of the power-supply filter system).
C12—Coupling and blocking capacitor.
C13—IF filter capacitor.
T3—Audio output transformer.
V4—Pentode power-amplifier tube.
SP1—Speaker.

Identification of Currents

The following electron currents are at work in this power-amplifier circuit:

1. Grid-driving current.
2. Plate current.
3. Screen-grid current.
4. Speaker current.
5. Plate and screen-grid currents from other tubes.

Details of Operation

The grid-driving current (shown in small dots) moves up and down through R8 at the audio frequencies being amplified. This current is in turn driven by the pulsations in plate current from the previous tube. Fig. 6-3 shows one such pulsation occurring, with plate current (a solid line) flowing onto the left-hand plate of C12. This action drives an equal number of electrons away from the right-hand plate of C12 and downward through grid resistor R8.

The grid-driving current flowing through R8 causes the voltage at the top of R8 to be negative. For this reason, Fig. 6-3 has been labeled as a negative half-cycle of operation. In Fig. 6-4, when the plate current pulsation is flowing out of the left-hand plate of C12, it draws the grid-driving current upward through R8. This indicates that the top of the resistor is more positive than the bottom.

In Fig. 6-3 when the top of R8 has its most negative voltage value, the plate-current electron stream flowing through V4 will be reduced to its minimum value. In Fig. 6-4 when the grid is positive, maximum plate current will flow.

Power Amplification—The construction of a power amplifier tube differs somewhat from that of a voltage-amplifier tube. A power-amplifier tube is constructed so that it will conduct a much heavier plate current at full conduction than a voltage-amplifier tube delivers. Electrical power varies as to the *square* of the current; therefore an amplifier tube which is capable of delivering wide extremes of electron current has been given the name of power amplifier. This title can be misleading because *all* amplifier tubes deliver their plate currents into some kind of a load, and consequently, some power is developed in each of these loads by these currents. A tube is called a power amplifier when it delivers a heavy enough current into its load to develop an appreciable amount of power.

The basic formula for computing power across a resistive load is:

$$P = I^2R$$

where,

P is the power developed in watts,
I is the current flowing through the resistor in amperes,
R is the resistance of the load in ohms.

The formula for computing power developed across an inductive load, such as the primary winding of T3, is:

$$P = I^2X$$

where,

P is the power in watts,
I is the current through the inductor in amperes,
X is the inductive reactance of the transformer primary in ohms.

Plate Current—The plate-current path starts at ground below R9. The electrons which make up this current flow upward

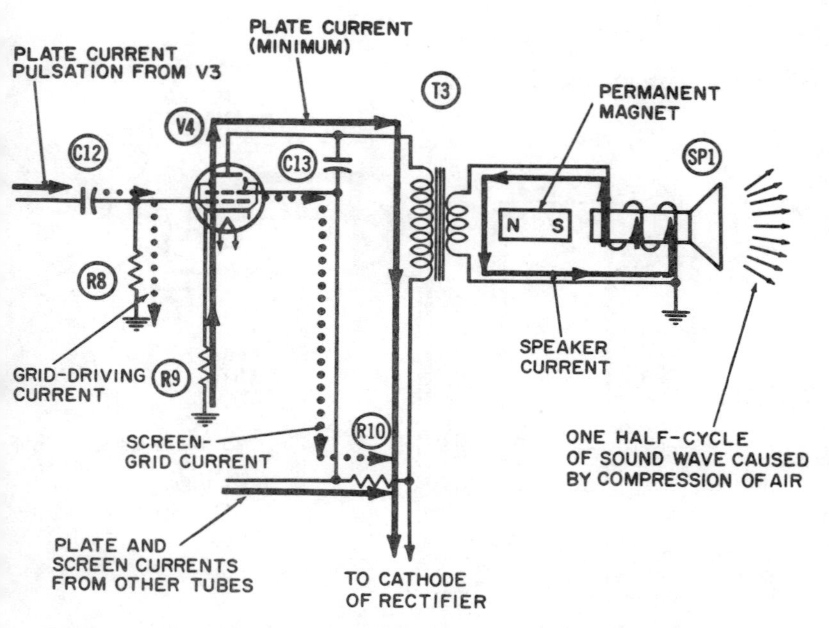

Fig. 6-3. Operation of an audio-output circuit—negative half-cycle.

through R9, through the vacuum tube from cathode to plate, downward through the primary winding of T3, from where it flows directly to the cathode of the rectifier tube. All of the other plate and screen-grid currents join with this plate current at the right-hand end of R10, to be drawn eventually through the rectifier tube and delivered back to ground.

Output Transformer—Transformer action can be a very difficult physical action to visualize. Stated in the simplest terms, when a current, such as the plate current of V4, can be made to pulsate through one winding of the transformer, it will cause another current to flow back and forth in the other winding. The pulsations of plate current flowing downward through the primary winding of T3 cause the speaker current (a solid line) to flow in the secondary winding.

When this plate current is increasing, as it does during the positive half-cycle of Fig. 6-4, the speaker current increases in the upward direction through the secondary winding. When the plate current decreases during the negative half-cycles (Fig. 6-3), the speaker current increases in the downward direction through the secondary winding. Since the plate current consists of continuous pulsations at the various audio frequencies, the speaker current will flow up and down through the secondary winding

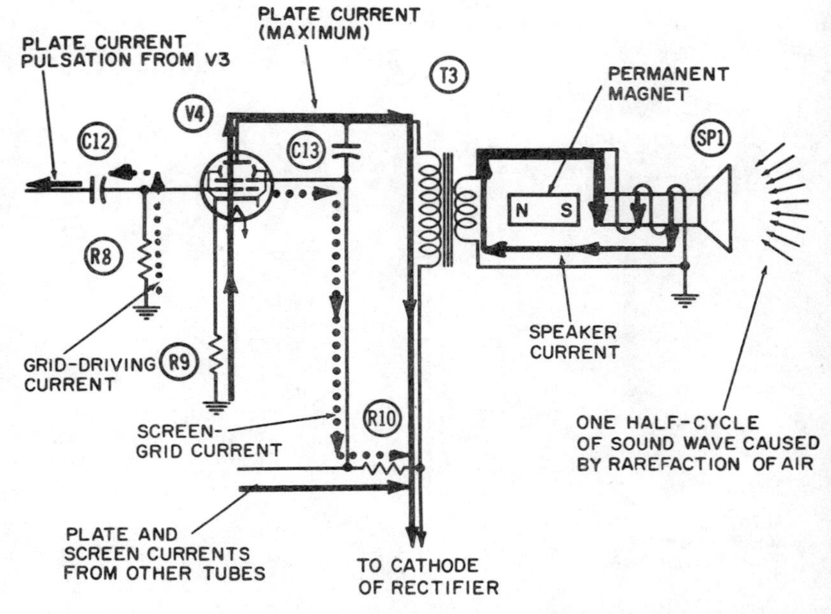

Fig. 6-4. Operation of an audio-output circuit—positive half-cycle.

(and through the moving coil of the speaker) at these same audio frequencies.

Like all similar output transformers, T3 has more turns in its primary winding than its secondary winding. This means that it is a current step-up transformer. (A current step-up transformer is the same thing as a voltage step-down transformer.)

The determination as to whether a transformer will step the current up or down is governed by a simple formula, which states:

$$\frac{I_p\,N_s}{I_s\,N_p}$$

where,

I_p is the current flowing through the primary winding,
I_s is the current flowing through the secondary winding,
N_s is the number of turns of wire in the secondary winding,
N_p is the number of turns of wire in the primary winding.

Speaker Action—The speaker is a typical moving-coil arrangement, which involves the use of two magnets—a permanent magnet and a temporary magnet (formed by the speaker voice-coil winding). The permanent magnet has been shown with its south pole adjacent to the left-hand end of the temporary magnet.

During the negative half-cycles shown in Fig. 6-3, the speaker current flows through the speaker voice coil in such a direction as to make the left end of the temporary magnet have a *south* magnetic pole. This causes it to be repelled by the adjacent south pole of the permanent magnet, and it moves to the right. Since the speaker cone is connected to the voice coil, it also moves. The movement of the speaker cone *compresses* the air in front of it and causes a half-cycle of a sound wave.

During the positive half-cycles depicted by Fig. 6-4, the speaker current flows in such a direction as to create a *north* magnetic pole at the left end of temporary magnet. This north magnetic pole will be attracted by the south magnetic pole of the permanent magnet, and its movement will cause a rarefaction of the air in front of the speaker diaphragm. This rarefaction of the air constitutes a second half-cycle of the sound wave.

Lenz's Law—The rule for relating the direction of electron flow around an iron core to the resulting magnetic polarity of that core is known as Lenz's law. If the iron core is grasped by the left hand so that the fingers point in the direction that the electron current is flowing through the coil, then the thumb points toward the temporary *north* magnetic pole of the electromagnet.

A moving-coil speaker is a current-operated device. The amount of movement of the iron core and the diaphragm to which it is attached depend on the amount of current flowing through the moving coil each half-cycle. The amount of diaphragm movement determines the loudness of the sound coming from your radio. Output transformers, such as T3, have more turns in their primary winding than in their secondary winding so that large fluctuations in plate current will be increased even more by the current step-up relationship between the windings.

VOLTAGE CHECKING THE SUPERHET RADIO

When the superhet radio becomes inoperative, or "dead," there are a number of tests, progressing from very simple ones to more complex ones, which the technician can use to locate and remedy the source of trouble. Probably the first of these tests is a visual check to see if all the tubes are lighted. Some tubes may have metal envelopes, but in simple table model radios, most if not all tubes will have glass envelopes. When the radio is turned on, we can look inside the envelope (after approximately 18 seconds have elapsed) and see a red glow that indicates that the cathode is heating normally. In tubes with metal cases, we can usually tell whether the filament is heating by feeling the metal envelope to see if it is warm.

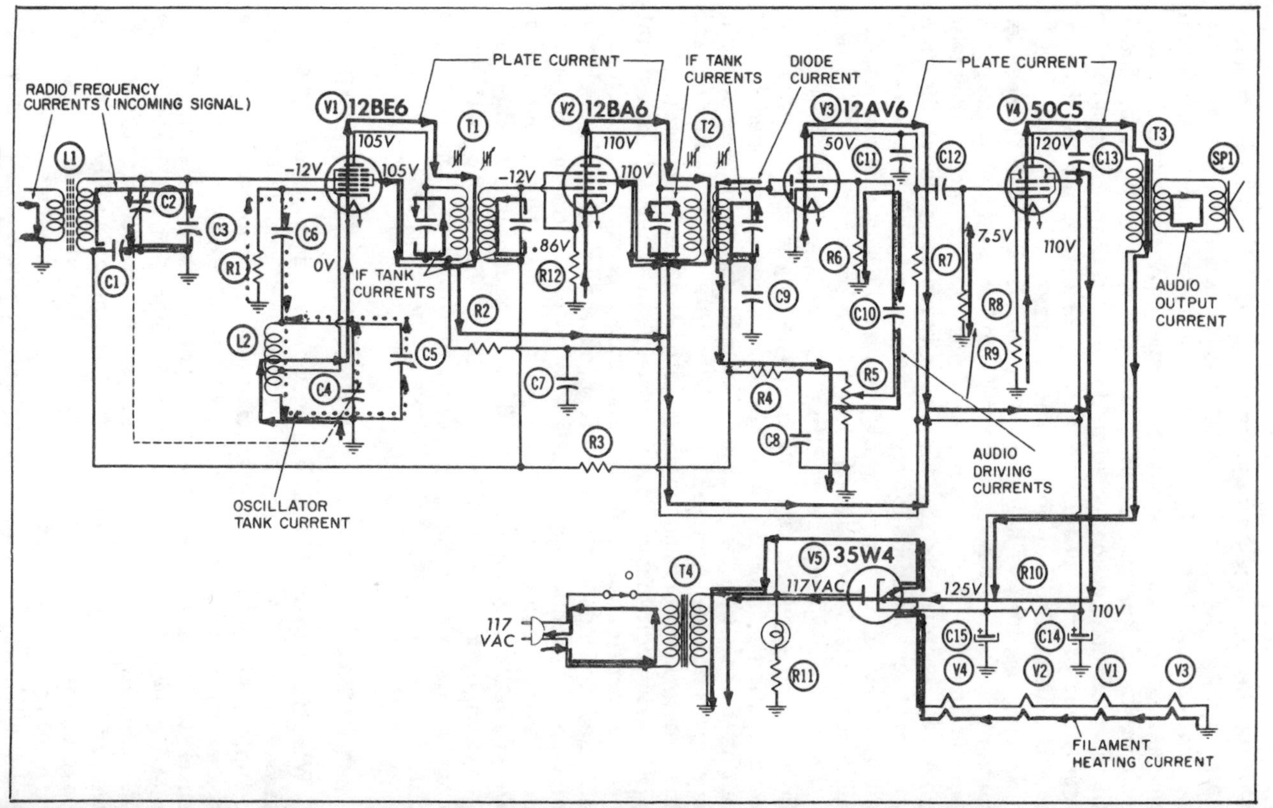

Fig. 6-5. Operation of a typical radio receiver.

Fig. 6-5 shows the circuit diagram of a typical radio, with the filament heating current (a solid line) flowing through its intended path. (Partial schematics of the various stages in this radio have been presented in the previous chapters.) The filaments for all five tubes are connected in series, so the same heating current must flow through every filament. If a single tube filament fails, no current can flow in any of them, so each must be checked on a tube tester. When the faulty tube is located in this manner, and replaced, the radio will usually operate again.

Once it is determined that the tubes are heating normally, the simplest method of isolating a faulty component is by the process known as voltage checking. This process requires only a single piece of test equipment—a standard voltmeter, along with a diagram showing voltages which can be expected at each tube electrode during normal operation. Each and every one of these electrode voltages has a value which is determined by one or more electron currents flowing through or along a certain resistive path. When each such current is understood and visualized, and when its complete path is recognized, the student or technician can infer a great deal about that current and about the components through which it flows by noting the voltage at the electrode where each current enters or exits from the tube.

The plate currents are superimposed on the circuit diagram in Fig. 6-5 in a line. Each of these currents is drawn up from ground below the respective cathodes of the tubes, and across the tube by the high positive voltage at the plate. The point of highest positive DC voltage in the radio is the cathode of the half-wave rectifier tube V5; this positive voltage draws all of the plate currents through their respective plate-load circuits (primary of T1, primary of T2, R7, and primary of T3) to the filter circuit (R10, C14, and C15) and to the cathode of V5.

The DC voltages which exist at the principal electrodes of the vacuum tubes are all directly associated with the flow of these currents. A great deal of servicing information can be inferred by observing these DC voltages.

Chapter 7

TYPICAL TRANSISTOR RECEIVER

In this chapter a typical transistor broadcast receiver will be analyzed. First, the operation of the receiver and each electron current flow will be explained. Then the methods of checking the various voltages will be outlined. Finally, signal substitution tests will be given.

TRANSISTOR BROADCAST RECEIVER

Figs. 7-1, 7-2, and 7-3 show three identical circuit diagrams of a typical transistor broadcast receiver. All of the electron currents which flow in this receiver during normal operation have been shown and identified in Fig. 7-1. In Figs. 7-2 and 7-3 these currents have been separated so that their actions may be more easily analyzed.

This receiver utilizes the superheterodyne circuit. Two stages of IF amplication are employed followed by a solid-state diode detector. The audio signal developed at the detector is then amplified by two voltage-amplifier stages, and push-pull power-amplifier stage. An adidtional stage is provided for overload protection.

Identification of Components

The individual components which make up this radio and their functions are as follows:

R1—Volume control.
R2, R3—Base-biasing resistors for X1.

R4—Emitter-biasing resistor for X1.
R5—Oscillator-tank damping resistor.
R6—Emitter-biasing resistor for X3.
R7, R8—Base-biasing resistors for X2.
R9—Emitter-biasing resistor for X2.
R10—Base-biasing resistor for X3.
R11—Collector resistor for X3.
R12, R13—Base-biasing resistors for X4.
R14—Emitter-biasing resistor for X4.
R15, R16,—Base-biasing resistors for X5.
R17—Collector-load resistor for X5 and base-biasing resistor for X6.
R18—Emitter-biasing resistor for X6.
R19—Power-supply decoupling resistor.
R20—Emitter-biasing resistor for X7 and X8.
R21—Base-biasing resistor for X7.
R22—Base-biasing resistor for X8.
C1—AVC capacitor.
C2A—Emitter bypass capacitor for X6.
C2B—Power-supply decoupling capacitor.
C3—Input coupling and blocking capacitor.
C4—Oscillator-tank coupling capacitor.
C5—Bypass capacitor for R6.
C6—Feedback and neutralizing capacitor for X2.
C7—Emitter bypass capacitor for X2.
C8—Bypass capacitor for R10.
C9—Bypass capacitor for R12.
C10—Feedback and neutralizing capacitor for X4.
C12—IF filter capacitor.
C13—Audio coupling and blocking capacitor.
C14—Parasitic suppression capacitor.
C15, C16—RF tank capacitors.
C17, C18—Oscillator tank capacitors.
L1—Antenna-coupling transformer.
L2—Oscillator-tank transformer.
L3—First IF transformer.
L4—Second IF transformer.
L5—Third IF transformer.
T1—Audio interstage transformer.
SP1—Speaker.
X1—2N412 Converter transistor.
X2, X4—2N410 IF amplifier transistors.
X3—3458 Overload transistor.
X5, X6—2N406 Audio-amplifier transistors.
X7, X8—2N408 Audio output transistors.

M1—Battery power supply.
M2—Diode detector.

Note that all of the foregoing transistors are PNP type. It is standard practice to use one type of transistor—either PNP or NPN—in small systems, such as this one. This practice provides simplification in biasing and power-supply requirements.

Identification of Currents

There are approximately 50 different electron currents at work in this radio during normal operation. These currents may be subdivided into seven main types, as follows:

1. Base-emitter current within each transistor—there is one such current for each transistor, making eight altogether.
2. Collector-emitter current within each transistor—there is one such current for each transistor, making eight in all.
3. Voltage-divider currents—there are six voltage-divider currents. All of them originate at a ground connection and are drawn through certain resistors to the positive terminal of the power supply. These resistor networks, and the important function provided by the current through each one, are:

   First Voltage-Divider Current (R2, R3, and R19)—Provides a small positive base-biasing voltage for X1 at junction of R2 and R3, and a more positive emitter-biasing voltage at junction of R3, R4, and R19.

   Second Voltage-Divider Current (R7, R8, R1, and R19)—Provides a small positive base-biasing voltage for X2 at junction of R7 and R8, and a more positive voltage at junction of R8 and R1 for the biasing cathode of M2.

   Third Voltage-Divider Current (R12, R13, and R19)—Provides a small positive base-biasing voltage for X4 at junction of R12 and R13.

   Fourth Voltage-Divider Current (R15, R16, and R19)—Provides small positive base-biasing voltage for X5 at junction of R15 and R16.

   Fifth Voltage-Divider Current (SP1, R21, and R20)—Provides a low positive base-biasing voltage for X7 at the junction of the two resistors.

   Sixth Voltage-Divider Current (SP1, R22, and R20)—Provides a low positive base-biasing voltage for X8 at the junction of the two resistors.

4. Driving current for each transistor—these currents have been shown in the same lines as the currents which excite them, (large dots for X1 and X5; large dots for X2, X3, X4, X6, X7, and X8).
5. Oscillating currents in the five tank circuits (large dots in RF tank, small dots in oscillator tank, and large dots in the 5 IF tank circuits).
6. Diode detector current through M2.
7. Filter currents—these currents have not been shown in the diagrams, but they flow through the following capacitors to ground: C5; C1 and C2B; C7, C9 and C11; C8, C12 and C2B; C2A, C14.

Details of Operation

In any transistor, two different electron currents—the base-emitter current and the collector-emitter current—must flow during normal operation. In the PNP transistor, such as those employed exclusively in this particular receiver, both of these currents *exit* from the emitter terminal of the transistor. The base-emitter current, which is frequently referred to merely as *emitter current,* enters the base terminal and flows through the emitter portion of the transistor before leaving at the emitter terminal.

The collector-emitter current, which is commonly referred to as *collector current,* enters the transistor at the collector terminal and flows through the collector, base, and emitter, within the transistor (in that order) before leaving via the emitter terminal.

In NPN-type transistors, both of these flow directions are reversed. The currents have the same names and they flow along the same two paths, except that they flow in opposite directions than for a PNP transistor.

The most important physical characteristic of the transistor, the one which permits it to function as an amplifier, is that property which permits the base-emitter current to control the flow of collector-emitter current. A small amount of base-emitter current flowing will permit a large amount of collector-emitter current to flow, and a small change in the amount of base-emitter current will cause a large change in the amount of collector-emitter current flowing. These current changes will always have the same sign, or phase, meaning that an increase in the base-emitter current brings an increase in the collector-emitter current, and vice versa.

Because of the foregoing characteristics, transistors are considered to be *current-operated* devices. However, the amount of base-emitter current which flows at any instant is precisely de-

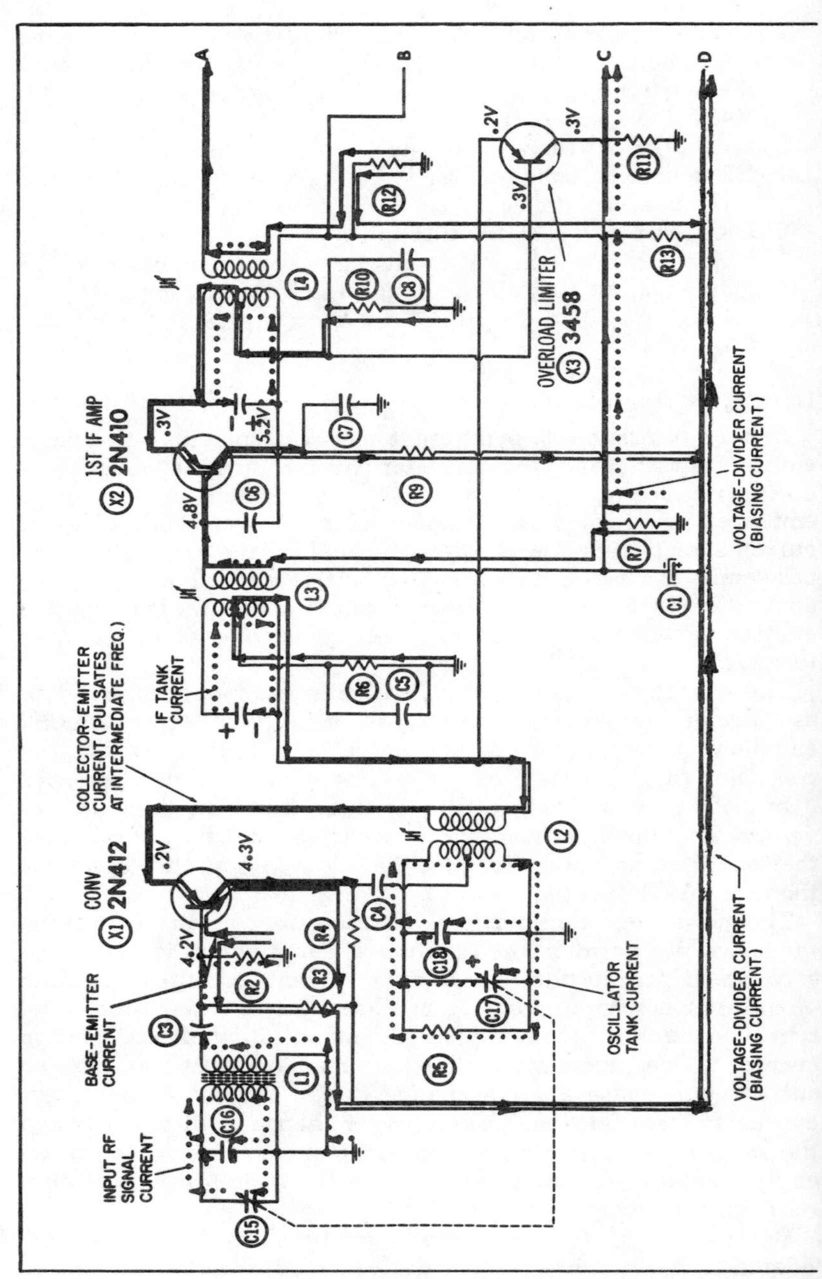

Fig. 7-1. Operation of a typical transistor broadcast

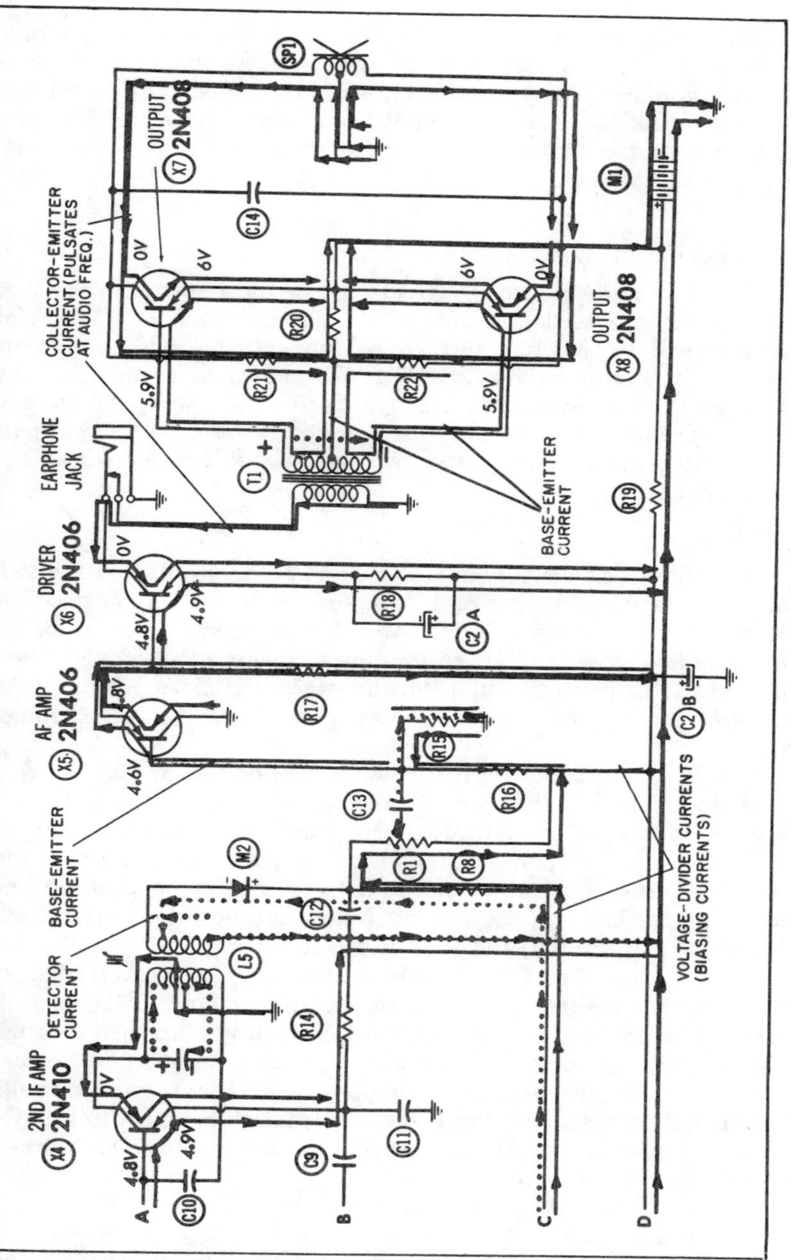

eceiver—all significant currents identified.

termined by the voltage difference existing between the base and the emitter. In other words, these two terminals are "biased" by the voltages which exist on them. It is important, therefore, that the reader be able to understand how these voltages are achieved and see what factors cause them to vary, and in what manner and degree.

OPERATION WHEN NO SIGNAL IS BEING RECEIVED

Consider the conditions which exist in the receiver when it is not tuned to a station. None of the signal currents (RF, IF, or audio) will be flowing, but all of the voltage-divider currents (solid line) will be flowing. Fig. 7-2 shows these currents by themselves for additional clarity. Each of these currents will create an *initial voltage difference* of appropriate polarity to permit some electron current to flow from base to emitter within the transistor.

X1 Biasing

The voltage at the junction of R2 and R3 is positive as a result of the voltage-divider current flow, and the base of X1 is positive. However, the voltage at the lower end of R3 is even more positive, because it is closer to the power supply. The emitter of X1 is connected to this point through R4; therefore it is more positive than the base, so that base-emitter current will begin flowing through the transistor.

This base-emitter current (shown as a thin line in Figs. 7-1 and 7-3) flows upward through R2, through the transistor in the direction indicated, then through R4 from right to left, where it joins the voltage-divider current and is drawn through R19 and into the positive terminal of the battery. From the positive terminal of the battery, the electrons flow through the battery, out the negative terminal, and back to ground.

Once a small amount of base-emitter current begins to flow, the collector-emitter current also begins to flow. This current (shown as a line in Figs. 7-1 and 7-3) is drawn upward through R6, through the lower portion of the primary winding of L3, upward through the secondary winding of L2, downward through the collector, base, and emitter of the transistor, then through R4 where it also joins the voltage-divider current on its journey to and through the power supply.

Each of these two transistor currents will create a separate component of voltage drop across R4. Each voltage drop will tend to make the left end of R4 more positive than the right-hand terminal. Since the transistor base is connected through R3 to the left end of R4, these two components of positive voltage will partially neutralize or nullify the original biasing condition caused by voltage-divider current flowing downward through R3. The net result will be a reduction in the amount of base-emitter current, and an accompanying reduction in collector-emitter current, from those amounts which would otherwise flow if R4 were not in the circuit.

The base-emitter and the collector-emitter currents will quickly stabilize at values that will allow the base voltage to remain slightly more negative than the emitter voltage. (A tenth or two-tenths of a volt difference between these two terminals is typical.) When the voltages at the base and emitter of a transistor are of such magnitudes that base-emitter current flows through the transistor, it is said to be *forward-biased*. If these voltages are such that base-emitter current cannot flow, then the transistor is *reverse-biased*.

In addition to their usefulness in describing the *total* voltage difference between base and emitter, these terms are also used to describe individual components of this total voltage difference, or bias. When used in this sense, the voltage-divider current flowing downward through R3 is said to contribute a substantial component of forward bias to the transistor, and the two transistor currents flowing through R4 contribute, or *add*, some reverse bias which reduces the forward bias. In the absence of any received signal, each of the transistor currents will be a pure DC.

The base-emitter and collector-emitter currents can never cut off the transistor entirely because, with no electron current flowing through R4, the emitter voltage would immediately rise to the high positive value existing on the main power-supply line. This would create so much forward bias that the transistor currents would again begin flowing.

Similar stories about DC operating conditions can be told about transistors X2, X4, X5, X7, and X8. Each of these transistor circuits is initially biased in the forward direction by the flow of one of the voltage-divider currents previously discussed. This initial forward bias starts the flow of base-emitter current which in turn starts the flow of collector-emitter current. In each case, the flow of the two transistor currents through an adjacent resistor will alter, or modify, the initial biasing conditions by contributing some reverse bias. And in each case, the amounts of the two transistor currents will settle down or stabilize at values which will permit the total bias (voltage difference between base and emitter) to remain at about a tenth of a volt in the forward direction. In the PNP transistor, this means that the emitter must be more positive than the base.

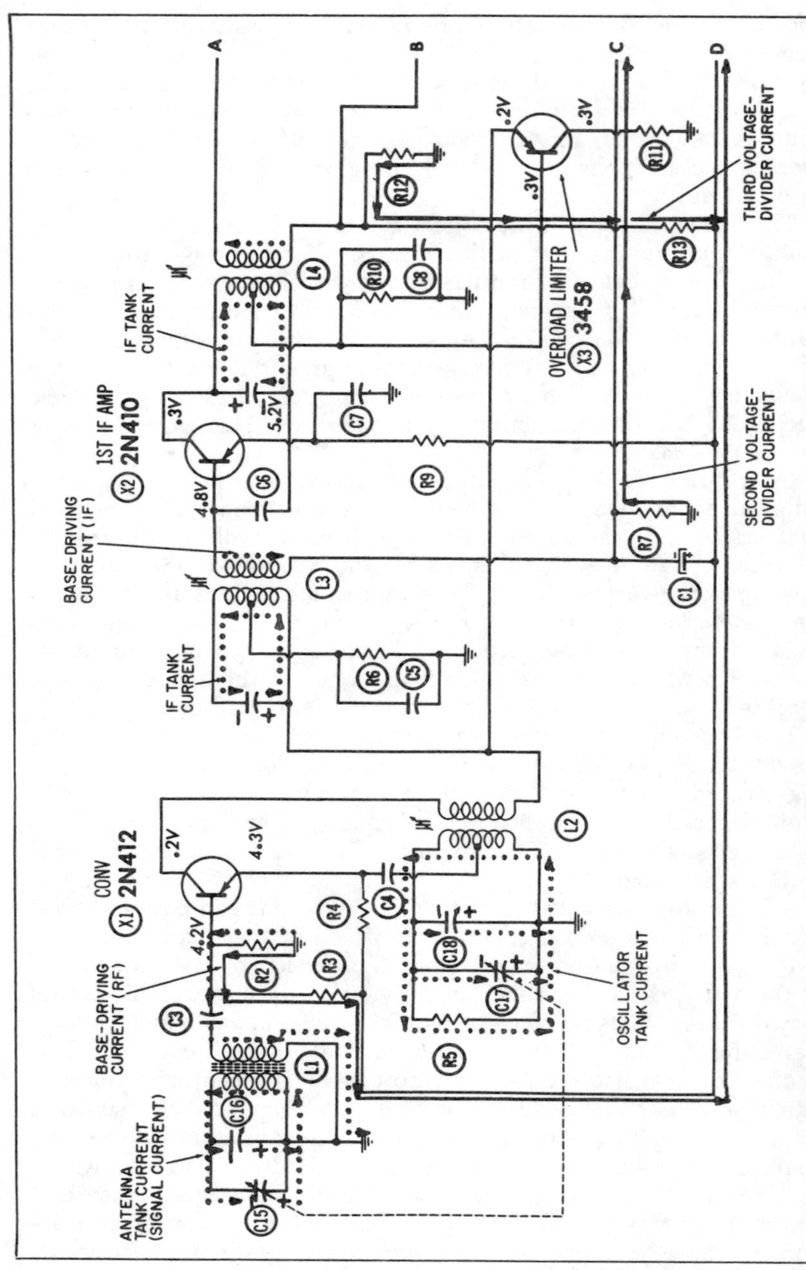

Fig. 7-2. Operation of a typical transistor broadcast

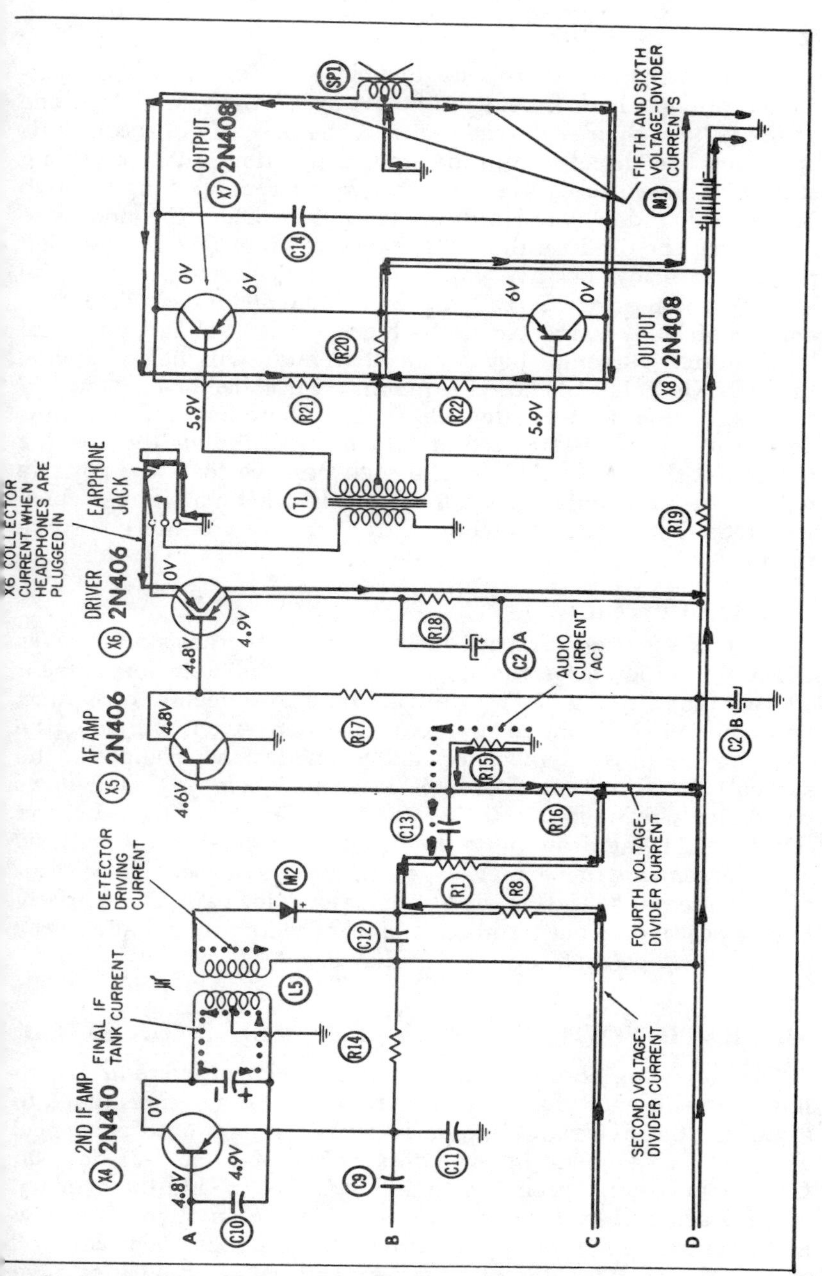

receiver—voltage-divider and signal currents.

X6 Biasing

The manner in which this stage is biased differs from those discussed previously. Whereas the others are initially biased by one of the voltage-divider current actions, the base of X6 receives its operating bias directly from the emitter junction of the preceding stage. We have already seen how the two currents flowing through X5 must flow downward through emitter resistor R17, and in so doing how they reduce the voltage at the top of R17 from a high positive to a low positive value. This voltage becomes both the emitter voltage for X5 and the base voltage for X6. Since the emitter of X6 is connected to the high positive voltage, an initial base-emitter current, shown in a thin line, will begin flowing through X6. This will cause the much larger collector-emitter current (solid line) to begin flowing. Both of these currents will flow downward through R18, and in so doing will reduce the positive voltage at the top of R18 from a high positive to a low positive value. These currents will stabilize or settle down at values which will keep the emitter of X6 only a fraction of a volt more positive than the base.

Detector Current

The final electron current which flows in this transistor receiver when no station or signal is tuned in is the detector current (dotted line in Fig. 7-1) through M2. The detector, of course, will conduct electrons in only one direction—from cathode (the straight line in the symbol for M2) to anode (the triangle in the symbol). This current originates at ground below R7 and flows successively through R7, R8, M2, the secondary winding of L5, and R19 before being drawn into the positive terminal of the battery. This current might very easily be classed as one of the voltage-divider currents, and at each point along its path is a slightly higher positive voltage than any point preceding it, and at a lower positive voltage than any point following it.

OPERATION WHEN A SIGNAL IS BEING RECEIVED

The foregoing accounts for all of the electron currents which flow in the absence of a received signal. When the radio is tuned to a station, the five radio-frequency currents come into existence. The first one (shown in large dots in Figs. 7-1 and 7-2) flows in the RF tuned-tank circuit composed of C15, C16, and the primary of L1. Each half-cycle of it induces a companion current to flow in the secondary circuit, which includes the secondary winding of L1, C3, and R2. Current directions and voltage polarities have been chosen in Fig. 7-1 so that this secondary current flows *downward* through R2 during this particular half-cycle, developing a small component of negative voltage at the top of R2 which must be subtracted from the normal positive base voltage of 4.2 volts. In a PNP transistor, a less positive voltage at the base constitutes "forward bias" and drives an additional amount of base-emitter current through the transistor. This, in turn, causes an increase in the amount of collector-emitter current.

A completely separate oscillation of electrons will meanwhile be occurring at a higher frequency in the oscillator tank circuit. This tank circuit (L2, C17, C18, and R5) is designed so that the oscillator frequency will always be 455 kc higher than the carrier frequency. Part of the oscillatory voltage is coupled via C4 to the emitter of X1. An instantaneous voltage polarity for this tank oscillation has been chosen in Fig. 7-1 so that the emitter has been made temporarily more positive than its normal voltage of 4.3 volts. This also constitutes forward bias in the PNP transistor so that another additional component of base-emitter current is encouraged to flow, drawing another additional component of collector-emitter current through the transistor.

This additional component of collector current must first flow upward through the secondary winding of L2; in so doing, it induces a feedback current to flow downward in the primary winding. Since this feedback current is approximately in phase with the tank current, the oscillation will be sustained or replenished during each cycle of operation.

Since the biasing conditions are being simultaneously varied by two separate frequencies, the collector current will be caused to pulsate through the external circuit at each of these two frequencies. As in the case with vacuum tube mixing and converting circuits, the collector current also pulsates at other frequencies, such as the sum and difference of the two applied frequencies. The tank circuit at the primary of L3 is tuned to this difference frequency of 455 kc so that each pulsation which occurs in the collector current at this frequency will surge downward through the lower half of the inductor and sustain or reinforce one cycle of the oscillation.

The IF oscillating current in L3 and its associated capacitor is shown in large dots. Current directions in Fig. 7-1 are chosen as downward in the primary winding, thereby inducing an upward current in the secondary. This secondary current delivers electrons to the base of X2, thereby increasing its forward bias during this half-cycle and causing a momentary increase in the flow of both base-emitter and collector-emitter current through the first IF amplifier transistor X2.

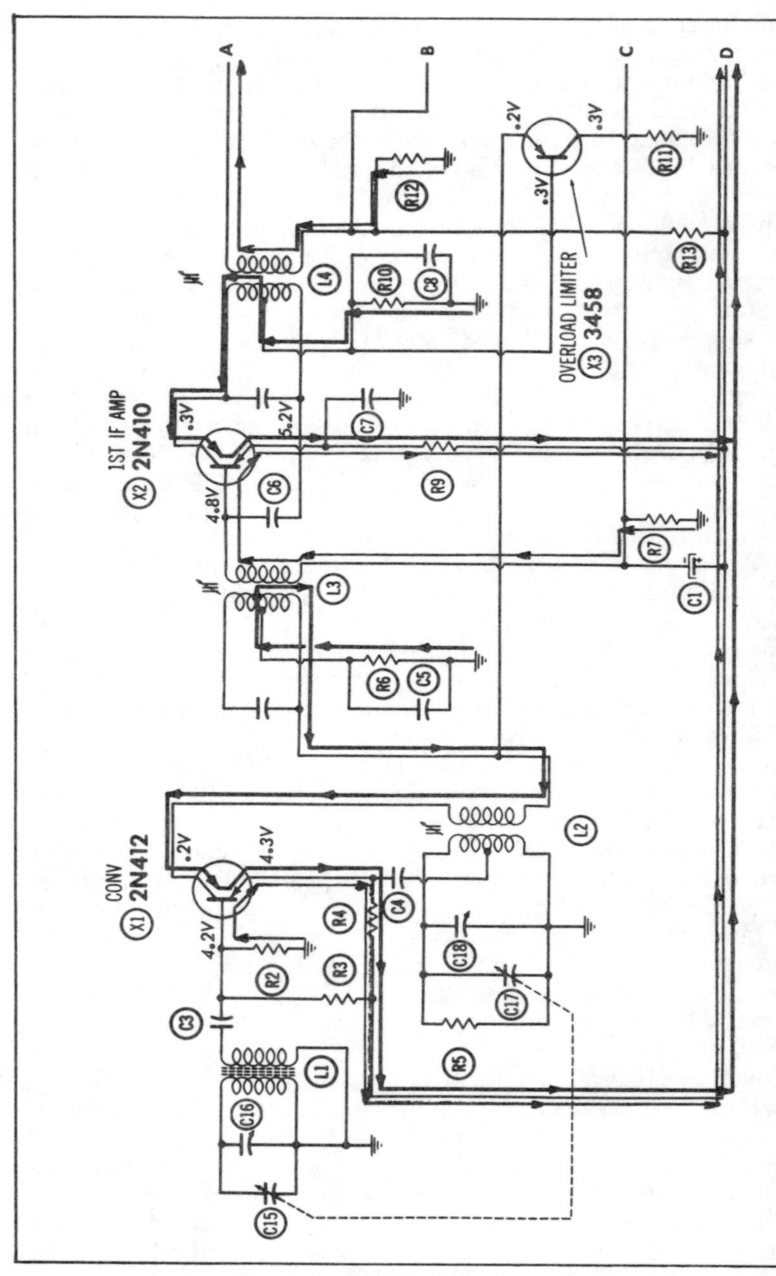

Fig. 7-3. Operation of a typical transistor broadcast

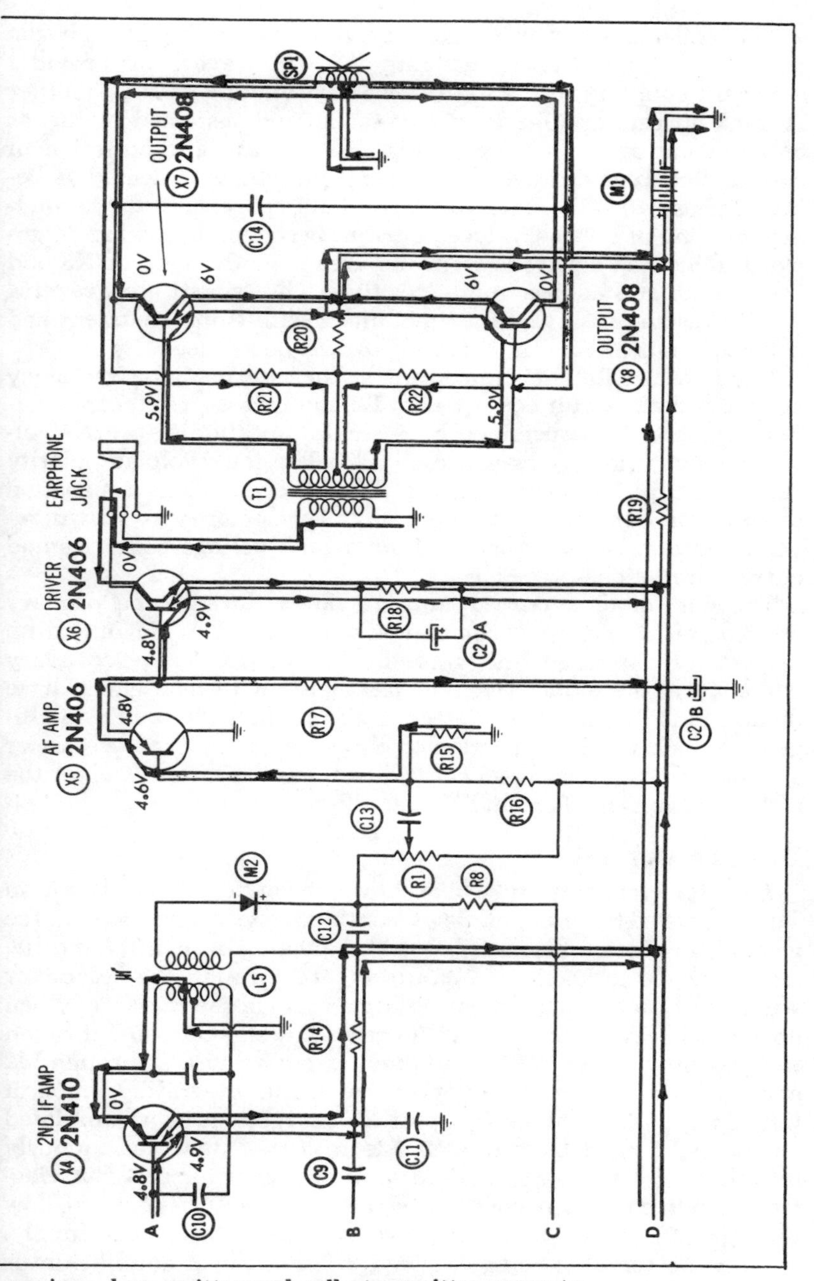

receiver—base-emitter and collector-emitter currents.

The collector-emitter current of X2 flows only through the upper half of the primary winding of L4. However, this provides sufficient coupling to the entire primary winding so that another IF tank current oscillation will be set up and sustained by means of autotransformer action. In Fig. 7-2 the upward pulsation of collector current through the primary winding is indicated as being in phase with the upward flow of tank current. In the secondary winding of L4, the induced current is shown as flowing downward, removing electrons from the area near the base of X3 and thus making the base more positive. This constitutes reverse bias, with the result that the amounts of base-emitter current and collector-emitter current through X4 will be reduced.

The final oscillation of electrons at the intermediate frequency occurs in tank circuit composed of L5 and its associated capacitor. Each cycle of it is sustained by a single pulsation of the X4 collector current as it surges through L5. The tank voltage polarity shown in Fig. 7-1 corresponds to the half-cycle when a pulsation of collector current is not occurring. This polarity is shown reversed in Fig. 7-2 along with all other tank voltage polarities and current directions.

The flow of tank current up and down through the primary winding of L5 induces a companion current to flow down and up respectively through the secondary winding. This secondary current and the induced voltage associated with it in effect drive diode M2 and cause it to detect or demodulate the audio intelligence which has been carried to the antenna by the RF carrier signal, and which has been carried through the "front end" of the radio by the converter and IF amplifiers.

Detector Current

The detector current which flows through M2 is shown in large dots. This current flows continuously, originating at the ground connection below R7, flowing upward through R7 and R8, from cathode to anode of M2, downward through the secondary winding of L5 to the main power supply line of the receiver. When no signal is being received, this current flows as pure DC through the resistive portion of the path, and as pulsating DC through M2 and L5. This feature is made possible by the integrating action of C12 along with R8 and R7. A positive charge will be accumulated on the right hand plate of C12. Electrons will flow continuously into this point from R8, tending to discharge it to zero. But electrons are also drawn continually away from this point to flow through M2 and on to the positive voltage of the power supply.

When no signal is being received, a condition of equilibrium is established. The quantity of electrons flowing into C12 equals the quantity being drawn out, and a positive voltage exists on the right-hand plate. When a signal is being received, the voltage at the upper end of the secondary winding of L5 fluctuates between higher and lower positive values because of the movements of the IF current (shown in large dots) induced in this winding. In Fig. 7-1, this IF current and associated voltage polarity are such that the upper electrode (the anode) is made less positive than it was before. This restricts the flow of electron current through M2.

In the alternate half-cycle shown in Fig. 7-2, the anode of M2 is made more positive by the IF current/voltage combination in the secondary winding of L5. This momentarily increases the flow of electrons through M2. Because of the large size of C12, the fluctuations in detector current are drawn directly from the right hand plate of C12 without causing a significant change in its total voltage.

When a modulated carrier signal is being received, the strength of the IF oscillation in the primary of L5 varies from cycle to cycle. A modulation peak is characterized by a succession of relatively strong individual cycles of IF. A modulation trough is characterized by a succession of relatively weak IF cycles. One cycle of audio voltage consists of one modulation peak and one modulation trough; together they will encompass many hundreds or even several thousand cycles of the IF oscillation.

The audio or modulating voltage, which is the intelligence we seek to hear from our radio, makes its first appearance in the radio on the right-hand plate of C12. The positive voltage at this point rises and falls at an audio rate. During modulation peaks, the strong IF cycles will draw an increased number of electrons away from C12, and the positive voltage at this point must increase. During modulation troughs, the weaker IF cycles will draw a reduced number of electrons away from C12, and the positive voltage at C12 will go up again. Thus, the positive voltage on C12 rises and falls in accordance with the strength and the frequency of the modulating voltage (the audio).

The second voltage-divider current flows downward through R1. When no modulated signal is being received, this current is a pure DC. When an audio voltage appears on C12, this voltage-divider current will pulsate at an audio rate. When the voltage on C12 reaches its most positive value during the modulation troughs, it approaches more nearly the value of the voltage existing at the bottom of R1. Thus, there will be a reduction in the amount of current flowing downward through the volume control during modulation troughs.

When the voltage on C12 reaches its least positive value during modulation peaks, it differs by a greater amount from the voltage which exists at the bottom of R1. Consequently, the amount of electron current flowing downward through R1 must increase during modulation peaks.

As a result of these pulsations in current flowing downward through R1, the voltage at any point along R1 will also pulsate at the same audio frequency. This pulsating voltage is coupled to C13, where it drives a small component of current up and down through R15. R15 functions as a base-driving resistor in much the same manner that a grid driving resistor functions in a vacuum tube circuit. During modulation peaks when the voltage divider current flowing downward through R1 increases, electrons will flow onto the left hand plate of C13, driving an equal number out of the right hand plate and downward through R15. This action develops a small component of negative voltage at the top of R15 which must be subtracted from the positive voltage developed at that point by the upward flow of the fourth voltage-divider current and the base-emitter current through this same resistor.

The reduction in positive voltage at the base of X5 constitutes forward bias in the PNP transistor, with the result that both the base-emitter and the collector-emitter currents through X5 will increase. This constitutes a half-cycle of the audio signal. The base driving current has been shown in large dots. It flows downward in Fig. 7-1.

Fig. 7-2 shows the base-driving current flowing upward through R15. This action constitutes the half of the audio cycle corresponding to a modulation trough. When it flows upward through R15, it adds an additional component of positive voltage to the other positive voltages already there. An increase in the positive voltage at the base of a PNP transistor constitutes reverse bias, so that both the base-emitter and collector-emitter currents will decrease during this half-cycle.

During the modulation peak, the increase in transistor currents through X5 will develop an increased voltage drop across R17, which can be observed as a decrease in the positive voltage existing at the emitter of X5, the base of X6, and the upper end of R17. Thus, an additional component of forward bias is applied to the base of X6 so that its two transistor currents also increase during a modulation peak.

Push-Pull Output Amplifier

The two output transistors (X7 and X8) are connected in push-pull arrangement to provide a heavy audio current to drive speaker SP1. Operating, or bias, voltages for these transistors are selected so that when one of them conducts, the other one will be cut off.

Both transistors are driven from the secondary winding of T1. The collector current for X6 pulsates upward through the primary winding of T1 at the audio frequency, and each pulsation induces a half-cycle of current to flow downward in the secondary winding. The induced voltage associated with this induced current is indicated in Fig. 7-1 as positive at the top of the secondary winding and negative at the bottom. This applies reverse bias to the base of X7, making the base more positive than the emitter and cutting off the flow of both transistor currents through X7.

It also applies forward bias to the base of X8, making its voltage significantly less positive than the 6 volts applied to the emitter, and causing an increase in base-emitter current, and a consequent heavy surge of collector-emitter current through X8.

In the succeeding half-cycle an opposite set of conditions prevail. The collector current through X6 decreases, the voltage induced across the secondary winding of T1 is negative at the top and positive at the obttom. This cuts off both of the transistor currents through X8, and causes an increase in base-emitter current through X7, and a resultant heavy collector-emitter current through X7.

The collector currents for X7 and X8 originate at the center-tapped ground connection of speaker SP1. Each of these currents flow through only half of this winding, and of course they flow in opposite directions. Since they flow on alternate half-cycles and in opposite directions through this winding, the speaker diaphragm will be alternately driven to the right and left at the frequency of the audio modulating voltage carried by the original carrier signal.

The headphone jack above X6 in the diagrams can divert the collector current of X6 so that it will flow either through the phones or through the primary winding of T1, but not through both. Headphones require a much smaller power than a speaker; therefore the pulsations in the collector current of X6 will be adequate to operate the phones without the additional amplification provided by X7 and X8. The headphone current path has been indicated in Fig. 7-2. The headphone current is, of course, identical to the collector-emitter current of X6.

Chapter 8

TUNED RADIO-FREQUENCY RECEIVER

The tuned radio-frequency (TRF) receiver differs from the vast majority of receivers in use today in that it does not utilize the heterodyne principle. As is pointed out in Chapter 1, the heterodyne principle is a frequency-changing process, in which two signals are "beat" against each other to obtain a new third frequency, called the "intermediate frequency." The manifold advantages inherent in this process have led to its adoption in virtually all receiver functions—AM and FM broadcast receivers, TV and radar receivers, and many communications and special purpose receivers.

The principal advantage of generating a fixed value of intermediate frequency in a heterodyne receiver is that all IF amplifier stages can be fixed-tuned, rather than variable-tuned. This enables each such amplifier circuit to be engineered for peak performance at the chosen fixed frequency, with little opportunity for or possibility of maladjustment by an operator.

TYPICAL TRF RECEIVER

The TRF receiver found its greatest popularity in the early days of radio, before frequency-converting circuits or principles were highly developed. Figs. 8-1 and 8-2 show a typical TRF receiver circuit. There are three tuned circuits in the receiver, one being connected to the control grid of each of the first three amplifier tubes, V1, V2, and V3. The tuning elements, usually one capacitor from each tank, must be mechanically connected together, or ganged, so that when one circuit is tuned to a new frequency, the other two tank circuits will also be tuned to the same frequency.

The third amplifier, V3, utilizes the principle of grid-leak detection to demodulate the audio signal directly from the RF carrier signal. From this point on the audio section of the receiver functions exactly as a comparable audio section in a conventional heterodyne receiver.

Fig. 8-1 differs from 8-2 in that a series diode limiting circuit, constructed around diode V5, has been added to Fig. 8-2 to illustrate a typical application of the noise limiting function.

Identification of Components

The individual circuit components and their principal functions are as follows:

R1, R4, R9, R12, R14 (Fig. 8-2)—Cathode-biasing resistors.
R2, R5, R13—Screen-grid voltage-dropping resistors.
R3, R6—Power-supply decoupling resistors.
R7—Grid-leak biasing and driving resistor for V3.
R8—AVC resistor.
R10—Plate-load resistor for V3.
R11—Volume control potentiometer.
R15 (Fig. 8-2)—Noise-limiter control potentiometer.
R16 (Fig. 8-2)—Grid-driving resistor for V4.
C1, C2—Tuning capacitors for first RF tank.
C3, C8, C17—Cathode-bypass capacitors.
C4, C9, C16—Screen-grid filter capacitors.
C5, C10—Power-supply decoupling capacitors.
C6, C7—Tuning capacitors for second RF tank.
C11—Tuning capacitor for third RF tank.
C12, C14, C18 (Fig. 8-2), C19 (Fig. 8-2)—Coupling and blocking capacitors.
C13—Plate RF bypass capacitor for V3.
C15—AVC storage capacitor.
L1 (Fig. 8-2)—Radio-frequency choke.
T1, T2, T3—Radio-frequency transformers.
T4—Audio output transformer.
V1, V2—Radio-frequency amplifier tubes.
V3—Grid-leak detector and audio-amplifier tube.
V4—Audio power-amplifier tube.
V5 (Fig. 8-2)—Diode noise-limiter tube.
M1—Power supply.

Identification of Currents

The several "families" of currents which flow in the TRF receiver all have familiar counterparts in the superhet receiver discussed in Chapter 6. These current families are:

1. Cathode heating current (not shown).
2. Three RF tank currents.
3. Five tube plate currents.
4. Three screen-grid currents.
5. Three cathode filter currents.
6. Three screen-grid filter currents.
7. Two power-supply filter currents.
8. One plate-filtering current.
9. Three audio-signal currents
10. AVC current.

Details of Operation

The cathode heating currents are not shown in Fig. 8-1 and 8-2. It is common practice to omit the filament windings in circuit diagram, because the filament circuit is isolated both electrically and functionally from the remainder of the circuits in a radio. It is universally taken for granted that tube cathodes must be heated before the tubes can perform their normal function of conducting electrons.

Each of the three RF tank currents flows up and down through the secondary winding of its respective RF transformers (T1, T2, or T3); each is sustained in oscillation by the RF current flowing in the associated primary winding. Note that unlike the transformers employed in superheterodyne receivers, the primaries of the coupling transformers are not tuned by a capacitor. In the case of T1, an RF alternating current flows back and forth from the antenna to the primary winding. In the case of T2 and T3, the primary winding currents are pulsating direct currents, since they are the plate currents of V1 and V2, respectively, and these pulsations occur at the radio frequency being received.

All of the amplifier tubes have been biased with grid and cathode voltages so that the tubes will operate under Class-A conditions, which means that each tube conducts electrons continuously throughout an entire cycle of RF or audio voltage. Each plate current (shown a solid line) starts at the ground below the cathode and flows up through the cathode resistor, through the tube from cathode to plate, out the plate and through the

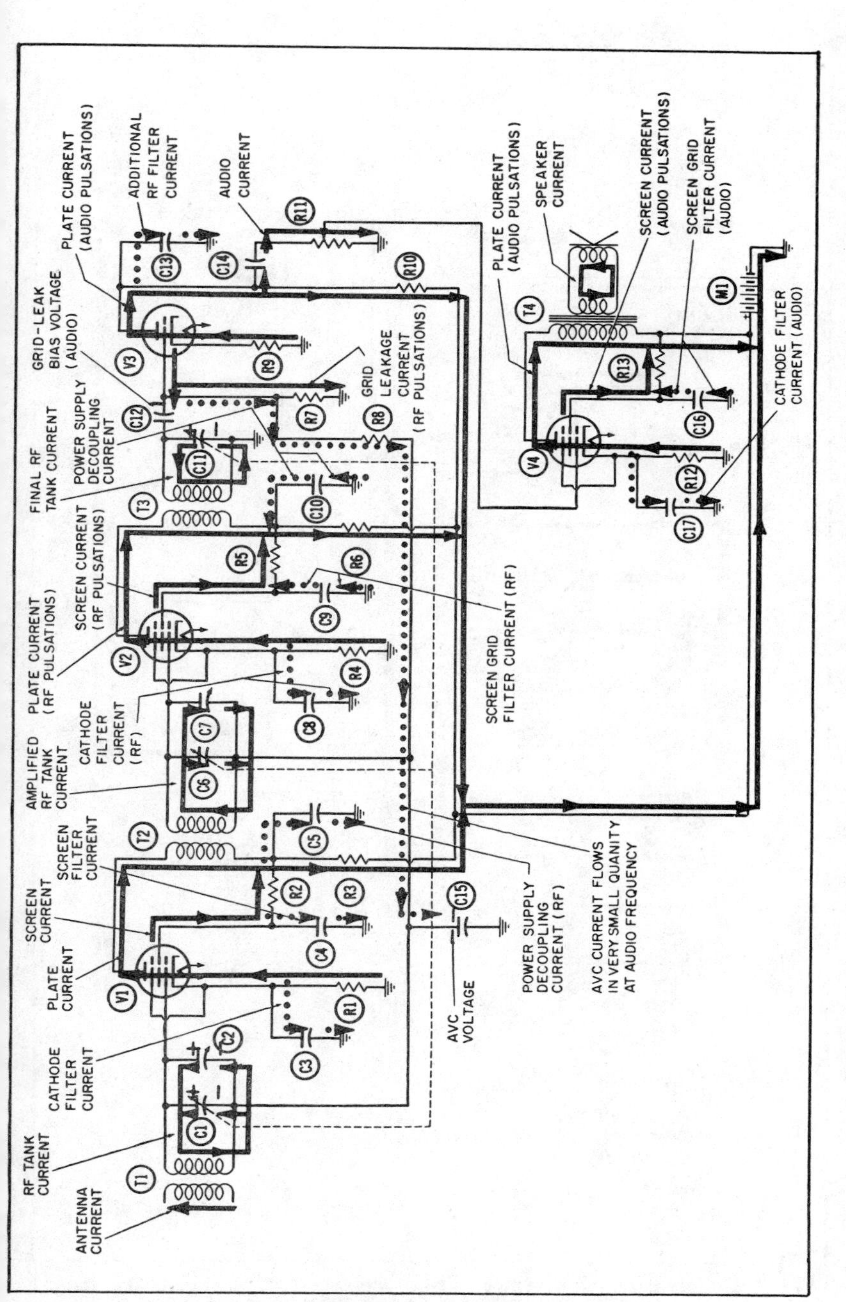

Fig. 8-1. Operation of a typical TRF receiver.

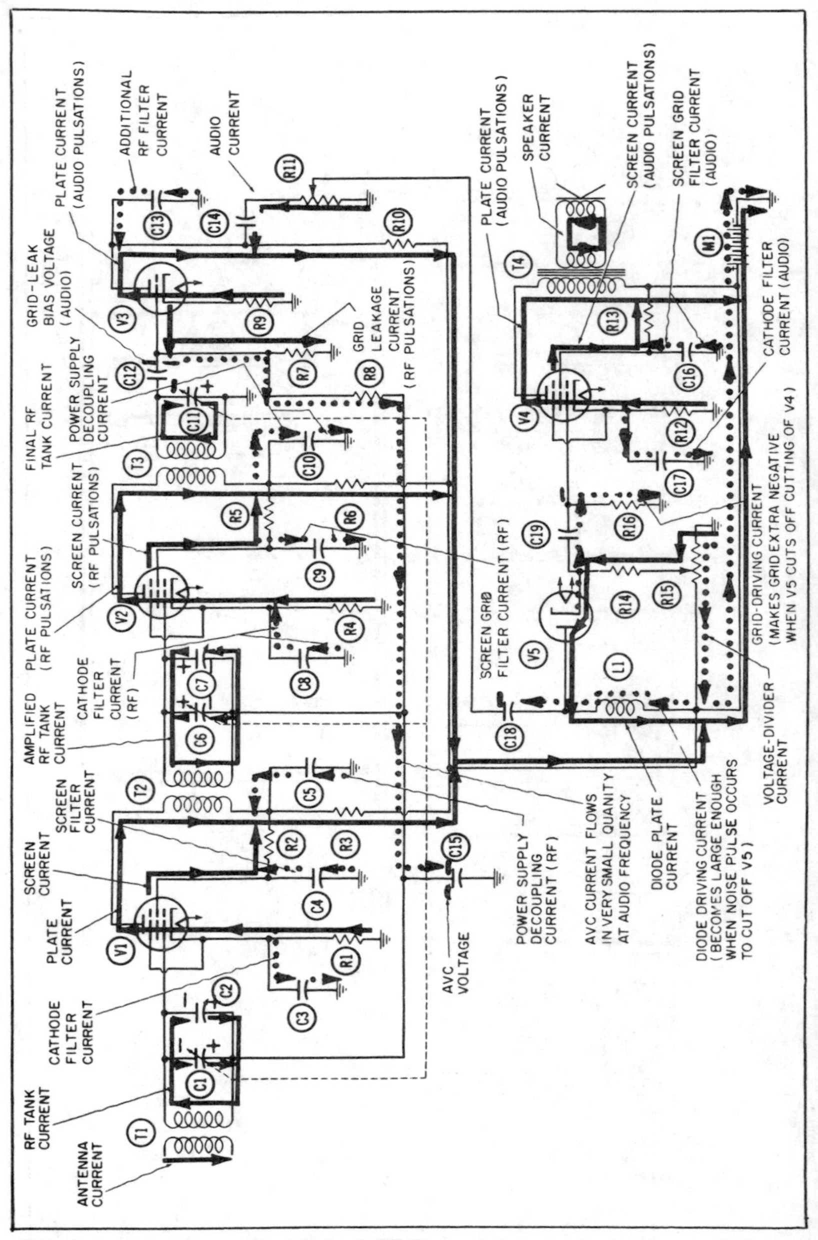

Fig. 8-2. Operation of a typical TRF receiver—currents reversed from those in Fig. 8-1 and a noise limiter added.

plate load to the positive terminal of the power supply, then through the power supply to ground.

The three pentode tubes (V1, V2, and V4) also have screen-grid currents. These currents, which have also been shown as a line, flow out the screen-grid terminal and through the screen-grid resistor where they join the plate currents and flow through the B+ line to the positive terminal of the power supply.

The plate currents through V1 and V2 pulsate at the radio-frequency being received, while the plate currents through V3 and V4 pulsate at the frequency of the audio intelligence which is carried by the carrier signal. The current through V5 flows continuously as long as the incoming signal strength is not made excessive by unwanted noise pulses. When the diode current is flowing, it also pulsates at the audio frequency being demodulated from the carrier.

There are nine separate filter currents in the circuit of Figs. 8-1 and 8-2. Filter currents have been explained many times previously, so they will not be repeated here. Each pulsates back and forth, alternately storing and drawing electrons away from the top plate according to the needs of the circuit to which it is connected. A corresponding electron current flows up from ground to the lower capacitor plate or from the capacitor to ground, in step with the electron flow to or from the upper plate.

The first two cathode filter currents flow at the RF rate in and out of C3 and C8 (shown in large dots). The third cathode filter current (shown in small dots) is at the audio rate, and flows in and out of C17. Likewise, the first two screen grid filter currents (also shown in large dots) are RF currents and flow in and out of C4 and C9. The other screen-grid filter current flows at the audio rate in and out of C16 (small dots).

The two power-supply decoupling currents are shown in large dots and flow in and out of C5 and C10 to prevent RF variations from existing on the B+ line. Another filtering current flows in and out of C13 in the plate circuit of V3 to remove the RF pulsations following detection.

The three audio-signal currents (shown as a solid line) carry the audio signal from the point of demodulation on C12 to the speaker. These are the grid leakage current from V3 to C12, which pulsates downward through R7, the two-way audio current which flows up and down through R11, and the two-way audio current flowing back and forth through the closed speaker circuit.

The AVC current, shown in small dots, flows back and forth through R8 at the basic audio frequency being demodulated. The amount of this current which flows during a single half cycle is very slight, because R8 has a very high resistive value. It is this current which delivers electrons to C15, and thereby builds up the negative AVC voltage. The AVC current can be looked upon as an "equalizing" current, since it attempts constantly to equalize the stored voltages at the opposite ends of R8. These voltages are the instantaneous audio voltage on the right hand plate of C12, and the AVC voltage on C15. When the voltage on C12 is more negative than that on C15 (during the modulation peaks), the AVC current flows downward through R8. When the voltage on C12 is less negative than that on C15 (during a modulation trough), the AVC current flows upward through R8.

A little reflection leads to the conclusion that the AVC voltage on C15 will always tend to stabilize at the *average* voltage existing on C12. In other words, this voltage will be midway between the trough and peak values. An example or two may serve to clarify this conclusion. First, imagine an instance where the audio voltage on C12 varies between −2 and −4 volts. (This is identified as the grid leak bias voltage in Fig. 8-1 and 8-2). This voltage will be (−2) volts during a modulation trough, and −4 volts during a modulation peak a half of an audio cycle later. The AVC voltage will tend to assume the average value of these two voltages, or −3 volts. This voltage will be applied directly to the control grids of V1 and V2 as part of their over-all "bias" voltages.

Now imagine that the signal strength increases due to some peculiar atmospheric condition. The three RF tank currents will all be proportionately increased in strength, and the amount of grid-leak detector current flowing out of V3 each cycle will also be increased. As a result, the electron accumulation on the right hand plate of C12 will be proportionately increased, so that the new trough and peak voltages will now be −3 and −6 volts, respectively. The average of these values is −4.5 volts; this is the amount of voltage which will build up on C15 under the new conditions. This increased negative voltage applied to the control grids of V1 and V2 will reduce their over-all gains, and will largely compensate for the unwanted increase in signal strength.

Noise-Limiting Diode Operation

V5 is connected as a series noise limiter, and functions in substantially the same manner as described in Chapter 3. R15 acts a voltage divider to "bias" the plate of V5 more positively than the cathode. This biasing action is accomplished by the voltage divider current shown in large dots in Fig. 8-2. Because the point midway along R15 where the cathode voltage is tapped off will always be more negative than the left-hand terminal of R15, the diode plate current shown in as a line will flow continuously along the indicated path unless interrupted by some other action.

During the modulation peaks and troughs which characterize all audio voltages, the amount of diode current through V5 will be increased and decreased. These variations in current flow through R14 will cause the positive voltage at the upper end of the resistor to increase and decrease at the same audio frequency. These audio voltage fluctuations will be coupled across C19 to R16, the grid driving resistor for V4.

When diode current through R14 and V5 increases during modulation peaks, electron current will be drawn upward through R16, making the control grid of V4 positive. When the diode current decreases during modulation troughs, electron current will be driven downward through R16, making the control grid of V4 negative.

Diode V5 can be cut off entirely only by an excessively strong noise pulse having a negative polarity when it reaches the plate of V5. Such a pulse makes the plate of V5 more negative than the cathode, and the upward current flow through R14 is stopped. The positive voltage at the cathode then decreases to the same value existing at the tap on R15. This drop in positive voltage at the upper end of R14 drives a large electron current downward through R16, making the grid of V4 sufficiently negative to cut off this tube entirely and noise limiting has been accomplished.

A strong noise pulse of the opposite polarity, namely positive, will not have this same effect. However, the listener is protected from such a noise pulse by the operating characteristics of V3. V3 itself will be cut off by such a noise pulse; the positive voltage of M1 represents the maximum or limiting value to which the plate of V3 can rise.

With the exception of the circuits just discussed—detector, AVC, and noise limiter—the circuits in the TRF receiver are amplifier circuits which are discussed more fully in other chapters of this book.

INDEX